George A. Anastassiou

Fuzzy Mathematics: Approximation Theory

T0180660

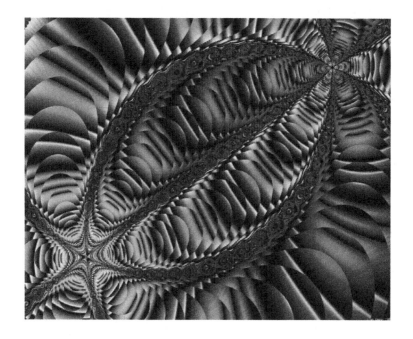

Studies in Fuzziness and Soft Computing, Volume 251

Editor-in-Chief

Prof. Janusz Kacprzyk
Systems Research Institute
Polish Academy of Sciences
ul. Newelska 6
01-447 Warsaw
Poland
E-mail: kacprzyk@ibspan.waw.pl

Further volumes of this series can be found on our homepage: springer.com

George A. Anastassiou

Fuzzy Mathematics: Approximation Theory

 Springer

Author

George A. Anastassiou,Ph.D
Professor of Mathematics
The University of Memphis
Department of Mathematical Sciences
TN 38152 Memphis
USA
E-mail: ganastss@memphis.edu

ISBN 978-3-642-26239-5   e-ISBN 978-3-642-11220-1

DOI 10.1007/978-3-642-11220-1

Studies in Fuzziness and Soft Computing   ISSN 1434-9922

Typeset & Cover Design: Scientific Publishing Services Pvt. Ltd., Chennai, India.

Printed in acid-free paper

9 8 7 6 5 4 3 2 1

springer.com

Dedicated to my daughters Angela and Peggy.

"No assertion is ever known with certainty ...
but that does not stop us making assertions."

Carneades, 214-129 BCE

"Any useful logic must concern itself with Ideas with a fringe of vagueness
and a Truth that is a matter of degree."

Norbert Wiener

"The facts were always fuzzy or vague or inexact ... Science treated the
gray or fuzzy facts as if they were the black-white facts of math. Yet no one
had put forth a single fact about the world that was 100% true or 100%
false."

Bart Kosko, Fuzzy Thinking, 1994, Preface.

Preface

This monograph is the first in Fuzzy Approximation Theory. It contains mostly the author's research work on fuzziness of the last ten years and relies a lot on [10]-[32] and it is a natural outgrowth of them. It belongs to the broader area of Fuzzy Mathematics.

Chapters are self-contained and several advanced courses can be taught out of this book.

We provide lots of applications but always within the framework of Fuzzy Mathematics. In each chapter is given background and motivations. A complete list of references is provided at the end. The topics covered are very diverse. In Chapter 1 we give an extensive basic background on Fuzziness and Fuzzy Real Analysis, as well a complete description of the book. In the following Chapters 2,3 we cover in deep Fuzzy Differentiation and Integration Theory, e.g. we present Fuzzy Taylor Formulae. It follows Chapter 4 on Fuzzy Ostrowski Inequalities. Then in Chapters 5, 6 we present results on classical algebraic and trigonometric polynomial Fuzzy Approximation. In Chapters 7-11 we develop completely the theory of convergence with rates of Fuzzy Positive linear operators to Fuzzy Unit operator, the so called Fuzzy Korovkin Theory. We include there the related topic of Fuzzy Global Smoothness, see Chapter 9. In Chapters 12-14 we deal with Fuzzy Wavelet type operators and their convergence with rates to Fuzzy Unit operator. In Chapters 15-16 we discuss similarly as above the Fuzzy Neural Network Operators. In Chapter 17 we deal with Fuzzy Random Korovkin type approximation theory. In Chapter 18 we deal with Fuzzy Random Neural Network approximations.

In Chapters 19, 20 we present Fuzzy Korovkin type approximations in the Sense of Summability.

Finally in Chapter 21 we estimate in the fuzzy sense differences of Fuzzy Wavelet type operators.

The monograph's approach is Quantitative and almost all main results are given through Fuzzy inequalities, involving fuzzy moduli of continuity, that is fuzzy Jackson type inequalities. Thus all fuzzy convergences are given with rates and the proofs are constructive.

The exposed theory is destined and expected to find applications to all aspects of Fuzziness from theoretical to practical in almost all sciences, technology and industry; in our real world we mostly perform fuzzy approximations. On the other hand our theory has its own theoretical merit and interest within the framework of Pure Mathematics. So this monograph is suitable for researchers, graduate students and seminars of theoretical and applied mathematics, computer science, statistics, engineering, etc., also suitable for all science libraries.

Fuzzy set theory and applications has experienced a rapid development since its discovery by L. Zadeh in 1965, see [103], its growth and applications now cover almost all kinds of mathematics and applied sciences with great applications to real life. We mention here only a few: finance and stock market, weather prediction, nuclear science, robotics, biomedicine, handwriting analysis, space exploration and satellites, radars, electronics, rheology, agriculture, elevators, ecology, geography and philosophy. For a much lenghtier list of applications of Fuzzy sets and Fuzzy logic see Chapter 1, Section 1.1.

I would like to thank Professor Sorin Gal, University of Oradea, Romania, for introducing me into Fuzzy Sets.

The final preparation of book took place during 2009 in Memphis, Tennessee, USA. I would like to thank my family for their dedication and love to me, which was the strongest support during the writing of the monograph.

I am also indebted and thankful to my graduate student Razvan Mezei for the typing preparation of the manuscript in a short time.

November 1, 2009
George A. Anastassiou
Department of Mathematical Sciences
The University of Memphis, TN, USA

Contents

1
INTRODUCTION

1.1 Basics

The concept of fuzziness was first discovered and introduced in the seminal article written by Lotfi A. Zadeh in 1965, see [103].

So in our description next we follow [103].

Frequently classes of objects encountered in the real natural world do not have exactly defined criteria of membership. For example, the class of animals clearly includes lions, tigers, horses, birds, fish, etc. as its members and obviously excludes objects such as trees, gases, cars, stones, houses, metals, etc. However there are objects such as starfish, bacteria, etc. that have an ambiguous status in comparison to the class of animals.

Similar ambiguity arises when we compare the number 20 to the class of real numbers much greater than zero. Clearly, "the class of real numbers much greater than zero", or "the class of beautiful women", or "the class of tall men" or "the class of smart students" are not defined precisely, thus they do not constitute sets of objects in the usual mathematical sense where each element of a set is 100% there. However such imprecisely considered "classes" of objects exist frequently and play an important role in every aspect of our lives,they show up a lot especially in engineering, computer science, pattern recognition, industry, etc. So the concept under consideration is the fuzzy set, which is a class of objects with a continuum of grades of membership. Such a set is characterized by a membership function which assigns to each object a grade of membership varying from zero to one. The notions of inclusion, union, intersection, complementation,

G.A. Anastassiou: Fuzzy Mathematics: Approximation The., STUDFUZZ 251, pp. 1–14.
springerlink.com

relation, convexity, etc., are extended to fuzzy sets, see [103]. So the above framework of consideration gives us a natural way of dealing with such imprecise phenomena, when classes of objects lack precise criteria of membership for their elements. Clearly the concept of fuzziness is distinctively different than the concepts of deterministic and random phenomena.

So random variables cannot describe the fuzzy sets.

Thus a fuzzy set A related to a space of objects X has a membership function $f_A(x)$ which assigns to each element $x \in X$ a number in the interval $[0, 1]$ that represents the "grade of membership" of x in A. Hence, the closer this value to 1, the higher the grade of membership of x in A. If A is an ordinary set, then $f_A(x) = 1$ or 0 according to if $x \in A$ or $x \notin A$. In this last case $f_A = \chi_A$ the characteristic function on A. As another example let $X = \mathbb{R}$ and A be the fuzzy set of real numbers much larger than 0. Then one can define $f_A(0) = f_A(1) = 0$, $f_A(10) = 0.02$, $f_A(1000) = 0.96$, $f_A(10000) = 1$. Again, the membership function though it resembles to a probability measure has nothing to do with it. They have totally different properties and are based on totally different concepts. So most clearly the notion of fuzziness is totally non-stochastic. For all the details of the above see again [103].

A fuzzy set is empty if and only if its membership function is identically equal to zero on X.

The fuzzy sets A and B are equal, $A = B$, if and only if their membership functions are identically equal on X.

The complement of a fuzzy set A is a fuzzy set denoted by A' and its membership function is defined by $f_{A'} = 1 - f_A$.

Also A is contained in B, $A \subset B$, if and only if $f_A \leq f_B$, where A, B are fuzzy sets.

The union $A \cup B$ is a fuzzy set with membership function

$$f_{A \cup B}(x) = \max \{f_A(x), f_B(x)\}, x \in X,$$

abbreviated as $f_{A \cup B} = f_A \vee f_B$.

The intersection $A \cap B$ is a fuzzy set with membership function

$$f_{A \cap B}(x) = \min \{f_A(x), f_B(x)\}, x \in X,$$

in short written as $f_{A \cap B} = f_A \wedge f_B$.

We say the fuzzy sets A and B are disjoint iff they are fuzzy-empty.

For the rest of definitions and all properties of the algebra of the fuzzy sets see [103], where were first given and described. L. A. Zadeh has done an extensive body of work on Fuzziness and he shaped this subject.

From all the above we derive that the concept of "membership function" plays a pivotal role in the study of fuzzy sets, in fact it is the most important component describing and defining a fuzzy set.

Therefore in the Fuzzy Mathematics usually we deal with membership functions from the sets of real or complex numbers to $[0, 1]$. These are the real or complex fuzzy numbers. For the description of the real fuzzy numbers see Section 1.2. Based on the above in formal mathematical calculations we identify fuzzy sets with membership functions. The extensive works on the fuzzy subject by D. Dubois and H. Prade are well known and play an important role there.

To Fuzzy set theory corresponds the Fuzzy logic which also was founded by Zadeh [103]. The Fuzzy logic is a generalization of classical and Boolean logic. It relaxes the Aristotelian law of the excluded middle. Fuzzy logic via the concept of "membership function" is a multivalued logic as opposed to the Aristotle's bi-valued logic: true or false / Yes or No only. Earlier than Zadeh was Jan Lukasiewicz to propose a 3-valued logic, with the third value "Possible".

Typical applications of Fuzzy logic include to control small devices, to control large systems, to learning computer programs and neural networks, to pattern recognition in speech and handwriting, etc.

It has been established that fuzzy mathematics calculations are much simpler, efficient and faster than numerical and other approximate calculations. In our natural world's computations there is not complete precision, so fuzzy mathematics have and will have a major effect to control complicated machines from robots to rockets which need quick computations due to quick changes.

So Fuzzy logic is reasoning with the use of Fuzzy sets, in particular with the use of membership functions. Fuzziness is a measure of completeness and has no relation to uncertainty. Probability is a measure of certainty and increases to 1 as more information is provided. So they are two completely different concepts.

Words themselves may be fuzzy values. Saying "airplane" or "boat" we cover all kinds of airplanes or boats. Assigning adjectives we reduce the size of fuzzy sets, e.g. "big airplane" or "small boat". So the concept of Fuzziness appears everywhere in real life including the natural languages we speak.

In engineering, triangular and trapezoidal fuzzy membership functions are examples of sufficiently accurate and with simple implementation such functions.

Let $a \leq b \leq c \leq d$ denote some characteristic numbers, we define

$$\text{"Triangular"} \quad : \quad f(x; a, b, c) = \max\left(\min\left(\frac{x-a}{b-a}, \frac{c-x}{c-b}\right), 0\right),$$

$$\text{"Trapezoidal"} \quad : \quad f(x; a, b, c, d) = \max\left(\min\left(\frac{x-a}{b-a}, 1, \frac{d-x}{d-c}\right), 0\right).$$

Multi-valued fuzzy logic mathematics belong to the wider umbrella of non-classical mathematics. They include fuzzy subgroupoids and fuzzy subgroups, fuzzy fields, fuzzy Galois theory, fuzzy topology, fuzzy geometry, fuzzy relations, fuzzy graphs, possibility theory, non-additive measures, fuzzy measure theory and fuzzy integrals.

This monograph is the first one in Fuzzy Approximation Theory. The fuzzy logic and fuzzy mathematics development so far has been through three important stages:

(i) the direct fuzzification of most mathematical concepts during 1965-80,

(ii) the development of all possible ways in fuzzy generalization process during 1980-90,

(iii) the axiomatic foundation and standardization of fuzziness during 1990-2000.

The applications of fuzziness are uncountable and varied. Additionally to consumer applications in Japanese electronics, auto industry in Germany, and home appliances; Fuzzy logic is applied to finance, stock market, biomedicine, ecology, philosophy, agriculture, geography, rheology, satellite remote control, nuclear science, weather prediction, elevators, robotics and rocket science, to mention a few of them. In general fuzziness is applied to engineering and control theory very widely.

Finally, we would like to emphasize that the fuzzy-neural network methodologies find lots of applications to learning strategies of radars, sonar signal processing and classification, speech recognition, classification of remotely sensed images, handwritten character recognition, semiconductor manufacturing processes, smart home appliances, hybrid fuzzy/expert systems for medical diagnosis, etc.

The fuzzy literature is vast and huge and goes to all possible directions. Thus we have decided the list of references of this book to contain only articles and books that are used here.

This monograph contains mainly the research work of the author in Fuzzy Approximation Theory that was conducted during the last ten years.

The chapters of the book are self-contained and each chapter can be read independently from the rest.

1.2   Background

We need the following

Definition 1.1 (see [53]). Let $\mu \colon \mathbb{R} \to [0,1]$ with the following properties.

(i) is *normal*, i.e., $\exists x_0 \in \mathbb{R}$; $\mu(x_0) = 1$.

(ii) $\mu(\lambda x + (1-\lambda)y) \geq \min\{\mu(x), \mu(y)\}$, $\forall x, y \in \mathbb{R}$, $\forall \lambda \in [0,1]$ (μ is called a convex fuzzy subset).

(iii) μ is *upper semicontinuous* on $\mathbb{R}$, i.e., $\forall x_0 \in \mathbb{R}$ and $\forall \varepsilon > 0$, $\exists$ neighborhood $V(x_0)$: $\mu(x) \leq \mu(x_0) + \varepsilon$, $\forall x \in V(x_0)$.

(iv) The set $\overline{\text{supp}(\mu)}$ is compact in $\mathbb{R}$ (where $\text{supp}(\mu) := \{x \in \mathbb{R}; \mu(x) > 0\}$).

We call μ a *fuzzy real number*. Denote the set of all μ with $\mathbb{R}_{\mathcal{F}}$.

E.g., $\mathcal{X}_{\{x_0\}} \in \mathbb{R}_{\mathcal{F}}$, for any $x_0 \in \mathbb{R}$, where $\mathcal{X}_{\{x_0\}}$ is the characteristic function at x_0.

For $0 < r \le 1$ and $\mu \in \mathbb{R}_{\mathcal{F}}$ define $[\mu]^r := \{x \in \mathbb{R}: \mu(x) \ge r\}$ and

$$[\mu]^0 := \overline{\{x \in \mathbb{R} : \mu(x) > 0\}}.$$

Then it is well known that for each $r \in [0,1]$, $[\mu]^r$ is a closed and bounded interval of $\mathbb{R}$ ([67]). For $u, v \in \mathbb{R}_{\mathcal{F}}$ and $\lambda \in \mathbb{R}$, we define uniquely the sum $u \oplus v$ and the product $\lambda \odot u$ by

$$[u \oplus v]^r = [u]^r + [v]^r, \quad [\lambda \odot u]^r = \lambda[u]^r, \quad \forall r \in [0,1],$$

where $[u]^r + [v]^r$ means the usual addition of two intervals (as subsets of $\mathbb{R}$) and $\lambda[u]^r$ means the usual product between a scalar and a subset of $\mathbb{R}$ (see, e.g., [53]). Notice $1 \odot u = u$ and it holds $u \oplus v = v \oplus u$, $\lambda \odot u = u \odot \lambda$. If $0 \le r_1 \le r_2 \le 1$ then $[u]^{r_2} \subseteq [u]^{r_1}$. Actually $[u]^r = [u_-^{(r)}, u_+^{(r)}]$, where $u_-^{(r)} \le u_+^{(r)}$, $u_-^{(r)}, u_+^{(r)} \in \mathbb{R}$, $\forall r \in [0,1]$. For $\lambda > 0$ one has $\lambda u_\pm^{(r)} = (\lambda \odot u)_\pm^{(r)}$, respectively.

Define

$$D: \mathbb{R}_{\mathcal{F}} \times \mathbb{R}_{\mathcal{F}} \to \mathbb{R}_+$$

by

$$D(u,v) \quad : \quad = \sup_{r \in [0,1]} \max\{|u_-^{(r)} - v_-^{(r)}|, |u_+^{(r)} - v_+^{(r)}|\}$$
$$= \sup_{r \in [0,1]} \text{Hausdorff distance} ([u]^r, [v]^r),$$

where $[v]^r = [v_-^{(r)}, v_+^{(r)}]$; $u, v \in \mathbb{R}_{\mathcal{F}}$. We have that D is a metric on $\mathbb{R}_{\mathcal{F}}$. Then $(\mathbb{R}_{\mathcal{F}}, D)$ is a complete metric space, see [50],[53], with the properties

$$D(u \oplus w, v \oplus w) = D(u,v), \quad \forall u, v, w \in \mathbb{R}_{\mathcal{F}},$$
$$D(k \odot u, k \odot v) = |k|D(u,v), \quad \forall u, v \in \mathbb{R}_{\mathcal{F}}, \forall k \in \mathbb{R},$$
$$D(u \oplus v, w \oplus e) \le D(u,w) + D(v,e), \quad \forall u, v, w, e \in \mathbb{R}_{\mathcal{F}}.$$

Let $f, g: \mathbb{R} \to \mathbb{R}_{\mathcal{F}}$ be *fuzzy number valued functions*. The distance between f, g is defined by

$$D^*(f,g) := \sup_{x \in \mathbb{R}} D(f(x), g(x)).$$

On $\mathbb{R}_{\mathcal{F}}$ we define a *partial order by* "$\le$"(or "$\preceq$"): $u, v \in \mathbb{R}_{\mathcal{F}}$, $u \le v$ (or $u \preceq v$) iff $u_-^{(r)} \le v_-^{(r)}$ and $u_+^{(r)} \le v_+^{(r)}$, $\forall r \in [0,1]$.

We mention

Lemma 1.2 ([31] and Lemma 5.3). *For any $a, b \in \mathbb{R}$: $a \cdot b \geq 0$ and any $u \in \mathbb{R}_{\mathcal{F}}$ we have*

$$D(a \odot u, b \odot u) \leq |a - b| \cdot D(u, \tilde{o}),$$

where $\tilde{o} \in \mathbb{R}_{\mathcal{F}}$ is defined by $\tilde{o} := \mathcal{X}_{\{0\}}$.

Lemma 1.3 ([31] and Lemma 5.6).

(i) *If we denote $\tilde{o} := \mathcal{X}_{\{0\}}$, then $\tilde{o} \in \mathbb{R}_{\mathcal{F}}$ is the neutral element with respect to $\oplus$, i.e., $u \oplus \tilde{o} = \tilde{o} \oplus u = u$, $\forall u \in \mathbb{R}_{\mathcal{F}}$.*

(ii) *With respect to $\tilde{o}$, none of $u \in \mathbb{R}_{\mathcal{F}}$, $u \neq \tilde{o}$ has opposite in $\mathbb{R}_{\mathcal{F}}$.*

(iii) *Let $a, b \in \mathbb{R}$: $a \cdot b \geq 0$, and any $u \in \mathbb{R}_{\mathcal{F}}$, we have $(a + b) \odot u = a \odot u \oplus b \odot u$. For general $a, b \in \mathbb{R}$, the above property is false.*

(iv) *For any $\lambda \in \mathbb{R}$ and any $u, v \in \mathbb{R}_{\mathcal{F}}$, we have $\lambda \odot (u \oplus v) = \lambda \odot u \oplus \lambda \odot v$.*

(v) *For any $\lambda, \mu \in \mathbb{R}$ and $u \in \mathbb{R}_{\mathcal{F}}$, we have $\lambda \odot (\mu \odot u) = (\lambda \cdot \mu) \odot u$.*

If we denote $\|u\|_{\mathcal{F}} := D(u, \tilde{o})$, $\forall u \in \mathbb{R}_{\mathcal{F}}$, then $\| \cdot \|_{\mathcal{F}}$ has the properties of a usual norm on $\mathbb{R}_{\mathcal{F}}$, i.e.,

$$\|u\|_{\mathcal{F}} = 0 \text{ iff } u = \tilde{o}, \|\lambda \odot u\|_{\mathcal{F}} = |\lambda| \cdot \|u\|_{\mathcal{F}},$$
$$\|u \oplus v\|_{\mathcal{F}} \leq \|u\|_{\mathcal{F}} + \|v\|_{\mathcal{F}}, \|u\|_{\mathcal{F}} - \|v\|_{\mathcal{F}} \leq D(u, v).$$

Notice that $(\mathbb{R}_{\mathcal{F}}, \oplus, \odot)$ is *not* a linear space over $\mathbb{R}$, and consequently $(\mathbb{R}_{\mathcal{F}}, \| \cdot \|_{\mathcal{F}})$ is *not* a normed space. Here $\sum^*$ denotes the fuzzy summation.

Definition 1.4. Let $a_1, a_2, b_1, b_2 \in \mathbb{R}$ such that $a_1 \leq b_1$ and $a_2 \leq b_2$. Then we define

$$[a_1, b_1] + [a_2, b_2] = [a_1 + a_2, b_1 + b_2]. \tag{1.1}$$

Let $a, b \in \mathbb{R}$ such that $a \leq b$ and $k \in \mathbb{R}$, then we define,

$$\text{if } k \geq 0, \quad k[a, b] = [ka, kb],$$
$$\text{if } k < 0, \quad k[a, b] = [kb, ka]. \tag{1.2}$$

Here we use

Lemma 1.5. Let $f : [a, b] \to \mathbb{R}_{\mathcal{F}}$ be fuzzy continuous and let $g : [a, b] \to \mathbb{R}_+$ be continuous. Then $f(x) \odot g(x)$ is fuzzy continuous function $\forall x \in [a, b]$.
Proof. The same as of Lemma 2 ([14]), using Lemma 2 of [11]. □

Let $f, g : U \to \mathbb{R}_{\mathcal{F}}, U \subseteq (M, d)$ metric space, be *fuzzy real number valued functions*. The distance between f, g is also defined by

$$D^*(f, g) := \sup_{x \in U} D(f(x), g(x)).$$

A subset $K \subseteq \mathbb{R}_{\mathcal{F}}$ is called *fuzzy bounded*, iff $D(u, v) \leq M, M > 0$, $\forall u, v \in K$.

We need

Definition 1.6. (see [53]). Let $x, y \in \mathbb{R}_{\mathcal{F}}$. If there exists a $z \in \mathbb{R}_{\mathcal{F}}$ such that $x = y \oplus z$, then we call z the *H-difference* of x and y, denoted by $z := x - y$.

Definition 1.7 ([53]). Let $T := [x_0, x_0 + \beta] \subset \mathbb{R}$, with $\beta > 0$. A function $f : T \to \mathbb{R}_{\mathcal{F}}$ is *H-differentiable* at $x \in T$ if there exists an $f'(x) \in \mathbb{R}_{\mathcal{F}}$ such that the limits (with respect to metric D)

$$\lim_{h \to 0^+} \frac{f(x + h) - f(x)}{h}, \quad \lim_{h \to 0^+} \frac{f(x) - f(x - h)}{h}$$

exist and are equal to $f'(x)$. We assume that the $H-$differences $f(x + h) - f(x)$, $f(x) - f(x - h) \in \mathbb{R}_{\mathcal{F}}$ in a neighborhood of x. We call f' the *derivative* or *H-derivative* of f at x. If f is H-differentiable at any $x \in T$, we call f *differentiable* or *H-differentiable* and it has *H-derivative over T* the function f'.

The last definition was given first by M. Puri and D. Ralescu [93].

E x a m p l e 1.8. Let $f : \mathbb{R}_+ \to \mathbb{R}_{\mathcal{F}}$ be such that for any $\lambda, \mu \geq 0$ it holds

$$f(\lambda x + \mu y) = \lambda \odot f(x) \oplus \mu \odot f(y), \quad \forall x, y \in \mathbb{R}_+.$$

Then the H-derivative $f'(x) = f(1), \forall x \in \mathbb{R}_+$.

Proof. By $f(x + h) = f(x) \oplus f(h)$, that is the H-difference

$$f(x + h) - f(x) = f(h) \in \mathbb{R}_{\mathcal{F}}.$$

Thus

$$\frac{f(x + h) - f(x)}{h} = f(1), \quad h > 0.$$

Similarly, $f(x) = f(x - h) \oplus f(h)$, for $h > 0$ small, that is the H-difference $f(x) - f(x - h) = f(h) \in \mathbb{R}_{\mathcal{F}}$. Hence

$$\frac{f(x) - f(x - h)}{h} = f(1).$$

But

$$\lim_{h \to 0^+} D(f(1), f(1)) = 0.$$

Clearly for $f'(0)$ we take the right-hand side H-derivative.   □

We need also a particular case of the *Fuzzy Henstock integral* $(\delta(x) = \frac{\delta}{2})$ introduced in [53], Definition 2.1, there.

That is,

Definition 1.9 ([66], p. 644). Let $f: [a, b] \to \mathbb{R}_{\mathcal{F}}$. We say that f is *Fuzzy-Riemann integrable* to $I \in \mathbb{R}_{\mathcal{F}}$ if for any $\varepsilon > 0$, there exists $\delta > 0$ such that for any division $P = \{[u, v]; \xi\}$ of $[a, b]$ with the norms $\Delta(P) < \delta$, we have

$$D\left(\sum_{P}{}^{*}(v - u) \odot f(\xi), I\right) < \varepsilon,$$

where $\sum^{*}$ denotes the fuzzy summation. We choose to write

$$I := (FR)\int_{a}^{b} f(x)dx.$$

We also call an f as above (FR)-integrable.

We need

Theorem 1.10 ([67]). *If* $f, g: [c, d] \to \mathbb{R}_{\mathcal{F}}$ *are* (FR)-*integrable fuzzy functions, and* α, β *are real numbers, then*

$$(FR)\int_{c}^{d}(\alpha f(x) \oplus \beta g(x))dx \;=\; \alpha(FR)\int_{c}^{d}f(x)dx$$

$$\oplus\; \beta(FR)\int_{c}^{d}g(x)dx.$$

Corollary 1.11 (Corollary 13.2 of [66]). *If* $f \in C([a, b], \mathbb{R}_{\mathcal{F}})$ *then* f *is* (FR) *integrable on* $[a, b]$.

We use the following fundamental theorem of Fuzzy Calculus:

Corollary 1.12 ([11]). *If* $f: [a, b] \to \mathbb{R}_{\mathcal{F}}$ *has a fuzzy continuous H-derivative* f' *on* $[a, b]$, *then* $f'(x)$ *is* (FR)-*integrable over* $[a, b]$ *and*

$$f(s) = f(t) \oplus (FR)\int_{t}^{s} f'(x)dx, \quad \text{for any } s \geq t, \; s, t \in [a, b].$$

Note. In Corollary 1.12 when $s < t$ the formula is invalid! since fuzzy real numbers correspond to closed intervals etc.

We mention

Lemma 1.13 ([11]). *If* $f, g: [a, b] \subseteq \mathbb{R} \to \mathbb{R}_{\mathcal{F}}$ *are fuzzy continuous functions, then the function* $F: [a, b] \to \mathbb{R}_{+}$ *defined by* $F(x) := D(f(x), g(x))$ *is continuous on* $[a, b]$, *and*

$$D\left((FR)\int_{a}^{b}f(x)dx, (FR)\int_{a}^{b}g(x)dx\right) \leq \int_{a}^{b}D(f(x), g(x))dx.$$

Lemma 1.14 ([11]). *Let $f\colon [a,b] \to \mathbb{R}_\mathcal{F}$ fuzzy continuous (with respect to metric D), then $D(f(x),\tilde{o}) \leq M$, $\forall x \in [a,b]$, $M > 0$, that is f is fuzzy bounded. Equivalently we get $\chi_{-M} \leq f(x) \leq \chi_M$, $\forall x \in [a,b]$.*

Lemma 1.15 ([11]). *Let $f\colon [a,b] \subseteq \mathbb{R} \to \mathbb{R}_\mathcal{F}$ be fuzzy continuous. Then*

$$(FR)\int_a^x f(t)dt \quad \text{is a fuzzy continuous function in } x \in [a,b].$$

Lemma 1.16 ([11]). *Let $f\colon [a,b] \subset \mathbb{R} \to \mathbb{R}_\mathcal{F}$ fuzzy continuous, $r \in \mathbb{N}$. Then the following integrals*

$$(FR)\int_a^{s_{r-1}} f(s_r)ds_r, (FR)\int_a^{s_{r-2}}\left(\int_a^{s_{r-1}} f(s_r)ds_r\right)ds_{r-1},$$

$$\ldots,(FR)\int_a^s\left(\int_a^{s_1}\cdots\left(\int_a^{s_{r-2}}\left(\int_a^{s_{r-1}} f(s_r)ds_r\right)ds_{r-1}\right)\cdots\right)ds_1,$$

are fuzzy continuous functions in $s_{r-1}, s_{r-2}, \ldots, s$, respectively. Here $a \leq s_{r-1} \leq s_{r-2} \leq \cdots \leq s \leq b$.

Lemma 1.17 ([17]). *Let $f\colon [a,b] \to \mathbb{R}_\mathcal{F}$ have an existing H-fuzzy derivative f' at $c \in [a,b]$. Then f is fuzzy continuous at c.*

We need

Theorem 1.18 ([67]). *Let $f\colon [a,b] \to \mathbb{R}_\mathcal{F}$ be fuzzy continuous. Then $(FR)\int_a^b f(x)dx$ exists and belongs to $\mathbb{R}_\mathcal{F}$, furthermore it holds*

$$\left[(FR)\int_a^b f(x)dx\right]^r = \left[\int_a^b (f)_-^{(r)}(x)dx, \int_a^b (f)_+^{(r)}(x)dx\right], \quad \forall r \in [0,1]. \tag{1.3}$$

Clearly $f_\pm^{(r)}\colon [a,b] \to \mathbb{R}$ are continuous functions.

We also need

Theorem 1.19 ([71]). *Let $f\colon [a,b] \subseteq \mathbb{R} \to \mathbb{R}_\mathcal{F}$ be H-fuzzy differentiable. Let $t \in [a,b]$, $0 \leq r \leq 1$. (Clearly*

$$[f(t)]^r = [(f(t))_-^{(r)}, (f(t))_+^{(r)}] \subseteq \mathbb{R}, \text{ denoted by } [f]^r = [f_-^{(r)}, f_+^{(r)}]). \tag{1.4}$$

Then $(f(t))_\pm^{(r)}$ are differentiable and

$$[f'^r = [((f(t))_-^{(r)})', (f(t))_+^{(r)})']. \tag{1.5}$$

The last can be used to find f'.

Here $C^n([a,b], \mathbb{R}_{\mathcal{F}})$, $n \geq 1$ denotes the space of n-times fuzzy continuously H-differentiable functions from $[a,b] \subseteq \mathbb{R}$ into $\mathbb{R}_{\mathcal{F}}$. By Theorem 5.2 of [71] (here Theorem 1.19), for $f \in C^n([a,b], \mathbb{R}_{\mathcal{F}})$ we obtain

$$[f^{(i)}(t)]^r = \left[((f(t))_{-}^{(r)})^{(i)}, ((f(t))_{+}^{(r)})^{(i)} \right], \qquad (1.6)$$

for $i = 0, 1, 2, \ldots, n$ and in particular we have

$$(f^{(i)})_{\pm}^{(r)} = (f_{\pm}^{(r)})^{(i)}, \quad \forall r \in [0,1]. \qquad (1.7)$$

1.3   Chapters description

Here we describe our monograph's chapters.

In Chapter 2 the concept of H-fuzzy differentiation is discussed thoroughly in the univariate and multivariate cases. Basic H-derivatives are calculated and then important theorems are presented on the topic, such as, the H-mean value theorem, the univariate and multivariate H-chain rules, and the interchange of the order of H-fuzzy differentiation. Finally is given a multivariate H-fuzzy Taylor formula.

In Chapter 3 we produce Fuzzy Taylor formulae with integral remainder in the univariate and multivariate cases, analogs of the real setting.

In Chapter 4 we present optimal upper bounds for the deviation of a fuzzy continuous function from its fuzzy average over $[a, b] \subset \mathbb{R}$, error is measured in the D-fuzzy metric. The established fuzzy Ostrowski type inequalities are sharp, in fact attained by simple fuzzy real number valued functions. These inequalities are given for fuzzy Hölder and fuzzy differentiable functions and these facts are reflected in their right-hand sides.

In Chapter 5 we establish that any 2π-periodic fuzzy continuous function from $\mathbb{R}$ to the fuzzy number space $\mathbb{R}_{\mathcal{F}}$, can be uniformly approximated by some fuzzy trigonometric polynomials.

About Chapter 6: In [31] was proved that any 2π-periodic continuous fuzzy-number-valued function can be uniformly approximated by sequences of generalized fuzzy trigonometric polynomials, but without giving any estimate for the approximation error. In this chapter, connected to the best approximation problem we present Jackson-type estimates. For the algebraic case we also give a Jackson-type estimate, using the Szabados-type polynomials. Finally, as an application we study the convergence of fuzzy Lagrange interpolation polynomials.

In Chapter 7 we present the basic fuzzy Korovkin theorem via a fuzzy Shisha–Mond inequality given here. This determines the degree of convergence with rates of a sequence of fuzzy positive linear operators to the fuzzy unit operator. The surprising fact is that only the real case Korovkin assumptions are enough for the validity of the fuzzy Korovkin theorem, along with a natural realization condition fulfilled by the sequence of fuzzy positive linear operators. The last condition is fulfilled by almost all operators defined via fuzzy summation or fuzzy integration.

In Chapter 8 we present the fuzzy Korovkin trigonometric theorem via a fuzzy Shisha–Mond trigonometric inequality presented here too. This determines the degree of approximation with rates of a sequence of fuzzy positive linear operators to the fuzzy unit operator. The astonishing fact is that only the real case trigonometric assumptions are enough for the validity of the fuzzy trigonometric Korovkin theorem, along with a very natural realization condition fulfilled by the sequence of fuzzy positive linear operators. The latter condition is satisfied by almost all operators defined via fuzzy summation or fuzzy integration.

In Chapter 9 we present the *property of global smoothness preservation* for fuzzy linear operators acting on spaces of fuzzy continuous functions. Basically we transfer the property of *real global smoothness preservation* into the fuzzy setting, via some *natural realization* condition fulfilled by almost all example-fuzzy linear operators. The derived inequalities involve fuzzy moduli of continuity and we give examples.

In Chapter 10 we study the fuzzy positive linear operators acting on fuzzy continuous functions. We prove the fuzzy Riesz representation theorem, the fuzzy Shisha–Mond type inequalities and fuzzy Korovkin type theorems regarding the fuzzy convergence of fuzzy positive linear operators to the fuzzy unit in various cases. Special attention is paid to the study of fuzzy weak convergence of finite positive measures to the unit Dirac measure. All convergences are with rates and are given via fuzzy inequalities involving the fuzzy modulus of continuity of the engaged fuzzy valued function. The assumptions for the Korovkin theorems are minimal and of natural realization, fulfilled by almost all example – fuzzy positive linear operators. The surprising fact is that the real Korovkin test functions assumptions carry over here in the fuzzy setting and they are the only enough to impose the conclusions of fuzzy Korovkin theorems. We give a lot of examples and applications to our theory, namely: to fuzzy Bernstein operators, to fuzzy Shepard operators, to fuzzy Szasz–Mirakjan and fuzzy Baskakov-type operators and to fuzzy convolution type operators.

We work in general, basically over real normed vector space domains that are compact and convex or just convex. On the way to prove the main theorems we establish a lot of other interesting and important side results.

In Chapter 11 is studied with rates the fuzzy uniform and L_p, $p \geq 1$, convergence of a sequence of fuzzy positive linear operators to the fuzzy unit operator acting on spaces of fuzzy differentiable functions. This is done quantitatively via fuzzy Korovkin type inequalities involving the fuzzy modulus of continuity of a fuzzy derivative of the engaged function. From there we deduce general fuzzy Korovkin type theorems with high rate of convergence. The surprising fact is that basic real positive linear operator simple assumptions enforce here the fuzzy convergences. At the end we give applications. The results are univariate and multivariate. The assumptions are minimal and natural fulfilled by almost all example—fuzzy positive linear operators.

About Chapter 12: The basic wavelet type operators A_k, B_k, C_k, D_k, $k \in \mathbb{Z}$ were studied extensively in the real case, e.g., see [9]. Here they are extended to the fuzzy setting and are defined similarly via a real valued scaling function. Their pointwise and uniform convergence with rates to the fuzzy unit operator I is presented. The produced Jackson type inequalities involve the fuzzy first modulus of continuity and usually are proved to be sharp, in fact attained. Furthermore all fuzzy wavelet like operators A_k, B_k, C_k, D_k preserve monotonicity in the fuzzy sense. Here we do not assume any kind of orthogonality condition on the scaling function φ, and the operators act on fuzzy valued continuous functions over $\mathbb{R}$.

About Chapter 13: The basic fuzzy wavelet type operators A_k, B_k, C_k, D_k, $k \in \mathbb{Z}$ were first introduced in [14], see also Chapter 12, where they were studied among others for their pointwise/uniform convergence with rates to the fuzzy unit operator I. Here we continue this study by estimating the fuzzy distances between these operators. We give the pointwise convergence with rates of these distances to zero. The related approximation is of higher order since we involve these higher order fuzzy derivatives of the engaged fuzzy continuous function f. The derived Jackson type inequalities involve the fuzzy (first) modulus of continuity. Some comparison inequalities are also given so we get better upper bounds to the distances we study. The defining of these operators scaling function φ is of compact support in $[-a, a]$, $a > 0$ and is not assumed to be orthogonal.

About Chapter 14: Here we study four sequences of naturally arising fuzzy integral operators of convolution type that are integral analogs of known fuzzy wavelet type operators, defined via a scaling function. Their fuzzy convergence with rates to the fuzzy unit operator is established through fuzzy inequalities involving the fuzzy modulus of continuity. Also their high order fuzzy approximation is given similarly by involving the fuzzy modulus of continuity of the Nth order ($N \geq 1$) H-fuzzy derivative of the engaged fuzzy number valued function. The *fuzzy global smoothness preservation property* of these operators is presented too.

In Chapter 15 we study the rate of convergence to the unit operator of very specific well described univariate Fuzzy neural network operators of Cardaliaguet–Euvrard and "Squashing" types. These Fuzzy operators arise in a very natural and common way among Fuzzy neural networks. These rates are given through *Jackson type inequalities* involving the Fuzzy modulus of continuity of the engaged Fuzzy valued function or its derivative in the Fuzzy sense. Also several interesting results in Fuzzy real analysis are presented to be used in the proofs of the main results.

In Chapter 16 are studied in terms of fuzzy high approximation to the unit several basic sequences of fuzzy wavelet type operators and fuzzy neural network operators. These operators are fuzzy analogs of earlier studied real ones. The produced results generalize earlier real ones into the fuzzy setting. Here the high order fuzzy pointwise convergence with rates to the fuzzy unit operator is established through fuzzy inequalities involving the fuzzy modulus of continuity of the Nth order ($N \geq 1$) H-fuzzy derivative of the engaged fuzzy number valued function. At the end we present a related L_p result for fuzzy neural network operators.

In Chapter 17 we study the fuzzy random positive linear operators acting on fuzzy random continuous functions. We establish a series of fuzzy random Shisha–Mond type inequalities of L^q-type $1 \leq q < \infty$ and related fuzzy random Korovkin type theorems, regarding the fuzzy random q-mean convergence of fuzzy random positive linear operators to the fuzzy random unit operator for various cases. All convergences are with rates and are given using the above fuzzy random inequalities involving the fuzzy random modulus of continuity of the engaged fuzzy random function. The assumptions for the Korovkin theorems are minimal and of natural realization, fulfilled by almost all example fuzzy random positive linear operators. The astonishing fact is that the real Korovkin test functions assumptions are enough for the conclusions of the fuzzy random Korovkin theory. We give at the end applications.

In Chapter 18 we study the rate of pointwise convergence in the q-mean to the Fuzzy-Random unit operator of very precise univariate Fuzzy-Random neural network operators of Cardaliaguet–Euvrard and "Squashing" types. These Fuzzy-Random operators arise in a natural and common way among Fuzzy-Random neural networks. These rates are given through *Probabilistic-Jackson type inequalities* involving the Fuzzy-Random modulus of continuity of the engaged Fuzzy-Random function or its Fuzzy derivatives. Also several interesting results in Fuzzy-Random Analysis are given of independent merit, which are used then in the proofs of the main results of the chapter.

The aim of Chapter 19 is to present a fuzzy Korovkin-type approximation theorem by using a matrix summability method. We also study the rates of convergence of fuzzy positive linear operators.

The aim of Chapter 20 is to present a fuzzy trigonometric Korovkin-type approximation theorem by using a matrix summability method. We also study the rates of convergence of fuzzy positive linear operators in trigonometric environment.

Finally about Chapter 21: The basic fuzzy wavelet type operators A_k, B_k, C_k, D_k, $k \in \mathbb{Z}$ were studied in [3], [5], for their pointwise and uniform convergence with rates to the fuzzy unit operator. Also they were studied in [6], in terms of estimating their fuzzy differences and giving their pointwise convergence with rates to zero. For prior related and similar study of convergence to the unit of real analogs of these wavelet type operators see [1], section II. Here in Section 1 we present the complete study of finding uniform estimates for the distances between the real Wavelet type operators A_k, B_k, C_k, D_k, $k \in \mathbb{Z}$. Their differences converge to zero with rates. This is done via elegant tight Jackson type inequalities involving the modulus of continuity of the higher order derivative of the engaged real function. Based on these real analysis results in Section 2 we establish the corresponding fuzzy results regarding uniform estimates for the fuzzy differences between the fuzzy wavelet type operators. These fuzzy differences converge to zero with rates given via fuzzy Jackson type tight inequalities. The last inequalities involve the fuzzy modulus of continuity of the higher order fuzzy derivative of the engaged fuzzy function. The defining all these operators real scaling function is not assumed to be orthogonal and is of compact support.

2

ABOUT H-FUZZY DIFFERENTIATION

The concept of H-fuzzy differentiation is discussed thoroughly in the univariate and multivariate cases. Basic H-derivatives are calculated and then important theorems are presented on the topic, such as, the H-mean value theorem, the univariate and multivariate H-chain rules, and the interchange of the order of H-fuzzy differentiation. Finally is given a multivariate H-fuzzy Taylor formula. This treatment relies in [10].

2.1   Introduction

Fuzziness was first introduced in the celebrated paper [103]. For the notion of H-fuzzy derivative see [93] and [53]. First we give some background from Fuzziness, motivation and justification, necessary for the results to follow. In Propositions 2.5, 2.7, 2.8, 2.10 we calculate basic H-fuzzy derivatives. In Lemmas 1.13 and 1.14 we give results on fuzzy continuity, and in Propositions 2.13 and 2.14 we give basic properties of H-fuzzy differentiation. Then come the main results.

Theorem 2.15 is on H-Fuzzy Mean Value Theorem, Lemmas 2.16, 2.17 and 2.20 are auxiliary on fuzzy convergence and fuzzy continuity, Theorem 2.18 is on univariate H-fuzzy chain rule, and Theorem 2.19 is on multivariate H-fuzzy chain rule.

We conclude with Theorem 2.21 on the interchange of the order of H-fuzzy differentiation, and the development of a multivariate H-fuzzy Taylor formula with integral remainder, see Theorem 2.22 and Corollary 2.23.

G.A. Anastassiou: Fuzzy Mathematics: Approximation The., STUDFUZZ 251, pp. 15–50.
springerlink.com   © Springer-Verlag Berlin Heidelberg 2010

2.2 Background

We need the Fuzzy Taylor formula

Theorem 2.1 ([11], see also Theorem 15.14). *Let $T := [x_0, x_0 + \beta] \subset \mathbb{R}$, with $\beta > 0$. We assume that $f^{(i)} : T \to \mathbb{R}_{\mathcal{F}}$ are H-differentiable for all $i = 0, 1, \ldots, n - 1$, for any $x \in T$. (I.e., there exist in $\mathbb{R}_{\mathcal{F}}$ the H-differences $f^{(i)}(x + h) - f^{(i)}(x)$, $f^{(i)}(x) - f^{(i)}(x - h)$, $i = 0, 1, \ldots, n - 1$ for all small $h : 0 < h < \beta$. Furthermore there exist $f^{(i+1)}(x) \in \mathbb{R}_{\mathcal{F}}$ such that the limits in D-distance exist and*

$$f^{(i+1)}(x) = \lim_{h \to 0^+} \frac{f^{(i)}(x + h) - f^{(i)}(x)}{h} = \lim_{h \to 0^+} \frac{f^{(i)}(x) - f^{(i)}(x - h)}{h},$$

for all $i = 0, 1, \ldots, n - 1$.) Also we assume that $f^{(n)}$, is fuzzy continuous on T. Then for $s \geq a$; $s, a \in T$ we obtain

$$\begin{aligned}
f(s) &= f(a) \oplus f'(a) \odot (s - a) \oplus f''(a) \odot \frac{(s - a)^2}{2!} \\
&\oplus \cdots \oplus f^{(n-1)}(a) \odot \frac{(s - a)^{n-1}}{(n - 1)!} \oplus R_n(a, s),
\end{aligned}$$

where

$$R_n(a, s) := (FR) \int_a^s \left(\int_a^{s_1} \cdots \left(\int_a^{s_{n-1}} f^{(n)}(s_n) ds_n \right) ds_{n-1} \right) \cdots \right) ds_1.$$

Here $R_n(a, s)$ is fuzzy continuous on T as a function of s.

N o t e . This formula is invalid when $s < a$, as it is totally based on Corollary 1.12.

Next $\overline{C}[0, 1]$ stands for the class of all real-valued bounded functions f on $[0, 1]$ such that f is left continuous for any $x \in (0, 1]$ and f has a right limit for any $x \in [0, 1)$, especially f is right continuous at 0. With the norm $\|f\| = \sup\limits_{x \in [0,1]} |f(x)|$, $\overline{C}[0, 1]$ is a Banach space [50].

We mention

Theorem 2.2 (Wu and Ma [50]). *For $u \in \mathbb{R}_{\mathcal{F}}$, denote $j : j(u) := (u_-, u_+)$, where $u_{\pm} = u_{\pm}(r) := u_{\pm}^{(r)}$, $0 \leq r \leq 1$. Then $j(\mathbb{R}_{\mathcal{F}})$ is a closed convex cone with vertex 0 in $\overline{C}[0, 1] \times \overline{C}[0, 1]$ (here $\overline{C}[0, 1] \times \overline{C}[0, 1]$ is a Banach space with the norm defined by $\|(f, g)\| := \max(\|f\|, \|g\|)$), and $j : \mathbb{R}_{\mathcal{F}} \to \overline{C}[0, 1] \times \overline{C}[0, 1]$ satisfies*

(1) *for all $u, v \in \mathbb{R}_{\mathcal{F}}$, $s \geq 0$, $t \geq 0$, $j(su + tv) = sj(u) + tj(v)$,*

(2) *$D(u, v) = \|j(u) - j(v)\|$, i.e., j embeds $\mathbb{R}_{\mathcal{F}}$ into $\overline{C}[0, 1] \times \overline{C}[0, 1]$ isometrically and isomorphically.*

We finally mention the important connections of the H-fuzzy derivative to the Fréchet derivative.

Lemma 2.3 (Wu and Ma [51]). *If* $f\colon [a,b] \subseteq \mathbb{R} \to \mathbb{R}_{\mathcal{F}}$ *satisfies the condition* H: *for any* $x \in [a,b]$, *there exists* $\beta > 0$ *such that the* H-*differences of* $f(x+h) - f(x)$, $f(x) - f(x-h)$ *exist for all* $0 < h < \beta$, *then the* H-*differentiability of* $f(x)$ *implies the differentiability of* $(j \circ f)(x)$ *and* $(j \circ f)'(x) \in j(\mathbb{R}_{\mathcal{F}})$, *where the differentiability of* $(j \circ f)(x)$ *on* $\overline{C}[0,1] \times \overline{C}[0,1]$ *is in the Fréchet's sense.*

Lemma 2.4 (Wu and Ma [51]). *If* $(j \circ f)(x)$ *is Fréchet differentiable and* $(j \circ f)'(x) \in j(\mathbb{R}_{\mathcal{F}})$, *then* $f(x)$ *is* H-*differentiable, and* $f'(x) = j^{-1}((j \circ f)'(x))$. *Here* $f\colon [a,b] \to \mathbb{R}_{\mathcal{F}}$, $j\colon \mathbb{R}_{\mathcal{F}} \to (\overline{C}[0,1])^2$, *and* $(j \circ f)\colon [a,b] \to (\overline{C}[0,1])^2$.

2.3   Basic Results

We present

Proposition 2.5. Let $F(t) := t^n \odot u$, $t \geq 0$, $n \in \mathbb{N}$, and $u \in \mathbb{R}_{\mathcal{F}}$ be fixed. Then (the H-derivative)

$$F'(t) = nt^{n-1} \odot u. \tag{2.1}$$

In particular when $n = 1$ then $F'(t) = u$.

Proof. We need to establish that

$$F'(t) = F'_+(t) = F'_-(t),$$

where

$$F'_+(t) := \lim_{h \to 0^+} \frac{(t+h)^n \odot u - t^n \odot u}{h},$$

and

$$F'_-(t) := \lim_{h \to 0^+} \frac{t^n \odot u - (t-h)^n \odot u}{h},$$

the limits are taken with respect to the D-metric.

First we take care of the case $t > 0$, $n \geq 2$. Here h is a small positive quantity approaching zero. By Lemma 4.1 (iii) of [31] we notice that

$$(t+h)^n \odot u = t^n \odot u \oplus \left(\sum_{k=1}^{n} \binom{n}{k} t^{n-k} h^k \right) \odot u,$$

where

$$t^n, \sum_{k=1}^{n} \binom{n}{k} t^{n-k} h^k > 0.$$

That is the H-difference

$$(t+h)^n \odot u - t^n \odot u = \left(\sum_{k=1}^{n} \binom{n}{k} t^{n-k} h^k \right) \odot u$$

exists, and

$$\frac{(t+h)^n \odot u - t^n \odot u}{h} = \left(\sum_{k=1}^{n} \binom{n}{k} t^{n-k} h^{k-1} \right) \odot u.$$

Then we observe that

$$\lim_{h \to 0^+} D \left(\frac{(t+h)^n \odot u - t^n \odot u}{h}, \, nt^{n-1} \odot u \right)$$

$$= \lim_{h \to 0^+} D \left(\left(\sum_{k=1}^{n} \binom{n}{k} t^{n-k} h^{k-1} \right) \odot u, nt^{n-1} \odot u \right)$$

$$\leq \text{ (by Lemma 2.2 of [31])}$$

$$\lim_{h \to 0^+} \left| \left(\sum_{k=1}^{n} \binom{n}{k} t^{n-k} h^{k-1} \right) - nt^{n-1} \right| D(u, \tilde{o})$$

$$= \lim_{h \to 0^+} \left(\sum_{k=2}^{n} \binom{n}{k} t^{n-k} h^{k-1} \right) D(u, \tilde{o}) = 0 D(u, \tilde{o}) = 0.$$

That is

$$F'_+(t) = nt^{n-1} \odot u, \quad t > 0, \quad n \geq 2.$$

Furthermore we notice that

$$F'_-(t) = \lim_{h \to 0^+} \frac{((t-h)+h)^n \odot u - (t-h)^n \odot u}{h}.$$

We set $\beta := t - h$, which for sufficiently small $h > 0$ is positive, i.e., $\beta > 0$. Thus

$$F'_-(t) = \lim_{h \to 0^+} \frac{(\beta+h)^n \odot u - \beta^n \odot u}{h}.$$

Again we have

$$(\beta+h)^n \odot u = \beta^n \odot u \oplus \left(\sum_{k=1}^{n} \binom{n}{k} \beta^{n-k} h^k \right) \odot u,$$

where

$$\beta^n, \, \sum_{k=1}^{n} \binom{n}{k} \beta^{n-k} h^k > 0.$$

That is the H-difference

$$(\beta + h)^n \odot u - \beta^n \odot u = \left(\sum_{k=1}^{n} \binom{n}{k} \beta^{n-k} h^k \right) \odot u$$

exists, and

$$\frac{(\beta + h)^n \odot u - \beta^n \odot u}{h} = \left(\sum_{k=1}^{n} \binom{n}{k} \beta^{n-k} h^{k-1} \right) \odot u.$$

Then we observe that

$$\lim_{h \to 0^+} D \left(\frac{t^n \odot u - (t - h)^n \odot u}{h}, n t^{n-1} \odot u \right)$$

$$= \lim_{h \to 0^+} D \left(\left(\sum_{k=1}^{n} \binom{n}{k} \beta^{n-k} h^{k-1} \right) \odot u, n t^{n-1} \odot u \right)$$

$$\leq \lim_{h \to 0^+} \left| \sum_{k=1}^{n} \binom{n}{k} \beta^{n-k} h^{k-1} - n t^{n-1} \right| D(u, \tilde{o})$$

$$= \lim_{h \to 0^+} \left| n(t - h)^{n-1} + \sum_{k=2}^{n} \binom{n}{k} (t - h)^{n-k} h^{k-1} - n t^{n-1} \right| D(u, \tilde{o})$$

$$= 0 D(u, \tilde{o}) = 0.$$

Hence $F'_-(t) = n t^{n-1} \odot u,\ t > 0,\ n \geq 2$. That is,

$$F'(t) = n t^{n-1} \odot u, \quad t > 0, \quad n \geq 2.$$

Next we treat separately the case of $n = 1,\ t > 0$ for the sake of clarity. Here

$$\lim_{h \to 0^+} D \left(\frac{(t + h) \odot u - t \odot u}{h}, u \right) = \lim_{h \to 0^+} D \left(\frac{h \odot u}{h}, u \right)$$

$$= \lim_{h \to 0^+} D(u, u) = 0.$$

I.e., $F'_+(t) = u,\ t > 0,\ n = 1$. And we see that

$$\lim_{h \to 0^+} D \left(\frac{t \odot u - (t - h) \odot u}{h}, u \right)$$

$$= \lim_{h \to 0^+} D \left(\frac{((t - h) + h) \odot u - (t - h) \odot u}{h}, u \right)$$

$$= \lim_{h \to 0^+} D \left(\frac{(\beta + h) \odot u - \beta \odot u}{h}, u \right)$$

$$= \lim_{h \to 0^+} D \left(\frac{h \odot u}{h}, u \right) = \lim_{h \to 0^+} D(u, u) = 0,$$

where $\beta := t - h > 0$, for sufficiently small $h > 0$. I.e., $F'_-(t) = u$, $t > 0$, $n = 1$. That is

$$F'(t) = u, \quad t > 0, \quad n = 1.$$

At last we do the case of $t = 0$. Here we need to find

$$F'_+(0) = \lim_{h \to 0^+} \frac{h^n \odot u}{h} = \lim_{h \to 0^+} h^{n-1} \odot u.$$

For $n = 1$, we see that

$$\lim_{h \to 0^+} D(h^{n-1} \odot u, u) = \lim_{h \to 0^+} D(u, u) = 0.$$

Thus

$$F'(0) = F'_+(0) = u, \quad \text{for } n = 1.$$

For $n \geq 2$ we see that

$$\lim_{h \to 0^+} D(h^{n-1} \odot u, \tilde{o}) = D(\tilde{o}, \tilde{o}) = 0.$$

Therefore

$$F'(0) = F'_+(0) = \tilde{o}, \quad \text{for } n \geq 2.$$

That is

$$F'(t) = nt^{n-1} \odot u \quad \text{is true for } t = 0.$$

$\square$

R e m a r k 2.6. Let a_i, $i = 1, \ldots$, be a sequence of real numbers all of the same sign such that $\left| \sum_{i=1}^{\infty} a_i \right| < +\infty$. Then

$$\left(\sum_{i=1}^{n} \alpha_i \right) \odot u = \sum_{i=1}^{n}{}^{*} (a_i \odot u), \quad u \in \mathbb{R}_{\mathcal{F}}, \quad \forall n \in \mathbb{N},$$

by Lemma 4.1 (iii) of [31]. Since

$$D\left(\left(\sum_{i=1}^{n} a_i \right) \odot u, \sum_{i=1}^{n}{}^{*} (\alpha_i \odot u) \right) = 0,$$

one obtains

$$\lim_{n \to +\infty} D\left(\left(\sum_{i=1}^{n} a_i \right) \odot u, \sum_{i=1}^{n}{}^{*} (a_i \odot u) \right) = 0.$$

That is

$$\left(\sum_{i=1}^{\infty} a_i \right) \odot u = \sum_{i=1}^{\infty}{}^{*} (a_i \odot u) \in \mathbb{R}_{\mathcal{F}}.$$

Next we give

Proposition 2.7. *Let* $F(x) = x^p \odot u$, $x \geq 0$, $u \in \mathbb{R}_{\mathcal{F}}$, *and* $p > 0$ *not an integer. Then*

$$F'(x) = px^{p-1} \odot u, \quad p > 0, \quad x > 0, \tag{2.2}$$

and

$$F'(o) = \tilde{o}, \quad \text{for } p > 1. \tag{2.3}$$

Proof. When $p > 0$ and $-1 \leq x \leq 1$ from [98], p. 232 we obtain the Binomial series, which converges absolutely

$$(1+x)^p = 1 + px + \frac{p(p-1)}{2!}x^2 + \cdots + \frac{p(p-1)\cdots(p-n+1)}{n!}x^n + \cdots .$$

In the last we plug in instead of x, $\frac{h}{x}$ for $h, x > 0$ and $h \leq x$. Clearly $-1 \leq \frac{h}{x} \leq 1$ is automatically fulfilled and $x + h > 0$. That is

$$\left(1 + \frac{h}{x}\right)^p = 1 + p\frac{h}{x} + \frac{p(p-1)}{2!}\frac{h^2}{x^2} + \cdots$$
$$+ \frac{p(p-1)\cdots(p-n+1)}{n!}\frac{h^n}{x^n} + \cdots .$$

And

$$(x+h)^p = x^p + phx^{p-1} + \frac{p(p-1)}{2!}h^2x^{p-2} + \cdots$$
$$+ \frac{p(p-1)\cdots(p-n+1)}{n!}h^n x^{p-n} + \cdots .$$

By $x + h > x$ we have $(x+h)^p > x^p > 0$ and $(x+h)^p - x^p > 0$. Consequently it holds

$$\Delta := \quad phx^{p-1} + \frac{p(p-1)}{2!}h^2x^{p-2} + \cdots$$
$$+ \frac{p(p-1)\cdots(p-n+1)}{n!}h^n x^{p-n} + \cdots > 0.$$

Therefore

$$(x+h)^p \odot u - x^p \odot u = \Delta \odot u \quad \text{exists in } \mathbb{R}_{\mathcal{F}}.$$

Hence

$$\lim_{h \to 0^+} D\left(\frac{(x+h)^p \odot u - x^p \odot u}{h}, px^{p-1} \odot u\right)$$

$$= \lim_{h \to 0^+} D\left(\frac{\Delta}{h} \odot u, px^{p-1} \odot u\right) \leq \lim_{h \to 0^+} \left|\frac{\Delta}{h} - px^{p-1}\right| D(u, \tilde{o})$$

$$= \lim_{h \to 0^+} \left|px^{p-1} + \frac{p(p-1)}{2!}hx^{p-2} + \cdots\right.$$
$$\left. + \frac{p(p-1)\cdots(p-n+1)}{n!}h^{n-1}x^{p-n} + \cdots - px^{p-1}\right| D(u, \tilde{o})$$

$$= 0D(u, \tilde{o}) = 0.$$

That is,

$$F'_+(x) = (x^p \odot u)'_+ = px^{p-1} \odot u, \quad p > 0, \quad x > 0.$$

Next we evaluate in D-metric

$$
\begin{aligned}
F'_-(t) &= \lim_{h \to 0+} \frac{x^p \odot u - (x-h)^p \odot u}{h} \\
&= \lim_{h \to 0+} \frac{((x-h)+h)^p \odot u - (x-h)^p \odot u}{h} \\
&= \lim_{h \to 0+} \frac{(\beta + h)^p \odot u - \beta^p \odot u}{h},
\end{aligned}
$$

where $\beta := x - h > 0$, for $h > 0$ small enough. In fact we choose h such that $2h < x$, that is, $h < x - h = \beta$. I.e., $0 < h < \beta$. Next we apply the Binomial series for $\frac{h}{\beta}$. Thus

$$
\begin{aligned}
(\beta + h)^p &= \beta^p + ph\beta^{p-1} + \frac{p(p-1)}{2!}h^2\beta^{p-2} + \cdots \\
&+ \frac{p(p-1)\cdots(p-n+1)}{n!}h^n\beta^{p-n} + \cdots.
\end{aligned}
$$

Clearly $\beta + h > \beta$ and $(\beta+h)^p > \beta^p > 0$, by $p > 0$. And $(\beta+h)^p - \beta^p > 0$. Hence

$$
\begin{aligned}
\Delta^* := \quad & ph\beta^{p-1} + \frac{p(p-1)}{2!}h^2\beta^{p-2} + \cdots \\
& + \frac{p(p-1)\cdots(p-n+1)}{n!}h^n\beta^{p-n} + \cdots > 0.
\end{aligned}
$$

Therefore

$$(\beta + h)^p \odot u - \beta^p \odot u = \Delta^* \odot u \quad \text{exists in } \mathbb{R}_{\mathcal{F}}.$$

Furthermore we have

$$
\begin{aligned}
&\lim_{h \to 0+} D\left(\frac{(\beta + h)^p \odot u - \beta^p \odot u}{h}, px^{p-1} \odot u \right) \\
&= \lim_{h \to 0+} D\left(\frac{\Delta^*}{h} \odot u, px^{p-1} \odot u \right) \\
&= \lim_{h \to 0+} D\left(\left(p\beta^{p-1} + \frac{p(p-1)}{2!}h\beta^{p-2} + \cdots \right.\right. \\
&\quad \left.\left. + \frac{p(p-1)\cdots(p-n+1)}{n!}h^{n-1}\beta^{p-n} \right) \odot u, px^{p-1} \odot u \right) \\
&= D(px^{p-1} \odot u, px^{p-1} \odot u) = 0.
\end{aligned}
$$

I.e., $F'_-(x) = (x^p \odot u)'_- = px^{p-1} \odot u$, $p > 0$, $x > 0$. That is

$$F'(x) = (x^p \odot u)' = px^{p-1} \odot u, \quad p > 0, \quad x > 0.$$

Finally at $x = 0$ we get

$$F_+''(0) = \lim_{h \to 0^+} \frac{(o + h)^p \odot u}{h} = \lim_{h \to 0^+} h^{p-1} \odot u.$$

Hence

$$\lim_{h \to 0^+} D(h^{p-1} \odot u, \tilde{o}) = D(\tilde{o}, \tilde{o}) = 0, \quad p > 1.$$

I.e., $F'(0) = (x^p \odot u)'\big|_{x=0} = \tilde{o}, p > 1$.   □

It follows

Proposition 2.8. *Let $u \in \mathbb{R}_{\mathcal{F}}$ be fixed. Then*

$$(e^x \odot u)' = e^x \odot u, \quad \text{any } x \in \mathbb{R}. \tag{2.4}$$

Proof. We have

$$e^x = 1 + x + \frac{x^2}{2!} + \frac{x^3}{3!} + \cdots + \frac{x^n}{n!} + \cdots, \quad -\infty < x < +\infty.$$

Then

$$e^{x+h} = 1 + (x + h) + \frac{(x+h)^2}{2!} + \frac{(x+h)^3}{3!} + \cdots + \frac{(x+h)^n}{n!} + \cdots, \quad h > 0.$$

Consequently we get

$$
\begin{aligned}
e^{x+h} - e^x &= h + \left(\frac{2xh + h^2}{2!} \right) + \left(\frac{3x^2h + 3xh^2 + h^3}{3!} \right) \\
&\quad + \cdots + \left(\frac{\sum\limits_{k=1}^{n} \binom{n}{k} x^{n-k} h^k}{n!} \right) + \cdots =: \Delta.
\end{aligned}
$$

Here $x \in \mathbb{R}$ and $x + h > x$. Since e^x is increasing then $e^{x+h} > e^x > 0$ and $e^{x+h} - e^x > 0$. I.e., $\Delta > 0$.

Therefore the next H-difference and quotient makes sense in $\mathbb{R}_{\mathcal{F}}$,

$$
\begin{aligned}
\frac{e^{x+h} \odot u - e^x \odot u}{h} &= \frac{\Delta}{h} \odot u \\
&= \left\{ 1 + \left(\frac{2x + h}{2!} \right) + \left(\frac{3x^2 + 3xh + h^2}{3!} \right) + \cdots \right. \\
&\quad \left. + \left(\frac{\sum\limits_{k=1}^{n} \binom{n}{k} x^{n-k} h^{k-1}}{n!} \right) + \cdots \right\} \odot u =: K \odot u, \quad K > 0.
\end{aligned}
$$

Thus

$$\lim_{h\to 0^+} D(K\odot u, e^x\odot u) \;\le\; \lim_{h\to 0^+} |K - e^x| D(u,\tilde{o})$$

$$= \left| 1 + x + \frac{x^2}{2!} + \cdots + \frac{x^n}{n!} + \cdots - e^x \right|.$$

$$D(u,\tilde{o}) = |e^x - e^x| D(u,\tilde{o}) = 0.$$

We prove that $(e^x \odot u)'_+ = e^x \odot u$.

Next we evaluate

$$(e^x \odot u)'_- = \lim_{h\to 0^+} \frac{e^x \odot u - e^{x-h} \odot u}{h}, \quad x \in \mathbb{R}, \; u \in \mathbb{R}_{\mathcal{F}}.$$

By setting $\beta := x - h$ we get

$$(e^x \odot u)'_- = \lim_{h\to 0^+} \frac{e^{\beta+h} \odot u - e^{\beta} \odot u}{h}.$$

Again we have $\beta + h > \beta$ and $e^{\beta+h} > e^{\beta} > 0$, and $e^{\beta+h} - e^{\beta} > 0$. Furthermore it holds

$$e^{\beta+h} - e^{\beta} = h + \left(\frac{2\beta h + h^2}{2!}\right) + \left(\frac{3\beta^2 h + 3\beta h^2 + h^3}{3!}\right)$$

$$+ \cdots + \left(\frac{\sum\limits_{k=1}^{n} \binom{n}{k}\beta^{n-k} h^k}{n!}\right) + \cdots =: \Delta^*.$$

Clearly $0 < \Delta^* < +\infty$.

The next make sense in $\mathbb{R}_{\mathcal{F}}$

$$\frac{e^{\beta+h} \odot u - e^{\beta} \odot u}{h} = \frac{\Delta^*}{h} \odot u$$

$$= \left\{ 1 + \left(\frac{2\beta + h}{2!}\right) + \left(\frac{3\beta^2 + 3\beta h + h^2}{3!}\right) + \cdots \right.$$

$$\left. + \left(\frac{\sum\limits_{k=1}^{n} \binom{n}{k}\beta^{n-k} h^{k-1}}{n!}\right) + \cdots \right\} \odot u$$

$$=: K^* \odot u, \quad K^* > 0.$$

Thus

$$\lim_{h\to 0^+} D(K^* \odot u, e^x \odot u) \le \lim_{h\to 0^+} |K^* - e^x| D(u,\tilde{o})$$

$$= \left| 1 + x + \frac{x^2}{2} + \cdots + \frac{x^n}{n!} + \cdots - e^x \right| D(u,\tilde{o}) = 0.$$

We have established
$$(e^x \odot u)'_- = e^x \odot u,$$
and finally proved (2.4). □

Note. Clearly $(e^x \odot u)^{(\ell)} = e^x \odot u$, $\ell \in \mathbb{N}$, $u \in \mathbb{R}_{\mathcal{F}}$ is fixed, $x \in \mathbb{R}$.

Next we need

Bernstein's Theorem 2.9 (see [37], p. 418). *Assume that $f \in C^\infty$ on an open interval of the form $(a - \delta, b)$, where $\delta > 0$, and suppose that f and all its derivatives are non-negative in the half-open interval $[a, b)$. Then, for every x_0 in $[a, b)$, we have*
$$f(x) = \sum_{n=0}^{\infty} \frac{f^{(n)}(x_0)}{n!}(x - x_0)^n, \quad \text{if } x_0 \leq x < b.$$

We present

Proposition 2.10. *Let $u \in \mathbb{R}_{\mathcal{F}}$ be fixed, and $f \in C^\infty(-\varepsilon, r)$, $\varepsilon > 0$, $r > 0$ and assume that $f, f', f'', \ldots \geq 0$ on $[0, r)$, with $f(0) = 0$. Then*
$$(f(x) \odot u)' = f'(x) \odot u, \quad \text{for } 0 \leq x < r. \tag{2.5}$$

Clearly
$$(f(x) \odot u)^{(\ell)} = f^{(\ell)}(x) \odot u, \quad \text{for } 0 \leq x < r, \quad \ell \in \mathbb{N}. \tag{2.6}$$

E.g., $f(x) = \sin hx$.

Proof. By Bernstein's Theorem we have
$$f(x) = \sum_{n=0}^{\infty} \frac{f^{(n)}(0)}{n!} x^n,$$
and
$$f(x + h) = \sum_{n=0}^{\infty} \frac{f^{(n)}(0)}{n!}(x + h)^n, \quad x \in [0, r)$$
and $h > 0$ such that $x + h \in [0, r)$. Since f is non-decreasing we have $f(x + h) \geq f(x) \geq 0$, and $f(x + h) - f(x) \geq 0$. Consequently we see that
$$f(x + h) - f(x) = \sum_{n=0}^{\infty} \frac{f^{(n)}(0)}{n!}((x + h)^n - x^n)$$
$$= \sum_{n=0}^{\infty} \frac{f^{(n)}(0)}{n!}\left(\sum_{k=1}^{n} \binom{n}{k} x^{n-k} h^k\right) \geq 0.$$

Thus
$$\frac{f(x + h) - f(x)}{h} = \sum_{n=0}^{\infty} \frac{f^{(n)}(0)}{n!}\left(\sum_{k=1}^{n} \binom{n}{k} x^{n-k} h^{k-1}\right) \geq 0.$$

Therefore the next makes sense in $\mathbb{R}_{\mathcal{F}}$

$$\frac{f(x+h)\odot u - f(x)\odot u}{h} = \sum_{n=0}^{\infty}\frac{f^{(n)}(0)}{n!}\left(\sum_{k=1}^{n}\binom{n}{k}x^{n-k}h^{k-1}\right)\odot u.$$

Then

$$\lim_{h\to 0^+} D\left(\left(\sum_{n=0}^{\infty}\frac{f^{(n)}(0)}{n!}\left(\sum_{k=1}^{n}\binom{n}{k}x^{n-k}h^{k-1}\right)\right)\odot u, f'(x)\odot u\right)$$

$$\leq \lim_{h\to 0^+}\left|\sum_{n=0}^{\infty}\frac{f^{(n)}(0)}{n!}\left(\sum_{k=1}^{n}\binom{n}{k}x^{n-k}h^{k-1}\right) - f'(x)\right| D(u,\tilde{o})$$

$$= \left|\sum_{n=1}^{\infty}\frac{f^{(n)}(0)}{n!}(nx^{n-1}) - f'(x)\right| D(u,\tilde{o})$$

$$= |f'(x) - f'(x)| D(u,\tilde{o}) = 0.$$

That is,

$$(f(x)\odot u)'_+ = f'(x)\odot u, \quad 0 \leq x < r.$$

Call $\beta := x - h$, $x > 0$, $x > h$ as $h \to 0^+$. Clearly $\beta > 0$. Here

$$f(x) = \sum_{n=0}^{\infty}\frac{f^{(n)}(0)}{n!}(\beta + h)^n,$$

and

$$f(x-h) = f(\beta) = \sum_{n=0}^{\infty}\frac{f^{(n)}(0)}{n!}\beta^n.$$

Also $f(x)$, $f(x-h) \geq 0$ and $f(x) \geq f(x-h)$. Thus

$$f(x) - f(x-h) = \sum_{n=0}^{\infty}\frac{f^{(n)}(0)}{n!}((\beta+h)^n - \beta^n)$$

$$= \sum_{n=0}^{\infty}\frac{f^{(n)}(0)}{n!}\left(\sum_{k=1}^{n}\binom{n}{k}\beta^{n-k}h^k\right) \geq 0.$$

Furthermore

$$\frac{f(x) - f(x-h)}{h} = \sum_{n=0}^{\infty}\frac{f^{(n)}(0)}{n!}\left(\sum_{k=1}^{n}\binom{n}{k}\beta^{n-k}h^{k-1}\right) \geq 0.$$

Consequently

$$\lim_{h \to 0^+} D\left(\frac{f(x) \odot u - f(x-h) \odot u}{h}, f'(x) \odot u \right)$$

$$= \lim_{h \to 0^+} D\left(\left(\sum_{n=0}^{\infty} \frac{f^{(n)}(0)}{n!} \left(\sum_{k=1}^{n} \binom{n}{k} \beta^{n-k} h^{k-1} \right) \odot u, f'(x) \odot u \right) \right)$$

$$\leq \lim_{h \to 0^+} \left| \sum_{n=0}^{\infty} \frac{f^{(n)}(0)}{n!} \left(\sum_{k=1}^{n} \binom{n}{k} \beta^{n-k} h^{k-1} \right) - f'(x) \right|.$$

$$D(u, \tilde{o}) = \left| \sum_{n=1}^{\infty} \frac{f^{(n)}(0)}{n!} (nx^{n-1}) - f'(x) \right| D(u, \tilde{o})$$

$$= |f'(x) - f'(x)| D(u, \tilde{o}) = 0.$$

I.e.,

$$(f(x) \odot u)'_- = f'(x) \odot u, \quad 0 < x < r.$$

We have established (2.5).   □

Note. One can do other examples of calculation of H-derivatives of basic fuzzy functions, working as above with power series over appropriate intervals.

We mention

Lemma 2.11. *Let $f, g: (a, b) \subseteq \mathbb{R} \to \mathbb{R}_{\mathcal{F}}$ be fuzzy continuous functions. Assume that the H-difference function $f - g$ exists on (a, b). Then $f - g$ is a fuzzy continuous function on (a, b).*

Proof. Let $x_n, x \in (a, b)$ such that $x_n \to x$, as $n \to +\infty$. We observe that

$$D(f(x_n) - g(x_n), f(x) - g(x))$$
$$= D(f(x_n) - g(x_n) \oplus g(x_n), g(x_n) \oplus f(x) - g(x))$$
$$= D(f(x_n), g(x_n) \oplus f(x) - g(x))$$
$$= D(f(x_n) \oplus g(x), g(x_n) \oplus f(x) - g(x) \oplus g(x))$$
$$= D(f(x_n) \oplus g(x), g(x_n) \oplus f(x))$$
$$\leq D(f(x_n), f(x)) + D(g(x_n), g(x)) \to 0.$$

□

Lemma 2.12. *Let U be an open subset of $\mathbb{R}^2$ and let $f, g: U \to \mathbb{R}_{\mathcal{F}}$ be fuzzy continuous (jointly) in $(x, y) \in U$. Then $D(f(x, y), g(x, y))$ is continuous (jointly) in (x, y).*

Proof. It is similar to [66], p. 644, Lemma 13.2 (ii). It goes as follows: Let $U \ni z_n := (x_n, y_n) \to z := (x, y)$, as $n \to +\infty$. We have

$$D(f(z_n), g(z_n)) \leq D(f(z_n), f(z)) + D(f(z), g(z)) + D(g(z), g(z_n)),$$

and

$$D(f(z), g(z)) \leq D(f(z), f(z_n)) + D(f(z_n), g(z_n)) + D(g(z_n), g(z)).$$

Passing to the limit as $n \to +\infty$, from the continuity of f and g we obtain

$$\lim_{n \to +\infty} D(f(z_n), g(z_n)) = D(f(z), g(z)).$$

□

We give

Proposition 2.13. *Let I be an open interval of $\mathbb{R}$ and let $f, g \colon I \to \mathbb{R}_{\mathcal{F}}$ be fuzzy differentiable functions with H-derivatives f', g'. Then $(f \oplus g)'$ exists and*

$$(f \oplus g)' = f' \oplus g'. \tag{2.7}$$

Proof. Let $h \to 0^+$, then by assumption

$$\alpha := f(x + h) - f(x), \quad \beta := g(x + h) - g(x) \in \mathbb{R}_{\mathcal{F}}.$$

Hence $f(x + h) = \alpha \oplus f(x)$, $g(x + h) = \beta \oplus g(x)$. Thus

$$(f \oplus g)(x + h) = \alpha \oplus \beta \oplus (f \oplus g)(x),$$

i.e.,

$$(f \oplus g)(x + h) - (f \oplus g)(x) = \alpha \oplus \beta.$$

Therefore

$$D\left(\frac{(f \oplus g)(x + h) - (f \oplus g)(x)}{h}, f'(x) \oplus g'(x)\right)$$

$$= D\left(\frac{\alpha}{h} \oplus \frac{\beta}{h}, f'(x) \oplus g'(x)\right)$$

$$\leq D\left(\frac{\alpha}{h}, f'(x)\right) + D\left(\frac{\beta}{h}, g'(x)\right) \to 0, \quad \text{as } h \to 0^+.$$

Next we set

$$\gamma := f(x) - f(x - h), \quad \delta := g(x) - g(x - h).$$

Clearly $\gamma, \delta \in \mathbb{R}_{\mathcal{F}}$. Then $f(x) = \gamma \oplus f(x - h)$, $g(x) = \delta \oplus g(x - h)$. Hence

$$(f \oplus g)(x) = (\gamma \oplus \delta) \oplus (f \oplus g)(x - h),$$

i.e.,

$$(f \oplus g)(x) - (f \oplus g)(x - h) = \gamma \oplus \delta.$$

Therefore

$$D\left(\frac{(f \oplus g)(x) - (f \oplus g)(x - h)}{h}, f'(x) \oplus g'(x)\right)$$

$$= D\left(\frac{\gamma \oplus \delta}{h}, f'(x) \oplus g'(x)\right)$$

$$\leq D\left(\frac{\gamma}{h}, f'(x)\right) + D\left(\frac{\delta}{h}, g'(x)\right) \to 0, \quad \text{as } h \to 0^+.$$

That is, proving the claim. □

The counterpart of the above follows.

Proposition 2.14. *Let I be an open interval of $\mathbb{R}$ and let $f: \to \mathbb{R}_{\mathcal{F}}$ be H-fuzzy differentiable, $c \in \mathbb{R}$. Then*

$$(c \odot f)' \text{ exists and } (c \odot f)' = c \odot f'(x). \tag{2.8}$$

Proof. We see

$$D\left(\frac{(c \odot f)(x + h) - (c \odot f)(x)}{h}, c \odot f'(x)\right)$$

$$= D\left(\frac{c \odot f(x + h) - c \odot f(x)}{h}, c \odot f'(x)\right) =: (*).$$

Here $\alpha := f(x + h) - f(x) \in \mathbb{R}_{\mathcal{F}}$, so that $f(x + h) = \alpha \oplus f(x)$. Then

$$c \odot f(x + h) = c \odot \alpha \oplus c \odot f(x).$$

I.e., $c \odot f(x + h) - c \odot f(x) = c \odot a$. Therefore

$$(*) = D\left(\frac{c \odot a}{h}, c \odot f'(x)\right)$$

$$= |c|D\left(\frac{a}{h}, f'(x)\right) \to 0, \quad \text{as } h \to 0^+.$$

Next let $\beta := f(x) - f(x - h) \in \mathbb{R}_{\mathcal{F}}$, so that $f(x) = \beta \oplus f(x - h)$. Hence

$$c \odot f(x) = c \odot \beta \oplus c \odot f(x - h),$$

i.e.,

$$c \odot f(x) - c \odot f(x - h) = c \odot \beta.$$

Therefore

$$D\left(\frac{(c \odot f)(x) - (c \odot f)(x - h)}{h}, c \odot f'(x)\right)$$

$$= D\left(\frac{c \odot f(x) - c \odot f(x - h)}{h}, c \odot f'(x)\right)$$

$$= D\left(\frac{c \odot \beta}{h}, c \odot f'(x)\right) = |c|D\left(\frac{\beta}{h}, f'(x)\right) \to 0, \quad \text{as } h \to 0^+.$$

That is establishing the claim.   □

N o t e. Linearity is true in *H*-fuzzy differentiation, that is

$$(\lambda \odot f \oplus \mu \odot g)' = \lambda \odot f' \oplus \mu \odot g',$$

when $\lambda, \mu \in \mathbb{R}$ and f, g are *H*-fuzzy differentiable.

2.4   Main Results

We present the "Fuzzy Mean Value Theorem".

Theorem 2.15. *Let* $f \colon [a, b] \to \mathbb{R}_{\mathcal{F}}$ *be a fuzzy differentiable function on* $[a, b]$ *with H-fuzzy derivative* f' *which is assumed to be fuzzy continuous. Then*

$$D(f(d), f(c)) \le (d - c) \sup_{t \in [c,d]} D(f'(t), \tilde{o}), \qquad (2.9)$$

for any $c, d \in [a, b]$ *with* $d \ge c$.

Proof. By Corollary A of [11] it holds that

$$f(c) = f(a) \oplus (FR) \int_a^c f'(t) dt,$$

and

$$f(d) = f(a) \oplus (FR) \int_a^d f'(t) dt.$$

Then

$$
\begin{aligned}
D(f(d), f(c)) &= D\left(f(a) \oplus (FR) \int_a^d f'(t) dt, f(a) \oplus (FR) \int_a^c f'(t) dt \right) \\
&= D\left((FR) \int_a^d f'(t) dt, (FR) \int_a^c f'(t) dt \right) \\
&= D\left((FR) \int_a^c f'(t) dt \oplus (FR) \int_c^d f'(t) dt, (FR) \int_a^c f'(t) dt \right) \\
&= D\left((FR) \int_c^d f'(t) dt, \tilde{o} \right) =: (*).
\end{aligned}
$$

Clearly $k \odot \tilde{o} = \tilde{o}$ for $k \in \mathbb{R}$. And

$$\tilde{o} = \tilde{o} \odot (d - c) = \tilde{o} \odot \int_c^d 1 \, dt = (FR) \int_c^d (\tilde{o} \odot 1) dt = (FR) \int_c^d \tilde{o} \, dt.$$

Hence

$$(*) = D\left((FR)\int_c^d f'(t)dt, (FR)\int_c^d \tilde{o}\,dt\right)$$

$$\text{(by Lemma 1, [11])} \le \int_c^d D(f'(t), \tilde{o})dt \le (d - c)\sup_{t\in[c,d]} D(f'(t), \tilde{o}) < +\infty,$$

by Lemma 2 of [11]. □

We need

Lemma 2.16. *Let* $u_n, v_n, u, v \in \mathbb{R}_{\mathcal{F}}$, $n \in \mathbb{N}$. *Let* $u_n \to u$, $v_n \to v$, *as* $n \to +\infty$. *Then* $D(u_n, v_n) \to D(u, v)$, *as* $n \to +\infty$ *(i.e.,* $D(u, v)$ *is continuous in* (u, v)*). In particular* $D(u_n, v) \to D(u, v)$, *as* $n \to +\infty$. *We write*

$$\lim_{n\to+\infty} D(u_n, v_n) = D\left(\lim_{n\to+\infty} u_n, \lim_{n\to+\infty} v_n\right) = D(u, v).$$

Lemma 2.17. *Let* $u_n, u \in \mathbb{R}_{\mathcal{F}}$; $c_n, c \in \mathbb{R}_+$, *such that* $u_n \to u$ *and* $c_n \to c$, *as* $n \to +\infty$. *Then in* D-*metric*

$$u_n \odot c_n \to u \odot c, \quad as\ n \to +\infty,$$

i.e.,

$$\lim_{n\to+\infty}(u_n \odot c_n) = \left(\lim_{n\to+\infty} u_n\right) \odot \left(\lim_{n\to+\infty} c_n\right) = u \odot c.$$

Proof. We notice that

$$D(u_n \odot c_n, u \odot c) \le D(u_n \odot c_n, u_n \odot c) + D(u_n \odot c, u \odot c)$$

$$\text{(by Lemma 2.2, [31])} \le |c_n - c|D(u_n, \tilde{o}) + cD(u_n, u)$$

$$\text{(by Lemma 2.16)} \to 0D(u, \tilde{o}) + c0 = 0.$$

That is

$$\lim_{n\to+\infty} D(u_n \odot c_n, u \odot c) = 0.$$

We present the "Univariate Fuzzy Chain Rule".

Theorem 2.18. *Let* I *be a closed interval in* $\mathbb{R}$. *Here* $g: I \to \zeta := g(I) \subseteq \mathbb{R}$ *is differentiable, and* $f: \zeta \to \mathbb{R}_{\mathcal{F}}$ *is H-fuzzy differentiable. Assume that* g *is strictly increasing. Then* $(f \circ g)'(x)$ *exists and*

$$(f \circ g)'(x) = f'(g(x)) \odot g'(x), \quad \forall x \in I. \tag{2.10}$$

Proof. Call $u := g(x)$. Let $\Delta x > 0$, such that $\Delta x \to 0^+$.

i) Let $\Delta u := g(x + \Delta x) - g(x)$. Then $\Delta u > 0$, and as $\Delta x \to 0^+$ we get $\Delta u \to 0^+$ by continuity of g. See that $g(x + \Delta x) = u + \Delta u$. We observe that

$$\lim_{\Delta x \to 0^+} D\left(\frac{f(g(x + \Delta x)) - f(g(x))}{\Delta x}, f'(g(x)) \odot g'(x) \right)$$

$$= \lim_{\Delta x \to 0^+} D\left(\left(\frac{f(g(x + \Delta x)) - f(g(x))}{g(x + \Delta x) - g(x)} \right) \right.$$

$$\odot \left. \left(\frac{g(x + \Delta x) - g(x)}{\Delta x} \right), f'(g(x)) \odot g'(x) \right)$$

$$= \lim_{\Delta x \to 0^+} D\left(\left(\frac{f(u + \Delta u) - f(u)}{\Delta u} \right) \right.$$

$$\odot \left. \left(\frac{g(x + \Delta x) - g(x)}{\Delta x} \right), f'(g(x)) \odot g'(x) \right)$$

$$= D(f'(u) \odot g'(x), f'(g(x)) \odot g'(x)) = 0,$$

by Lemmas 2.16 and 2.17. I.e.,

$$(f \odot g)'_+ = f'(g(x)) \odot g'(x).$$

ii) Let $\Delta u := g(x) - g(x - \Delta x)$. Then $\Delta u > 0$, and as $\Delta x \to 0^+$ we get $\Delta u \to 0^+$ by continuity of g. Notice that $g(x - \Delta x) = u - \Delta u$. We observe that

$$\lim_{\Delta x \to 0^+} D\left(\frac{f(g(x)) - f(g(x - \Delta x))}{\Delta x}, f'(g(x)) \odot g'(x) \right)$$

$$= \lim_{\Delta x \to 0^+} D\left(\left(\frac{f(g(x)) - f(g(x - \Delta x))}{g(x) - g(x - \Delta x)} \right) \right.$$

$$\odot \left. \left(\frac{g(x) - g(x - \Delta x)}{\Delta x} \right), f'(g(x)) \odot g'(x) \right)$$

$$= \lim_{\Delta x \to 0^+} D\left(\left(\frac{f(u) - f(u - \Delta u)}{\Delta u} \right) \right.$$

$$\odot \left. \left(\frac{g(x) - g(x - \Delta x)}{\Delta x} \right), f'(g(x)) \odot g'(x) \right)$$

$$= D(f'(u) \odot g'(x), f'(g(x)) \odot g'(x)) = 0,$$

by Lemmas 2.16 and 2.17. I.e.,

$$(f \circ g)'_- = f'(g(x)) \odot g'(x).$$

At the endpoints of I we take one-sided derivatives.   □

Next follows the multivariate fuzzy chain rule.

Theorem 2.19. Let $\phi_i \colon [a, b] \subseteq \mathbb{R} \to \phi_i([a, b]) := I_i \subseteq \mathbb{R}$, $i = 1, \ldots, n$, $n \in \mathbb{N}$, are strictly increasing and differentiable functions. Denote $x_i :=$

$x_i(t) := \phi_i(t)$, $t \in [a, b]$, $i = 1, \ldots, n$. Consider U an open subset of $\mathbb{R}^n$ such that $\times_{i=1}^n I_i \subseteq U$. Consider $f : U \to \mathbb{R}_{\mathcal{F}}$ a fuzzy continuous function. Assume that $f_{x_i} : U \to \mathbb{R}_{\mathcal{F}}$, $i = 1, \ldots, n$, the H-fuzzy partial derivatives of f, exist and are fuzzy continuous. Call $z := z(t) := f(x_1, \ldots, x_n)$. Then $\frac{dz}{dt}$ exists and

$$\frac{dz}{dt} = \sum_{i=1}^{n}{}^{*} \frac{dz}{dx_i} \odot \frac{dx_i}{dt}, \quad \forall t \in [a, b] \tag{2.11}$$

where $\frac{dz}{dt}$, $\frac{dz}{dx_i}$, $i = 1, \ldots, n$ are the H-fuzzy derivatives of f with respect to t, x_i, respectively.

Proof. Let first $t \in (a, b)$. Let a general $(x_1, x_2, \ldots, x_n) \in U$ be fixed and let $\Delta x_i > 0$, $i = 1, \ldots, n$, be small.

I) Call

$$\begin{aligned} \alpha_1 \quad := \quad & f(x_1 + \Delta x_1, x_2 + \Delta x_2, \ldots, x_n + \Delta x_n) \\ & - f(x_1, x_2 + \Delta x_2, \ldots, x_n + \Delta x_n) \in \mathbb{R}_{\mathcal{F}}. \end{aligned}$$

That is

$$f(x_1 + \Delta x_1, x_2 + \Delta x_2, \ldots, x_n + \Delta x_n) = \alpha_1 \oplus f(x_1, x_2 + \Delta x_2, \ldots, x_n + \Delta x_n).$$

Call

$$\begin{aligned} \alpha_2 \quad := \quad & f(x_1, x_2 + \Delta x_2, \ldots, x_n + \Delta x_n) \\ & - f(x_1, x_2, x_3 + \Delta x_3, \ldots, x_n + \Delta x_n) \in \mathbb{R}_{\mathcal{F}}. \end{aligned}$$

That is

$$f(x_1, x_2 + \Delta x_2, \ldots, x_n + \Delta x_n) = \alpha_2 \oplus f(x_1, x_2, x_3 + \Delta x_3, \ldots, x_n + \Delta x_n).$$

Call

$$\begin{aligned} \alpha_3 \quad := \quad & f(x_1, x_2, x_3 + \Delta x_3, \ldots, x_n + \Delta x_n) \\ & - f(x_1, x_2, x_3, x_4 + \Delta x_4, \ldots, x_n + \Delta x_n) \in \mathbb{R}_{\mathcal{F}}. \end{aligned}$$

That is

$$f(x_1, x_2, x_3 + \Delta x_3, \ldots, x_n + \Delta x_n) = \alpha_3 \oplus f(x_1, x_2, x_3, x_4 + \Delta x_4, \ldots, x_n + \Delta x_n).$$

Etc. Call

$$a_n := f(x_1, x_2, \ldots, x_{n-1}, x_n + \Delta x_n) - f(x_1, x_2, \ldots, x_n) \in \mathbb{R}_{\mathcal{F}}.$$

That is

$$f(x_1, x_2, \ldots, x_{n-1}, x_n + \Delta x_n) = \alpha_n \oplus f(x_1, x_2, \ldots, x_n).$$

I.e., it holds

$$\mathbb{R}_{\mathcal{F}} \in f(x_1 + \Delta x_1, x_2 + \Delta x_2, \ldots, x_n + \Delta x_n) - f(x_1, x_2, \ldots, x_n) = \sum_{i=1}^{n}{}^{*} \alpha_i.$$

Since the partial derivatives f_{x_i} exist, the above H-differences α_i, $i = 1, \ldots, n$ exist in $\mathbb{R}_{\mathcal{F}}$ for small $\Delta x_i > 0$. In particular we define

$$\Delta x_i := \phi_i(t + \Delta t) - \phi_i(t), \quad \Delta t > 0, \quad i = 1, \ldots n$$

(i.e.,

$$\phi_i(t + \Delta t) = x_i + \Delta x_i, \quad x_i := \phi_i(t)).$$

Since ϕ_i, $i = 1, \ldots, n$ are strictly increasing we have that $\Delta x_i > 0$. So as $\Delta t \to 0^+$, then $\Delta x_i \to 0^+$ by continuity of ϕ_i.

We observe that

$$\lim_{\Delta t \to 0^+} D\left(\frac{f(\phi_1(t + \Delta t), \ldots, \phi_n(t + \Delta t)) - f(\phi_i(t), \ldots, \phi_n(t))}{\Delta t}, \right.$$

$$\left. \sum_{i=1}^{n}{}^{*} f_{x_i}(x_1, \ldots, x_n) \odot x_i'(t) \right)$$

$$= \lim_{\Delta t \to 0^+} D\left(\frac{f(x_1 + \Delta x_1, \ldots, x_n + \Delta x_n) - f(x_1, \ldots, x_n)}{\Delta t}, \right.$$

$$\left. \sum_{i=1}^{n}{}^{*} f_{x_i}(x_1, \ldots, x_n) \odot x_i'(t) \right)$$

$$= \lim_{\Delta t \to 0^+} D\left(\frac{\sum_{i=1}^{n}{}^{*} \alpha_i}{\Delta t}, \sum_{i=1}^{n}{}^{*} f_{x_i}(x_1, \ldots, x_n) \odot x_i'(t) \right)$$

$$\leq \lim_{\Delta t \to 0^+} D\left(\frac{f(x_1 + \Delta x_1, x_2 + \Delta x_2, \ldots, x_n + \Delta x_n)}{\Delta t} \right.$$

$$\left. - \frac{f(x_1, x_2 + \Delta x_2, \ldots, x_n + \Delta x_n)}{\Delta t}, \; f_{x_1}(x_1, \ldots, x_n) \odot x_1'(t) \right)$$

$$+ \lim_{\Delta t \to 0^+} D\left(\frac{f(x_1, x_2 + \Delta x_2, \ldots, x_n + \Delta x_n)}{\Delta t} \right.$$

$$\left. - \frac{f(x_1, x_2, x_3 + \Delta x_3, \ldots, x_n + \Delta x_n)}{\Delta t}, \; f_{x_2}(x_1, \ldots, x_n) \odot x_2'(t) \right)$$

$$+ \lim_{\Delta t \to 0^+} D\left(\frac{f(x_1, x_2, x_3 + \Delta x_3, \ldots, x_n + \Delta x_n)}{\Delta t} \right.$$

$$\left. - \frac{f(x_1, x_2, x_3, x_4 + \Delta x_4, \ldots, x_n + \Delta x_n)}{\Delta t}, \; f_{x_3}(x_1, \ldots, x_n) \odot x_3'(t) \right)$$

$$
+ \cdots + \lim_{\Delta t \to 0^+} D\left(\frac{f(x_1, x_2, \ldots, x_{n-1}, x_n + \Delta x_n) - f(x_1, x_2, \ldots, x_n)}{\Delta t}, \right.
$$

$$
\left. f_{x_n}(x_1, \ldots, x_n) \odot x_n'(t) \right)
$$

$$
= \lim_{\Delta t \to 0^+} D\left(\left(\frac{f(x_1 + \Delta x_1, x_2 + \Delta x_2, \ldots, x_n + \Delta x_n)}{\Delta x_1} \right. \right.
$$

$$
\left. \left. - \frac{f(x_1, x_2 + \Delta x_2, \ldots, x_n + \Delta x_n)}{\Delta x_1} \right) \odot \frac{\Delta x_1}{\Delta t}, \; f_{x_1}(x_1, \ldots, x_n) \odot x_1'(t) \right)
$$

$$
+ \lim_{\Delta t \to 0^+} D\left(\left(\frac{f(x_1, x_2 + \Delta x_2, \ldots, x_n + \Delta x_n)}{\Delta x_2} \right. \right.
$$

$$
\left. \left. - \frac{f(x_1, x_2, x_3 + \Delta x_3, \ldots, x_n + \Delta x_n)}{\Delta x_2} \right) \odot \frac{\Delta x_2}{\Delta t}, \; f_{x_2}(x_1, \ldots, x_n) \odot x_2'(t) \right)
$$

$$
+ \lim_{\Delta t \to 0^+} D\left(\left(\frac{f(x_1, x_2, x_3 + \Delta x_3, \ldots, x_n + \Delta x_n)}{\Delta x_3} \right. \right.
$$

$$
\left. \left. - \frac{f(x_1, x_2, x_3, x_4 + \Delta x_4, \ldots, x_n + \Delta x_n)}{\Delta x_3} \right) \odot \frac{\Delta x_3}{\Delta t}, \; f_{x_3}(x_1, \ldots, x_n) \odot x_3'(t) \right)
$$

$$
+ \cdots + \lim_{\Delta t \to 0^+} D\left(\left(\frac{f(x_1, x_2, \ldots, x_{n-1}, x_n + \Delta x_n) - f(x_1, \ldots, x_n)}{\Delta x_n} \right) \right.
$$

$$
\left. \odot \frac{\Delta x_n}{\Delta t}, \; f_{x_n}(x_1, \ldots, x_n) \odot x_n'(t) \right)
$$

(by Corollary A, [11])

$$
= \lim_{\Delta t \to 0^+} D\left(\left(\frac{(FR)\int_{x_1}^{x_1 + \Delta x_1} f_{x_1}(t, x_2 + \Delta x_2, \ldots, x_n + \Delta x_n)dt}{\Delta x_1} \right) \right.
$$

$$
\left. \odot \frac{\Delta x_1}{\Delta t}, \; f_{x_1}(x_1, \ldots, x_n) \odot x_1'(t) \right)
$$

$$
+ \lim_{\Delta t \to 0^+} D\left(\left(\frac{(FR)\int_{x_2}^{x_2 + \Delta x_2} f_{x_2}(x_1, t, x_3 + \Delta x_3, \ldots, x_n + \Delta x_n)dt}{\Delta x_2} \right) \right.
$$

$$
\left. \odot \frac{\Delta x_2}{\Delta t}, \; f_{x_2}(x_1, \ldots, x_n) \odot x_2'(t) \right)
$$

$$
+ \lim_{\Delta t \to 0^+} D\left(\left(\frac{(FR)\int_{x_3}^{x_3 + \Delta x_3} f_{x_3}(x_1, x_2, t, x_4 + \Delta x_4, \ldots, x_n + \Delta x_n)dt}{\Delta x_3} \right) \right.
$$

$$\odot \frac{\Delta x_3}{\Delta t}, f_{x_3}(x_1, \ldots, x_n) \odot x_3'(t) \Bigg)$$

$$+ \cdots + \lim_{\Delta t \to 0^+} D\Bigg(\Bigg(\frac{(FR) \int_{x_{n-1}}^{x_{n-1}+\Delta x_{n-1}} f_{x_{n-1}}(x_1, \ldots, x_{n-2}, t, x_n + \Delta x_n) dt}{\Delta x_{n-1}} \Bigg)$$

$$\odot \frac{\Delta x_{n-1}}{\Delta t}, f_{x_{n-1}}(x_1, \ldots, x_n) \odot x_{n-1}'(t) \Bigg)$$

$$+ D(f_{x_n}(x_1, \ldots, x_n) \odot x_n'(t), f_{x_n}(x_1, \ldots, x_n) \odot x_n'(t))$$

(by Lemmas 2.16 and 2.17)

$$= \quad x_1'(t) \lim_{\Delta t \to 0^+} \frac{1}{\Delta x_1} D\Bigg((FR) \int_{x_1}^{x_1 + \Delta x_1} f_{x_1}(t, x_2 + \Delta x_2, \ldots, x_n + \Delta x_n) dt,$$

$$\Delta x_1 \odot f_{x_1}(x_1, \ldots, x_n) \Bigg)$$

$$+ x_2'(t) \lim_{\Delta t \to 0^+} \frac{1}{\Delta x_2} D\Bigg((FR) \int_{x_2}^{x_2 + \Delta x_2} f_{x_2}(x_1, t, x_3 + \Delta x_3,$$

$$\ldots, x_n + \Delta x_n) dt, \ \Delta x_2 \odot f_{x_2}(x_1, \ldots, x_n) \Bigg)$$

$$+ x_3'(t) \lim_{\Delta t \to 0^+} \frac{1}{\Delta x_3} D\Bigg((FR) \int_{x_3}^{x_3 + \Delta x_3} f_{x_3}(x_1, x_2, t, x_4 + \Delta x_4,$$

$$\ldots, x_n + \Delta x_n) dt, \ \Delta x_3 \odot f_{x_3}(x_1, \ldots, x_n) \Bigg) + \cdots$$

$$+ x_{n-1}'(t) \lim_{\Delta t \to 0^+} \frac{1}{\Delta x_{n-1}} D\Bigg((FR) \int_{x_{n-1}}^{x_{n-1}+\Delta x_{n-1}}$$

$$f_{x_{n-1}}(x_1, x_2, \ldots, x_{n-2}, t, x_n + \Delta x_n) dt, \Delta x_{n-1} \odot f_{x_{n-1}}(x_1, \ldots, x_n) \Bigg)$$

$$= \quad \sum_{i=1}^{n-1} x_i'(t) \lim_{\Delta t \to 0^+} \frac{1}{\Delta x_i} D\Bigg((FR) \int_{x_i}^{x_i + \Delta x_i} f_{x_i}(x_1, x_2, \ldots, x_{i-1},$$

$$t, x_{i+1} + \Delta x_{i+1}, \ldots, x_n + \Delta x_n) dt, (FR) \int_{x_i}^{x_i + \Delta x_i} f_{x_i}(x_1, \ldots, x_n) dt \Bigg)$$

(by Lemma 1 of [11]) $\leq \sum_{i=1}^{n-1} x_i'(t) \lim_{\Delta t \to 0^+} \frac{1}{\Delta x_i} \left(\int_{x_i}^{x_i + \Delta x_i} D(f_{x_i}(x_1, \right.$

$$x_2, \ldots, x_{i-1}, t, x_{i+1} + \Delta x_{i+1}, \ldots, x_n + \Delta x_n), f_{x_i}(x_1, \ldots, x_n))dt \Big)$$

$$\leq \sum_{i=1}^{n-1} x_i'(t) \lim_{\Delta t \to 0^+} \frac{1}{\Delta x_i} \left(\sup_{\tau \in [x_i, x_i + \Delta x_i]} (D(f_{x_i}(x_1, x_2, \ldots, x_{i-1}, \tau, \right.$$

$$x_{i+1} + \Delta x_{i+1}, \ldots, x_n + \Delta x_n), f_{x_i}(x_1, \ldots, x_n))) \Big) \Delta x_i$$

(by Lemma 1 of [11])
$$= \qquad \text{(for some } \tau_i^* \in [x_i, x_i + \Delta x_i])$$

$$\sum_{i=1}^{n-1} x_i'(t) \lim_{\Delta t \to 0^+} D(f_{x_i}(x_1, x_2, \ldots, x_{i-1}, \tau_i^*, x_{i+1} + \Delta x_{i+1},$$

$$\ldots, x_n + \Delta x_n), f_{x_i}(x_1, \ldots, x_n))$$

(as $\Delta t \to 0^+$, then all $\Delta x_i \to 0^+$ and thus $\tau_i^* \to x_i$, for all $i = 1, \ldots, n$)

$$= \sum_{i=1}^{n-1} x_i'(t) D(f_{x_i}(x_1, \ldots, x_n), f_{x_i}(x_1, \ldots, x_n))$$

$$= \sum_{i=1}^{n-1} x_i'(t) \cdot 0 = 0,$$

by continuity of f_{x_i}, $i = 1, \ldots, n-1$. I.e., we have proved that

$$\left(\frac{dz}{dt} \right)_+ = \sum_{i=1}^{n} {}^* \frac{dz}{dx_i} \odot \frac{dx_i}{dt}.$$

II) Call

$$\beta_1 := f(x_1, x_2, \ldots, x_n) - f(x_1, x_2, \ldots, x_{n-1}, x_n - \Delta x_n) \in \mathbb{R}_{\mathcal{F}}.$$

That is

$$f(x_1, x_2, \ldots, x_n) = \beta_1 \oplus f(x_1, x_2, \ldots, x_{n-1}, x_n - \Delta x_n).$$

Call

$$\beta_2 := \quad f(x_1, x_2, \ldots, x_{n-1}, x_n - \Delta x_n)$$
$$- f(x_1, x_2, \ldots x_{n-2}, x_{n-1} - \Delta x_{n-1}, x_n - \Delta x_n) \in \mathbb{R}_{\mathcal{F}}.$$

That is

$$f(x_1, x_2, \ldots, x_{n-1}, x_n - \Delta x_n) = \beta_2 \oplus f(x_1, x_2, \ldots, x_{n-2}, x_{n-1} - \Delta x_{n-1},$$
$$x_n - \Delta x_n).$$

Call

$$\beta_3 := \quad f(x_1, x_2, \ldots, x_{n-2}, x_{n-1} - \Delta x_{n-1}, x_n - \Delta x_n)$$
$$- f(x_1, x_2, \ldots, x_{n-3}, x_{n-2} - \Delta x_{n-2}, x_{n-1} - \Delta x_{n-1}, x_n - \Delta x_n)$$
$$\in \mathbb{R}_{\mathcal{F}}.$$

That is

$$f(x_1, x_2, \ldots, x_{n-2}, x_{n-1} - \Delta x_{n-1}, x_n - \Delta x_n)$$
$$= \beta_3 \oplus f(x_1, x_2, \ldots, x_{n-3}, x_{n-2} - \Delta x_{n-2}, x_{n-1} - \Delta x_{n-1}, x_n - \Delta x_n).$$

Etc. Call

$$\beta_n := f(x_1, x_2 - \Delta x_2, \ldots, x_n - \Delta x_n) - f(x_1 - \Delta x_1, x_2 - \Delta x_2, \ldots, x_n - \Delta x_n) \in \mathbb{R}_{\mathcal{F}}.$$

That is

$$f(x_1, x_2 - \Delta x_2, \ldots, x_n - \Delta x_n) = \beta_n \oplus f(x_1 - \Delta x_1, x_2 - \Delta x_2, \ldots, x_n - \Delta x_n).$$

I.e., it holds

$$\mathbb{R}_{\mathcal{F}} \ni f(x_1, x_2, \ldots, x_n) - f(x_1 - \Delta x_1, x_2 - \Delta x_2, \ldots, x_n - \Delta x_n) = \sum_{i=1}^{n}{}^{*} \beta_i.$$

Since the partial derivatives f_{x_i} exist, the above H-differences β_i, $i = 1, \ldots, n$ exist in $\mathbb{R}_{\mathcal{F}}$ for small $\Delta x_i > 0$. In particular we define $\Delta x_i := \phi_i(t) - \phi_i(t - \Delta t)$, $\Delta t > 0$, $i = 1, \ldots, n$ (i.e., $\phi_i(t - \Delta t) = x_i - \Delta x_i$, $x_i := \phi_i(t)$). Since ϕ_i, $i = 1, \ldots, n$ are strictly increasing we have that $\Delta x_i > 0$. So as $\Delta t \to 0^+$, then $\Delta x_i \to 0^+$ by continuity of ϕ_i.

We observe that

$$\lim_{\Delta t \to 0^+} D\left(\frac{f(\phi_1(t), \ldots, \phi_n(t)) - f(\phi_1(t - \Delta t), \ldots, \phi_n(t - \Delta t))}{\Delta t}, \right.$$

$$\left. \sum_{i=1}^{n}{}^{*} f_{x_i}(x_1, \ldots, x_n) \odot x_i'(t) \right)$$

$$= \lim_{\Delta t \to 0^+} D\left(\frac{f(x_1, \ldots, x_n) - f(x_1 - \Delta x_1, \ldots, x_n - \Delta x_n)}{\Delta t}, \right.$$

$$\left. \sum_{i=1}^{n}{}^{*} f_{x_i}(x_1, \ldots, x_n) \odot x_i'(t) \right)$$

$$= \lim_{\Delta t \to 0^+} D\left(\frac{\sum_{i=1}^{*n} \beta_i}{\Delta t}, \sum_{i=1}^{*n} f_{x_i}(x_1, x_2, \ldots, x_n) \odot x_i'(t) \right)$$

$$\leq \lim_{\Delta t \to 0^+} D\left(\left(\frac{f(x_1, x_2, \ldots, x_n) - f(x_1, x_2, \ldots, x_{n-1}, x_n - \Delta x_n)}{\Delta x_n} \right), \right.$$

$$f_{x_n}(x_1, \ldots, x_n) \odot x_n'(t)) + \lim_{\Delta t \to 0^+}$$

$$D\left(\left(\frac{f(x_1, x_2, \ldots, x_{n-1}, x_n - \Delta x_n)}{\Delta t} \right.\right.$$

$$\left.- \frac{f(x_1, x_2, \ldots, x_{n-2}, x_{n-1} - \Delta x_{n-1}, x_n - \Delta x_n)}{\Delta t}, f_{x_{n-1}}(x_1, \ldots, x_n) \odot x_{n-1}'(t) \right) +$$

$$\lim_{\Delta t \to 0^+} D\left(\left(\frac{\begin{matrix} f(x_1, x_2, \ldots, x_{n-2}, x_{n-1} - \Delta x_{n-1}, x_n - \Delta x_n) \\ -f(x_1, x_2, \ldots, x_{n-3}, x_{n-2} - \Delta x_{n-2}, x_{n-1} - \Delta x_{n-1}, x_n - \Delta x_n) \end{matrix}}{\Delta t} \right.\right.$$

$$\left., f_{x_{n-2}}(x_1, \ldots, x_n) \odot x_{n-2}'(t) \right) + \cdots + \lim_{\Delta t \to 0^+}$$

$$D\left(\left(\frac{f(x_1, x_2 - \Delta x_2, \ldots, x_n - \Delta x_n)}{\Delta t} \right.\right.$$

$$\left.- \frac{f(x_1 - \Delta x_1, x_2 - \Delta x_2, \ldots, x_n - \Delta x_n)}{\Delta t}, f_{x_1}(x_1, \ldots, x_n) \odot x_1'(t) \right)$$

$$\lim_{\Delta t \to 0^+} D\left(\left(\frac{f(x_1, x_2, \ldots, x_n) - f(x_1, x_2, \ldots, x_{n-1}, x_n - \Delta x_n)}{\Delta x_n} \right.\right.$$

$$\odot \frac{\Delta x_n}{\Delta t}, f_{x_n}(x_1, \ldots, x_n) \odot x_n'(t) \right) + \lim_{\Delta t \to 0^+}$$

$$D\left(\left(\frac{f(x_1, x_2, \ldots, x_{n-1}, x_n - \Delta x_n) - f(x_1, x_2, \ldots, x_{n-2}, x_{n-1} - \Delta x_{n-1}, x_n - \Delta x_n)}{\Delta x_{n-1}} \right.\right.$$

$$\odot \frac{\Delta x_{n-1}}{\Delta t}, f_{x_{n-1}}(x_1, \ldots, x_n) \odot x_{n-1}'(t) \right) + \lim_{\Delta t \to 0^+}$$

$$D\left(\left(\frac{\begin{matrix} f(x_1, x_2, \ldots, x_{n-2}, x_{n-1} - \Delta x_{n-1}, x_n - \Delta x_n) \\ -f(x_1, x_2, \ldots, x_{n-3}, x_{n-2} - \Delta x_{n-2}, x_{n-1} - \Delta x_{n-1}, x_n - \Delta x_n) \end{matrix}}{\Delta x_{n-2}} \right.\right.$$

$$\odot \frac{\Delta x_{n-2}}{\Delta t}, f_{x_{n-2}}(x_1,...,x_n) \odot x'_{n-2}(t)\Bigg) + \lim_{\Delta t \to 0^+}$$

$$D\Bigg(\Bigg(\frac{f(x_1, x_2 - \Delta x_2, ..., x_n - \Delta x_n) - f(x_1 - \Delta x_1, x_2 - \Delta x_2, ..., x_n - \Delta x_n)}{\Delta x_1}\Bigg)$$

$$\odot \frac{\Delta x_1}{\Delta t}, f_{x_1}(x_1,...,x_n) \odot x'_1(t)\Bigg)$$

$$\underset{\text{(by Corollary A, [11])}}{=}$$

$$D\big(f_{x_n}(x_1,\ldots,x_n) \odot x'_n(t), f_{x_n}(x_1,\ldots,x_n) \odot x'_n(t)\big)$$

$$+ \lim_{\Delta t \to 0^+} D\Bigg(\Bigg(\frac{(FR)\int_{x_{n-1}-\Delta x_{n-1}}^{x_{n-1}} f_{x_{n-1}}(x_1, x_2, \ldots, x_{n-2}, t, x_n - \Delta x_n)dt}{\Delta x_{n-1}}\Bigg)$$

$$\odot \frac{\Delta x_{n-1}}{\Delta t}, f_{x_{n-1}}(x_1,\ldots,x_n) \odot x'_{n-1}(t)\Bigg) + \lim_{\Delta t \to 0^+}$$

$$D\Bigg(\frac{(FR)\int_{x_{n-2}-\Delta x_{n-2}}^{x_{n-2}} f_{x_{n-2}}(x_1, x_2, \ldots, x_{n-3}, t, x_{n-1} - \Delta x_{n-1}, x_n - \Delta x_n)dt}{\Delta x_{n-2}}\Bigg)$$

$$\odot \frac{\Delta x_{n-2}}{\Delta t}, f_{x_{n-2}}(x_1,\ldots,x_n) \odot x'_{n-2}(t)\Bigg) + \cdots + \lim_{\Delta t \to 0^+}$$

$$D\Bigg(\Bigg(\frac{(FR)\int_{x_1-\Delta x_1}^{x_1} f_{x_1}(t, x_2 - \Delta x_2, \ldots, x_n - \Delta x_n)dt}{\Delta x_1}\Bigg)$$

$$\odot \frac{\Delta x_1}{\Delta t}, f_{x_1}(x_1,\ldots,x_n) \odot x'_1(t)\Bigg)$$

$$\underset{\text{(by Lemmas 2.16, 2.17)}}{=} x'_{n-1}(t) \lim_{\Delta t \to 0^+} \frac{1}{\Delta x_{n-1}} D\Bigg((FR)$$

$$\int_{x_{n-1}-\Delta x_{n-1}}^{x_{n-1}} f_{x_{n-1}}(x_1, x_2, \ldots, x_{n-2}, t, x_n - \Delta x_n)dt,$$

$$\left. \Delta x_{n-1} \odot f_{x_{n-1}}(x_1, \ldots, x_n)\right)$$

$$+ x_{n-2}'(t) \lim_{\Delta t \to 0^+} \frac{1}{\Delta x_{n-2}} D\left((FR) \int_{x_{n-2}-\Delta x_{n-2}}^{x_{n-2}} f_{x_{n-2}}(x_1, x_2, \ldots, x_{n-3}, \right.$$

$$\left. t, x_{n-1} - \Delta x_{n-1}, x_n - \Delta x_n)dt, \Delta x_{n-2} \odot f_{x_{n-2}}(x_1, \ldots, x_n)\right)$$

$$+ \cdots + x_1'(t) \lim_{\Delta t \to 0^+} \frac{1}{\Delta x_1}$$

$$D\left((FR) \int_{x_1-\Delta x_1}^{x_1} f_{x_1}(t, x_2 - \Delta x_2, \ldots, x_n - \Delta x_n)dt, \Delta x_1 \odot f_{x_1}(x_1, \ldots, x_n)\right)$$

$$= \sum_{i=1}^{n-1} x_i'(t) \lim_{\Delta t \to 0^+} \frac{1}{\Delta x_i} D\left((FR) \int_{x_i-\Delta x_i}^{x_i} f_{x_i}(x_1, x_2, \ldots, x_{i-1}, \right.$$

$$\left. t, x_{i+1} - \Delta x_{i+1}, \ldots, x_{n-1} - \Delta x_{n-1}, x_n - \Delta x_n)\right)dt,$$

$$\left. (FR) \int_{x_i-\Delta x_i}^{x_i} f_{x_i}(x_1, \ldots, x_n)dt\right)$$

(by Lemma 1 of [11])
$$\leq \sum_{i=1}^{n-1} x_i'(t) \lim_{\Delta t \to 0^+} \frac{1}{\Delta x_i} \left(\int_{x_i-\Delta x_i}^{x_i} D(f_{x_i}(x_1, x_2, \ldots, x_{i-1}, t, \right.$$

$$\left. x_{i+1} - \Delta x_{i+1}, \ldots, x_{n-1} - \Delta x_{n-1}, x_n - \Delta x_n), f_{x_i}(x_1, \ldots, x_n))dt\right)$$

$$\leq \sum_{i=1}^{n-1} x_i'(t) \lim_{\Delta t \to 0^+} \frac{1}{\Delta x_i} \left(\sup_{\tau \in [x_i-\Delta x_i, x_i]} (D(f_{x_i}(x_1, x_2, \ldots, x_{i-1}, \tau, x_{i+1} - \Delta x_{i+1}, \right.$$

$$\left. \ldots, x_{n-1} - \Delta x_{n-1}, x_n - \Delta x_n), f_{x_i}(x_1, \ldots, x_n)))\right) \Delta x_i$$

$$\text{(for some } \tau_i^* \in [x_i - \Delta x_i, x_i])$$

(by Lemma 1 of [11])
$$= \sum_{i=1}^{n-1} x_i'(t) \lim_{\Delta t \to 0^+} D\left(f_{x_i}(x_1, x_2, \ldots, x_{i-1}, \right.$$

$$\left. \tau_i^*, x_{i+1} - \Delta x_{i+1}, \ldots, x_{n-1} - \Delta x_{n-1}, x_n - \Delta x_n), f_{x_i}(x_1, \ldots, x_n))\right.$$

(as $\Delta t \to 0^+$, then all $\Delta x_i \to 0^+$ and thus $\tau_i^* \to x_i$, for all $i = 1, \ldots, n$)

$$= \sum_{i=1}^{n-1} x_i'(t) D\left(f_{x_i}(x_1, \ldots, x_n), f_{x_i}(x_1, \ldots, x_n)\right)$$

$$= \sum_{i=1}^{n-1} x_i'(t) \cdot 0 = 0,$$

by continuity of f_{x_i}, $i = 1, \ldots, n - 1$. I.e., we have proved that

$$\left(\frac{dz}{dt}\right)_{-} = \sum_{i=1}^{n}{}^{*} \frac{dz}{dx_i} \odot \frac{dx_i}{dt}.$$

When $t = a$, or b, then $\frac{dz}{dt}$ equals $\left(\frac{dz}{dt}\right)_{+}$, or $\left(\frac{dz}{dt}\right)_{-}$, respectively. Clearly here

$$\left.\frac{dx_i}{dt}\right|_{t=a} = \left.\left(\frac{dx_i}{dt}\right)_{+}\right|_{t=a}, \quad \text{and} \quad \left.\frac{dx_i}{dt}\right|_{t=b} = \left.\left(\frac{dx_i}{dt}\right)_{-}\right|_{t=b}.$$

Etc., the same proof as before. The theorem basically is proved.

A further explanation follows.
If $\Delta t \to 0^{+}$, then all $\Delta x_i \to 0^{+}$, $i = 1, \ldots, n$.
We notice that

$$\lim_{\Delta t \to 0^{+}} D\left(\frac{(FR)\int_{x_1}^{x_1+\Delta x_1} f_{x_1}(t, x_2 + \Delta x_2, \ldots, x_n + \Delta x_n)dt}{\Delta x_1}, \right.$$

$$\left. \frac{(FR)\int_{x_1}^{x_1+\Delta x_1} f_{x_1}(t, x_2, \ldots, x_n)dt}{\Delta x_1}\right)$$

$$\leq \lim_{\Delta t \to 0^{+}} \frac{1}{\Delta x_1} \int_{x_1}^{x_1+\Delta x_1} D\left(f_{x_1}(t, x_2 + \Delta x_2, \ldots, x_n + \Delta x_n),\right.$$

$$f_{x_1}(t, x_2, \ldots, x_n))\, dt$$

(the integrand just above is continuous in t)

$$= \lim_{\Delta t \to 0^{+}} \frac{1}{\Delta x_1} D\left(f_{x_1}(t^{*}, x_2 + \Delta x_2, \ldots, x_n + \Delta x_n),\right.$$

$$f_{x_1}(t^{*}, x_2, \ldots, x_n)) \cdot \Delta x_1 \quad (t^{*} \text{ in } [x_1, x_1 + \Delta x_1])$$

$$= \lim_{\Delta t \to 0^{+}} D\left(f_{x_1}(t^{*}, x_2 + \Delta x_2, \ldots, x_n + \Delta x_n), f_{x_1}(t^{*}, x_2, \ldots, x_n)\right)$$

(i.e. $\Delta x_1 \to 0$, and $t^{*} \to x_1$)

$$= D\left(f_{x_1}(x_1, x_2, \ldots, x_n), f_{x_1}(x_1, x_2, \ldots, x_n)\right) = 0,$$

by continuity of f_{x_1}, etc.

We further see that

$$f(x_1 + \Delta x_1, x_2, \ldots, x_n) - f(x_1, x_2, \ldots, x_n) = \int_{x_1}^{x_1+\Delta x_1} f_{x_1}(t, x_2, \ldots, x_n)\, dt.$$

That is

$$\frac{f(x_1 + \Delta x_1, x_2, \ldots, x_n) - f(x_1, x_2, \ldots, x_n)}{\Delta x_1} = \frac{\int_{x_1}^{x_1+\Delta x_1} f_{x_1}(t, x_2, \ldots, x_n)\, dt}{\Delta x_1}.$$

Also it holds

$$\lim_{\Delta t \to 0^+} D\left(\frac{f(x_1 + \Delta x_1, x_2, \ldots, x_n) - f(x_1, x_2, \ldots, x_n)}{\Delta x_1}, f_{x_1}(x_1, x_2, \ldots, x_n) \right)$$
$$= 0.$$

Therefore

$$\lim_{\Delta t \to 0^+} D\left(\frac{(FR)\int_{x_1}^{x_1 + \Delta x_1} f_{x_1}(t, x_2 + \Delta x_2, \ldots, x_n + \Delta x_n)dt}{\Delta x_1}, \right.$$

$$f_{x_1}(x_1, x_2, \ldots, x_n))$$

$$\leq \lim_{\Delta t \to 0^+} D\left(\frac{(FR)\int_{x_1}^{x_1 + \Delta x_1} f_{x_1}(t, x_2 + \Delta x_2, \ldots, x_n + \Delta x_n)dt}{\Delta x_1}, \right.$$

$$\left. \frac{(FR)\int_{x_1}^{x_1 + \Delta x_1} f_{x_1}(t, x_2, \ldots, x_n)dt}{\Delta x_1} \right)$$

$$+ \lim_{\Delta t \to 0^+} D\left(\frac{(FR)\int_{x_1}^{x_1 + \Delta x_1} f_{x_1}(t, x_2, \ldots, x_n)dt}{\Delta x_1}, f_{x_1}(x_1, x_2, \ldots, x_n) \right)$$

$$= 0.$$

Therefore we get
$$\left(\frac{(FR)\int_{x_1}^{x_1 + \Delta x_1} f_{x_1}(t, x_2 + \Delta x_2, \ldots, x_n + \Delta x_n)dt}{\Delta x_1} \right) \xrightarrow[as\ \Delta t \to 0^+]{D} f_{x_1}(x_1, x_2, \ldots, x_n), \text{ etc.}$$
The proof of the theorem now is clear and completed. □

We need

Lemma 2.20. *Let f be a fuzzy continuous function from the open set $U \subseteq \mathbb{R}^n$, $n \in \mathbb{N}$, into $\mathbb{R}_{\mathcal{F}}$. Then $f_{\pm}^{(r)}$ are continuous functions from U into $\mathbb{R}$, for all $r \in [0, 1]$.*

Proof. Let $x_m, x \in U$, $m \in \mathbb{N}$, be such that $x_m \to x$ as $m \to +\infty$. Then by continuity of f we get $D(f(x_m), f(x)) \to 0$, as $m \to +\infty$. Hence we have

$$D(f(x_m), f(x)) = \sup_{r \in [0,1]} \max\{|(f(x_m))_-^{(r)} - (f(x))_-^{(r)}|,$$

$$|(f(x_m))_+^{(r)} - (f(x))_+^{(r)}|\} \to 0.$$

Therefore $|(f(x_m))_-^{(r)} - (f(x))_-^{(r)}| \to 0$ and $|(f(x_m))_+^{(r)} - (f(x))_+^{(r)}| \to 0$, as $m \to +\infty$, for all $r \in [0, 1]$. Consequently $(f(x_m))_{\pm}^{(r)} \to (f(x))_{\pm}^{(r)}$, proving that $f_{\pm}^{(r)} \in C(U, \mathbb{R})$, for all $0 \leq r \leq 1$. □

We present the interchange of the order of H-fuzzy differentiation.

Theorem 2.21. *Let U be an open subset of $\mathbb{R}^n$, $n \in \mathbb{N}$, and $f : U \to \mathbb{R}_{\mathcal{F}}$ be a fuzzy continuous function. Assume that all H-fuzzy partial derivatives of f up to order $m \in \mathbb{N}$ exist and are fuzzy continuous. Let $x :=$ $(x_1, \ldots, x_n) \in U$. Then the H-fuzzy mixed partial derivative of order k, $D_{x_{\ell_1}, \ldots, x_{\ell_k}} f(x)$ is unchanged when the indices $\ell_1, \ldots, \ell_k$ are permuted. Each ℓ_i is a positive integer $\le n$. Here some or all of ℓ_i's can be equal. Also $k = 2, \ldots, m$ and there are n^k partials of order k.*

Proof. We only need to demonstrate the proof for the case $n = k = 2$. The rest is true by induction on k, and similarly true for $n > 2$. So here $z = f(x, y) : U \subseteq \mathbb{R}^2 \to \mathbb{R}_{\mathcal{F}}$ and $\frac{\partial^2 f}{\partial x^2}, \frac{\partial^2 f}{\partial y^2}, \frac{\partial^2 f}{\partial x \partial y}, \frac{\partial^2 f}{\partial y \partial x}$ exist and are fuzzy continuous functions from U into $\mathbb{R}_{\mathcal{F}}$. We make use of Theorem 5.2 from [71] repeatedly. Here we have

$$[f(x,y)]^r = \left[(f(x,y))_-^{(r)}, (f(x,y))_+^{(r)}\right], \quad 0 \le r \le 1.$$

By that theorem and the above assumptions $\frac{\partial}{\partial x}(f(x,y))_{\pm}^{(r)}$ exist and

$$\left[\frac{\partial}{\partial x} f(x,y)\right]^r = \left[\frac{\partial}{\partial x}(f(x,y))_-^{(r)}, \frac{\partial}{\partial x}(f(x,y))_+^{(r)}\right],$$

for all $0 \le r \le 1$ and all $(x,y) \in U$. Furthermore, the same way $\frac{\partial^2}{\partial y \partial x}(f(x,y))_{\pm}^{(r)}$ exist and

$$\left[\frac{\partial^2}{\partial y \partial x} f(x,y)\right]^r = \left[\frac{\partial^2}{\partial y \partial x}(f(x,y))_-^{(r)}, \frac{\partial^2}{\partial y \partial x}(f(x,y))_+^{(r)}\right],$$

for all $0 \le r \le 1$ and all $(x,y) \in U$. Similarly we obtain

$$\left[\frac{\partial^2}{\partial x \partial y} f(x,y)\right]^r = \left[\frac{\partial^2}{\partial x \partial y}(f(x,y))_-^{(r)}, \frac{\partial^2}{\partial x \partial y}(f(x,y))_+^{(r)}\right],$$

for all $0 \le r \le 1$ and all $(x,y) \in U$.

Clearly it also holds that

$$\left[\frac{\partial^2}{\partial x^2} f(x,y)\right]^r = \left[\frac{\partial^2}{\partial x^2}(f(x,y))_-^{(r)}, \frac{\partial^2}{\partial x^2}(f(x,y))_+^{(r)}\right],$$

and

$$\left[\frac{\partial^2}{\partial y^2} f(x,y)\right]^r = \left[\frac{\partial^2}{\partial y^2}(f(x,y))_-^{(r)}, \frac{\partial^2}{\partial y^2}(f(x,y))_+^{(r)}\right],$$

for all $0 \le r \le 1$ and all $(x,y) \in U$. By Lemma 2.20 we find that

$$\frac{\partial^2}{\partial x^2}(f(x,y))_{\pm}^{(r)}, \frac{\partial^2}{\partial y^2}(f(x,y))_{\pm}^{(r)}, \frac{\partial^2}{\partial x \partial y}(f(x,y))_{\pm}^{(r)}, \frac{\partial^2}{\partial y \partial x}(f(x,y))_{\pm}^{(r)}$$

are all continuous for any $r \in [0,1]$. But by basic real analysis, Theorem 6-20, p. 121 of [37] we have

$$\frac{\partial^2}{\partial x \partial y} (f(x,y))_{\pm}^{(r)} = \frac{\partial^2}{\partial y \partial x} (f(x,y))_{\pm}^{(r)},$$

for any $r \in [0,1]$. Thus we get

$$\left[\frac{\partial^2}{\partial x \partial y} f(x,y) \right]^r = \left[\frac{\partial^2}{\partial y \partial x} f(x,y) \right]^r,$$

for all $0 \le r \le 1$. That is the H-fuzzy partial derivatives are equal, $\frac{\partial^2}{\partial x \partial y} f(x,y) = \frac{\partial^2 f(x,y)}{\partial y \partial x}$ for all $(x,y) \in U$.　　　　□

Finally it follows a multivariate Fuzzy Taylor's formula.

Theorem 2.22. *Let U be an open convex subset of $\mathbb{R}^n$, $n \in \mathbb{N}$ and $f : U \to \mathbb{R}_{\mathcal{F}}$ be a fuzzy continuous function. Assume that all H-fuzzy partial derivatives of f up to order $m \in \mathbb{N}$ exist and are fuzzy continuous. Let $z := (z_1, \ldots, z_n)$, $x_0 := (x_{01}, \ldots, x_{0n}) \in U$ such that $z_i \ge x_{0i}$, $i = 1, \ldots, n$. Let $0 \le t \le 1$, we define $x_i := x_{0i} + t(z_i - x_{0i})$, $i = 1, 2, \ldots, n$ and $g_z(t) := f(x_0 + t(z - x_0))$. (Clearly $x_0 + t(z - x_0) \in U$.) Then for $N = 1, \ldots, m$ we obtain*

$$g_z^{(N)}(t) = \left[\left(\sum_{i=1}^{n} {}^* (z_i - x_{0i}) \odot \frac{\partial}{\partial x_i} \right)^N f \right] (x_1, x_2, \ldots, x_n). \qquad (2.12)$$

Furthermore it holds the following fuzzy multivariate Taylor formula

$$f(z) = f(x_0) \oplus \sum_{N=1}^{m-1} {}^* \frac{g_z^{(N)}(0)}{N!} \oplus \mathcal{R}_m(0,1), \qquad (2.13)$$

where

$$\mathcal{R}_m(0,1) := (FR) \int_0^1 \left(\int_0^{s_1} \cdots \left(\int_0^{s_{m-1}} g_z^{(m)}(s_m) ds_m \right) ds_{m-1} \right) \cdots \right) ds_1. \qquad (2.14)$$

N o t e (Explaining formula (2.12)). When $N = n = 2$ we have ($z_i \ge x_{0i}$, $i = 1, 2$)

$$g_z(t) = f(x_{01} + t(z_1 - x_{01}), x_{02} + t(z_2 - x_{02})), \quad 0 \le t \le 1.$$

We apply Theorems 2.19 and 2.21 repeatedly, etc. Thus we have

$$g_z'(t) = (z_1 - x_{01}) \odot \frac{\partial f}{\partial x_1}(x_1, x_2) \oplus (z_2 - x_{02}) \odot \frac{\partial f}{\partial x_2}(x_1, x_2).$$

Furthermore it holds

$$g_z''(t) = (z_1 - x_{01})^2 \odot \frac{\partial^2 f}{\partial x_1^2}(x_1, x_2) \oplus 2(z_1 - x_{01}) \cdot (z_2 - x_{02})$$
$$\odot \frac{\partial^2 f(x_1, x_2)}{\partial x_1 \partial x_2} \oplus (z_2 - x_{02})^2 \odot \frac{\partial^2 f}{\partial x_2^2}(x_1, x_2). \tag{2.15}$$

When $n = 2$ and $N = 3$ we get

$$g_z'''(t) = (z_1 - x_{01})^3 \odot \frac{\partial^3 f}{\partial x_1^3}(x_1, x_2) \oplus 3(z_1 - x_{01})^2(z_2 - x_{02}) \tag{2.16}$$
$$\odot \frac{\partial^3 f(x_1, x_2)}{\partial x_1^2 \partial x_2} \oplus 3(z_1 - x_{01})(z_2 - x_{02})^2 \cdot \frac{\partial^3 f(x_1, x_2)}{\partial x_1 \partial x_2^2}$$
$$\oplus (z_2 - x_{02})^3 \odot \frac{\partial^3 f}{\partial x_2^3}(x_1, x_2).$$

When $n = 3$ and $N = 2$ we obtain ($z_i \geq x_{0i}$, $i = 1, 2, 3$)

$$g_z''(t) = (z_1 - x_{01})^2 \odot \frac{\partial^2 f}{\partial x_1^2}(x_1, x_2, x_3) \oplus (z_2 - x_{02})^2 \odot \frac{\partial^2 f}{\partial x_2^2}(x_1, x_2, x_3)$$
$$\oplus (z_3 - x_{03})^2 \odot \frac{\partial^2 f}{\partial x_3^2}(x_1, x_2, x_3) \oplus 2(z_1 - x_{01})(z_2 - x_{02})$$
$$\odot \frac{\partial^2 f(x_1, x_2, x_3)}{\partial x_1 \partial x_2} \oplus 2(z_2 - x_{02})(z_3 - x_{03}) \tag{2.17}$$
$$\odot \frac{\partial^2 f(x_1, x_2, x_3)}{\partial x_2 \partial x_3} \oplus 2(z_3 - x_{03})(z_1 - x_{01}) \odot \frac{\partial^2 f}{\partial x_3 \partial x_1}(x_1, x_2, x_3).$$

Etc.

Proof of Theorem 2.22. Let $z := (z_1, \ldots, z_n)$, $x_0 := (x_{01}, \ldots, x_{0n}) \in U$, $n \in \mathbb{N}$, such that $z_i > x_{0i}$, $i = 1, 2, \ldots, n$. We define

$$x_i := \phi_i(t) := x_{0i} + t(z_i - x_{0i}), \quad 0 \leq t \leq 1; \quad i = 1, 2, \ldots, n.$$

Thus $\frac{dx_i}{dt} = z_i - x_{0i} > 0$. Consider

$$Z := g_z(t) := f(x_0 + t(z - x_0)) = f(x_{01} + t(z_1 - x_{01}), \ldots, x_{0n} + t(z_n - x_{0n}))$$
$$= f(\phi_1(t), \ldots, \phi_n(t)).$$

Since by assumptions $f \colon U \to \mathbb{R}_\mathcal{F}$ is fuzzy continuous, also f_{x_i} exist and are fuzzy continuous, by Theorem 2.19 (2.11) we get

$$\frac{dZ(x_1, \ldots, x_n)}{dt} = \sum_{i=1}^{n}{}^* \frac{\partial Z(x_1, \ldots, x_n)}{\partial x_i} \odot \frac{dx_i}{dt}$$
$$= \sum_{i=1}^{n}{}^* \frac{\partial f(x_1, \ldots, x_n)}{\partial x_i} \odot (z_i - x_{0i}).$$

That is,

$$g'_z(t) = \sum_{i=1}^{n}{}^* \frac{\partial f(x_1, \ldots, x_m)}{\partial x_i} \odot (z_i - x_{0i}).$$

Next we see

$$
\begin{aligned}
\frac{d^2 Z}{dt^2} &= g''_z(t) = \frac{d}{dt}\left(\sum_{i=1}^{n}{}^* \frac{\partial f(x_1, \ldots, x_n)}{\partial x_i} \odot (z_i - x_{0i})\right) \\
&= \sum_{i=1}^{n}{}^* (z_i - x_{0i}) \odot \frac{d}{dt}\left(\frac{\partial f(x_1, \ldots, x_n)}{\partial x_i}\right) \\
&= \sum_{i=1}^{n}{}^* (z_i - x_{0i}) \odot \left[\sum_{j=1}^{n}{}^* \frac{\partial^2 f(x_1, \ldots, x_n)}{\partial x_j \partial x_i} \odot (z_j - x_{0j})\right] \\
&= \sum_{i=1}^{n}{}^* \sum_{j=1}^{n}{}^* \frac{\partial^2 f(x_1, \ldots, x_n)}{\partial x_j \partial x_i} \odot (z_i - x_{0i}) \cdot (z_j - x_{0j}).
\end{aligned}
$$

That is

$$g''_z(t) = \sum_{i=1}^{n}{}^* \sum_{j=1}^{n}{}^* \frac{\partial^2 f(x_1, \ldots, x_m)}{\partial x_j \partial x_i} \odot (z_i - x_{0i}) \cdot (z_j - x_{0j}).$$

The last is true by Theorem 2.19 (2.11) under the additional assumptions that f_{x_i}; $\frac{\partial^2 f}{\partial x_j \partial x_i}$, $i, j = 1, 2, \ldots, n$ exist and are fuzzy continuous.

Working similarly we find

$$
\begin{aligned}
\frac{d^3 Z}{dt^3} &= g'''_z(t) = \frac{d}{dt}\left(\sum_{i=1}^{n}{}^* \sum_{j=1}^{n}{}^* \frac{\partial^2 f(x_1, \ldots, x_n)}{\partial x_j \partial x_i} \odot (z_i - x_{0i}) \cdot (z_j - x_{0j})\right) \\
&= \sum_{i=1}^{n}{}^* \sum_{j=1}^{n}{}^* (z_i - x_{0i}) \cdot (z_j - x_{0j}) \frac{d}{dt}\left(\frac{\partial^2 f(x_1, \ldots, x_n)}{\partial x_j \partial x_i}\right) \\
&= \sum_{i=1}^{n}{}^* \sum_{j=1}^{n}{}^* (z_i - x_{0i}) \cdot (z_j - x_{0j}) \left[\sum_{k=1}^{n}{}^* \frac{\partial^3 f(x_1, \ldots, x_n)}{\partial x_k \partial x_j \partial x_i} \odot (z_k - x_{0k})\right] \\
&= \sum_{i=1}^{n}{}^* \sum_{j=1}^{n}{}^* \sum_{k=1}^{n}{}^* \frac{\partial^3 f(x_1, \ldots, x_n)}{\partial x_k \partial x_j \partial x_i} \odot (z_i - x_{0i}) \cdot (z_j - x_{0j}) \cdot (z_k - x_{0k}).
\end{aligned}
$$

That is

$$g'''_z(t) = \sum_{i=1}^{n}{}^* \sum_{j=1}^{n}{}^* \sum_{k=1}^{n}{}^* \frac{\partial^3 f(x_1, \ldots, x_n)}{\partial x_k \partial x_j \partial x_i} \odot (z_i - x_{0i}) \cdot (z_j - x_{0j}) \cdot (z_k - x_{0k}).$$

That last is true by Theorem 2.19 (2.11) under the additional assumptions that

$$\frac{\partial^3 f(x_1, \ldots, x_n)}{\partial x_k \partial x_j \partial x_i}, \quad i, j, k = 1, \ldots, n$$

do exist and are fuzzy continuous. Etc. In general one obtains that for $N = 1, \ldots, m \in \mathbb{N}$,

$$g_z^{(N)}(t) = \sum_{i_1=1}^{n} {}^* \sum_{i_2=1}^{n} {}^* \cdots \sum_{i_N=1}^{n} {}^* \frac{\partial^N f(x_1, \ldots, x_n)}{\partial x_{i_N} \partial x_{i_{N-1}} \cdots \partial x_{i_1}} \odot \prod_{r=1}^{N} (z_{i_r} - x_{0i_r}),$$

which by Theorem 2.21 is the same as (2.12) for the case $z_i > x_{0i}$, see also (2.15), (2.16), and (2.17). The last is true by Theorem 2.19 (2.11) under the assumptions that all H-partial derivatives of f up to order m exist and they are all fuzzy continuous including f itself.

Next let $t_{\tilde{m}} \to \tilde{t}$, as $\tilde{m} \to +\infty$, $t_{\tilde{m}}, \tilde{t} \in [0, 1]$. Consider

$$x_{i\tilde{m}} := x_{0i} + t_{\tilde{m}}(z_i - x_{0i})$$

and

$$\tilde{x}_i := x_{0i} + \tilde{t}(z_i - x_{0i}), \quad i = 1, 2, \ldots, n.$$

That is

$$x_{\tilde{m}} = (x_{1\tilde{m}}, x_{2\tilde{m}}, \ldots, x_{n\tilde{m}}) \quad \text{and} \quad \tilde{x} = (\tilde{x}_1, \ldots, \tilde{x}_n) \quad \text{in } U.$$

Then $x_{\tilde{m}} \to \tilde{x}$, as $\tilde{m} \to +\infty$. Clearly using the properties of D-metric and under the theorem's assumptions, we obtain that

$$g_z^{(N)}(t) \quad \text{is fuzzy continuous for } N = 0, 1, \ldots, m.$$

Then by Theorem 2.1, from the univariate fuzzy Taylor formula, we obtain

$$g_z(1) = g_z(0) \oplus g_z'(0) \oplus \frac{g_z''(0)}{2!} \oplus \cdots \oplus \frac{g_z^{(m-1)}(0)}{(m-1)!} \oplus \mathcal{R}_m(0, 1),$$

where

$$\mathcal{R}_m(0, 1) := (FR) \int_0^1 \left(\int_0^{s_1} \cdots \left(\int_0^{s_{m-1}} g_z^{(m)}(s_m) ds_m \right) ds_{m-1} \right) \cdots ds_1.$$

By Lemma 4, [11] and Corollary 13.2, p. 644, [66], the remainder $\mathcal{R}_m(0, 1)$ exist in $\mathbb{R}_{\mathcal{F}}$. I.e., we get the multivariate fuzzy Taylor formula

$$f(z) = f(x_0) \oplus g_z'(0) \oplus \frac{g_z''(0)}{2!} \oplus \cdots \oplus \frac{g_z^{(m-1)}(0)}{(m-1)!} \oplus \mathcal{R}_m(0, 1),$$

when $z_i > x_{0i}$, $i = 1, 2, \ldots, n$.

Finally we would like to take care of the case that some $x_{0i} = z_i$. Without loss of generality we may assume that $x_{01} = z_1$, and $z_i > x_{0i}$, $i = 2, \ldots, n$. In this case we define

$$\tilde{Z} := \tilde{g}_z(t) := f(x_{01}, x_{02} + t(z_2 - x_{02}), \ldots, x_{0n} + t(z_n - x_{0n})).$$

Therefore one has

$$\tilde{g}_z'(t) = \sum_{i=2}^{n}{}^{*} \frac{\partial f(x_{01}, x_2, \ldots, x_n)}{\partial x_i} \odot (z_i - x_{0i}),$$

and in general we find

$$\tilde{g}_z^{(N)}(t) = \sum_{i_2=2,\ldots,i_N=2}^{n}{}^{*} \frac{\partial^N f(x_{01}, x_2, \ldots, x_n)}{\partial x_{i_N} \partial x_{N-1} \cdots \partial x_{i_2}} \odot \prod_{r=2}^{N} (z_{i_r} - x_{0i_r}),$$

for $N = 1, \ldots, m \in \mathbb{N}$. Notice that all $\tilde{g}_z^{(N)}$, $N = 0, 1, \ldots, m$ are fuzzy continuous and

$$\tilde{g}_z(0) = f(x_{01}, x_{02}, \ldots, x_{0n}), \quad \tilde{g}_z(1) = f(x_{01}, z_2, z_3, \ldots, z_n).$$

Then one can write down a fuzzy Taylor formula, as above, for $\tilde{g}_z$. But $\tilde{g}_z^{(N)}(t)$ coincides with $g_z^{(N)}(t)$ formula at $z_1 = x_{01} = x_1$. That is both Taylor formulae in that case coincide.

At last we remark that if $z = x_0$, then we define $Z^* := g_z^*(t) := f(x_0) =: c \in \mathbb{R}_{\mathcal{F}}$ a constant. Since $c = c + \tilde{o}$, that is $c - c = \tilde{o}$, we obtain the H-fuzzy derivative $(c)' = \tilde{o}$. Consequently we have that

$$g_z^{*(N)}(t) = \tilde{o}, \quad N = 1, \ldots, m.$$

The last coincide with the $g_z^{(N)}$ formula, established earlier, if we apply there $z = x_0$. And, of course, the fuzzy Taylor formula now can be applied trivially for g_z^*. Furthermore in that case it coincides with the Taylor formula proved earlier for g_z. We have established a multivariate fuzzy Taylor formula for the case of $z_i \geq x_{0i}$, $i = 1, 2, \ldots, n$. That is (2.12)–(2.14) are true. $\qquad\square$

At last we give the useful

Corollary 2.23. *Let U be an open convex subset of $\mathbb{R}^n$, $n \in \mathbb{N}$, and $f \colon U \to \mathbb{R}_{\mathcal{F}}$ be a fuzzy continuous function. Assume that all the first H-fuzzy partial derivatives f_{x_i} of f exist and are fuzzy continuous. Let $z := (z_1, \ldots, z_n)$, $x_0 := (x_{01}, \ldots, x_{0n}) \in U$ such that $z_i \geq x_{0i}$, $i = 1, \ldots, n$. Let $0 \leq t \leq 1$, we define $x_i := x_{0i} + t(z_i - x_{0i})$, $i = 1, 2, \ldots, n$ and $g_z(t) := f(x_0 + t(z - x_0))$. Then*

$$g_z'(t) = \sum_{i=1}^{n}{}^{*} \frac{\partial f(x_1, \ldots, x_n)}{\partial x_i} \odot (z_i - x_{0i}). \qquad (2.18)$$

Furthermore it holds

$$f(z) \;=\; f(x_0) \oplus (FR) \int_0^1 g_z'(s)\,ds \qquad\qquad (2.19)$$

$$\;=\; f(x_0) \oplus \sum_{i=1}^{n}{}^{*}(z_i - x_{0i}) \odot (FR) \int_0^1 \frac{\partial f(x_1(s), \ldots, x_n(s))}{\partial x_i}\,ds.$$

Proof. By Theorem 2.22, case of $m = 1$. The second part of (2.19) is valid by Theorem 2.6 of [53]. Here $x_i(s) = x_{0i} + s(z_i - x_{0i})$, $s \in [0, 1]$, $i = 1, \ldots, n$ with $z_i \geq x_{0i}$. $\qquad\square$

Comment. Theorem 2.22 and Corollary 2.23 are still valid when U is a compact convex subset of $\mathbb{R}^n$ such that $U \subseteq W$, where W is an open subset of $\mathbb{R}^n$. Now $f \colon W \to \mathbb{R}_{\mathcal{F}}$ and it has all the properties of f as in Theorem 2.22 and Corollary 2.23. Clearly here $x_0, z \in U$.

3

ON FUZZY TAYLOR FORMULAE

We present Fuzzy Taylor formulae with integral remainder in the univariate and multivariate cases, analogs of the real setting. This chapter is based on [19].

3.1   Main Results

We present the following fuzzy Taylor theorem in one dimension.

Theorem 3.1. Let $f \in C^n([a,b], \mathbb{R}_{\mathcal{F}})$, $n \geq 1$, $[\alpha, \beta] \subseteq [a,b] \subseteq \mathbb{R}$. Then

$$f(\beta) = f(\alpha) \quad \oplus \quad f'^{(n-1)}(\alpha) \odot \frac{(\beta - \alpha)^{n-1}}{(n-1)!}$$

$$\oplus \quad \frac{1}{(n-1)!} \odot (FR) \int_{\alpha}^{\beta} (\beta - t)^{n-1} \odot f^{(n)}(t)\, dt. \quad (3.1)$$

The integral remainder is a fuzzy continuous function in β.

Proof. Let $r \in [0,1]$. We have here $[f(\beta)]^r = [f_-^{(r)}(\beta), f_+^{(r)}(\beta)]$, and by Theorem 5.2 ([71]) $f_{\pm}^{(r)}$ is n-times continuously differentiable on $[a,b]$. By (1.7) we get

$$(f_{\pm}^{(i)}(\alpha))^{(r)} = (f_{\pm}^{(r)}(\alpha))^{(i)}, \quad \text{all } i = 0, 1, \ldots, n, \quad (3.2)$$

and

$$[f^{(i)}(\alpha)]^r = [(f_-^{(r)}(\alpha))^{(i)}, (f_+^{(r)}(\alpha))^{(i)}].$$

G.A. Anastassiou: Fuzzy Mathematics: Approximation The., STUDFUZZ 251, pp. 51–63.
springerlink.com

Thus by Taylor's theorem we obtain

$$f_\pm^{(r)}(\beta) = f_\pm^{(r)}(\alpha) + (f_\pm^{(r)}(\alpha))'(\beta - \alpha)$$
$$+ \cdots + (f_\pm^{(r)}(\alpha))^{(n-1)}\frac{(\beta - \alpha)^{n-1}}{(n-1)!} + \frac{1}{(n-1)!}\int_\alpha^\beta (\beta - t)^{n-1}(f_\pm^{(r)})^{(n)}(t)d$$

Furthermore by (3.2) we have

$$f_\pm^{(r)}(\beta) = f_\pm^{(r)}(\alpha) + (f_\pm'(\alpha))^{(r)}(\beta - \alpha)$$
$$+ \cdots + (f_\pm^{(n-1)}(\alpha))^{(r)}\frac{(\beta - \alpha)^{n-1}}{(n-1)!} + \frac{1}{(n-1)!}\int_\alpha^\beta (\beta - t)^{n-1}(f_\pm^{(n)})^{(r)}(t)d$$

Here it holds $\beta - \alpha \geq 0$, $\beta - t \geq 0$ for $t \in [\alpha, \beta]$, and

$$(f_-^{(i)}(t))^{(r)} \leq (f_+^{(i)}(t))^{(r)}, \quad \forall t \in [a, b]$$

all $i = 0, 1, \ldots, n$, and any $r \in [0, 1]$.

We see that

$$\left[f_-^{(r)}(\beta), f_+^{(r)}(\beta)\right] = \Big[f_-^{(r)}(\alpha) + (f_-'(\alpha))^{(r)}(\beta - \alpha) + \cdots + (f_-^{(n-1)}(\alpha))^{(r)}\frac{(\beta - \alpha)^{n-}}{(n-1)!}$$
$$+ \frac{1}{(n-1)!}\int_\alpha^\beta (\beta - t)^{n-1}(f_-^{(n)})^{(r)}(t)dt, \; f_+^{(r)}(\alpha)$$
$$+ (f_+'(\alpha))^{(r)}(\beta - \alpha) + \cdots + (f_+^{(n-1)}(\alpha))^{(r)}\frac{(\beta - \alpha)^{n-1}}{(n-1)!}$$
$$+ \frac{1}{(n-1)!}\int_\alpha^\beta (\beta - t)^{n-1}(f_+^{(n)})^{(r)}(t)\,dt\Big].$$

To split the above closed interval into a sum of smaller closed intervals is where we use $\beta - \alpha \geq 0$. So we get

$$[f(\beta)]^r = [f_-^{(r)}(\beta), f_+^{(r)}(\beta)] = [f_-^{(r)}(\alpha), f_+^{(r)}(\alpha)] + [(f_-'(\alpha))^{(r)}, (f_+'(\alpha))^{(r)}](\beta - \alpha)$$
$$+ \cdots + [(f_-^{(n-1)}(\alpha))^{(r)}, (f_+^{(n-1)}(\alpha))^{(r)}]\frac{(\beta - \alpha)^{n-1}}{(n-1)!}$$
$$+ \frac{1}{(n-1)!}\left[\int_\alpha^\beta (\beta - t)^{n-1}(f_-^{(n)})^{(r)}(t)dt, \int_\alpha^\beta (\beta - t)^{n-1}(f_+^{(n)})^{(r)}(t)dt\right]$$
$$= [f(\alpha)]^r + [f'^r(\beta - \alpha) + \cdots + [f^{(n-1)}(\alpha)]^r\frac{(\beta - \alpha)^{n-1}}{(n-1)!}$$
$$+ \frac{1}{(n-1)!}\left[\int_\alpha^\beta ((\beta - t)^{n-1} \odot f^{(n)}(t))_-^{(r)}dt,\right.$$
$$\left.\int_\alpha^\beta ((\beta - t)^{n-1} \odot f^{(n)}(t))_+^{(r)}dt\right].$$

By Theorem 3.2 ([67]) we next get

$$[f(\beta)]^r = [f(\alpha)]^r + [f'^r(\beta - \alpha) + \cdots + [f^{(n-1)}(\alpha)]^r \frac{(\beta - \alpha)^{n-1}}{(n-1)!}$$

$$+ \frac{1}{(n-1)!} \left[(FR) \int_\alpha^\beta (\beta - t)^{n-1} \odot f^{(n)}(t)dt \right]^r .$$

Finally we obtain

$$[f(\beta)]^r = \left[f(\alpha) \oplus f'^{(n-1)}(\alpha) \odot \frac{(\beta - \alpha)^{n-1}}{(n-1)!} \right.$$

$$\left. \oplus \frac{1}{(n-1)!} \odot (FR) \int_\alpha^\beta (\beta - t)^{n-1} \odot f^{(n)}(t)dt \right]^r , \quad \text{all } r \in [0,1].$$

By Theorem 3.2 of [67] and Lemma 1.5 we get that the remainder of (3.1) is in $\mathbb{R}_\mathcal{F}$, and by Lemma 3 ([11]) is a fuzzy continuous function in β. The theorem has been proved. □

Next we present another multivariate fuzzy Taylor theorem.

We need the following multivariate fuzzy chain rule. Here the H-fuzzy partial derivatives are defined according to the Definition 3.3 of [53], see Section 1 there, and the analogous way to the real case.

Theorem 3.2 ([10]). *Let $\phi_i \colon [a,b] \subseteq \mathbb{R} \to \phi_i([a,b]) := I_i \subseteq \mathbb{R}$, $i = 1, \ldots, n$, $n \in \mathbb{N}$, are strictly increasing and differentiable functions. Denote $x_i := x_i(t) := \phi_i(t)$, $t \in [a,b]$, $i = 1, \ldots, n$. Consider U an open subset of $\mathbb{R}^n$ such that $\times_{i=1}^n I_i \subseteq U$. Consider $f \colon U \to \mathbb{R}_\mathcal{F}$ a fuzzy continuous function. Assume that $f_{x_i} \colon U \to \mathbb{R}_\mathcal{F}$, $i = 1, \ldots, n$, the H-fuzzy partial derivatives of f, exist and are fuzzy continuous. Call $z := z(t) := f(x_1, \ldots, x_n)$. Then $\frac{dz}{dt}$ exists and*

$$\frac{dz}{dt} = \sum_{i=1}^n {}^* \frac{dz}{dx_i} \odot \frac{dx_i}{dt}, \quad \forall t \in [a,b] \tag{3.3}$$

where $\frac{dz}{dt}$, $\frac{dz}{dx_i}$, $i = 1, \ldots, n$ are the H-fuzzy derivatives of f with respect to t, x_i, respectively.

The interchange of the order of H-fuzzy differentiation is needed too.

Theorem 3.3 ([10]). *Let U be an open subset of $\mathbb{R}^n$, $n \in \mathbb{N}$, and $f \colon U \to \mathbb{R}_\mathcal{F}$ be a fuzzy continuous function. Assume that all H-fuzzy partial derivatives of f up to order $m \in \mathbb{N}$ exist and are fuzzy continuous. Let $x := (x_1, \ldots, x_n) \in U$. Then the H-fuzzy mixed partial derivative of order k, $D_{x_{\ell_1}, \ldots, x_{\ell_k}} f(x)$ is unchanged when the indices $\ell_1, \ldots, \ell_k$ are permuted. Each ℓ_i is a positive integer $\leq n$. Here some or all of ℓ_i's can be equal. Also $k = 2, \ldots, m$ and there are n^k partials of order k.*

We present

Theorem 3.4. *Let U be an open convex subset of $\mathbb{R}^n$, $n \in \mathbb{N}$ and $f\colon U \to \mathbb{R}_{\mathcal{F}}$ be a fuzzy continuous function. Assume that all H-fuzzy partial derivatives of f up to order $m \in \mathbb{N}$ exist and are fuzzy continuous. Let $z := (z_1, \ldots, z_n)$, $x_0 := (x_{01}, \ldots, x_{0n}) \in U$ such that $x_i \geq x_{0i}$, $i = 1, \ldots, n$. Let $0 \leq t \leq 1$, we define $x_i := x_{0i} + t(z_i - z_{0i})$, $i = 1, 2, \ldots, n$ and $g_z(t) := f(x_0 + t(z - x_0))$. (Clearly $x_0 + t(z - x_0) \in U$.) Then for $N = 1, \ldots, m$ we obtain*

$$g_z^{(N)}(t) = \left[\left(\sum_{i=1}^{n}{}^{*} (z_i - x_{0i}) \odot \frac{\partial}{\partial x_i} \right)^N f \right] (x_1, x_2, \ldots, x_n). \qquad (3.4)$$

Furthermore it holds the following fuzzy multivariate Taylor formula

$$f(z) = f(x_0) \oplus \sum_{N=1}^{m-1}{}^{*} \frac{g_z^{(N)}(0)}{N!} \oplus \mathcal{R}_m(0, 1), \qquad (3.5)$$

where

$$\mathcal{R}_m(0, 1) := \frac{1}{(m-1)!} \odot (FR) \int_0^1 (1 - s)^{m-1} \odot g_z^{(m)}(s) ds. \qquad (3.6)$$

Comment. (Explaining formula (3.4)). When $N = n = 2$ we have ($z_i \geq x_{0i}$, $i = 1, 2$)

$$g_z(t) = f(x_{01} + t(z_1 - x_{01}), x_{02} + t(z_2 - x_{02})), \quad 0 \leq t \leq 1.$$

We apply Theorems 3 and 4 of [10] repeatedly, etc. Thus we find

$$g_z'(t) = (z_1 - x_{01}) \odot \frac{\partial f}{\partial x_1}(x_1, x_2) \oplus (z_2 - x_{02}) \odot \frac{\partial f}{\partial x_2}(x_1, x_2).$$

Furthermore it holds

$$\begin{aligned}
g_z''(t) &= (z_1 - x_{01})^2 \odot \frac{\partial^2 f}{\partial x_1^2}(x_1, x_2) \oplus 2(z_1 - x_{01}) \cdot (z_2 - x_{02}) \quad (3.7) \\
&\quad \odot \frac{\partial^2 f(x_1, x_2)}{\partial x_1 \partial x_2} \oplus (z_2 - x_{02})^2 \odot \frac{\partial^2 f}{\partial x_2^2}(x_1, x_2).
\end{aligned}$$

When $n = 2$ and $N = 3$ we obtain

$$\begin{aligned}
g_z'''(t) &= (z_1 - x_{01})^3 \odot \frac{\partial^3 f}{\partial x_1^3}(x_1, x_2) \oplus 3(z_1 - x_{01})^2(z_2 - x_{02}) \\
&\quad \odot \frac{\partial^3 f(x_1, x_2)}{\partial x_1^2 \partial x_2} \oplus 3(z_1 - x_{01})(z_2 - x_{02})^2 \cdot \frac{\partial^3 f(x_1, x_2)}{\partial x_1 \partial x_2^2} \\
&\quad \oplus (z_2 - x_{02})^3 \odot \frac{\partial^3 f}{\partial x_2^3}(x_1, x_2). \qquad (3.8)
\end{aligned}$$

When $n = 3$ and $N = 2$ we get $(z_i \geq x_{0i}, \; i = 1, 2, 3)$

$$g_z''(t) = (z_1 - x_{01})^2 \odot \frac{\partial^2 f}{\partial x_1^2}(x_1, x_2, x_3) \oplus (z_2 - x_{02})^2 \odot \frac{\partial^2 f}{\partial x_2^2}(x_1, x_2, x_3)$$

$$\oplus (z_3 - x_{03})^2 \odot \frac{\partial^2 f}{\partial x_3^2}(x_1, x_2, x_3) \oplus 2(z_1 - x_{01})(z_2 - x_{02})$$

$$\odot \frac{\partial^2 f(x_1, x_2, x_3)}{\partial x_1 \partial x_2} \oplus 2(z_2 - x_{02})(z_3 - x_{03}) \tag{3.9}$$

$$\odot \frac{\partial^2 f(x_1, x_2, x_3)}{\partial x_2 \partial x_3} \oplus 2(z_3 - x_{03})(z_1 - x_{01}) \odot \frac{\partial^2 f}{\partial x_3 \partial x_1}(x_1, x_2, x_3),$$

etc.

Proof of Theorem 3.4. Let $z := (z_1, \ldots, z_n)$, $x_0 := (x_{01}, \ldots, x_{0n}) \in U$, $n \in \mathbb{N}$, such that $z_i > x_{0i}$, $i = 1, 2, \ldots, n$. We define

$$x_i := \phi_i(t) := x_{0i} + t(z_i - x_{0i}), \quad 0 \leq t \leq 1; \quad i = 1, 2, \ldots, n.$$

Thus $\frac{dx_i}{dt} = z_i - x_{0i} > 0$. Consider

$$Z := g_z(t) := f(x_0 + t(z - x_0)) = f(x_{01} + t(z_1 - x_{01}), \ldots, x_{0n} + t(z_n - x_{0n}))$$
$$= f(\phi_1(t), \ldots, \phi_n(t)).$$

Since by assumptions $f: U \to \mathbb{R}_{\mathcal{F}}$ is fuzzy continuous, also f_{x_i} exist and are fuzzy continuous, by Theorem 3 (10) of [10] we get

$$\frac{dZ(x_1, \ldots, x_n)}{dt} = \sum_{i=1}^{n}{}^* \frac{\partial Z(x_1, \ldots, x_n)}{\partial x_i} \odot \frac{dx_i}{dt}$$

$$= \sum_{i=1}^{n}{}^* \frac{\partial f(x_1, \ldots, x_n)}{\partial x_i} \odot (z_i - x_{0i}).$$

Thus

$$g_z'(t) = \sum_{i=1}^{n}{}^* \frac{\partial f(x_1, \ldots, x_n)}{\partial x_i} \odot (z_i - x_{0i}).$$

Next we observe that

$$\frac{d^2 Z}{dt^2} = g_z''(t) = \frac{d}{dt}\left(\sum_{i=1}^{n}{}^* \frac{\partial f(x_1, \ldots, x_n)}{\partial x_i} \odot (z_i - x_{0i})\right)$$

$$= \sum_{i=1}^{n}{}^* (z_i - x_{0i}) \odot \frac{d}{dt}\left(\frac{\partial f(x_1, \ldots, x_n)}{\partial x_i}\right)$$

$$= \sum_{i=1}^{n}{}^* (z_i - x_{0i}) \odot \left[\sum_{j=1}^{n}{}^* \frac{\partial^2 f(x_1, \ldots, x_n)}{\partial x_j \partial x_i} \odot (z_j - x_{0j})\right]$$

$$= \sum_{i=1}^{n}{}^* \sum_{j=1}^{n}{}^* \frac{\partial^2 f(x_1, \ldots, x_n)}{\partial x_j \partial x_i} \odot (z_i - x_{0i}) \cdot (z_j - x_{0j}).$$

That is

$$g_z''(t) = \sum_{i=1}^{n}{}^* \sum_{j=1}^{n}{}^* \frac{\partial^2 f(x_1,\ldots,x_n)}{\partial x_j \partial x_i} \odot (z_i - x_{0i}) \cdot (z_j - x_{0j}).$$

The last is true by Theorem 3 (10) of [10] under the additional assumptions that f_{xi}; $\frac{\partial^2 f}{\partial x_j \partial x_i}$, $i,j = 1,2,\ldots,n$ exist and are fuzzy continuous.

Working the same way we find

$$
\begin{aligned}
\frac{d^3 Z}{dt^3} &= g_z'''(t) = \frac{d}{dt}\left(\sum_{i=1}^{n}{}^* \sum_{j=1}^{n}{}^* \frac{\partial^2 f(x_1,\ldots,x_n)}{\partial x_j \partial x_i} \odot (z_i - x_{0i}) \cdot (z_j - x_{0j}) \right)\\
&= \sum_{i=1}^{n}{}^* \sum_{j=1}^{n}{}^* (z_i - x_{0i}) \cdot (z_j - x_{0j}) \frac{d}{dt}\left(\frac{\partial^2 f(x_1,\ldots,x_n)}{\partial x_j \partial x_i} \right)\\
&= \sum_{i=1}^{n}{}^* \sum_{j=1}^{n}{}^* (z_i - x_{0i}) \cdot (z_j - x_{0j}) \left[\sum_{k=1}^{n}{}^* \frac{\partial^3 f(x_1,\ldots,x_n)}{\partial x_k \partial x_j \partial x_i} \odot (z_k - x_{0k}) \right]\\
&= \sum_{i=1}^{n}{}^* \sum_{j=1}^{n}{}^* \sum_{k=1}^{n}{}^* \frac{\partial^3 f(x_1,\ldots,x_n)}{\partial x_k \partial x_j \partial x_i} \odot (z_i - x_{0i}) \cdot (z_j - x_{0j}) \cdot (z_k - x_{0k}).
\end{aligned}
$$

Therefore,

$$g_z'''(t) = \sum_{i=1}^{n}{}^* \sum_{j=1}^{n}{}^* \sum_{k=1}^{n}{}^* \frac{\partial^3 f(x_1,\ldots,x_n)}{\partial x_k \partial x_j \partial x_i} \odot (z_i - x_{0i}) \cdot (z_j - x_{0j}) \cdot (z_k - x_{0k}).$$

That last is true by Theorem 3 (10) of [10] under the additional assumptions that

$$\frac{\partial^3 f(x_1,\ldots,x_n)}{\partial x_k \partial x_j \partial x_i}, \quad i,j,k = 1,\ldots,n$$

do exist and are fuzzy continuous. Etc. In general one obtains that for $N = 1,\ldots, m \in \mathbb{N}$,

$$g_z^{(N)}(t) = \sum_{i_1=1}^{n}{}^* \sum_{i_2=1}^{n}{}^* \cdots \sum_{i_N=1}^{n}{}^* \frac{\partial^N f(x_1,\ldots,x_n)}{\partial x_{i_N} \partial x_{i_{N-1}} \cdots \partial x_{i_1}} \odot \prod_{r=1}^{N}(z_{i_r} - x_{0i_r}),$$

which by Theorem 4 of [10] is the same as (3.4) for the case $z_i > x_{0i}$, see also (3.7), (3.8), and (3.9). The last is true by Theorem 3 (10) of [10] under the assumptions that all H-partial derivatives of f up to order m exist and they are all fuzzy continuous including f itself.

Next let $t_{\tilde{m}} \to \tilde{t}$, as $\tilde{m} \to +\infty$, $t_{\tilde{m}}, \tilde{t} \in [0,1]$. Consider

$$x_{i\tilde{m}} := x_{0i} + t_{\tilde{m}}(z_i - x_{0i})$$

and

$$\tilde{x}_i := x_{0i} + \tilde{t}(z_i - x_{0i}), \quad i = 1, 2, \ldots, n.$$

That is

$$x_{\tilde{m}} = (x_{1\tilde{m}}, x_{2\tilde{m}}, \ldots, x_{n\tilde{m}}) \text{ and } \tilde{x} = (\tilde{x}_1, \ldots, \tilde{x}_n) \text{ in } U.$$

Then $x_{\tilde{m}} \to \tilde{x}$, as $\tilde{m} \to +\infty$. Clearly using the properties of D-metric and under the theorem's assumptions, we obtain that

$$g_z^{(N)}(t) \text{ is fuzzy continuous for } N = 0, 1, \ldots, m.$$

Then by Theorem 3.1, from the univariate fuzzy Taylor formula (3.1), we find

$$g_z(1) = g_z(0) \oplus g_z'(0) \oplus \frac{g_z''(0)}{2!} \oplus \cdots \oplus \frac{g_z^{(m-1)}(0)}{(m-1)!} \oplus \mathcal{R}_m(0, 1),$$

where $\mathcal{R}_m(0, 1)$ comes from (3.6).

By Theorem 3.2 of [67] and Lemma 1.5 we get that $\mathcal{R}_m(0, 1) \in \mathbb{R}_{\mathcal{F}}$. That is we get the multivariate fuzzy Taylor formula

$$f(z) = f(x_0) \oplus g_z'(0) \oplus \frac{g_z''(0)}{2!} \oplus \cdots \oplus \frac{g_z^{(m-1)}(0)}{(m-1)!} \oplus \mathcal{R}_m(0, 1),$$

when $z_i > x_{0i}$, $i = 1, 2, \ldots, n$.

Finally we would like to take care of the case that some $x_{0i} = z_i$. Without loss of generality we may assume that $x_{01} = z_1$, and $z_i > x_{0i}$, $i = 2, \ldots, n$. In this case we define

$$\tilde{Z} := \tilde{g}_z(t) := f(x_{01}, x_{02} + t(z_2 - x_{02}), \ldots, x_{0n} + t(z_n - x_{0n})).$$

Therefore one has

$$\tilde{g}_z'(t) = \sum_{i=2}^{n}{}^* \frac{\partial f(x_{01}, x_2, \ldots, x_n)}{\partial x_i} \odot (z_i - x_{0i}),$$

and in general we find

$$\tilde{g}_z^{(N)}(t) = \sum_{i_2=2,\ldots,i_N=2}^{n}{}^* \frac{\partial^N f(x_{01}, x_2, \ldots, x_n)}{\partial x_{i_N} \partial x_{N-1} \cdots \partial x_{i_2}} \odot \prod_{r=2}^{N}(z_{i_r} - x_{0i_r}),$$

for $N = 1, \ldots, m \in \mathbb{N}$. Notice that all $\tilde{g}_z^{(N)}$, $N = 0, 1, \ldots, m$ are fuzzy continuous and

$$\tilde{g}_z(0) = f(x_{01}, x_{02}, \ldots, x_{0n}), \quad \tilde{g}_z(1) = f(x_{01}, z_2, z_3, \ldots, z_n).$$

Then one can write down a fuzzy Taylor formula, as above, for $\tilde{g}_z$. But $\tilde{g}_z^{(N)}(t)$ coincides with $g_z^{(N)}(t)$ formula at $z_1 = x_{01} = x_1$. That is both Taylor formulae in that case coincide.

At last we remark that if $z = x_0$, then we define $Z^* := g_z^*(t) := f(x_0) =: c \in \mathbb{R}_{\mathcal{F}}$ a constant. Since $c = c + \tilde{o}$, that is $c - c = \tilde{o}$, we obtain the H-fuzzy derivative $(c)' = \tilde{o}$. Consequently we have that

$$g_z^{*(N)}(t) = \tilde{o}, \quad N = 1, \ldots, m.$$

The last coincide with the $g_z^{(N)}$ formula, established earlier, if we apply there $z = x_0$. And, of course, the fuzzy Taylor formula now can be applied trivially for g_z^*. Furthermore in that case it coincides with the Taylor formula proved earlier for g_z. We have established a multivariate fuzzy Taylor formula for the case of $z_i \geq x_{0i}$, $i = 1, 2, \ldots, n$. That is (3.4)–(3.6) are true. □

Note. Theorem 3.4 is still valid when U is a compact convex subset of $\mathbb{R}^n$ such that $U \subseteq W$, where W is an open subset of $\mathbb{R}^n$. Now $f \colon W \to \mathbb{R}_{\mathcal{F}}$ and it has all the properties of f as in Theorem 3.4. Clearly here we take $x_0, z \in U$.

3.2   Addendum

As related material we give

Theorem 3.5. Let $f : [a, b] \subseteq \mathbb{R} \to \mathbb{R}_{\mathcal{F}}$ be fuzzy continuous. Then

$$G(t) = (FR) \int_a^t f(s)ds \in \mathbb{R}_{\mathcal{F}},$$

any $t \in [a, b]$. Also $G(t)$ is $H-$differentiable and $G'(t) = f(t)$, $t \in [a, b]$.

Proof. Clearly $G(t)$ is fuzzy continuous in t, see Lemma 1.15.

Here

$$
\begin{aligned}
G(t + h) &= (FR) \int_a^{t+h} f(s)ds \\
&= (FR) \int_a^t f(s)ds \oplus (FR) \int_t^{t+h} f(s)ds \\
&= G(t) \oplus (FR) \int_t^{t+h} f(s)ds.
\end{aligned}
$$

Thus the $H-$ difference

$$G(t + h) - G(t) = (FR) \int_t^{t+h} f(s)ds \in \mathbb{R}_{\mathcal{F}},$$

exists. And furthermore we have

$$\frac{G(t+h) - G(t)}{h} = \frac{1}{h} \odot (FR) \int_t^{t+h} f(s)ds \in \mathbb{R}_{\mathcal{F}}.$$

Thus

$$
\begin{aligned}
D\left(\frac{G(t+h) - G(t)}{h}, f(t)\right) &= \frac{1}{h}D\left((FR)\int_t^{t+h} f(s)ds, h \odot f(t)\right) \\
&= \frac{1}{h}D\left((FR)\int_t^{t+h} f(s)ds, \left(\int_t^{t+h} 1ds\right)\right. \\
&\quad \odot f(t)) \\
&= \frac{1}{h}D\left((FR)\int_t^{t+h} f(s)ds,\right. \\
&\quad \left.(FR)\int_t^{t+h} f(t)ds\right) \\
&\leq \frac{1}{h}\int_t^{t+h} D\left(f(s), f(t)\right)ds \\
&\quad \text{(the integrand just above is continuous in } s) \\
&\leq \frac{1}{h}h \sup_{s \in [t,t+h]} D\left(f(s), f(t)\right) \\
&= D\left(f(s^*), f(t)\right) \to 0,
\end{aligned}
$$

where $s^* \in [t, t+h]$, and as $h \to 0^+$ we get $s^* \to t$, and we assumed that f is fuzzy continuous.

Similarly we have

$$G(t-h) = (FR)\int_a^{t-h} f(s)ds, \ (h > 0),$$

and

$$G(t) = (FR)\int_a^{t-h} f(s)ds \oplus (FR)\int_{t-h}^t f(s)ds.$$

Also it holds

$$\frac{G(t) - G(t-h)}{h} = \frac{(FR)\int_{t-h}^t f(s)ds}{h},$$

$h \to 0^+$.

Hence we have

$$D\left(\frac{G(t)-G(t-h)}{h}, f\left(t\right)\right) = \frac{1}{h}D\left((FR)\int_{t-h}^{t} f(s)ds, h \odot f(t)\right)$$

$$= \frac{1}{h}D\left((FR)\int_{t-h}^{t} f(s)ds,\right.$$

$$\left.(FR)\int_{t-h}^{t} f(t)ds\right)$$

$$\leq \frac{1}{h}\int_{t-h}^{t} D\left(f(s), f(t)\right)ds$$

$$\leq \frac{1}{h}h \sup_{s\in[t-h,t]} D\left(f(s), f(t)\right)$$

$$= D\left(f\left(s_*\right), f\left(t\right)\right) \to 0,$$

where $s_* \in [t-h,t]$, by f−continuity; as $h \to 0^+$, then $s_* \to t$.
So that G is fuzzy differentiable and $G'(t) = f(t)$. $\qquad \square$

Comment 3.6. Let $f : [a,b] \to \mathbb{R}_{\mathcal{F}}$ be fuzzy continuous. Then $G_1(t_1) = (FR)\int_a^{t_1} f(s)ds$ is H−differentiable and $G_1'(t_1) = f(t_1)$. Here $G_1(t_1)$ is fuzzy continuous. Let $G_2(t_2) := (FR)\int_a^{t_2} G_1(t_1)dt_1 \in \mathbb{R}_{\mathcal{F}}$, it is also fuzzy continuous, and $G_2'(t_2) = G_1(t_2) = \int_a^{t_2} f(s)ds$.
Hence $(G_2'(t_2))' = (G_1(t_2))' = f(t_2)$. I.e. $G_2''(t_2) = f(t_2)$, etc. That is, there are non-trivial higher order H−fuzzy derivatives.
We continue with

Theorem 3.7. Let f be fuzzy continuous on $R := [a,b] \times [c,d]$. Then

$$F(x_1,x_2) := (FR)\int_a^{x_1}\left((FR)\int_c^{x_2} f(x,y)dy\right)dx \in \mathbb{R}_{\mathcal{F}},$$

any $(x_1,x_2) \in R$, and

$$D_{2,1}F\left(x_1,x_2\right) = f\left(x_1,x_2\right).$$

Proof. The function

$$G(x,x_2) = (FR)\int_c^{x_2} f(x,y)dy,$$

is fuzzy continuous in x_2, and belongs to $\mathbb{R}_{\mathcal{F}}$ for each $x \in [a,b]$. We have

$$F(x_1,x_2) = (FR)\int_a^{x_1} G(x,x_2)dx.$$

We would like to prove that $G(x,x_2)$ is fuzzy continuous in $x \in [a,b]$. Here $f(x,y)$ is fuzzy continuous, and thus uniformly fuzzy continuous on R.

Let h be such that $x + h \in [a, b]$. Uniform fuzzy continuity implies $\forall \epsilon > 0, \exists \delta > 0$ such that, whenever $\|(x + h, y) - (x, y)\| < \delta$, where $\|\cdot\|$ is a norm in $\mathbb{R}^2$, we get $D(f(x + h, y), f(x, y)) < \epsilon$.

Notice that

$$\|(x + h, y) - (x, y)\| = \|(h, 0)\| = |h| \, \|(1, 0)\| \, .$$

Thus

$$
\begin{aligned}
D(G(x + h, x_2), G(x, x_2)) \; &= \; D\left((FR) \int_c^{x_2} f(x + h, y) dy, \right. \\
& \qquad \left. (FR) \int_c^{x_2} f(x, y) dy \right) \\
& \text{(both integrands just above are fuzzy} \\
& \text{continuous in } y, \text{ hence by Lemma 1.13 we get)} \\
& \leq \; \int_c^{x_2} D(f(x + h, y), f(x, y)) \, dy \\
& \leq \; \epsilon (x_2 - c),
\end{aligned}
$$

proving fuzzy continuity of $G(x, x_2)$ in $x \in [a, b]$.

Clearly then $F(x_1, x_2) \in \mathbb{R}_{\mathcal{F}}$, by Corollary 1.11.

Therefore

$$\frac{\partial F(x_1, x_2)}{\partial x_1} = G(x_1, x_2) = (FR) \int_c^{x_2} f(x_1, y) dy,$$

by Theorem 3.5.

Thus

$$\frac{\partial^2 F(x_1, x_2)}{\partial x_2 \partial x_1} = \frac{\partial G(x_1, x_2)}{\partial x_2} = f(x_1, x_2),$$

proving the claim.   □

We finish with

Theorem 3.8. Let f be fuzzy continuous on $R = [a, b] \times [c, d]$. Then

$$M(x_1, x_2) = (FR) \int_a^{x_1} \left((FR) \int_c^{x_2} \left((FR) \int_c^y f(x, s) ds \right) dy \right) dx \in \mathbb{R}_{\mathcal{F}},$$

any $(x_1, x_2) \in R$, and

$$D_{2,2,1} M(x_1, x_2) = f(x_1, x_2) \, .$$

Proof. Call

$$\varphi(x, y) = (FR) \int_c^y f(x, s) ds \in \mathbb{R}_{\mathcal{F}}.$$

Notice φ is fuzzy continuous in x and y. Then

$$M\left(x_1, x_2\right) = (FR) \int_a^{x_1} \left((FR) \int_c^{x_2} \varphi(x,y)dy\right) dx.$$

Call

$$G(x, x_2) = (FR) \int_c^{x_2} \varphi(x,y)dy \in \mathbb{R}_{\mathcal{F}},$$

notice G is fuzzy continuous in x_2 and

$$M(x_1, x_2) = (FR) \int_a^{x_1} G(x, x_2)dx.$$

We need to prove that $\varphi(x,y)$ is (simultaneously) fuzzy continuous in (x, y). Indeed for letting $y_n \geq y$ we observe

$$
\begin{aligned}
D\left(\varphi(x_n, y_n), \varphi(x,y)\right) &= D\left((FR) \int_c^{y_n} f(x_n, s)ds, (FR) \int_c^{y} f(x, s)ds\right) \\
&= D\left((FR) \int_c^{y} f(x_n, s)ds \oplus (FR) \int_y^{y_n} f(x_n, s)ds,\right. \\
&\qquad \left.(FR) \int_c^{y} f(x, s)ds\right) \\
&\leq D\left((FR) \int_c^{y} f(x_n, s)ds, (FR) \int_c^{y} f(x, s)ds\right) + \\
&\qquad D\left((FR) \int_y^{y_n} f(x_n, s)ds, \tilde{o}\right)
\end{aligned}
$$

$(\epsilon > 0 \text{ small})$
$$
\begin{aligned}
&\overset{\leq}{} \epsilon \\
&+ D\left((FR) \int_y^{y_n} f(x_n, s)ds, (FR) \int_y^{y_n} \tilde{o}ds\right)
\end{aligned}
$$

(since the integrands just above are fuzzy continuous in s, by Lemma 1.13 we get)

$$\leq \epsilon + \int_y^{y_n} D\left(f(x_n, s), \tilde{o}\right) ds$$

$$\leq \epsilon + \tau(y_n - y) \to 0,$$

(here $D\left(f(x_n, s), \tilde{o}\right) \leq \tau, \tau > 0$, as in Lemma 2, [11]), i.e.

$$D\left(\varphi(x_n, y_n), \varphi(x,y)\right) \to 0,$$

as $(x_n, y_n) \to (x, y)$, with $y_n \geq y$.

Let now $y_n \leq y$, then

$$
\begin{aligned}
D\left(\varphi(x_n, y_n), \varphi(x, y)\right) &= D\left((FR)\int_c^{y_n} f(x_n, s)ds, (FR)\int_c^y f(x, s)ds\right) \\
&= D\left((FR)\int_c^{y_n} f(x_n, s)ds \oplus (FR)\int_{y_n}^y f(x_n, s)ds,\right. \\
&\qquad \left. (FR)\int_{y_n}^y f(x_n, s)ds \oplus (FR)\int_c^y f(x, s)ds\right) \\
&= D\left((FR)\int_c^y f(x_n, s)ds,\right. \\
&\qquad \left. (FR)\int_{y_n}^y f(x_n, s)ds \oplus (FR)\int_c^y f(x, s)ds\right) \\
&\leq D\left((FR)\int_c^y f(x_n, s)ds, (FR)\int_c^y f(x, s)ds\right) + \\
&\qquad D\left((FR)\int_{y_n}^y f(x_n, s)ds, \tilde{o}\right) \\
&\overset{(\epsilon>0\ \text{small})}{\leq} \epsilon + \\
&\qquad D\left((FR)\int_{y_n}^y f(x_n, s)ds, (FR)\int_{y_n}^y \tilde{o}ds\right) \\
&\leq \epsilon + \int_{y_n}^y D\left(f(x_n, s), \tilde{o}\right)ds \\
&\leq \epsilon + \tau\left(y - y_n\right) \to 0.
\end{aligned}
$$

So as $(x_n, y_n) \to (x, y)$, with $y_n \leq y$, we get again

$$
D\left(\varphi(x_n, y_n), \varphi(x, y)\right) \to 0.
$$

Therefore we have proved that $\varphi(x, y)$ is (jointly) fuzzy continuous in (x, y) over R.

Consequently $G(x, x_2)$ is fuzzy continuous in $x \in [a, b]$, i.e. $M(x_1, x_2) \in \mathbb{R}_{\mathcal{F}}$.

Thus

$$
\frac{\partial^2 M(x_1, x_2)}{\partial x_2 \partial x_1} = \frac{\partial G(x_1, x_2)}{\partial x_2} = \varphi(x_1, x_2),
$$

and

$$
\frac{\partial^3 M(x_1, x_2)}{\partial x_2^2 \partial x_1} = \frac{\partial^2 G(x_1, x_2)}{\partial x_2^2} = \frac{\partial \varphi(x_1, x_2)}{\partial x_2} = f(x_1, x_2).
$$

$\square$

Conclusion. There are higher order nontrivial $H-$fuzzy partial derivatives.

4

FUZZY OSTROWSKI INEQUALITIES

We present optimal upper bounds for the deviation of a fuzzy continuous function from its fuzzy average over $[a, b] \subset \mathbb{R}$, error is measured in the D-fuzzy metric. The established fuzzy Ostrowski type inequalities are sharp, in fact attained by simple fuzzy real number valued functions. These inequalities are given for fuzzy Hölder and fuzzy differentiable functions and these facts are reflected in their right-hand sides. This chapter relies on [13].

4.1 Introduction

Ostrowski inequality (see [92]) has as follows

$$\left| \frac{1}{b-a} \int_a^b f(y)dy - f(x) \right| \leq \left(\frac{1}{4} + \frac{\left(x - \frac{a+b}{2}\right)^2}{(b-a)^2} \right) (b-a)\|f'\|_\infty,$$

where $f \in C^1([a, b])$, $x \in [a, b]$. The last inequality is sharp, see [5].

Since 1938 when A. Ostrowski proved his famous inequality, see [92], many people have been working about and around it, in many different directions and with a lot of applications in Numerical Analysis and Probability, etc.

One of the most notable works extending Ostrowski's inequality is the work of A.M. Fink, see [64]. The author in [5] continued that tradition.

G.A. Anastassiou: Fuzzy Mathematics: Approximation The., STUDFUZZ 251, pp. 65–73.
springerlink.com   © Springer-Verlag Berlin Heidelberg 2010

This chapter is mainly motivated by [5], [64], [92], [103] and extends Ostrowski type inequalities into the fuzzy setting, as fuzziness is a natural reality genuine feature different than randomness and determinism.

4.2 Background

We use the Fuzzy Taylor formula.

Theorem 4.1 (Theorem 1 of [11]). *Let* $T := [x_0, x_0 + \beta] \subset \mathbb{R}$, *with* $\beta > 0$. *We assume that* $f^{(i)} \colon T \to \mathbb{R}_{\mathcal{F}}$ *are* H-*differentiable for all* $i = 0, 1, \ldots, n-1$, *for any* $x \in T$. *(I.e., there exist in* $\mathbb{R}_{\mathcal{F}}$ *the* H-*differences* $f^{(i)}(x + h) - f^{(i)}(x)$, $f^{(i)}(x) - f^{(i)}(x - h)$, $i = 0, 1, \ldots, n - 1$ *for all small* $h \colon 0 < h < \beta$. *Furthermore there exist* $f^{(i+1)}(x) \in \mathbb{R}_{\mathcal{F}}$ *such that the limits in* D-*distance exist and*

$$f^{(i+1)}(x) = \lim_{h \to 0^+} \frac{f^{(i)}(x + h) - f^{(i)}(x)}{h} = \lim_{h \to 0^+} \frac{f^{(i)}(x) - f^{(i)}(x - h)}{h},$$

for all $i = 0, 1, \ldots, n - 1$.) *Also we assume that* $f^{(n)}$, *is fuzzy continuous on* T. *Then for* $s \geq a$, $s, a \in T$ *we obtain*

$$
\begin{aligned}
f(s) &= f(a) \oplus f'(a) \odot (s - a) \oplus f''(a) \odot \frac{(s - a)^2}{2!} \\
&\oplus \cdots \oplus f^{(n-1)}(a) \odot \frac{(s - a)^{n-1}}{(n - 1)!} \oplus R_n(a, s),
\end{aligned}
$$

where

$$R_n(a, s) := (FR) \int_a^s \left(\int_a^{s_1} \cdots \left(\int_a^{s_{n-1}} f^{(n)}(s_n) ds_n \right) ds_{n-1} \cdots \right) ds_1.$$

Here $R_n(a, s)$ *is fuzzy continuous on* T *as a function of* s.
    We use

Proposition 4.2 (Proposition 1 of [10]). *Let* $F(t) := t^n \odot u$, $t \geq 0$, $n \in \mathbb{N}$, *and* $u \in \mathbb{R}_{\mathcal{F}}$ *be fixed. Then (the* H-*derivative)*

$$F'(t) = nt^{n-1} \odot u.$$

In particular when $n = 1$ *then* $F'(t) = u$.
    We mention

Proposition 4.3 (Proposition 6 of [10]). *Let* I *be an open interval of* $\mathbb{R}$ *and let* $f \colon I \to \mathbb{R}_{\mathcal{F}}$ *be* H-*fuzzy differentiable,* $c \in \mathbb{R}$. *Then*

$$(c \odot f)' \text{ exists and } (c \odot f)' = c \odot f'(x).$$

We use the "Fuzzy Mean Value Theorem".

Theorem 4.4 (Theorem 1 of [10]). *Let $f\colon [a,b] \to \mathbb{R}_{\mathcal{F}}$ be a fuzzy differentiable function on $[a,b]$ with H-fuzzy derivative f' which is assumed to be fuzzy continuous. Then*

$$D(f(d), f(c)) \le (d-c) \sup_{t \in [c,d]} D(f'(t), \tilde{o}),$$

for any $c, d \in [a,b]$ with $d \ge c$.

We finally need the "Univariate Fuzzy Chain Rule".

Theorem 4.5 (Theorem 2 of [10]). *Let I be a closed interval in $\mathbb{R}$. Here $g\colon I \to \zeta := g(I) \subseteq \mathbb{R}$ is differentiable, and $f\colon \zeta \to \mathbb{R}_{\mathcal{F}}$ is H-fuzzy differentiable. Assume that g is strictly increasing. Then $(f \circ g)'(x)$ exists and*

$$(f \circ g)'(x) = f'(g(x)) \odot g'(x), \qquad \forall x \in I.$$

4.3 Results

We present the following

Theorem 4.6. *Let $f \in C([a,b], \mathbb{R}_{\mathcal{F}})$, the space of fuzzy continuous functions, $x \in [a,b]$ be fixed. We assume that f fulfills the Hölder condition*

$$D(f(y), f(z)) \le L_f \cdot |y - z|^\alpha, \quad 0 < \alpha \le 1, \ \forall y, z \in [a,b],$$

for some $L_f > 0$. Then

$$D\left(\frac{1}{b-a} \odot (FR) \int_a^b f(y) dy, f(x)\right) \le L_f \left(\frac{(x-a)^{\alpha+1} + (b-x)^{\alpha+1}}{(\alpha+1)(b-a)}\right).$$

$$(4.1)$$

Proof. We have that

$$D\left(\frac{1}{b-a} \odot (FR) \int_a^b f(y) dy, f(x)\right)$$

$$= D\left(\frac{1}{b-a} \odot (FR) \int_a^b f(y) dy, \frac{1}{b-a} \odot (FR) \int_a^b f(x) dy\right)$$

$$= \frac{1}{b-a} D\left(\int_a^b f(y) dy, \int_a^b f(x) dy\right)$$

$$\underset{\le}{\text{(by Lemma 1.13)}} \ \frac{1}{b-a} \int_a^b D(f(y), f(x)) dy$$

$$\le \frac{L_f}{b-a} \int_a^b |y - x|^\alpha dy = \left(\frac{L_f}{b-a}\right) \left(\frac{(x-a)^{\alpha+1} + (b-x)^{\alpha+1}}{\alpha+1}\right).$$

□

Optimality of (4.1) comes next.

Proposition 4.7. *Inequality (4.1) is sharp, in fact, attained by* $f^*(y) :=$ $|y - x|^\alpha \odot u$, $0 < \alpha \leq 1$, *with* $u \in \mathbb{R}_\mathcal{F}$ *fixed. Here* $x, y \in [a, b]$.

Proof. Clearly $f^* \in C([a, b], \mathbb{R}_\mathcal{F})$: for letting $y_n \to y$, $y_n \in [a, b]$, then

$$D(f^*(y_n), f^*(y)) = D(|y_n - x|^\alpha \odot u, |y - x|^\alpha \odot u)$$

(by Lemma 1.2)
$$\leq \quad ||y_n - x|^\alpha - |y - x|^\alpha| D(u, \tilde{o}) \to 0, \quad \text{as } n \to +\infty.$$

Furthermore

$$D(f^*(y), f^*(z)) = D(|y - x|^\alpha \odot u, |z - x|^\alpha \odot u)$$

(by Lemma 1.2)
$$\leq \quad ||y - x|^\alpha - |z - x|^\alpha| D(u, \tilde{o})$$
$$\leq ||y - x| - |z - x||^\alpha D(u, \tilde{o}) \leq |y - z|^\alpha D(u, \tilde{o}).$$

That is, for $L_{f^*} := D(u, \tilde{o})$ we get

$$D(f^*(y), f^*(z)) \leq L_{f^*} |y - z|^\alpha, \quad 0 < \alpha \leq 1, \text{ any } y, z \in [a, b].$$

So that f^* is a Hölder function.

Finally we have

$$D\left(\frac{1}{b-a} \odot (FR) \int_a^b f^*(y) dy, f^*(x)\right)$$

$$= D\left(\frac{1}{b-a} \odot (FR) \int_a^b (|y - x|^\alpha \odot u) dy, \tilde{o}\right)$$

$$= \frac{1}{b-a} \cdot D\left((FR) \int_a^b (|y - x|^\alpha \odot u) dy, \tilde{o}\right)$$

$$= \frac{1}{b-a} D\left(\left(\int_a^b |y - x|^\alpha dy\right) \odot u, \tilde{o}\right)$$

$$= \frac{1}{b-a} D\left(\left(\frac{(x-a)^{\alpha+1} + (b-x)^{\alpha+1}}{\alpha+1}\right) \odot u, \tilde{o}\right)$$

$$= \frac{L_{f^*}}{b-a} \left(\frac{(x-a)^{\alpha+1} + (b-x)^{\alpha+1}}{\alpha+1}\right). \quad \square$$

Next comes the basic Ostrowski type fuzzy result in

Theorem 4.8 *let* $f \in C^1([a, b], \mathbb{R}_\mathcal{F})$, *the space of one time continuously differentiable functions in the fuzzy sense. Then for* $x \in [a, b]$,

$$D\left(\frac{1}{b-a} \odot (FR) \int_a^b f(y) dy, f(x)\right) \leq \left(\sup_{t \in [a,b]} D(f'(t), \tilde{o})\right) \left(\frac{(x-a)^2 + (b-x)^2}{2(b-a)}\right).$$
$$(4.2)$$

Inequality (4.2) is sharp at $x = a$, in fact attained by $f^*(y) := (y - a)(b - a) \odot u$, $u \in \mathbb{R}_{\mathcal{F}}$ being fixed.

Proof. We observe that

$$D\left(\frac{1}{b-a} \odot (FR) \int_a^b f(y)dy, f(x)\right)$$

$$= D\left(\frac{1}{b-a} \odot (FR) \int_a^b f(y)dy, \frac{1}{b-a} \odot (FR) \int_a^b f(x)dy\right)$$

$$= \frac{1}{b-a} D\left((FR) \int_a^b f(y)dy, (FR) \int_a^b f(x)dy\right)$$

$$\begin{array}{c}\text{(by Lemma 1.13)}\\ \leq \end{array} \quad \frac{1}{b-a} \int_a^b D(f(y), f(x))dy$$

$$\begin{array}{c}\text{(by Theorem 4.4)}\\ \leq \end{array} \quad \frac{1}{b-a} \int_a^b |y - x| \left(\sup_{t\in[a,b]} D(f'(t), \tilde{o})\right) dy$$

$$= \frac{\left(\sup_{t\in[a,b]} D(f'(t), \tilde{o})\right)}{b-a} \left(\frac{(x-a)^2 + (b-x)^2}{2}\right),$$

proving (4.2).

By Propositions 4.2, 4.3 and Theorem 4.5 we get that $f^{*\prime}(y) = (b-a) \odot u$. We have that

$$\text{L.H.S.(4.2)} = D\left(\frac{1}{b-a} \odot (FR) \int_a^b ((y-a)(b-a) \odot u)dy, \tilde{o}\right)$$

$$= D\left((FR) \int_a^b ((y-a) \odot u)dy, \tilde{o}\right)$$

$$= D\left(\left(\int_a^b (y-a)dy\right) \odot u, \tilde{o}\right)$$

$$= D\left(\frac{(b-a)^2}{2} \odot u, \tilde{o}\right) = \frac{(b-a)^2}{2} D(u, \tilde{o}).$$

And

$$\text{R.H.S.(4.2)} = \sup_{t\in[a,b]} D((b-a) \odot u, \tilde{o}) \frac{(b-a)}{2} = \frac{(b-a)^2}{2} D(u, \tilde{o}).$$

That is equality in (4.2) is attained. □

We conclude with the following Ostrowski type inequality fuzzy generalization in

Theorem 4.9. *Let $f \in C^{n+1}([a, b], \mathbb{R}_{\mathcal{F}})$, $n \in \mathbb{N}$, the space of $(n+1)$ times continuously differentiable functions on $[a, b]$ in the fuzzy sense. Call*

$$M := \sum_{i=1}^{n} \frac{(b-a)^i}{(i+1)!} D(f^{(i)}(a), \tilde{o}).$$

Then

$$D\left(\frac{1}{b-a} \odot (FR) \int_a^b f(x)dx, f(a) \right) \leq \qquad (4.3)$$

$$\left[M + \left(\sup_{t \in [a,b]} D(f^{(n+1)}(t), \tilde{o}) \right) \frac{(b-a)^{n+1}}{(n+2)!} \right].$$

If $f^{(i)}(a) = \tilde{o}$, $i = 1, \ldots, n$. Then

$$D\left(\frac{1}{b-a} \odot (FR) \int_a^b f(x)dx, f(a) \right) \leq \left(\sup_{t \in [a,b]} D(f^{(n+1)}(t), \tilde{o}) \right) \frac{(b-a)^{n+1}}{(n+2)!}.$$
$$(4.4)$$

Inequalities (4.3) and (4.4) are sharp, in fact attained by

$$f^*(x) := (b-a)(x-a)^{n+1} \odot u, \qquad u \in \mathbb{R}_{\mathcal{F}} \text{ being fixed.}$$

Corollary 4.10. *Let $f \in C^2([a, b], \mathbb{R}_{\mathcal{F}})$. Then*

$$D\left(\frac{1}{b-a} \odot (FR) \int_a^b f(x)dx, f(a) \right) \qquad (4.5)$$

$$\leq \left[\frac{(b-a)}{2} D(f'(a), \tilde{o}) + \left(\sup_{t \in [a,b]} D(f''(t), \tilde{o}) \right) \frac{(b-a)^2}{6} \right].$$

When $f'(a) = \tilde{o}$, then

$$D\left(\frac{1}{b-a} \odot (FR) \int_a^b f(x)dx, f(a) \right) \leq \left(\sup_{t \in [a,b]} D(f''(t), \tilde{o}) \right) \frac{(b-a)^2}{6}.$$
$$(4.6)$$

Proof of Theorem 4.9. Let $x \in [a, b]$, then by Theorem 4.1 we get

$$f(x) = \sum_{i=1}^{n-1} {}^* f^{(i)}(a) \odot \frac{(x-a)^i}{i!} \oplus \mathcal{R}_n(a, x),$$

where

$$\mathcal{R}_n(a, x) := (FR) \int_a^x \left(\int_a^{x_1} \cdots \left(\int_a^{x_{n-1}} f^{(n)}(x_n)dx_n \right) dx_{n-1} \cdots \right) dx_1$$

(here we need $x \geq a$). We observe that

$$D\left(\frac{1}{b-a} \odot (FR)\int_a^b f(x)dx, f(a)\right)$$

$$= \frac{1}{b-a}D\left((FR)\int_a^b f(x)dx, (FR)\int_a^b f(a)dx\right)$$

$$= \frac{1}{b-a}D\left((FR)\int_a^b \left(\sum_{i=0}^{n-1}{}^* f^{(i)}(a) \odot \frac{(x-a)^i}{i!} \oplus R_n(a,x)\right)dx,\right.$$
$$\left.(FR)\int_a^b f(a)dx\right)$$

$$= \frac{1}{b-a}\cdot D\left((FR)\int_a^b \left(\sum_{i=1}^{n-1}{}^* f^{(i)}(a) \odot \frac{(x-a)^i}{i!} \oplus R_n(a,x)\right)dx, \tilde{o}\right)$$

$$= \frac{1}{b-a}D\left((FR)\int_a^b \left(\sum_{i=1}^{n}{}^* f^{(i)}(a) \odot \frac{(x-a)^i}{i!} \oplus R_n(a,x)\right)dx,\right.$$
$$\left.(FR)\int_a^b f^{(n)}(a) \odot \frac{(x-a)^n}{n!}dx\right)$$

$$= \frac{1}{b-a}D\left(\sum_{i=1}^{n}{}^* (FR)\int_a^b f^{(i)}(a) \odot \frac{(x-a)^i}{i!}dx\right.$$
$$\left.\oplus (FR)\int_a^b R_n(a,x)dx, (FR)\int_a^b f^{(n)}(a) \odot \frac{(x-a)^n}{n!}dx\right)$$

$$= \frac{1}{b-a}D\left(\sum_{i=1}^{n}{}^* f^{(i)}(a) \odot \frac{(b-a)^{i+1}}{(i+1)!} \oplus (FR)\int_a^b R_n(a,x)dx,\right.$$
$$\left.(FR)\int_a^b f^{(n)}(a) \odot \frac{(x-a)^n}{n!}dx\right)$$

$$\leq \frac{1}{b-a}\left[\sum_{i=1}^{n}\frac{(b-a)^{i+1}}{(i+1)!}D(f^{(i)}(a), \tilde{o})\right.$$
$$\left.+ D\left((FR)\int_a^b R_n(a,x)dx, (FR)\int_a^b f^{(n)}(a) \odot \frac{(x-a)^n}{n!}dx\right)\right]$$

$$= M + \frac{1}{b-a}D\left((FR)\int_a^b R_n(a,x)dx, (FR)\int_a^b f^{(n)}(a) \odot \frac{(x-a)^n}{n!}dx\right)$$

$$= M + \frac{1}{b-a} D\Bigg((FR) \int_a^b \Bigg(\int_a^x \Bigg(\int_a^{x_1} \cdots \Bigg(\int_a^{x_{n-1}} f^{(n)}(x_n) dx_n\Bigg) dx_{n-1}\Bigg)$$

$$\cdots\Bigg) dx_1\Bigg) dx,$$

$$(FR) \int_a^b \Bigg(\int_a^x \Bigg(\int_a^{x_1} \cdots \Bigg(\int_a^{x_{n-1}} f^{(n)}(a) dx_n\Bigg) dx_{n-1}\Bigg) \cdots\Bigg) dx_1\Bigg) dx\Bigg)\Bigg]$$

(by Lemmas 1.13, 1.15)
$$\leq \quad M + \frac{1}{b-a}\Bigg[\int_a^b \Bigg(\int_a^x \Bigg(\int_a^{x_1} \cdots \Bigg(\int_a^{x_{n-1}}$$

$$D(f^{(n)}(x_n), f^{(n)}(a)\Bigg) dx_n\Bigg) dx_{n-1}\Bigg) \cdots\Bigg) dx_1\Bigg) dx\Bigg]$$

(by Theorem 4.4)
$$\leq \quad M + \frac{1}{b-a}\Bigg[\int_a^b \Bigg(\int_a^x \Bigg(\int_a^{x_1} \cdots \Bigg(\int_a^{x_{n-1}} (x_n - a)$$

$$\cdot \Bigg(\sup_{t\in[a,b]} D(f^{(n+1)}(t), \tilde{o})\Bigg) dx_n\Bigg) dx_{n-1}\Bigg) \cdots\Bigg) dx_1\Bigg) dx\Bigg]$$

$$= M + \frac{\Big(\sup_{t\in[a,b]} D(f^{(n+1)}(t), \tilde{o})\Big)}{b-a} \frac{(b-a)^{n+2}}{(n+2)!}$$

$$= M + \Bigg(\sup_{t\in[a,b]} D(f^{(n+1)}(t), \tilde{o})\Bigg) \frac{(b-a)^{n+1}}{(n+2)!}.$$

We have established inequalities (4.3) and (4.4).

Consider $g(x) := c(x-a)^\ell \odot u$, $x \in [a,b]$, $c > 0$, $\ell \in \mathbb{Z}_+$, $u \in \mathbb{R}_\mathcal{F}$ fixed. We prove that g is fuzzy continuous. Let $x_n \in [a,b]$ such that $x_n \to x$ as $n \to +\infty$. Then

$$\begin{aligned} D(g(x_n), g(x)) &= D(c(x_n - a)^\ell \odot u, c(x-a)^\ell \odot u) \\ &\leq c|(x_n - a)^\ell - (x-a)^\ell| D(u, \tilde{o}) \to 0. \end{aligned}$$

Hence by the last argument, Propositions 4.2, 4.3 and Theorem 4.5 we obtain that $f^* \in C^{n+1}([a,b], \mathbb{R}_\mathcal{F})$.

We see that

$$f^{*(i)}(a) = \tilde{o}, \quad \text{for } i = 1, \ldots, n.$$

That is $M = 0$. Furthermore it holds

$$f^{*(n+1)}(x) = (b-a)(n+1)! \odot u.$$

Finally, we notice that

$$
\begin{aligned}
\text{L.H.S.}((4.3),(4.4)) \; &= \; D\left(\frac{1}{b-a} \odot (FR) \int_a^b ((b-a)(x-a)^{n+1} \odot u)dx, \tilde{o}\right) \\
&= \; D\left(u \odot \int_a^b (x-a)^{n+1} dx, \tilde{o}\right) = D\left(u \odot \frac{(b-a)^{n+2}}{n+2}, \tilde{o}\right) \\
&= \; \frac{(b-a)^{n+2}}{n+2} D(u, \tilde{o}).
\end{aligned}
$$

Also we find

$$
\text{R.H.S.}((4.3),(4.4)) = (b-a)(n+1)! D(u, \tilde{o}) \frac{(b-a)^{n+1}}{(n+2)!} = \frac{(b-a)^{n+2}}{n+2} D(u, \tilde{o}).
$$

Proving (4.3) and (4.4) sharp, in fact attained inequalities.   □

5

A FUZZY TRIGONOMETRIC APPROXIMATION THEOREM OF WEIERSTRASS-TYPE

In this chapter we show that any 2π-periodic fuzzy continuous function from $\mathbb{R}$ to the fuzzy number space $\mathbb{R}_{\mathcal{F}}$, can be uniformly approximated by some fuzzy trigonometric polynomials. This chapter is based on [31].

5.1   Introduction

A fuzzy valued function $f\colon [a,b] \to \mathbb{R}_{\mathcal{F}}$ is said to be continuous at $x_0 \in [a,b]$, if for each $\varepsilon > 0$ there is $\delta > 0$ such that $D(f(x), f(x_0)) < \varepsilon$, whenever $x \in [a,b]$ and $|x - x_0| < \delta$. We say that f is fuzzy continuous on $[a,b]$ if f is continuous at each $x_0 \in [a,b]$, and denote the space of all such functions by $C_{\mathcal{F}}[a,b]$.

For $f \in C_{\mathcal{F}}[0,1]$, let us consider the Bernstein-type fuzzy polynomials

$$B_n^{(\mathcal{F})}(f)(x) = \sum_{k=0}^{n}{}^{*} f(k/n) \odot p_{k,n}(x), \quad n \in \mathbb{N}, \ x \in [0,1] \qquad (5.1)$$

where $p_{k,n}(x) = \binom{n}{k} x^k (1-x)^{n-k}$ and $\sum^{*}$ means addition with respect to $\oplus$ in $\mathbb{R}_{\mathcal{F}}$. It is obvious that $p_{n,k}(x) \geq 0$, $\forall x \in [0,1]$ and $p_{n,0}(x)$, $p_{n,1}(x), \ldots, p_{n,n}(x)$ are linearly independent algebraic polynomials of degree $\leq n$.

Concerning these fuzzy polynomials recently was proved the following

G.A. Anastassiou: Fuzzy Mathematics: Approximation The., STUDFUZZ 251, pp. 75–82.
springerlink.com

Theorem 5.1 (see [49]). *If* $f \in C_{\mathcal{F}}[0,1]$, *then for any* $\varepsilon > 0$, *there exists* $n_0 \in \mathbb{N}$, *such that*

$$D\big(B_{n_0}^{(\mathcal{F})}(f)(x), f(x)\big) < \varepsilon, \quad \forall x \in [0,1]. \tag{5.2}$$

Moreover, in [66, p. 642, Theorem 13.13] *it is proved the following quantitative estimate*

$$D\big(B_n^{(\mathcal{F})}(f)(x), f(x)\big) \le \frac{3}{2}\omega_1^{(\mathcal{F})}\left(f; \frac{1}{\sqrt{n}}\right), \quad \forall n \in \mathbb{N}, \ \forall x \in [0,1], \tag{5.3}$$

where $\omega_1^{(\mathcal{F})}(f; \delta) = \sup\{D(f(x), f(y)); |x - y| \le \delta, \ x, y \in [0,1]\}$ *is the modulus of continuity of* f.

It is the main aim of this chapter to present an analogue of Theorem 5.1 in the case of approximation of 2π-periodic fuzzy continuous functions, by some trigonometric fuzzy polynomials.

5.2 Preliminaries

The proof of the main result requires some auxiliary results and concepts.

A function $f: \mathbb{R} \to \mathbb{R}_{\mathcal{F}}$ will be called 2π-periodic if $f(x + 2\pi) = f(x)$, $\forall x \in \mathbb{R}$.

A generalized fuzzy trigonometric polynomial of degree $\le n$, will be defined as a finite sum of the form $\sum^* T_k(x) \odot c_k$, where $c_k \in \mathbb{R}_{\mathcal{F}}$ and $T_k(x)$ are trigonometric polynomials of degree $\le n$. Here, from approximation theory's point of view, that is to approximate a function by other simpler functions; the polynomials $T_k(x)$, are not necessarily supposed to be linearly independent on $\mathbb{R}$.

We need a particular case of the Henstock integral introduced in [52].

Definition 5.2 ([66, p. 644]). Let $f: [a, b] \to \mathbb{R}_{\mathcal{F}}$. We say that f is Riemann integrable to $I \in \mathbb{R}_{\mathcal{F}}$, if for any $\varepsilon > 0$, there exists $\delta > 0$, such that for any division $P = \{[u, v]; \xi\}$ of $[a, b]$ with the norm $\Delta(P) < \delta$, we have

$$D\left(\sum_P {}^*(v - u) \odot f(\xi), I\right) < \varepsilon. \tag{5.4}$$

We write $I = (R) \int\limits_a^b f(x)dx$.

We also need the following:

Lemma 5.3. *For any* $a, b \in \mathbb{R}$, $a, b \ge 0$, *and any* $u \in \mathbb{R}_{\mathcal{F}}$ *we have*

$$D(a \odot u, b \odot u) \le |b - a| \cdot D(u, \tilde{o}), \tag{5.5}$$

where $\tilde{o} \in \mathbb{R}_{\mathcal{F}}$ is defined by $\tilde{o} = \chi_{\{0\}}$.

Proof. If $a = b$ then the inequality becomes $o = o$.

Let us suppose $0 \le a < b$ (the case $0 \le b < a$ is similar). We get (denoting $\lambda = a/(b-a) \ge 0$)

$$
\begin{aligned}
D(a \odot u, b \odot u) &= D\left((b-a) \odot \left(\frac{a}{b-a} \odot u\right), (b-a) \odot \left(\frac{b}{b-a} \odot u\right)\right) \\
&= (b-a) \cdot D\left(\frac{a}{b-a} \odot u, \frac{b}{b-a} \odot u\right) \\
&= (b-a) D(\lambda \odot u, (1-\lambda) \odot u) \\
&\le (b-a)[D(\lambda \odot u, u) + D(u, (1-\lambda) \odot u)] \\
&= (b-a)[D(\lambda \odot u \oplus \tilde{o}, \lambda \odot u \oplus (1-\lambda) \odot u) \\
&\quad + D(\lambda \odot u \oplus (1-\lambda) \odot u, (1-\lambda) \odot u \oplus \tilde{o})] \\
&= (b-a)[D(\tilde{o}, (1-\lambda) \odot u) + D(\lambda \odot u, \tilde{o})] \\
&= (b-a)[(1-\lambda) D(\tilde{o}, u) + \lambda D(\tilde{o}, u)] = (b-a) D(u, \tilde{o}).
\end{aligned}
$$

$\square$

Remark 5.4. If a, b are not both ≥ 0 then Lemma 5.3 is not in general valid, because the property $(\alpha + \beta) \odot u = \alpha \odot u \oplus \beta \odot u$, $(\alpha, \beta \in \mathbb{R}, u \in \mathbb{R}_{\mathcal{F}})$ is valid only when $\alpha, \beta \ge 0$, but in general it is not valid for arbitrary $\alpha, \beta \in \mathbb{R}$.

5.3 Main Result

Let us denote by $C_{2\pi}^{(\mathcal{F})}(\mathbb{R}) = \{f \colon \mathbb{R} \to \mathbb{R}_{\mathcal{F}}; f \text{ is fuzzy continuous and } 2\pi\text{-periodic on } \mathbb{R}\}$.

Theorem 5.5. *For any $\varepsilon > 0$, there exists a generalized fuzzy trigonometric polynomial $T(x)$, such that*

$$D(f(x), T(x)) < \varepsilon, \quad \forall x \in \mathbb{R}. \tag{5.6}$$

Proof. Let us define the fuzzy Jackson-type operators (see [66, p. 646])

$$\mathcal{J}_n(f)(x) = (R) \int_{-\pi}^{\pi} K_n(t) \odot f(x+t) dt = (R) \int_{-\pi}^{\pi} K_n(u-x) \odot f(u) du,$$

where $K_n(t) = L_{n'}(t)$, $n' = \left[\frac{n}{2}\right]$, $L_n(t) = \lambda_n^{-1} \left[\frac{\sin(nt/2)}{\sin(t/2)}\right]^4$, $\int_{-\pi}^{\pi} L_n(t) dt = 1$.

According to [66, p. 647, Theorem 13.14], we have

$$D(\mathcal{J}_n(f)(x), f(x)) \le C\omega_1^{(\mathcal{F})}\left(f; \frac{1}{n}\right), \quad \forall n \in \mathbb{N}, \forall x \in \mathbb{R},$$

where $\omega_1^{(\mathcal{F})}(f;\delta) = \sup\{D(f(x), f(y)); |x-y| \le \delta, x, y \in \mathbb{R}\}$ represents the modulus of continuity of f. Then, for fixed $\varepsilon > 0$, there exists $n_0 \in \mathbb{N}$ such that

$$D(\mathcal{J}_{n_0}(f)(x), f(x)) \le C\omega_1^{(\mathcal{F})}\left(f; \frac{1}{n_0}\right) < \frac{\varepsilon}{2}.$$

Now, divide $[-\pi, \pi]$ into m equal parts and consider the Riemann sum of the integral $\mathcal{J}_{n_0}(f)(x)$, i.e.,

$$\sum_{m}^{*}(f)(x) = \frac{2\pi}{m} \sum_{k=1}^{m} K_{n_0}\left(-\pi + \frac{2\pi k}{m} - x\right) \odot f\left(-\pi + \frac{2\pi}{m}k\right).$$

Obviously $\sum_{m}^{*}(f)(x)$ represents a generalized fuzzy trigonometric polynomial of degree $\le n_0$.

By [52, Theorem 2.5] we can write

$$\begin{aligned}
\mathcal{J}_{n_0}(f)(x) &= (R) \int_{-\pi}^{\pi} K_{n_0}(u - x) \odot f(u)du \\
&= \sum_{k=1}^{m} {}^{*} \int_{-\pi+(k-1)2\pi/m}^{-\pi+k\frac{2\pi}{m}} K_{n_0}(u - x) \odot f(u)du.
\end{aligned}$$

For any fixed $x \in \mathbb{R}$ we obtain

$$D\left(\mathcal{J}_{n_0}(f)(x), \sum_{m}^{*}(f)(x)\right)$$

$$= D\left(\sum_{k=1}^{m} {}^{*} \int_{-\pi+(k-1)\frac{2\pi}{m}}^{-\pi+k\frac{2\pi}{m}} K_{n_0}(u - x) \odot f(u)du,\right.$$

$$\left.\sum_{k=1}^{m} {}^{*} \left(\frac{2\pi}{m}\right) K_{n_0}\left(-\pi + k\frac{2\pi}{m} - x\right) \odot f\left(-\pi + k\frac{2\pi}{m}\right)\right)$$

$$\le \sum_{k=1}^{m} D\left(\int_{-\pi+(k-1)\frac{2\pi}{m}}^{-\pi+k\frac{2\pi}{m}} K_{n_0}(u - x) \odot f(u)du,\right.$$

$$\left.\frac{2\pi}{m} K_{n_0}\left(-\pi + k\frac{2\pi}{m} - x\right) \odot f\left(-\pi + k\frac{2\pi}{m}\right)\right)$$

$$= \sum_{k=1}^{m} D\left(\int_{-\pi+(k-1)\frac{2\pi}{m}}^{-\pi+k\frac{2\pi}{m}} K_{n_0}(u-x) \odot f(u)du, \right.$$

$$\int_{-\pi+(k-1)\frac{2\pi}{m}}^{-\pi+k\frac{2\pi}{m}} K_{n_0}\left(-\pi+k\frac{2\pi}{m}-x\right) \odot f\left(-\pi+k\frac{2\pi}{m}\right)\right)$$

$$\leq (\text{see [52, Remarks 3.2] or [66, p. 644, Lemma 13.2, (ii)]})$$

$$\leq \sum_{k=1}^{m} \int_{-\pi+(k-1)\frac{2\pi}{m}}^{-\pi+k\frac{2\pi}{m}} D(K_{n_0}(u-x) \odot f(u),$$

$$K_{n_0}\left(-\pi+k\frac{2\pi}{m}-x\right) \odot f\left(-\pi+k\frac{2\pi}{m}\right)\right)du$$

$$\leq 2\pi\omega_1^{(\mathcal{F})}\left(K_{n_0}(\cdot-x) \odot f(\cdot); \frac{2\pi}{m}\right).$$

Let $u_1, u_2 \in [-\pi, \pi]$, $|u_1 - u_2| \leq \frac{2\pi}{m}$ be fixed. We have

$$D(K_{n_0}(u_1-x) \odot f(u_1), K_{n_0}(u_2-x) \odot f(u_2))$$
$$\leq D(K_{n_0}(u_1-x) \odot f(u_1), K_{n_0}(u_1-x) \odot f(u_2))$$
$$+ D(K_{n_0}(u_1-x) \odot f(u_2), K_{n_0}(u_2-x) \odot f(u_2))$$
$$= K_{n_0}(u_1-x)D(f(u_1), f(u_2)) + D(K_{n_0}(u_1-x) \odot f(u_2),$$
$$K_{n_0}(u_2-x) \odot f(u_2))$$
$$\leq M_{n_0}\omega_1^{(\mathcal{F})}\left(f; \frac{2\pi}{m}\right) + D(K_{n_0}(u_1-x) \odot f(u_2), K_{n_0}(u_2-x) \odot f(u_2))$$

$$(\text{by Lemma 5.3})$$

$$\leq M_{n_0}\omega_1^{(\mathcal{F})}\left(f; \frac{2\pi}{m}\right) + |K_{n_0}(u_1-x) - K_{n_0}(u_2-x)| \cdot D(f(u_2), \tilde{o})$$
$$\leq M_{n_0}\omega_1^{(\mathcal{F})}\left(f; \frac{2\pi}{m}\right) + \omega_1\left(K_{n_0}; \frac{2\pi}{m}\right)$$
$$\cdot \sup\{D(f(u), \tilde{o}); u \in [0, 2\pi]\}$$
$$\leq M_{n_0}\omega_1^{(\mathcal{F})}\left(f; \frac{2\pi}{m}\right) + C\omega_1\left(K_{n_0}; \frac{2\pi}{m}\right),$$

because by e.g., the proof of Lemma 13.2 in [66, p. 644], we get that $F(u) = D(f(u), \tilde{o})$, $u \in \mathbb{R}$ is 2π-periodic and continuous on $\mathbb{R}$, that is $0 \leq F(u) \leq C$, $\forall u \in \mathbb{R}$. Now, choose $m_0 \in \mathbb{N}$ sufficiently large such that

$$M\omega_1^{(\mathcal{F})}\left(f; \frac{2\pi}{m}\right) + C\omega_1\left(K_{n_0}; \frac{2\pi}{m}\right) < \frac{\varepsilon}{2}$$

(this is possible because

$$\lim_{m \to +\infty} \omega_1^{(\mathcal{F})}\left(f; \frac{2\pi}{m}\right) = \lim_{m \to +\infty} \omega_1\left(K_{n_0}; \frac{2\pi}{m}\right) = 0).$$

Finally, we obtain

$$D\left(\sum_{m_0}^{*}(f)(x), f(x)\right) \leq D\left(\sum_{m_0}^{*}(f)(x), \mathcal{J}_{n_0}(f)(x)\right)$$

$$+ D(\mathcal{J}_{n_0}(f)(x), f(x)) < \frac{\varepsilon}{2} + \frac{\varepsilon}{2}, \quad \forall x \in \mathbb{R},$$

i.e., the theorem is proved with $T(x) = \sum_{m_0}^{*}(f)(x)$. $\square$

5.4   General Remarks

In this section we present some properties in the space $(\mathbb{R}_{\mathcal{F}}, D)$ that can be useful in the study of approximation of fuzzy valued functions. They can be summarized by the following:

Lemma 5.6. (i) *If we denote* $\tilde{o} = \chi_{\{0\}}$ *then* $\tilde{o} \in \mathbb{R}_{\mathcal{F}}$ *is neutral element with respect to* $\oplus$, *i.e.*

$$u \oplus \tilde{o} = \tilde{o} \oplus u = u, \quad \text{for all } u \in \mathbb{R}_{\mathcal{F}}.$$

(ii) *With respect to* $\tilde{o}$, *none of* $u \in \mathbb{R}_{\mathcal{F}}$, $u \neq \tilde{o}$, *has opposite in* $\mathbb{R}_{\mathcal{F}}$.
(iii) *For any* $a, b \in \mathbb{R}$ *with* $a, b \geq 0$ *or* $a, b \leq 0$, *and any* $u \in \mathbb{R}_{\mathcal{F}}$, *we have*

$$(a + b) \odot u = a \odot u \oplus b \odot u.$$

For general $a, b \in \mathbb{R}$, *the above property does not hold.*
(iv) *For any* $\lambda \in \mathbb{R}$ *and any* $u, v \in \mathbb{R}_{\mathcal{F}}$, *we have*

$$\lambda \odot (u \oplus v) = \lambda \odot u \oplus \lambda \odot v.$$

(v) *For any* $\lambda, \mu \in \mathbb{R}$ *and any* $u \in \mathbb{R}_{\mathcal{F}}$, *we have*

$$\lambda \odot (\mu \odot u) = (\lambda \cdot \mu) \odot u.$$

(vi) *If we denote* $\|u\|_{\mathcal{F}} = D(u, \tilde{o})$, $\forall u \in \mathbb{R}_{\mathcal{F}}$, *then* $\|\cdot\|_{\mathcal{F}}$ *has the properties of a usual norm on* $\mathbb{R}_{\mathcal{F}}$, *i.e.,* $\|u\|_{\mathcal{F}} = 0$ *iff* $u = \tilde{o}$, $\|\lambda \odot u\|_{\mathcal{F}} = |\lambda| \cdot \|u\|_{\mathcal{F}}$, $\|u \oplus v\|_{\mathcal{F}} \leq \|u\|_{\mathcal{F}} + \|v\|_{\mathcal{F}}$, $\left| \|u\|_{\mathcal{F}} - \|v\|_{\mathcal{F}} \right| \leq D(u, v)$.

Proof. (i) It is immediate by the definitions of $\tilde{o}$ and $\oplus$.
   (ii) Let $u \in \mathbb{R}_{\mathcal{F}}$ be fixed and let us suppose that there exists $v \in \mathbb{R}_{\mathcal{F}}$, such that $u \oplus v = \tilde{o}$.
   We get

$$[u]^r + [v]^r = [\tilde{o}]^r, \quad \forall r \in [0, 1],$$

that is $[u_-^r, u_+^r] + [v_-^r, v_+^r] = [0, 0] = \{0\}$, and finally $u_-^r + v_-^r = 0$, $u_+^r + v_+^r = 0$, $\forall r \in [0, 1]$. Because $u_-^r < u_+^r$ at least for one r (since $u \neq \tilde{o}$), the previous relation implies that for r the contradiction $v_+^r < v_-^r$.

(iii) Let $[u]^r = [u^r_-, u^r_+]$, $r \in [0,1]$. For $a, b \geq 0$ we get

$$
\begin{aligned}
(a+b) \odot [u^r_-, u^r_+] &= [(a+b)u^r_-, (a+b)u^r_+] = [au^r_- + bu^r_-, au^r_+ + bu^r_+] \\
&= [au^r_-, au^r_+] + [bu^r_-, bu^r_+] = a \cdot [u]^r + b \cdot [u]^r, \\
\forall r &\in [0,1].
\end{aligned}
$$

If $a, b \in \mathbb{R}$, $a, b \leq 0$, we get $(a+b) \leq 0$ and

$$
\begin{aligned}
(a+b) \cdot [u^r_-, u^r_+] &= [(a+b)u^r_+, (a+b)u^r_-] = [au^r_+ + bu^r_+, au^r_- + bu^r_-] \\
&= [au^r_+, au^r_-] + [bu^r_+, bu^r_-] = a[u^r_-, u^r_+] + b[u^r_-, u^r_+], \\
\forall r &\in [0,1].
\end{aligned}
$$

If $a, b \in \mathbb{R}$ are arbitrary, we cannot in general apply the above reasonings.
(iv) For any $r \in [0,1]$ we get by the definitions of $\odot$ and $\oplus$

$$
\begin{aligned}
[\lambda \odot (u \oplus v)]^r &= \lambda[u \oplus v]^r = \lambda([u]^r + [v]^r) \\
&= \lambda([u^r_-, u^r_+] + [v^r_-, v^r_+]) = \lambda([u^r_- + v^r_-, u^r_+ + v^r_+]).
\end{aligned}
$$

Now, if $\lambda \geq 0$, then

$$
\begin{aligned}
[\lambda \odot (u \oplus v)]^r &= [\lambda u^r_- + \lambda v^r_-, \lambda u^r_+ + \lambda v^r_+] \\
&= [\lambda u^r_-, \lambda u^r_+] + [\lambda v^r_-, \lambda v^r_+] = \lambda[u]^r + \lambda[v]^r.
\end{aligned}
$$

If $\lambda \leq 0$, then

$$
\begin{aligned}
[\lambda \odot (u \oplus v)]^r &= [\lambda(u^r_+ + v^r_+), \lambda(u^r_- + v^r_-)] \\
&= [\lambda u^r_+ + \lambda v^r_+, \lambda u^r_- + \lambda v^r_-] = [\lambda u^r_+, \lambda u^r_-] + [\lambda v^r_+, \lambda v^r_-] \\
&= \lambda[u]^r + \lambda[v]^r.
\end{aligned}
$$

(v) By definition of $\odot$, for any $r \in [0,1]$ we get

$$
[\lambda \odot (\mu \odot u)]^r = \lambda[\mu \odot u]^r = \lambda\mu[u]^r.
$$

(vi) The first three properties are immediate consequences of the properties of D. For the last property, by e.g. [69, p. 21, Exercise 1.2], in any metric space we have $|D(u,w) - D(v,w)| \leq D(u,v)$. Taking $w = \tilde{o}$ we get the proof. $\qquad\square$

Remark 5.7. The properties (ii) and (iii) in Lemma 5.6 show us that $(\mathbb{R}_{\mathcal{F}}, \oplus, \odot)$ is not a linear space over $\mathbb{R}$, and consequently $(\mathbb{R}_{\mathcal{F}}, \|\cdot\|_{\mathcal{F}})$ is not a normed space. But the properties of D and Lemma 5.6, (iv)–(vi), have as an effect that the metric properties of a function defined on $\mathbb{R}$ with values in a Banach space could be translated also to fuzzy functions in a similar manner.

Remark 5.8. The fact that the above Bernstein-type and Jackson-type (see the proof of Theorem 5.5) generalized fuzzy polynomials are of the form $\sum\limits_{k=1}^{n} c_k \odot P_k(x)$ and $\sum\limits_{k=1}^{n} c_k \odot T_k(x)$ (respectively), where $P_k(x)$ and $T_k(x)$ are positive polynomials on $[0,1]$ and $\mathbb{R}$ (respectively), is essential in applying the properties of the metric D. In case $P_k(x)$ (or $T_k(x)$) are not positive, by Lemma 5.6, (iii), the properties of D cannot be applied.

The concept of fuzzy polynomials can also be defined as in e.g. [66, p. 621, Definition 13.2]

$$\sum_{k=0}^{n}{}^{*} c_k \odot x^k, \quad x \in [a,b], \ c_k \in \mathbb{R}_{\mathcal{F}},$$

and

$$\sum_{k=0}^{n}{}^{*}(a_k \odot \cos kx + b_k \odot \sin kx), \quad a_k, b_k \in \mathbb{R}, \ k = \overline{0,n}.$$

Unfortunately, because of Lemma 5.6, (iii), we cannot apply the properties of D in order to obtain approximation results by such polynomials, so the problem of closure (in the uniform metric $D^*(f,g) = \sup\limits_{x}\{D(f(x), g(x))\}$) of the set of such polynomials remains open.

Obviously the generalized fuzzy polynomials of Bernstein-type and of Jackson-type mentioned earlier are not of the latter kind.

Remark 5.9. The property (iv) of Lemma 5.6 is useful to show the linearity of some approximation operators, like that of Bernstein-type

$$B_n^{(\mathcal{F})}(f)(x) = \sum_{k=0}^{n}{}^{*} f\left(\frac{k}{n}\right) \odot p_{n,k}(x),$$

for example. Indeed, we have

$$B_n^{(\mathcal{F})}(\lambda \odot f \oplus \mu \odot g)(x) = \sum_{k=0}^{n}{}^{*}[\lambda \odot f \oplus \mu \odot g]\left(\frac{k}{n}\right) \odot p_{n,k}(x)$$

$$= \sum_{k=0}^{n}{}^{*}\left[\lambda \odot f\left(\frac{k}{n}\right) \oplus \mu \odot g\left(\frac{k}{n}\right)\right] \odot p_{n,k}(x)$$

$$= \text{(by Lemma 5.6, (iv))}$$

$$= \sum_{k=0}^{n}{}^{*}\lambda \odot p_{n,k}(x) \odot f\left(\frac{k}{n}\right) + \sum_{k=0}^{n}{}^{*}\mu \odot p_{n,k}(x) \odot g\left(\frac{k}{n}\right)$$

$$= \lambda \odot \sum_{k=0}^{n}{}^{*} p_{n,k}(x) \odot f\left(\frac{k}{n}\right) + \mu \odot \sum_{k=0}^{n}{}^{*} p_{n,k}(x) \odot g\left(\frac{k}{n}\right)$$

$$= \lambda \odot B_n^{(\mathcal{F})}(f)(x) + \mu \odot B_n^{(\mathcal{F})}(g)(x).$$

$\square$

6

ON BEST APPROXIMATION AND JACKSON-TYPE ESTIMATES BY GENERALIZED FUZZY POLYNOMIALS

In [31] was proved that any 2π-periodic continuous fuzzy-number-valued function can be uniformly approximated by sequences of generalized fuzzy trigonometric polynomials, but without giving any estimates for the approximation error. In this chapter, connected to the best approximation problem we present Jackson-type estimates. For the algebraic case we also give a Jackson-type estimate, using the Szabados-type polynomials. Finally, as an application we study the convergence of fuzzy Lagrange interpolation polynomials. This chapter relies on [41].

6.1 Introduction

A function $f : \mathbb{R} \to \mathbb{R}_{\mathcal{F}}$ is called $2\pi-$ periodic if $f(x + 2\pi) = f(x)$, $\forall x \in \mathbb{R}$.

A generalized fuzzy trigonometric polynomial of degree $\leq n$ is defined as a finite sum of the form $T(x) = \sum_{k=0}^{n} t_k(x) \odot c_k$, where $c_k \in \mathbb{R}_{\mathcal{F}}$ and $t_k(x)$ are usual trigonometric polynomials of degree $\leq n$.

Let us denote $C_{2\pi}^{\mathcal{F}}(\mathbb{R}) = \{f : \mathbb{R} \to \mathbb{R}_{\mathcal{F}}; \ f \text{ is } 2\pi-\text{periodic and continuous on } \mathbb{R}\}$.

In [31], the following Weierstrass-type result is proved.

Theorem 6.1. For any $f \in C_{2\pi}^{\mathcal{F}}(\mathbb{R})$, there exists a sequence of generalized fuzzy trigonometric polynomials $(T_n(x))_{n\in\mathbb{N}}$ such that

$$\lim_{n\to\infty} \sup_{x\in\mathbb{R}} D(T_n(x), f(x)) = 0.$$

G.A. Anastassiou: Fuzzy Mathematics: Approximation The., STUDFUZZ 251, pp. 83–98.
springerlink.com

Other results concerning approximation and interpolation of fuzzy-number-valued functions can be found in: [11], [59], [57], [81], [84], [72], [80]. But the problems of existence of best approximation by fuzzy polynomials and of convergence of fuzzy Lagrange polynomials, were not discussed much in the fuzzy mathematical literature.

In Section 6.2 we consider some problems of best approximation by generalized fuzzy trigonometric polynomials (of degree $\leq n$) and a Jackson-type estimate is proved.

Section 6.3 contains the case of best approximation by generalized fuzzy algebraic polynomials.

In Section 6.4, as an application, we prove the convergence of Lagrange interpolating polynomials for the class of fuzzy Lipschitz functions of order $> \frac{1}{2}$.

6.2   Best approximation, trigonometric case

On $C_{2\pi}^{\mathcal{F}}(\mathbb{R})$ let us consider the uniform distance

$$D^*(f,g) = \sup\{D(f(x),g(x)); x \in \mathbb{R}\} = \sup\{D(f(x),g(x)); x \in [-\pi,\pi]\},$$

$\forall f,g \in C_{2\pi}^{\mathcal{F}}(\mathbb{R})$.

For an interval $I \subset \mathbb{R}$ and a subset $K \subset \mathbb{R}_{\mathcal{F}}$, $K \neq \emptyset$, let us consider $V_n^{K,I} = \{T_n; T_n(x) = \sum_{k=0}^n t_k(x) \odot c_k$, where all $c_k \in K$ and each $t_k(x)$ is an usual trigonometric polynomial of degree $\leq n$ with all its coefficients belonging to $I\}$.

For fixed $f \in C_{2\pi}^{\mathcal{F}}(\mathbb{R})$, $K \subset \mathbb{R}_{\mathcal{F}}$ and $I \subset \mathbb{R}$ and for each $n \in \mathbb{N}$, it is natural to consider the following problem of best approximation.
$E_n^{K,I}(f) = \inf\{D^*(f,T_n); T_n \in V_n^{K,I}\}$.

In the study of this problem, it is essential the following

Theorem 6.2. If $K \subset \mathbb{R}_{\mathcal{F}}$, $K \neq \emptyset$, is a compact and $I = [A,B]$ is compact subinterval of $\mathbb{R}$, then the set $V_n^{K,I}$ is sequentially compact in the metric space $(C_{2\pi}^{\mathcal{F}}(\mathbb{R}), D^*)$, for all $n \in \mathbb{N}$.

Proof. Let us denote by $\mathcal{T}_n^I = \{t_k; t_k$ is usual trigonometric polynomial of degree $\leq n$, with all coefficients belonging to $I\}$ and define $\varphi : K^{n+1} \times (\mathcal{T}_n^I)^{n+1} \to C_{2\pi}^{\mathcal{F}}(\mathbb{R})$, by $\varphi(c_0,...,c_n,t_0,...,t_n)(x) = \sum_{k=0}^n t_k(x) \odot c_k$.

First let us prove that φ is continuous.

Indeed, let another generalized fuzzy trigonometric polynomial of degree $\leq n$, $\sum_{k=0}^n s_k(x) \odot d_k$. By the properties of D in Introduction and by [31, Lemma 2.2], we get

$$D\left(\sum_{k=0}^n t_k(x) \odot c_k, \sum_{k=0}^n s_k(x) \odot d_k\right) \leq$$

$$\sum_{k=0}^{n} |t_k(x)| D(c_k, d_k) + \sum_{k=0}^{n} |t_k(x) - s_k(x)| \odot D(c_k, \widetilde{0}),$$

where $\widetilde{0} = \chi_{\{0\}} \in \mathbb{R}_{\mathcal{F}}$.

Because each $t_k(x)$ is of the form $\alpha_0 + \sum_{j=0}^{n} (\alpha_j \cos jx + \beta_j \sin jx)$, with $\alpha_j, \beta_j \in I = [A, B]$, it immediately follows that
$$|t_k(x)| \le (2n+1) \max\{|A|, |B|\} = M, \text{ for all } k = \overline{0, n} \text{ and all } x \in \mathbb{R}.$$

Also, because $c_k \in K-$ compact, $\forall k = \overline{0, n}$, we get that $K' = K \cup \{\widetilde{0}\}$ is compact too (in the metric space $\mathbb{R}_{\mathcal{F}}$), which implies that it is bounded and therefore there exists a constant $M' > 0$ such that $D(c_k, \widetilde{0}) \le M, \forall c_k \in K$.

As a conclusion, it follows

$$D\left(\sum_{k=0}^{n} t_k(x) \odot c_k, \sum_{k=0}^{n} s_k(x) \odot d_k\right) \le$$

$$M \sum_{k=0}^{n} D(c_k, d_k) + M' \sum_{k=0}^{n} \|t_k(x) - s_k(x)\|$$

(here $\|\cdot\|$ denotes the usual uniform norm on the set of real valued, $2\pi-$ periodic functions, denoted by $C_{2\pi}$).

This last inequality immediately shows that φ is continuous, if $K^{n+1} \times (T_n^I)^{n+1}$ is endowed with the box metric given by

$$\rho[(c_0, ..., c_n, t_0, ..., t_n), (d_0, ..., d_n, s_0, ..., s_n)] = \max_{k=\overline{0,n}} \{D(c_k, d_k), \|t_k - s_k\|\}.$$

Now we claim that T_n^I is compact in $(C_{2\pi}, \|\cdot\|)$. Indeed, if we consider $\psi : I^{2n+1} \to C_{2\pi}$ defined by

$$\psi(\alpha_0, ..., \alpha_n, \beta_1, ..., \beta_n)(x) = \alpha_0 + \sum_{k=0}^{n} [\alpha_k \cos kx + \beta_k \sin kx],$$

then it easily follows that ψ is continuous and therefore $T_n^I = \psi(I^{2n+1})$ is compact.

As a conclusion, $K^{n+1} \times (T_n^I)^{n+1}$ is compact which implies that $V_n^{K,I} = \varphi(K^{n+1} \times (T_n^I)^{n+1})$ is compact, and therefore as a compact subset of a metric space, $V_n^{K,I}$ is sequentially compact.   □

As an immediate consequence of Theorem 6.2, we get

Corollary 6.3. Let $f \in C_{2\pi}^{\mathcal{F}}(\mathbb{R})$. If $K \subset \mathbb{R}_{\mathcal{F}}$, $K \ne \emptyset$ is compact and $I = [A, B]$ is a compact interval of $\mathbb{R}$, then for each $n \in \mathbb{N}$, there exists $T^* \in V_n^{K,I}$ such that $E_n^{K,I}(f) = D^*(f, T^*)$, i.e. T^* is a generalized fuzzy trigonometric polynomial (of degree $\le n$) of best approximation for f.

Proof. Since $K \ne \emptyset$, it follows that $V_n^{K,I} \ne \emptyset$.

For $\varepsilon = \frac{1}{m}$, there exists $T_m \in V_n^{K,I}$ such that $E_n^{K,I}(f) \le D^*(f, T_m) \le E_n^{K,I}(f) + \frac{1}{m}$, $m = 1, 2, ...$. Since by Theorem 6.2, $V_n^{K,I}$ is sequentially compact, the sequence $(T_m)_m$ has a convergent subsequence $(T_{m_k})_k$ to an element $T^* \in V_n^{K,I}$. Passing above to limit, we get $E_n^{K,I}(f) = D^*(f, T^*)$, i.e. T^* is of best approximation, which proves the corollary. □

Note. If $f \in C_{2\pi}^{\mathcal{F}}(\mathbb{R})$, then by $f([-\pi, \pi]) = K$ compact, it follows that in Corollary 6.3 we can take $K = f([-\pi, \pi])$ (i.e. depending on f).

In what follows we will derive a Jackson-type estimate for $E_n^{K,I}(f)$ with $K = f([-\pi, \pi])$.

Theorem 6.4. If $f \in C_{2\pi}^{\mathcal{F}}(\mathbb{R})$ and $[-1, 1] \subset [A, B]$, then there exists a constant $C > 0$ (independent of f and n) and an index $n_0 \in \mathbb{N}$ (independent of f) such that for $K = f([-\pi, \pi])$ we have

$$E_n^{K,[A,B]}(f) \leq C\omega_1^{\mathcal{F}}\left(f, \frac{1}{n}\right), \quad \forall n \geq n_0$$

where $\omega_1^{\mathcal{F}}(f, \delta) = \sup\{D(f(x), f(y)); x, y \in \mathbb{R}, |x - y| \leq \delta\}$.

Proof. In [66, p.646] was introduced the following fuzzy Jackson operator

$$J_n(f)(x) = (R) \int_{-\pi}^{\pi} K_n(t) \odot f(x + t)dt = (R) \int_{-\pi}^{\pi} K_n(u - x) \odot f(u)du,$$

where $K_n(t) = L_{n'}(t)$, $n' = [n/2] + 1$,

$$L_{n'}(t) = \frac{3}{2\pi n'[2(n')^2 + 1]} \left[\frac{\sin(n't/2)}{\sin(t/2)}\right]^4, \quad \int_{-\pi}^{\pi} L_{n'}(t)dt = 1,$$

and it was proved [66, p.647, Theorem 13.14] the estimate

$$D(J_n(f)(x), f(x)) \leq C\omega_1^{\mathcal{F}}\left(f, \frac{1}{n}\right), \quad \forall n \in \mathbb{N}, \ x \in \mathbb{R}.$$

On the other hand by taking the Riemann sum of $J_n(f)(x)$ (on an equidistant division of $[-\pi, \pi]$), we get

$$T_n(x) = \frac{2\pi}{n'} \sum_{k=0}^{n'} L_{n'}\left(-\pi + \frac{2k\pi}{n'} - x\right) \odot f\left(-\pi + \frac{2k\pi}{n'}\right).$$

Obviously $T_n(x)$ is a generalized fuzzy trigonometric polynomial of degree $\leq n$ (since $n' \leq n$) and by [40, Corollary 3], for all $x \in [-\pi, \pi]$ and $n \in \mathbb{N}$, we have

$$D(J_n(f)(x), T_n(x)) \leq 2\pi\omega_1^{\mathcal{F}}\left(f, \frac{2\pi}{n'}\right)_{[-\pi,\pi]}$$

$$\leq 2\pi(2\pi + 1)\omega_1^{\mathcal{F}}\left(f, \frac{1}{n'}\right)_{[-\pi,\pi]} \leq 4\pi(2\pi + 1)\omega_1^{\mathcal{F}}\left(f, \frac{1}{n}\right)_{[-\pi,\pi]}$$

(since $n' = [n/2] + 1 > n/2$).

But reasoning exactly as in the usual case (see [30, p.75, Lemma 2.2.1], we have $\omega_1^{\mathcal{F}}(f, \delta)_{[-\pi,\pi]} \le \omega_1^{\mathcal{F}}(f, \delta) \le 2\omega_1^{\mathcal{F}}(f, \delta)_{[-\pi,\pi]}$.

As a consequence, we obtain the following Jackson-type estimate

$$D(T_n(x), f(x)) \le D(T_n(x), J_n(f)(x)) + D(J_n(f)(x), f(x)) \le$$

$$\le C\omega_1^{\mathcal{F}}\left(f, \frac{1}{n}\right), \quad \text{for all } n \in \mathbb{N}, \ x \in [-\pi, \pi].$$

(Note that above $\omega_1^{\mathcal{F}}\left(f, \frac{1}{n}\right)$ can be replaced by $\omega_1^{\mathcal{F}}\left(f, \frac{1}{n}\right)_{[-\pi,\pi]}$ too).

To finish the proof, we have to calculate the bounds for the coefficients of the usual trigonometric polynomials $L_{n'}\left(-\pi + \frac{2k\pi}{n'} - x\right)$ in the expression of $T_n(x)$.

First, it is well known the identity

$$\left(\frac{\sin\frac{mx}{2}}{\sin\frac{x}{2}}\right)^2 = m + 2\sum_{k=1}^{m-1}(m-k)\cos kx.$$

It follows

$$\left(\frac{\sin\frac{mx}{2}}{\sin\frac{x}{2}}\right)^4 = \left(m + 2\sum_{k=1}^{m-1}(m-k)\cos kx\right)\left(m + 2\sum_{k=1}^{m-1}(m-k)\cos kx\right)$$

$$= m^2 + 4m\sum_{k=1}^{m-1}(m-k)\cos kx + 4\sum_{i=1}^{m-1}\sum_{j=1}^{m-1}(m-i)(m-j)\cos(ix)\cos(jx) =$$

$$= m^2 + 4m\sum_{k=1}^{m-1}(m-k)\cos kx + 4\sum_{i=1}^{m-1}\sum_{j=1}^{m-1}(m-i)(m-j)\{\frac{1}{2}[\cos x(i+$$

$$j) + \cos x(i-j)]\} =$$

$$= m^2 + 4m\sum_{k=1}^{m-1}(m-k)\cos kx + 2\sum_{i=1}^{m-1}\sum_{j=1}^{m-1}(m-i)(m-j)\cos x(i+$$

$$j) + 2\sum_{i=1}^{m-1}\sum_{j=1}^{m-1}(m-i)(m-j)\cos x(i-j) =$$

$$= m^2 + 4m\sum_{k=1}^{m-1}(m-k)\cos kx + 2\sum_{\substack{i,j=1\\i+j\le m}}^{m-1}(m-i)(m-j)\cos x(i+j) +$$

$$2\sum_{\substack{i,j=1\\i+j>m}}^{m-1}(m-i)(m-j)\cos x(i+j) + 2\sum_{i,j=1}^{m-1}(m-i)(m-j)\cos x(i-j) :=$$

$$m^2 + S_1 + S_2 + S_3 + S_4.$$

By simple calculation we can write

$$S_2 = \sum_{k=2}^{m}[2\sum_{i=1}^{k-1}(m-i)(m-(k-i))]\cos(kx),$$

$$S_3 = \sum_{p=1}^{m-2}[2\sum_{i=1}^{m-p-1}(m-(p+i))(m-(m-i))]\cos((m+p)x),$$

$$S_4 = 4(1^2 + 2^2 + \ldots + (m-1)^2) + \sum_{k=1}^{m-2}[4\sum_{i=1}^{m-1-k}(m-i)(m-(k+i))]\cos(kx).$$

By the relations

$$\sum_{i=1}^{k-1}(m-i)(m-(k-i)) = \sum_{i=1}^{k-1}[-i^2 + ik + m(m-k)] \le \sum_{i=1}^{k-1}[k^2 + 4m(m-k)]/2 \le$$

$$(k-1)[k^2 + 4m(m-k)]/2 \le (m-1)[m^2 + 4m(m-2)]/2, k = 2, ..., m,$$

$$\sum_{i=1}^{m-p-1} (m-(p+i))(m-(m-i)) = \sum_{i=1}^{m-p-1} [-i^2 + i(m-p)] \le \sum_{i=1}^{m-p-1} (m-p)^2/2 \le$$

$$(m-p)^2(m-p-1)/2 \le (m-1)^2(m-2)/2, p = 1, ..., m-2,$$

$$\sum_{i=1}^{m-1-k} (m-i)(m-i-k) \le \sum_{i=1}^{m-1-k} (m-1)(m-1-k) \le$$

$$(m-1)(m-1-k)^2 \le (m-1)(m-2)^2, k = 1, ..., m-2,$$

$$1^2 + 2^2 + ... + (m-1)^2 = m(m-1)(2m-1)/6,$$

it follows that for $k \in \{0, ..., 2m-2\}$, the coefficients of $\cos kx$ in $\left(\frac{\sin \frac{mx}{2}}{\sin \frac{x}{2}}\right)^4$ are all positive and bounded by an algebraic polynomial of degree 3, with constant coefficients, independent of f, let us denote it by $H_3(m)$ (in S_1, obviously all the coefficients of $cos(kx)$ are bounded by $4m(m-1)$).

As a conclusion, it easily follows that in $(2\pi/n')L_{n'}\left(-\pi + \frac{2k\pi}{n'} - x\right)$ (which contains terms in $\cos kx$ and $\sin kx$), all the coefficients are bounded, in absolute value, by $F(n') = 3H_3(n')/[(n'^2(2(n'^2+1))]$, that is an $n_0 \in \mathbb{N}$ (independent of f) can be found (constructively), such that for all $n' \ge n_0$ we have $F(n') \le 1$ (since $F(n')$ converges to 0 when n' converges to infinity).

Therefore, for $n \ge 3n_0$, it follows that $T_n(x)$ belongs to $V_n^{K,[-1,1]}$. Now, for $[-1,1] \subset [A, B]$ it is obvious that $E_n^{K,[A,B]}(f) \le E_n^{K,[-1,1]}(f)$, which proves the theorem. $\qquad \square$

Remark 6.5. From the proof it is easily seen that an interval $[A, B]$ (independent of f and n) can be constructively determined such that the Jackson kind estimate in Theorem 6.4 holds for all $n = 1, 2, ...$

6.3 Best approximation, algebraic case

Let $C_{\mathcal{F}}[a, b] = \{f : [a, b] \to \mathbb{R}_{\mathcal{F}}; f \text{ continuous on } [a, b]\}$ where $[a, b]$ is a compact subinterval of $\mathbb{R}$. If we define the concept of generalized fuzzy algebraic polynomial of degree $\le n$ as in [49], i.e. as a finite sum of the form $\sum_{k=0}^{n} p_k(x) \odot c_k$, where $c_k \in \mathbb{R}_{\mathcal{F}}$ and $p_k(x)$ are algebraic polynomials of degree $\le n$, we can repeat the reasonings in the above Theorem 6.2 and Corollary 6.3 simply by replacing $[-\pi, \pi]$ there by $[a, b]$, $C_{2\pi}^{\mathcal{F}}(\mathbb{R})$ by $C_{\mathcal{F}}[a, b]$ and the generalized fuzzy trigonometric polynomials by generalized fuzzy algebraic polynomials. But if we consider probably the simplest generalized fuzzy algebraic polynomials, given by the fuzzy Bernstein polynomials

$$B_n(f)(x) = \sum_{k=0}^{n} p_{n,k}(x) \odot f(k/n), x \in [0, 1]$$

where $p_{n,k}(x) = \binom{n}{k}x^k(1-x)^{n-k}$, we easily see that the coefficients of x^s in $p_{n,k}(x)$ are in general unbounded, for $k = \overline{1,n-1}$, while however $|p_{n,k}(x)| \leq 1$, $\forall x \in [0,1]$, $n \in \mathbb{N}$, $k = \overline{0,n}$. Therefore, in algebraic case, would be more natural to consider the problem of best approximation as follows. For a constant $M > 0$ and a subset $K \subset \mathbb{R}_{\mathcal{F}}$, $K \neq \emptyset$, let us consider $A_n^{K,M}[a,b] = \{P_n;\ P_n(x) = \sum_{k=0}^n p_k(x) \odot c_k$, where all $c_k \in \mathbb{R}_{\mathcal{F}}$ and $p_k(x)$ are algebraic polynomials of degree $\leq n$, satisfying $|p_k(x)| \leq M$, for all k and all $x \in [a,b]\}$.

For fixed $f \in C_{\mathcal{F}}[a,b]$, $K \subset \mathbb{R}_{\mathcal{F}}$, $K \neq \emptyset$ and $M > 0$ and for each $n \in \mathbb{N}$, we can consider the following problem of best approximation

$$E_n^{K,M}(f) = \inf\{D^*(f,P_n);\ P_n \in A_n^{K,M}[a,b]\},$$

where $D^*(f,g) = \sup\{D(f(x),g(x);\ x \in [a,b]\}$, for $f,g \in C_{\mathcal{F}}[a,b]$.
We have:

Theorem 6.6. If $K \subset \mathbb{R}_{\mathcal{F}}$, $K \neq \emptyset$ is compact and $M > 0$, then the set $A_n^{K,M}[a,b]$ is sequentially compact in the metric space $(C_{\mathcal{F}}[a,b], D^*)$, for all $n \in \mathbb{N}$.

Proof. Let us denote by $\mathcal{P}_n^M = \{p;\ p$ usual algebraic polynomials of degree $\leq n$, satisfying $|p(x)| \leq M$, for all $x \in [a,b]\}$ and define $\varphi : K^{n+1} \times (\mathcal{P}_n^M)^{n+1} \to C_{\mathcal{F}}[a,b]$, by $\varphi(c_0,...,c_n,p_0,...,p_n)(x) = \sum_{k=0}^n p_k(x) \odot c_k$.

Reasoning exactly as in the proof of Theorem 6.2, we get that φ is continuous and because $\mathcal{P}_n^M$ is compact in $C[a,b]$ (endowed with the uniform norm $\|\cdot\|$), see e.g. [56, p16 Lemma 1], we get the desired conclusion. $\square$

Consequently, we obtain the following

Corollary 6.7. Let $f \in C_{\mathcal{F}}[a,b]$. If $K \subset \mathbb{R}_{\mathcal{F}}$, $K \neq \emptyset$ is compact and $M > 0$, then for all $n \in \mathbb{N}$, there exists $T^* \in A_n^{K,M}[a,b]$ such that $E_n^{K,M}(f) = D^*(f,T^*)$, i.e. T^* is a generalized fuzzy algebraic polynomial (of degree $\leq n$), of best approximation for f.

Remark 6.8. By [66, p. 642, Theorem 13.13], we immediately obtain

$$E_n^{K,1}(f) \leq C\omega_1^{\mathcal{F}}\left(f, \frac{1}{\sqrt{n}}\right)_{[0,1]}, \quad \forall n \in \mathbb{N},\ f \in C_{\mathcal{F}}[0,1],\ K = f([0,1]).$$

In what follows we deduce Jackson-type estimate for $E_n^{K,M}(f)$ by using some fuzzy analogous of Szabados-type polynomials (see e.g.[101]). For this aim we need the following lemmas.

Lemma 6.9. Let $f : \left[-\frac{1}{4}, \frac{1}{4}\right] \to \mathbb{R}_{\mathcal{F}}$, be continuous and

$$R_n(f,x) = \sum_{k=-n}^n r_{n,k}(x) \odot f(x_k),$$

with $r_{n,k}(x) = \frac{(x-x_k)^{-4}}{\sum_{j=-n}^{n}(x-x_j)^{-4}}$ and $x_k = \frac{k}{4n}$, $k = \overline{-n,n}$. Then the following estimate holds true:

$$D^*(f, R_n) \leq 5\omega_1^{\mathcal{F}}\left(f, \frac{1}{n}\right)_{\left[-\frac{1}{4}, \frac{1}{4}\right]}.$$

Proof. We follow the proof of Lemma 1 in [101] for $r = 0$ and $s = 4$. Thus, for fixed x, let i be an index such that

$$|x - x_i| = \min_{|k| \leq n} |x - x_k| \leq \frac{1}{8n}. \tag{6.1}$$

Then evidently

$$\frac{|i - k|}{8n} \leq |x - x_k| \leq \frac{|i - k|}{2n}, \text{ for } i \neq k. \tag{6.2}$$

Denote $I = \left[-\frac{1}{4}, \frac{1}{4}\right]$. Since $r_{n,k}(x) \geq 0$, $\forall k = \overline{-n,n}$ and $\sum_{k=-n}^{n} r_{n,k}(x) = 1$, by the properties of D we have:

$$D(f(x), R_n(f, x)) = D\left(\left(\sum_{k=-n}^{n} r_{n,k}(x)\right) \odot f(x), \sum_{k=-n}^{n} r_{n,k}(x) \odot f(x_k)\right)$$

$$\leq \sum_{k=-n}^{n} r_{n,k}(x) \odot D(f(x), f(x_k)) \leq \sum_{k=-n}^{n} r_{n,k}(x)\omega_1^{\mathcal{F}}(f, |x - x_k|)_I$$

$$\leq (x-x_i)^4 \sum_{k=-n}^{n} \frac{D(f(x), f(x_k))}{(x - x_k)^4} \leq (x-x_i)^4 \sum_{k=-n}^{n} |x-x_k|^{-4}\omega_1^{\mathcal{F}}(f, |x - x_k|)_I$$

$$\leq \omega_1^{\mathcal{F}}(f, |x - x_i|)_I + (8n)^{-4} \sum_{\substack{k=-n \\ k \neq i}}^{n} \omega_1^{\mathcal{F}}\left(f, \frac{|i - k|}{2n}\right)_I \left(\frac{8n}{|i - k|}\right)^4$$

$$\leq \left[1 + \sum_{\substack{k=-n \\ k \neq i}}^{n} |i - k|^{-2}\right] \omega_1^{\mathcal{F}}\left(f, \frac{1}{n}\right)_I \leq 5\omega_1^{\mathcal{F}}\left(f, \frac{1}{n}\right)_I.$$

$\square$

Remark 6.10. The result in the above Lemma 6.9 can be seen as a Jackson-type estimate for the error of the approximation by fuzzy generalized rational functions.

For the next results we need an embedding theorem.

Theorem 6.11. (see e.g. [52]) Let $\overline{C}[0, 1]$ be the class of all real valued bounded functions f on $[0, 1]$, such that f is left continuous on $(0, 1]$ and f

has right limit for $x \in [0,1)$, especially f is right continuous at 0. With the norm $\|f\| = \sup_{x \in [0,1]} |f(x)|$, $\overline{C}[0,1]$ is a Banach space. For $u \in \mathbb{R}_{\mathcal{F}}$, define $j : \mathbb{R}_{\mathcal{F}} \to \overline{C}[0,1]$, $j(u) = (u_-, u_+)$, where $u_-(r) = u_-^r$ and $u_+(r) = u_+^r$. Then $j(\mathbb{R}_{\mathcal{F}})$ is a closed convex cone in the Banach space $\overline{C}[0,1] \times \overline{C}[0,1]$ and:

(i) $j(s \odot u \oplus t \odot v) = s \cdot j(u) + t \cdot j(v)$, $\forall u, v \in \mathbb{R}_{\mathcal{F}}$ and $s, t \in \mathbb{R}_+$ (here "+" and "·"denote the addition and scalar multiplication in $\overline{C}[0,1] \times \overline{C}[0,1]$).

(ii) $D(u,v) = \|j(u) - j(v)\|$, $\forall u, v \in \mathbb{R}_{\mathcal{F}}$. i.e. j embeds $\mathbb{R}_{\mathcal{F}}$ in $\overline{C}[0,1] \times \overline{C}[0,1]$ isometrically and isomorphically ($\|\cdot\|$ being the usual product norm in $\overline{C}[0,1] \times \overline{C}[0,1]$).

The following lemmas give some approximation properties in Banach spaces.

Lemma 6.12. Let $(\mathbb{B}, \|\cdot\|)$ be a Banach space and $g : \left[-\frac{1}{4}, \frac{1}{4}\right] \to \mathbb{B}$ continuous. Let $R_n(g,x) = \sum_{k=-n}^{n} r_{n,k}(x) \cdot g(x_k)$, with $r_{n,k}$ as in Lemma 6.9. Then

$$\|R_n'(g,x)\| \leq 900 n \omega_1^{\mathbb{B}} \left(g, \frac{1}{n}\right)_{\left[-\frac{1}{4}, \frac{1}{4}\right]},$$

where $R_n'(g,x)$ is the Fréchet derivative of $R_n(g,x)$ in $\mathbb{B}$ and $\omega_1^{\mathbb{B}}(g,\delta)_{\left[-\frac{1}{4},\frac{1}{4}\right]} = \sup\left\{\|g(x) - g(y)\| ; \ x, y \in \left[-\frac{1}{4}, \frac{1}{4}\right], \ |x - y| \leq \delta \right\}$.

Proof. The proof is the same as that of [101, Lemma 2], written in the case of functions with values in a Banach space.

Thus by (6.1) and (6.2) we get

$$\|R_n'(g,x)\| = \left\| \frac{-4\sum_{k=-n}^{n} g(x_k)(x-x_k)^{-5} \sum_{k=-n}^{n}(x-x_k)^{-4}}{\left(\sum_{k=-n}^{n}(x-x_k)^{-4}\right)^2} + \right.$$

$$\left. + \frac{4\sum_{k=-n}^{n} g(x_k)(x-x_k)^{-4} \sum_{k=-n}^{n}(x-x_k)^{-5}}{\left(\sum_{k=-n}^{n}(x-x_k)^{-4}\right)^2} \right\|$$

$$= 2 \frac{\left\| \sum_{k=-n}^{n}(x-x_k)^{-5} \sum_{j=-n}^{n}[g(x_k) - g(x_j)](x_k - x_j)(x - x_j)^{-5} \right\|}{\left(\sum_{k=-n}^{n}(x-x_k)^{-4}\right)^2}$$

$$\leq 2(x-x_i)^8 \sum_{k=-n}^{n} |x - x_k|^{-5} \cdot \frac{1}{4n} \omega_1^{\mathbb{B}}\left(g, \frac{1}{n}\right)_{\left[-\frac{1}{4},\frac{1}{4}\right]} \sum_{\substack{j=-n \\ j \neq k}}^{n} \frac{|j-k|^2}{|x-x_j|^5}$$

$$\leq \frac{(8n)^{-3} \omega_1^{\mathbb{B}}\left(g, \frac{1}{n}\right)_{\left[-\frac{1}{4},\frac{1}{4}\right]}}{2n} \cdot$$

$$\cdot \left\{ \sum_{\substack{j=-n \\ j \neq i}}^{n} \left(\frac{8n}{|j-i|}\right)^5 |j-i|^2 + \sum_{\substack{k=-n \\ k \neq i}}^{n} \left(\frac{8n}{|k-i|}\right)^5 \left[|k-i|^2 + \sum_{\substack{j=-n \\ j \neq i}}^{n} \frac{|j-k|^2}{|j-i|^5}\right] \right\}$$

$$\leq 32n\omega_1^{\mathbb{B}}\left(g,\frac{1}{n}\right)_{[-\frac{1}{4},\frac{1}{4}]}\left\{\sum_{\substack{j=-n\\j\neq i}}^{n}|j-i|^{-3}+\sum_{\substack{k=-n\\k\neq i}}^{n}|k-i|^{-3}\left[1+4\sum_{\substack{j=-n\\j\neq i}}^{n}|j-i|^{-3}\right]\right\}$$

$$\leq 900n\omega_1^{\mathbb{B}}\left(g,\frac{1}{n}\right)_{[-\frac{1}{4},\frac{1}{4}]}.$$

$\square$

Lemma 6.13. Let $(\mathbb{B},\|\cdot\|)$ be a Banach space and $f:\left[-\frac{1}{2},\frac{1}{2}\right]\to\mathbb{B}$ continuous. Let $P_n(x)=c_n\left(\frac{\cos(2n\arccos x)}{x^2-\sin^2\frac{\pi}{4n}}\right)^2$ where c_n is chosen such that $\int_{-1}^{1}P_n(x)dx=1$ and let

$$K_n(f,x)=\int_{-\frac{1}{2}}^{\frac{1}{2}}[f(t)-f(0)]P_n(t-x)dt+f(0)$$

be the Bojanic-DeVore operator, where the integral is considered to be the usual Riemann integral for functions $g:\left[-\frac{1}{2},\frac{1}{2}\right]\to\mathbb{B}$. Then

$$\|f-K_n(f)\|_{C([-\frac{1}{4},\frac{1}{4}],\mathbb{B})}\leq C_4\omega_1^{\mathbb{B}}\left(f,\frac{1}{n}\right)_{[-\frac{1}{4},\frac{1}{4}]}.$$

Proof. The proof is the same as the proof of [56, p. 275-276, Proposition 3.4] but for functions with values in a Banach space. Indeed, firstly $P_n(x)$ is an even algebraic polynomial of degree $4n-4$, therefore $K_n(f,x)$ is a generalized (algebraic) polynomial of degree $4n-4$, with coefficients in the Banach space $\mathbb{B}$. Let us denote $I=[-1,1]$, $I'=\left[-\frac{1}{4},\frac{1}{4}\right]$ and for fixed $x\in I'$, $I_x=\left[-\frac{1}{2}+x,\frac{1}{2}-x\right]$. If we denote $g(x)=f(x)-f(0)$ and $L_n(g,x)=\int_{-\frac{1}{2}}^{\frac{1}{2}}g(t)P_n(t-x)dt$, then by [56, p.276, relation (3.10)], it follows

$$\|f-K_n(f)\|_{C(I',\mathbb{B})}=\|g-K_n(g)\|_{C(I',\mathbb{B})}, \qquad (6.3)$$

where $C(I',\mathbb{B})=\{f:I'\to\mathbb{B};f$ continuous on $I'\}$. Then for fixed $x\in I'$, as in [56, p.276] we get

$$L_n(g,x)-g(x)=\int_{I_x}[g(x+u)-g(x)]P_n(u)du-g(x)\int_{I\setminus I_x}P_n(u)du$$

where $\left\|g(x)\int_{I\setminus I_x}P_n(u)du\right\|_{C(I',\mathbb{B})}\leq C_2\|g\|_{C(I',\mathbb{B})}\cdot n^{-2}$ and

$$\left\|\int_{I_x}[g(x+u)-g(x)]P_n(u)du\right\|_{C(I',\mathbb{B})}\leq C_1\omega_1^{\mathbb{B}}\left(g,\frac{1}{n}\right)_{I'}=C_1\omega_1^{\mathbb{B}}\left(f,\frac{1}{n}\right)_{I'}.$$

Since $\|g\|_{C(I',\mathbb{B})}=\|f-f(0)\|_{C(I',\mathbb{B})}\leq\omega_1^{\mathbb{B}}\left(f,\frac{1}{4}\right)_{I'}$, by (6.3) we obtain

$$\|f-K_n(f)\|_{C(I',\mathbb{B})}\leq C_3\left[\omega_1^{\mathbb{B}}\left(f,\frac{1}{n}\right)_{I'}+\omega_1^{\mathbb{B}}\left(f,\frac{1}{4}\right)_{I'}\cdot n^{-2}\right]\leq C_4\omega_1^{\mathbb{B}}\left(f,\frac{1}{n}\right)_{I'},$$

taking into account that $\omega_1^{\mathbb{B}}\left(f, \frac{1}{4}\right)_{I'} \leq \omega_1^{\mathbb{B}}(f, 1)_{I'} = \omega_1^{\mathbb{B}}\left(f, n \cdot \frac{1}{n}\right)_{I'} \leq$
$\leq n\omega_1^{\mathbb{B}}\left(f, \frac{1}{n}\right)_{I'} \leq n^2\omega_1^{\mathbb{B}}\left(f, \frac{1}{n}\right)_{I'}$. The lemma is proved. $\qquad\square$

Now let us consider $h : \left[-\frac{1}{4}, \frac{1}{4}\right] \to \mathbb{B}$ and

$$\overline{R}_n(h, x) = \begin{cases} h\left(-\frac{1}{4}\right), & \text{if } -\frac{1}{2} \leq x \leq -\frac{1}{4} \\ R_n(h, x), & \text{if } -\frac{1}{4} \leq x \leq \frac{1}{4} \\ h\left(\frac{1}{4}\right), & \text{if } \frac{1}{4} \leq x \leq \frac{1}{2} \end{cases}.$$

For $f \in C_{\mathcal{F}}\left[-\frac{1}{4}, \frac{1}{4}\right]$, let $K_n(\overline{R}_n(j \circ f), x)$ be the Bojanic-DeVore operator associated to $\overline{R}_n(j \circ f)$, where j is the embedding in Theorem 6.1, i.e.

$$K_n(\overline{R}_n(j \circ f), x) = \int_{-\frac{1}{2}}^{\frac{1}{2}} [\overline{R}_n(j \circ f)(t) - \overline{R}_n(j \circ f)(0)]P_n(t - x)dt +$$

$$\overline{R}_n(j \circ f)(0).$$

Then we have

$$K_n(\overline{R}_n(j \circ f), x) = (j \circ f)(0)\left[1 - \int_{-\frac{1}{2}}^{\frac{1}{2}} P_n(t - x)dt\right]$$

$$+ (j \circ f)\left(-\frac{1}{4}\right)\int_{-\frac{1}{2}}^{-\frac{1}{4}} P_n(t - x)dt$$

$$+ \sum_{k=-n}^{n} (j \circ f)(x_k)\int_{-\frac{1}{4}}^{\frac{1}{4}} \frac{P_n(t - x)dt}{(t - x_k)^4 \sum_{j=-n}^{n}(t - x_j)^{-4}}$$

$$+ (j \circ f)\left(\frac{1}{4}\right)\int_{\frac{1}{4}}^{\frac{1}{2}} P_n(t - x)dt.$$

It is easy to see that all the terms in x associated to $(j \circ f)(x_k)$ are positive and therefore we obtain the form

$$K_n(\overline{R}_n(j \circ f), x) = \sum_{k=-n}^{n} (j \circ f)(x_k)p_{n,k}(x)$$

with $p_{n.k}(x) \geq 0$ for all $x \in \left[-\frac{1}{4}, \frac{1}{4}\right]$.

With the help of $p_{n.k}(x)$ given as above, we define the Szabados-type fuzzy generalized polynomial associated to $f : \left[-\frac{1}{4}, \frac{1}{4}\right] \to \mathbb{R}_{\mathcal{F}}$, by

$$S(x) = \sum_{k=-n}^{n} p_{n,k}(x) \odot f(x_k).$$

The following theorem gives Jackson-type estimate for the error of approximation by Szabados-type polynomials.

Theorem 6.14. Let $f : \left[-\frac{1}{4}, \frac{1}{4}\right] \to \mathbb{R}_{\mathcal{F}}$ be continuous and

$$S(x) = \sum_{k=-n}^{n} p_{n,k}(x) \odot f(x_k)$$

defined as above. Then

$$D^*(f, S) \leq C\omega_1^{\mathcal{F}}\left(f, \frac{1}{n}\right)_{\left[-\frac{1}{4}, \frac{1}{4}\right]}.$$

Proof. Since all $p_{n,k}(x) \geq 0$, we have

$$\sum_{k=-n}^{n} (j \circ f)(x_k)p_{n,k}(x) = j\left(\sum_{k=-n}^{n} p_{n,k}(x) \odot f(x_k)\right).$$

But j is an isometry, so we have:

$$D(f(x), S(x)) = D\left(f(x), \sum_{k=-n}^{n} p_{n,k}(x) \odot f(x_k)\right)$$

$$= \left\|(j \circ f)(x) - j\left(\sum_{k=-n}^{n} p_{n,k}(x) \odot f(x_k)\right)\right\|$$

$$= \left\|(j \circ f)(x) - \sum_{k=-n}^{n} (j \circ f)(x_k)p_{n,k}(x)\right\|,$$

where $\|\cdot\|$ is the norm in $\mathbb{B} = \overline{C}[0,1] \times \overline{C}[0,1]$.
We observe that the last sum is $K_n(\overline{R}_n(j \circ f), x)$. Also we have:

$$\left\|(j \circ f)(x) - K_n(\overline{R}_n(j \circ f), x)\right\| \leq \left\|(j \circ f)(x) - R_n(j \circ f, x)\right\| +$$

$$+ \left\|R_n(j \circ f, x) - K_n(\overline{R}_n(j \circ f), x)\right\|.$$

Since the coefficients $r_{n,k}(x)$ of R_n in Lemma 6.9 are all positive, we have

$$R_n(j \circ f, x) = \sum_{k=-n}^{n} r_{n,k}(x)(j \circ f)(x_k) = j\left(\sum_{k=-n}^{n} p_{n,k}(x) \odot f(x_k)\right)$$

and taking into account that j is an isometry, we obtain for all $x \in \left[-\frac{1}{4}, \frac{1}{4}\right]$

$$D(f(x), S(x)) \leq D(f(x), R_n(f, x)) + \left\|R_n(j \circ f, x) - K_n(\overline{R}_n(j \circ f), x)\right\| =$$
$$= D(f(x), R_n(f, x)) + \left\|\overline{R}_n(j \circ f, x) - K_n(\overline{R}_n(j \circ f), x)\right\|$$

(this last inequality is obvious by the definition of $\overline{R}_n$).

By Lemma 6.9 and Lemma 6.13 we obtain

$$D(f(x), S(x)) \leq 5\omega_1^{\mathcal{F}}\left(f, \frac{1}{n}\right)_{\left[-\frac{1}{4},\frac{1}{4}\right]} + C_4\omega_1^{\mathbb{B}}\left(\overline{R}_n(j \circ f), \frac{1}{n}\right)_{\left[-\frac{1}{4},\frac{1}{4}\right]}.$$

It is easy to see that $\omega_1^{\mathbb{B}}\left(\overline{R}_n(j \circ f), \frac{1}{n}\right)_{\left[-\frac{1}{4},\frac{1}{4}\right]} = \omega_1^{\mathbb{B}}\left(R_n(j \circ f), \frac{1}{n}\right)_{\left[-\frac{1}{4},\frac{1}{4}\right]}$.
By Lagrange theorem for functions with values in Banach spaces we obtain

$$\|R_n(j \circ f, y) - R_n(j \circ f, x)\| \leq \sup_{\xi \in [x,y]} \|R_n'(j \circ f, \xi)\|_{C(I', \mathbb{B})} \cdot |y - x|$$

with $I' = \left[-\frac{1}{4}, \frac{1}{4}\right]$ and for $|y - x| \leq \frac{1}{n}$, taking into account Lemma 6.12 we obtain

$$\omega_1^{\mathbb{B}}\left(\overline{R}_n(j \circ f), \frac{1}{n}\right)_{I'} \leq 900n \cdot \omega_1^{\mathbb{B}}\left(j \circ f, \frac{1}{n}\right)_{I'} \cdot \frac{1}{n} = 900\omega_1^{\mathbb{B}}\left(j \circ f, \frac{1}{n}\right)_{I'}.$$

It is easy to check that $\omega_1^{\mathbb{B}}\left(j \circ f, \frac{1}{n}\right)_{I'} = \omega_1^{\mathcal{F}}\left(f, \frac{1}{n}\right)_{I'}$ and we finally obtain

$$D(f(x), S(x)) \leq 5\omega_1^{\mathcal{F}}\left(f, \frac{1}{n}\right)_{I'} + C_4 \cdot 900\omega_1^{\mathcal{F}}\left(f, \frac{1}{n}\right)_{I'} = C\omega_1^{\mathcal{F}}\left(f, \frac{1}{n}\right)_{I'}$$

which completes the proof. □

As an immediate consequence we obtain the following Jackson-type estimate for the error in approximation by generalized fuzzy algebraic polynomials.

Corollary 6.15. For the best approximation by algebraic polynomials we have

$$E_n^{K,1} \leq C\omega_1^{\mathcal{F}}\left(f, \frac{1}{n}\right)_{\left[-\frac{1}{4},\frac{1}{4}\right]}, \quad \forall n \in \mathbb{N}, f \in C_{\mathcal{F}}\left[-\frac{1}{4},\frac{1}{4}\right], K = f\left(\left[-\frac{1}{4},\frac{1}{4}\right]\right).$$

where $C > 0$ is an absolute constant independent of n and f.
Proof. Since the polynomial $P_n(t - x) \geq 0$, $\forall t, x \in [-1, 1]$ and $\int_{-1}^{1} P_n(t - x)dt = 1$, we have

$$|p_{n,-n}(x)| = \int_{-\frac{1}{2}}^{-\frac{1}{4}} P_n(t - x)dt + \int_{-\frac{1}{4}}^{\frac{1}{4}} \frac{P_n(t - x)dt}{(t - x_{-n})^4 \sum_{j=-n}^{n}(t - x_j)^{-4}} \leq$$

$$\leq \int_{-\frac{1}{2}}^{\frac{1}{4}} P_n(t - x)dt \leq 1.$$

Also

$$|p_{n,n}(x)| = \int_{\frac{1}{4}}^{\frac{1}{2}} P_n(t - x)dt + \int_{-\frac{1}{4}}^{\frac{1}{4}} \frac{P_n(t - x)dt}{(t - x_n)^4 \sum_{j=-n}^{n}(t - x_j)^{-4}} \leq 1.$$

and

$$|p_{n,0}(x)| = 1 - \int_{-\frac{1}{2}}^{\frac{1}{2}} P_n(t-x)dt + \int_{-\frac{1}{4}}^{\frac{1}{4}} \frac{P_n(t-x)dt}{(t-x_0)^4 \sum_{j=-n}^{n}(t-x_j)^{-4}} \le$$

$$\le 1 - \int_{-\frac{1}{2}}^{\frac{1}{2}} P_n(t-x)dt + \int_{-\frac{1}{4}}^{\frac{1}{4}} P_n(t-x)dt \le 1.$$

For $k \notin \{-n, 0, n\}$ we have

$$|p_{n,k}(x)| = \int_{-\frac{1}{4}}^{\frac{1}{4}} \frac{P_n(t-x)dt}{(t-x_k)^4 \sum_{j=-n}^{n}(t-x_j)^{-4}} \le \int_{-\frac{1}{4}}^{\frac{1}{4}} P_n(t-x)dt \le 1.$$

and the proof is complete.   □

Remark 6.16. We can obtain the above results in any interval $[a, b]$ instead of $\left[-\frac{1}{4}, \frac{1}{4}\right]$ by mapping this interval in $[a, b]$ through a linear function which maps $-\frac{1}{4}$ to a and $\frac{1}{4}$ to b.

6.4   Application to fuzzy interpolation

In this section we prove the convergence of fuzzy Lagrange polynomials for some classes of fuzzy functions.

The fuzzy Lagrange polynomial is defined in [84], [72] as follows (see also [66, p.651]) $L_n(x) = \sum_{i=0}^{n} l_i(x) \odot f(x_i)$, where $l_i(x) = \frac{(x-x_0)...\diagup...(x-x_n)}{(x_i-x_0)...\diagup...(x_i-x_n)}$ are the usual fundamental Lagrange interpolation polynomials and the sign "/" means that the ith operand is missing.

Theorem 6.17. Let $f : [-1, 1] \to \mathbb{R}_{\mathcal{F}}$ be a Lipschitz mapping of order $\alpha > \frac{1}{2}$(i. e. there exists L such that $D(f(x), f(y)) \le L|x - y|^{\alpha}$ for all $x, y \in [-1, 1]$. Let $(x_{n,i})_{i=\overline{1,n}}$, $n \in \mathbb{N}$ be a normal matrix of nodes and $L_n(x)$ the fuzzy Lagrange polynomial which interpolates f on $\{x_{n,0}, ..., x_{n,n}\}$. Then

$$\lim_{n \to \infty} L_n(x) = f(x), \ \forall x \in [-1, 1].$$

The convergence is uniform in any interval $[-1 + h, 1 - h]$, $0 < h < 1$.

Proof. By Corollary 6.15 if we take $M = \max\left\{1, \sqrt{\frac{n}{2h}}\right\}$, $\forall h > 0$, then $E_n^{K,M}(f) \le C\omega_1^{\mathcal{F}}\left(f, \frac{1}{n}\right)_{[-1,1]}$, where $K = f([-1, 1])$ and also the best approximation polynomial in $A_n^{K,M}$ (denoted π_n) exists. By [43, Lemma 8.3.2, p. 351] for a normal matrix of nodes we have $|l_i(x)| \le \sum_{i=0}^{n} |l_i(x)| \le \sqrt{\frac{n}{2h}}$ for $x \in [-1 + h, 1 - h]$ and so $L_n \in A_n^{K,M}$. Then $D^*(L_n, f) \le D^*(L_n, \pi_n) + D^*(\pi_n, f)$. By Corollary 6.15 we obtain $D^*(\pi_n, f) \le C\omega_1^{\mathcal{F}}\left(f, \frac{1}{n}\right)_{[-1,1]}$.

Let $L_n(\pi_n)$ be the fuzzy Lagrange polynomial associated to π_n at $\{x_{n,0}, ..., x_{n,n}\}$. We prove that $L_n(\pi_n) = \pi_n$. We observe that $\pi_n(x_{n,j}) = \sum_{i=0}^{n} l_i(x) \odot \pi(x_{n,i})$ since $l_i(x_{n,j}) = \delta_{i,j}$ (Kronecker symbol $\delta_{i,j}$).

So $L_n(\pi_n)(x_{n,j}) = \pi_n(x_{n,j})$, $j = \overline{0,n}$. Since the Lagrange polynomial is unique (see [66, p.650]), we get $L_n(\pi_n) = \pi_n$. Then using the properties of the metric D we obtain:

$$D\left(L_n(x), \pi_n(x)\right) = D\left(\sum_{i=0}^{n} l_i(x) \odot f\left(x_{n,i}\right), \sum_{i=0}^{n} l_i(x) \odot \pi\left(x_{n,i}\right)\right) \le$$

$$\le \sum_{i=0}^{n} D(l_i(x) \odot f\left(x_{n,i}\right), l_i(x) \odot \pi\left(x_{n,i}\right)) \le \sum_{i=0}^{n} |l_i(x)| D(f\left(x_{n,i}\right), \pi\left(x_{n,i}\right)).$$

Using again Corollary 6.15 we obtain

$$D\left(L_n(x), \pi_n(x)\right) \le \sum_{i=0}^{n} |l_i(x)| C\omega_1^{\mathcal{F}}\left(f, \frac{1}{n}\right)_{[-1,1]}.$$

By [43, Lemma 8.3.2, p. 351], for a normal matrix of nodes we have

$$\sum_{i=0}^{n} |l_i(x)| \le \sqrt{\frac{(b-a)n}{h}}, \text{ for } x \in [a+h, b-h].$$

Then

$$D^*(L_n, f) \le C\omega_1^{\mathcal{F}}\left(f, \frac{1}{n}\right)_{[-1,1]} \left(1 + \sqrt{\frac{2}{h}}\sqrt{n}\right).$$

Since f is of Lipschitz-type, we have $\omega_1^{\mathcal{F}}\left(f, \frac{1}{n}\right)_{[-1,1]} \le L\frac{1}{n^\alpha}$, $\alpha > \frac{1}{2}$. Then

$$D^*\left(L_n, f\right) \le CL\frac{1}{n^\alpha} + CL\sqrt{\frac{2}{h}}\frac{1}{n^{\alpha-\frac{1}{2}}},$$

which completes the proof. $\qquad\square$

7
BASIC FUZZY KOROVKIN THEORY

We present the basic fuzzy Korovkin theorem via a fuzzy Shisha–Mond inequality given here. This determines the degree of convergence with rates of a sequence of fuzzy positive linear operators to the fuzzy unit operator. The surprising fact is that only the real case Korovkin assumptions are enough for the validity of the fuzzy Korovkin theorem, along with a natural realization condition fulfilled by the sequence of fuzzy positive linear operators. The last condition is fulfilled by almost all operators defined via fuzzy summation or fuzzy integration. This chapter relies on [18].

7.1 Basics

Motivation for this chapter are the references [4], [60], [57], [66], [78], [96], [102]. References [60], [102] are the first articles dealing with the fuzzy Korovkin issue and only one reference [102] provides a fuzzy Shisha–Mond inequality but to a very different and specialized direction for fuzzy random variables. Of course pre-existed some Korovkin type set valued literature not related at all to this chapter. The results of Theorems 7.7 and 7.8 are simple, basic and very general, directly transferring the real case of the convergence with rates of positive linear operators to the unit, to the fuzzy one. The same real assumptions are kept here in the fuzzy setting, and they are the only assumptions we make along with the very natural and general

G.A. Anastassiou: Fuzzy Mathematics: Approximation The., STUDFUZZ 251, pp. 99–104.
springerlink.com   © Springer-Verlag Berlin Heidelberg 2010

realization condition (7.5). Condition (7.5) is fulfilled by almost all example — fuzzy positive operators, that is, by most fuzzy summation and fuzzy integration operators. At each step of the chapter we provide an example to justify our method.

We use the following

Definition 7.1. Let $f\colon [a,b] \to \mathbb{R}_{\mathcal{F}}$ be a fuzzy real number valued function. We define the (first) *fuzzy modulus of continuity* of f by

$$\omega_1^{(\mathcal{F})}(f,\delta) := \sup_{\substack{x,y\in[a,b] \\ |x-y|\le\delta}} D(f(x),f(y)),$$

any $0 < \delta \le b - a$.

Definition 7.2. Let $f\colon [a,b] \subseteq \mathbb{R} \to \mathbb{R}_{\mathcal{F}}$. We say that f is *fuzzy continuous* at $x_0 \in [a,b]$ iff whenever $x_n \to x_0$, then $D(f(x_n),f(x_0)) \to 0$, as $n \to \infty$, $n \in \mathbb{N}$. We call f *fuzzy continuous* iff it is fuzzy continuous $\forall x \in [a,b]$ and we denote the space of fuzzy continuous functions by $C_{\mathcal{F}}([a,b])$.

Denote $[f]^r = [f_-^{(r)}, f_+^{(r)}]$ and we mean

$$[f(x)]^r = \left[f_-^{(r)}(x), f_+^{(r)}(x)\right], \quad \forall x \in [a,b], \text{ all } r \in [0,1].$$

Let $f,g \in C_{\mathcal{F}}([a,b])$ we say that f is *fuzzy larger* than g pointwise and we denote it by $f \succsim g$ iff $f(x) \succsim g(x)$ iff $f_-^{(r)}(x) \ge g_-^{(r)}(x)$ and $f_+^{(r)}(x) \ge g_+^{(r)}(x)$, $\forall x \in [a,b]$, $\forall r \in [0,1]$, iff $f_-^{(r)} \ge g_-^{(r)}$, $f_+^{(r)} \ge g_+^{(r)}$, $\forall r \in [0,1]$.

Let L be a map from $C_{\mathcal{F}}([a,b])$ into itself, we call it a *fuzzy linear operator* iff

$$L\big(c_1 \odot f_1 \oplus c_2 \odot f_2\big) = c_1 \odot L(f_1) \oplus c_2 \odot L(f_2),$$

for any $c_1, c_2 \in \mathbb{R}$, $f_1, f_2 \in C_{\mathcal{F}}([a,b])$. We say that L is a *fuzzy positive linear operator* iff for $f,g \in C_{\mathcal{F}}([a,b])$ with $f \succsim g$ we get $L(f) \succsim L(g)$ iff $(L(f))_-^{(r)} \ge (L(g))_-^{(r)}$ and $(L(f))_+^{(r)} \ge (L(g))_+^{(r)}$ on $[a,b]$ for all $r \in [0,1]$.

Example 7.3. Let $f \in C_{\mathcal{F}}([0,1])$, we define the *fuzzy Bernstein operator*

$$\big(B_n^{(\mathcal{F})}(f)\big)(x) = \sum_{k=0}^{n}{}^{*} \binom{n}{k} x^k (1-x)^{n-k} \odot f\left(\frac{k}{n}\right), \quad \forall x \in [0,1], \ n \in \mathbb{N}.$$

This is a fuzzy positive linear operator.

We mention the very interesting with rates approximation motivating this chapter.

Theorem 7.4 (see p. 642, [66], S. Gal). *If $f \in C_{\mathcal{F}}([0,1])$, then*

$$D^*\big(B_n^{(\mathcal{F})}(f),f\big) \le \frac{3}{2}\omega_1^{(\mathcal{F})}\left(f, \frac{1}{\sqrt{n}}\right), \quad \forall n \in \mathbb{N}$$

i.e.

$$\lim_{n\to\infty} D^*\big(B_n^{(\mathcal{F})}(f), f\big) = 0,$$

that is $B_n^{(\mathcal{F})} f \to f$, $n \to \infty$ in fuzzy uniform convergence.

The last fact comes by the property that $\omega_1^{(\mathcal{F})}(f,\delta) \to 0$ as $\delta \to 0$, whenever $f \in C_{\mathcal{F}}([a,b])$.

We need to use

Theorem 7.5 (Shisha and Mond (1968), [96]). *Let $[a,b] \subseteq \mathbb{R}$. Let $(\tilde{L}_n)_{n\in\mathbb{N}}$ be a sequence of positive linear operators from $C([a,b])$ into itself. For $n = 1, 2, \ldots$, suppose $\tilde{L}_n(1)$ is bounded. Let $f \in C([a,b])$. Then for $n = 1, 2, \ldots$, we have*

$$\|\tilde{L}_n f - f\|_\infty \le \|f\|_\infty \|\tilde{L}_n 1 - 1\|_\infty + \|\tilde{L}_n(1) + 1\|_\infty \omega_1(f, \mu_n),$$

where ω_1 is the standard real modulus of continuity and

$$\mu_n := \big\|(\tilde{L}_n((t-x)^2))(x)\big\|_\infty^{1/2},$$

and $\|\cdot\|_\infty$ stands for the sup-norm over $[a,b]$. In particular, if $L_n(1) = 1$ then

$$\|\tilde{L}_n f - f\|_\infty \le 2\omega_1(f, \mu_n).$$

Note. One can easily see ([96]), for $n = 1, 2, \ldots$,

$$\mu_n^2 \le \big\|(\tilde{L}_n(t^2))(x) - x^2\big\|_\infty + 2c\big\|(\tilde{L}_n(t))(x) - x\big\|_\infty + c^2\big\|(\tilde{L}_n(1))(x) - 1\big\|_\infty,$$

where $c := \max(|a|, |b|)$.

Assuming that $\tilde{L}_n(1) \overset{u}{\longrightarrow} 1$, $\tilde{L}_n(id) \overset{u}{\longrightarrow} id$, $\tilde{L}_n(id^2) \overset{u}{\longrightarrow} id^2$ (*id* is the identity map), $n \to \infty$, uniformly, then from Theorem 7.5's main inequality we get $\tilde{L}_n(f) \overset{u}{\longrightarrow} f$, $\forall f \in C([a,b])$, that is the famous Korovkin theorem in the real case.

We finally need

Lemma 7.6. *Let $f \in C_{\mathcal{F}}([a,b])$, $[a,b] \subseteq \mathbb{R}$. Then it holds*

$$\omega_1^{(\mathcal{F})}(f,\delta) = \sup_{r\in[0,1]} \max\{\omega_1(f_-^{(r)}, \delta), \omega_1(f_+^{(r)}, \delta)\},$$

for any $0 < \delta \le b - a$.

Proof. Let $x, y \in [a,b]$: $|x - y| \le \delta$, $0 < \delta \le b - a$. Then we have

$$
\begin{aligned}
D(f(x), f(y)) &= \sup_{r\in[0,1]} \max\{|(f(x))_-^{(r)} - (f(y))_-^{(r)}|, |(f(x))_+^{(r)} - (f(y))_+^{(r)}|\} \\
&\le \sup_{r\in[0,1]} \max\{\omega_1(f_-^{(r)}, \delta), \omega_1(f_+^{(r)}, \delta)\}.
\end{aligned}
$$

Thus
$$\omega_1^{(\mathcal{F})}(f,\delta) \leq \sup_{r \in [0,1]} \max\{\omega_1(f_-^{(r)},\delta), \omega_1(f_+^{(r)},\delta)\}.$$

For any $r \in [0,1]$ and any $x, y \in [a,b]$: $|x - y| \leq \delta$ we see that
$$\omega_1^{(\mathcal{F})}(f,\delta) \geq D(f(x), f(y)) \geq \left|(f(x))_-^{(r)} - (f(y))_-^{(r)}\right|, \left|(f(x))_+^{(r)} - (f(y))_+^{(r)}\right|.$$

Therefore
$$\omega_1(f_\pm^{(r)}, \delta) \leq \omega_1^{(\mathcal{F})}(f,\delta), \quad \forall r \in [0,1].$$

Hence
$$\sup_{r \in [0,1]} \max\{\omega_1(f_-^{(r)},\delta), \omega_1(f_+^{(r)},\delta)\} \leq \omega_1^{(\mathcal{F})}(f,\delta),$$

proving the claim. □

Note. For $f \in C_{\mathcal{F}}([a,b])$ we get that f is fuzzy bounded and $\omega_1^{(\mathcal{F})}(f,\delta)$ is finite for all $0 < \delta \leq b - a$. Also $f_\pm^{(r)}$ are continuous on $[a,b]$ and $\omega_1(f_\pm^{(r)},\delta)$ are finite too, all $r \in [0,1]$.

7.2   Main Results

We present the fuzzy analog of Shisha–Mond inequality of Theorem 7.5.

Theorem 7.7. *Let $\{L_n\}_{n \in \mathbb{N}}$ be a sequence of fuzzy positive linear operators from $C_{\mathcal{F}}([a,b])$ into itself, $[a,b] \subseteq \mathbb{R}$. We assume that there exists a corresponding sequence $\{\tilde{L}_n\}_{n \in \mathbb{N}}$ of positive linear operators from $C([a,b])$ into itself with the property*
$$(L_n(f))_\pm^{(r)} = \tilde{L}_n(f_\pm^{(r)}), \tag{7.1}$$

respectively, $\forall r \in [0,1]$, $\forall f \in C_{\mathcal{F}}([a,b])$. We assume that $\{\tilde{L}_n(1)\}_{n \in \mathbb{N}}$ is bounded. Then for $n \in \mathbb{N}$ we have
$$D^*(L_n f, f) \leq \|\tilde{L}_n 1 - 1\|_\infty D^*(f, \tilde{o}) + \|\tilde{L}_n(1) + 1\|_\infty \omega_1^{(\mathcal{F})}(f, \mu_n), \tag{7.2}$$

where
$$\mu_n := \left(\|(\tilde{L}_n((t - x)^2))(x)\|_\infty\right)^{1/2}, \tag{7.3}$$

$\forall f \in C_{\mathcal{F}}([a,b])$, $\tilde{o} := \mathcal{X}_{\{0\}}$ the neutral element for $\oplus$. If $\tilde{L}_n 1 = 1$, $n \in \mathbb{N}$, then
$$D^*(L_n f, f) \leq 2\omega_1^{(\mathcal{F})}(f, \mu_n). \tag{7.4}$$

Note. The fuzzy Bernstein operators $B_n^{(\mathcal{F})}$ and the real corresponding ones B_n acting on $C_{\mathcal{F}}([0,1])$ and $C([0,1])$, respectively, fulfill assumption (7.5).

We present now the Fuzzy Korovkin Theorem.

Theorem 7.8. *Let $\{L_n\}_{n\in\mathbb{N}}$ be a sequence of fuzzy positive linear operators from $C_{\mathcal{F}}([a,b])$ into itself, $[a,b]\subseteq\mathbb{R}$. We assume that there exists a corresponding sequence $\{\tilde{L}_n\}_{n\in\mathbb{N}}$ of positive linear operators from $C([a,b])$ into itself with the property*

$$\left(L_n(f)\right)_{\pm}^{(r)} = \tilde{L}_n(f_{\pm}^{(r)}), \tag{7.5}$$

respectively, $\forall r \in [0,1]$, $\forall f \in C_{\mathcal{F}}([a,b])$. Furthermore assume that

$$\tilde{L}_n(1) \xrightarrow{u} 1, \quad \tilde{L}_n(id) \xrightarrow{u} id, \quad \tilde{L}_n(id^2) \xrightarrow{u} id^2,$$

as $n \to \infty$, uniformly. Then

$$D^*(L_n f, f) \longrightarrow 0, \qquad \text{as } n \to \infty,$$

for any $f \in C_{\mathcal{F}}([a,b])$, i.e. $L_n f \xrightarrow{D^*} f$, that is $L_n \to I$ unit operator in the fuzzy sense, as $n \to \infty$.

Proof. Use of (7.2), property of (7.3), etc. □

Example for Theorem 7.8 the fuzzy Bernstein operators $B_n^{(\mathcal{F})}$.

Proof of Theorem 7.7. We would like to estimate

$$
\begin{aligned}
D^*(L_n f, f) &= \sup_{x\in[a,b]} D\big((L_n f)(x), f(x)\big) \\
&= \sup_{x\in[a,b]} \sup_{r\in[0,1]} \max\{|(L_n f)_{-}^{(r)}(x) - (f)_{-}^{(r)}(x)|, \\
&\qquad |(L_n f))_{+}^{(r)}(x) - (f)_{+}^{(r)}(x)|\} \\
&= \sup_{x\in[a,b]} \sup_{r\in[0,1]} \max\{|\tilde{L}_n(f_{-}^{(r)}(x) - (f)_{-}^{(r)}(x)|, \\
&\qquad |\tilde{L}_n(f_{+}^{(r)})(x) - (f)_{+}^{(r)}(x)|\} \\
&= \sup_{r\in[0,1]} \max\{\|\tilde{L}_n f_{-}^{(r)} - f_{-}^{(r)}\|_\infty, \|\tilde{L}_n f_{+}^{(r)} - f_{+}^{(r)}\|_\infty\} \\
&\qquad \text{(by Theorem 7.5)} \\
&\le \sup_{r\in[0,1]} \max\{\left(\|f_{-}^{(r)}\|_\infty\|\tilde{L}_n 1 - 1\|_\infty + \|\tilde{L}_n(1) + 1\|_\infty \omega_1(f_{-}^{(r)}, \mu_n)\right), \\
&\qquad \left(\|f_{+}^{(r)}\|_\infty\|\tilde{L}_n 1 - 1\|_\infty + \|\tilde{L}_n(1) + 1\|_\infty \omega_1(f_{+}^{(r)}, \mu_n)\right)\} \\
&\le \|\tilde{L}_n 1 - 1\|_\infty \sup_{r\in[0,1]} \max\left(\|f_{-}^{(r)}\|_\infty, \|f_{+}^{(r)}\|_\infty\right) \\
&\qquad + \|\tilde{L}_n(1) + 1\|_\infty \sup_{r\in[0,1]} \max\{\omega_1(f_{-}^{(r)}, \mu_n), \omega_1(f_{+}^{(r)}, \mu_n)\} \\
&\qquad \text{(by Lemma 7.6)} \\
&= \|\tilde{L}_n 1 - 1\|_\infty D^*(f, \tilde{o}) + \|\tilde{L}_n(1) + 1\|_\infty \omega_1^{(\mathcal{F})}(f, \mu_n),
\end{aligned}
$$

proving (7.2). □

Application 7.9. Let $f \in C_{\mathcal{F}}([0,1])$ then by applying (7.2) we obtain

$$D^*(B_n^{(\mathcal{F})}f, f) \leq 2\omega_1^{(\mathcal{F})}\left(f, \frac{1}{2\sqrt{n}}\right), \quad \forall n \in \mathbb{N}. \tag{7.6}$$

8

FUZZY TRIGONOMETRIC KOROVKIN THEORY

We present the fuzzy Korovkin trigonometric theorem via a fuzzy Shisha–Mond trigonometric inequality presented here too. This determines the degree of approximation with rates of a sequence of fuzzy positive linear operators to the fuzzy unit operator. The astonishing fact is that only the real case trigonometric assumptions are enough for the validity of the fuzzy trigonometric Korovkin theorem, along with a very natural realization condition fulfilled by the sequence of fuzzy positive linear operators. The latter condition is satisfied by almost all operators defined via fuzzy summation or fuzzy integration. This chapter is based on [32].

8.1 Basics

Motivation for this chapter are the references [102], [18], [24], [97], [96]. References [102], [18], [24] are the first articles dealing with the fuzzy Korovkin matter and inequalities, however [102] is very specialized and restrictive though very interesting dealing with fuzzy random variables and positive linearity. The main results here are Theorems 8.12 and 8.13. They are simple, basic and very general directly transferring the real trigonometric case, of the convergence with rates of positive linear operators to the unit under trigonometric assumptions, to the fuzzy one. The same real trigonometric

G.A. Anastassiou: Fuzzy Mathematics: Approximation The., STUDFUZZ 251, pp. 105–113.
springerlink.com   © Springer-Verlag Berlin Heidelberg 2010

assumptions are kept here in the fuzzy setting and they are the only convergence assumptions we make, along with the general realization condition
(8.5).

Condition (8.5) is satisfied by almost all example-fuzzy positive linear
operators of fuzzy summation or fuzzy integration form. At each step of
the development of our method we present an example that satisfies our
theory.

Let $f, g \colon I \subset \mathbb{R} \to \mathbb{R}_{\mathcal{F}}$ be *fuzzy real number valued functions*. The distance between f, g is defined by

$$D^*(f, g) := \sup_{x \in I} D(f(x), g(x)).$$

The function $f \colon \mathbb{R} \to \mathbb{R}_{\mathcal{F}}$ is 2π-periodic if $f(x) = f(x + 2\pi)$, $\forall x \in \mathbb{R}$. Here
$\sum^*$ stands for the fuzzy summation and $\tilde{o} = \mathcal{X}_{\{0\}}$.

We use the following

Definition 8.1. Let $f \colon \mathbb{R} \to \mathbb{R}_{\mathcal{F}}$ be a fuzzy real number valued function.
We define the (first) *fuzzy modulus of continuity* of f by

$$\omega_1^{(\mathcal{F})}(f, \delta) := \sup_{\substack{x, y \in \mathbb{R} \\ |x-y| \le \delta}} D(f(x), f(y)),$$

for any $\delta > 0$.

We have a similar obvious definition for subsets of $\mathbb{R}$.

Definition 8.2. Let $f \colon \mathbb{R} \to \mathbb{R}_{\mathcal{F}}$. We say that f is *fuzzy continuous at*
$x_0 \in \mathbb{R}$ iff whenever $x_n \to x_0$, then $D(f(x_n), f(x_0)) \to 0$, as $n \to \infty$,
$n \in \mathbb{N}$. We call f *fuzzy continuous* iff it is fuzzy continuous $\forall x \in \mathbb{R}$ and we
denote the space of fuzzy continuous functions by $C_{\mathcal{F}}(\mathbb{R})$. We call f *fuzzy*
uniformly continuous iff $\forall \varepsilon > 0 \; \exists \delta > 0$: whenever $x, y \in \mathbb{R}$ with $|x - y| \le \delta$
then $D(f(x), f(y)) \le \varepsilon$ and we denote the related space by $C_{\mathcal{F}}^U(\mathbb{R})$. Denote

$$[f]^r = [f_-^{(r)}, f_+^{(r)}]$$

and we mean

$$[f(x)]^r = [f_-^{(r)}(x), f_+^{(r)}(x)], \quad \forall x \in \mathbb{R},$$

all $r \in [0, 1]$. Let us denote $C_{2\pi}^{(\mathcal{F})}(\mathbb{R}) = \{f \colon \mathbb{R} \to \mathbb{R}_{\mathcal{F}}; f$ is fuzzy continuous
and 2π-periodic on $\mathbb{R}\}$.

Let $f, g \in C_{\mathcal{F}}(\mathbb{R})$ we say that f is *fuzzy larger than g pointwise* and we
denote it by $f \succsim g$ iff $f(x) \succsim g(x)$ iff $f_-^{(r)}(x) \ge g_-^{(r)}(x)$ and $f_+^{(r)}(x) \ge g_+^{(r)}(x)$,
$\forall x \in \mathbb{R}$, $\forall r \in [0, 1]$, iff $f_-^{(r)} \ge g_-^{(r)}$, $f_+^{(r)} \ge g_+^{(r)}$, $\forall r \in [0, 1]$.

Let L be a map from $C_{\mathcal{F}}(\mathbb{R})$ into itself, we call it a *fuzzy linear operator*
iff

$$L(c_1 \odot f_1 \oplus c_2 \odot f_2) = c_1 \odot L(f_1) \oplus c_2 \odot L(f_2),$$

for any $c_1, c_2 \in \mathbb{R}$, $f_1, f_2 \in C_{\mathcal{F}}(\mathbb{R})$. We say that L is a *fuzzy positive linear operator* iff for $f, g \in C_{\mathcal{F}}(\mathbb{R})$ with $f \succsim g$ we get $L(f) \succsim L(g)$, iff $(Lf)_-^{(r)} \geq (Lg)_-^{(r)}$ and $(Lf)_+^{(r)} \geq (Lg)_+^{(r)}$ on $\mathbb{R}$ for all $r \in [0, 1]$.

We need

Lemma 8.3. *Let the fuzzy trigonometric polynomial*

$$Q_n(x) = \sum_{k=0}^{n} {}^* \{ (\cos kx) \odot a_k \oplus (\sin kx) \odot b_k \},$$

where $x, y \in \mathbb{R}$; $a_k, b_k \in \mathbb{R}_{\mathcal{F}}$, $k = 0, 1, \ldots, n$. *Then* $Q_n(x)$ *is a fuzzy* 2π-*periodic continuous function in* $x \in \mathbb{R}$.

Proof. Clear.   □

We present

Lemma 8.4. *Let* $f : \mathbb{R} \to \mathbb{R}_{\mathcal{F}}$ *be a* 2π-*periodic and fuzzy continuous function, i.e.* $f \in C_{2\pi}^{(\mathcal{F})}(\mathbb{R})$. *Then for all* $\delta \in [0, \pi]$ *we have*

$$\omega_1^{(\mathcal{F})}\left(f\big|_{[0,2\pi]}, \delta\right) \leq \omega_1^{(\mathcal{F})}(f, \delta) \leq 2\omega_1^{(\mathcal{F})}\left(f\big|_{[0,2\pi]}, \delta\right).$$

Proof. The left hand side inequality is obvious. Now, let us denote $I_k = [2k\pi, 2(k+1)\pi]$, $\forall k \in \mathbb{Z}$. For $x, y \in \mathbb{R}$, $|x - y| \leq \delta$, there exist two possibilities:

(1) $\exists k \in \mathbb{Z}$ such that $x, y \in I_k$,

(2) $\exists k \in \mathbb{Z}$ such that $x \in I_k$, $y \in I_{k+1}$, or $x \in I_{k+1}$, $y \in I_k$.

Case (1). We have: $x' = x - 2k\pi$, $y' = y - 2k\pi \in [0, 2\pi]$, $|x' - y'| = |x - y| \leq \delta$ and

$$D(f(x), f(y)) = D\big(f(x'), f(y')\big) \leq \omega_1^{(\mathcal{F})}\left(f\big|_{[0,2\pi]}, \delta\right) \leq 2\omega_1^{(\mathcal{F})}\left(f\big|_{[0,2\pi]}, \delta\right).$$

Case (2). Let $x \in I_k$, $y \in I_{k+1}$ (the case $y \in I_k$, $x \in I_{k+1}$, as symmetric, is similar).

We have: $x' = x - 2k\pi \in [0, 2\pi]$, $y' = y - 2k\pi \in [2\pi, 4\pi]$, $|x' - y'| \leq \delta$, $x' \leq 2\pi \leq y'$. Thus

$$
\begin{aligned}
D(f(x), f(y)) &= D\big(f(x'), f(y')\big) \leq D\big(f(x'), f(2\pi)\big) + D\big(f(2\pi), f(y')\big) \\
&\leq \omega_1^{(\mathcal{F})}\left(f\big|_{[0,2\pi]}, \delta\right) + \omega_1^{(\mathcal{F})}\left(f\big|_{[2\pi,4\pi]}, \delta\right)
\end{aligned}
$$

(since $f \in C_{2\pi}^{(\mathcal{F})}(\mathbb{R})$ we obviously have

$$\omega_1^{(\mathcal{F})}\left(f\big|_{[0,2\pi]}, \delta\right) = \omega_1^{(\mathcal{F})}\left(f\big|_{[2\pi,4\pi]}, \delta\right)) = 2\omega_1^{(\mathcal{F})}\left(f\big|_{[0,2\pi]}, \delta\right).$$

Taking the supremum in the above with $x, y \in \mathbb{R}$, $|x - y| \leq \delta$ we establish the claim.   □

We give

Lemma 8.5. *Let* $f \in C_{2\pi}^{(\mathcal{F})}(\mathbb{R})$, *then* f *is fuzzy bounded and fuzzy uniformly continuous. I.e.* $C_{2\pi}^{(\mathcal{F})}(\mathbb{R}) = {}_{2\pi}C_{\mathcal{F}}^{U}(\mathbb{R})$; *the space of fuzzy uniformly continuous* 2π-*periodic functions.*

Proof. By Lemma 2 of [11] we have

$$D(f(x), \tilde{o}) \leq M, \quad \forall x \in [0, 2\pi], \quad M > 0.$$

For any $z \notin [0, 2\pi]$ there exists $x \in [0, 2\pi]$ such that $z = x + 2k\pi$, $k \in \mathbb{Z} - \{0\}$. Hence we have

$$D(f(z), \tilde{o}) = D(f(x), \tilde{o}) \leq M, \quad \forall z \in \mathbb{R} - [0, 2\pi],$$

proving that f is fuzzy bounded on $\mathbb{R}$.

By Proposition 2 of [24] we have that $\lim_{\delta \to 0} \omega_1^{(\mathcal{F})}\left(f|_{[0,2\pi]}, \delta\right) = 0$ because f is fuzzy uniformly continuous on $[0, 2\pi]$. Thus by Lemma 8.4 we get $\lim_{\delta \to 0} \omega_1^{(\mathcal{F})}(f, \delta) = 0$, equivalently by Proposition 2 of [11] we have $f \in {}_{2\pi}C_{\mathcal{F}}^{U}(\mathbb{R})$, proving the claim.   □

We also need

Proposition 8.6 ([24]). *Let* $f: \mathbb{R} \to \mathbb{R}_{\mathcal{F}}$ *be a fuzzy real number valued function. Assume that* $\omega_1^{(\mathcal{F})}(f, \delta)$, $\omega_1(f_-^{(r)}, \delta)$, $\omega_1(f_+^{(r)}, \delta)$ *are finite for* $\delta > 0$, *where* ω_1 *is the usual real modulus of continuity. Then it holds*

$$\omega_1^{(\mathcal{F})}(f, \delta) = \sup_{r \in [0,1]} \max\{\omega_1(f_-^{(r)}, \delta), \omega_1(f_+^{(r)}, \delta)\}.$$

Definition 8.7 ([66]). *Let* $f: [a, b] \to \mathbb{R}_{\mathcal{F}}$. *We say that* f *is fuzzy-Riemann integrable to* $I \in \mathbb{R}_{\mathcal{F}}$ *if, for any* $\varepsilon > 0$, $\exists \delta > 0$: *for any division* $P = \{[u, v]; \xi\}$ *of* $[a, b]$ *with the norms* $\Delta(P) < \delta$, *we have*

$$D\left(\sum_{P}{}^{*} (v - u) \odot f(\xi), I\right) < \varepsilon.$$

We write

$$I := (FR) \int_a^b f(x) \, dx,$$

we also call an f, as above, (FR)-integrable.

By Corollary 13.2 of [66], p. 644 we have that if $f \in C_{\mathcal{F}}([a, b])$ (fuzzy continuous on $[a, b]$), then f is fuzzy-Riemann integrable on $[a, b]$. Also, by Lemma 13.2 of [66], p. 644 for $f\colon \mathbb{R} \to \mathbb{R}_{\mathcal{F}}$ which fuzzy continuous 2π-periodic function we have that

$$(FR)\int_0^{2\pi} f(x)dx = (FR)\int_a^{a+2\pi} f(x)dx \left(= (FR)\int_{-\pi}^{\pi} f(x)dx\right), \quad \forall a \in \mathbb{R}.$$

We need the following.

Theorem 8.8 (see [67]). *Let $f\colon [a,b] \to \mathbb{R}_{\mathcal{F}}$ be fuzzy continuous function. Then*

$$\left[(FR)\int_a^b f(x)dx\right]^r = \left[\int_a^b (f)_{-}^{(r)}(x)dx, \int_a^b (f)_{+}^{(r)}(x)dx\right], \quad \forall r \in [0,1].$$

Clearly $f_{\pm}^{(r)}\colon [a,b] \to \mathbb{R}$ are continuous functions.

We are motivated by

Definition 8.9 (see [66], p. 646). *Let $f \in C_{2\pi}^{(\mathcal{F})}(\mathbb{R})$ we give the fuzzy Jackson operator*

$$(J_n(f))(x) = (FR)\int_{-\pi}^{\pi} K_n(t) \odot f(x+t)dt,$$

where

$$K_n(t) = L_{n'}(t), \quad n' = \left[\frac{n}{2}\right] + 1,$$

[ ] the integral part,

$$L_m(t) = \lambda_m^{-1} \left[\frac{\sin(mt/2)}{\sin(t/2)}\right]^4,$$

with λ_m being determined by

$$\int_{-\pi}^{\pi} L_m(t)dt = 1, \quad m \in \mathbb{N}.$$

It is noticed that $K_n(t) \geq 0$ being even trigonometric polynomial of degree n. By [66], p. 647 it is shown that $(J_n(f))(x)$ is a fuzzy continuous trigonometric polynomial (also by Lemma 8.3). Note by

$$(\tilde{J}_n(g))(x) = \int_{-\pi}^{\pi} K_n(t)g(x+t)dt, \quad g \in C_{2\pi}(\mathbb{R}).$$

the corresponding real Jackson operator.

We mention

Theorem 8.10 ([66], p. 647). *There exists a constant $C > 0$ (independent of n and f), such that for all $f \in C_{2\pi}^{(\mathcal{F})}(\mathbb{R})$ we have*

$$D((J_n(f))(x), f(x)) \leq C\omega_1^{(\mathcal{F})}\left(f, \frac{1}{n}\right), \quad \forall n \in \mathbb{N}, \ x \in \mathbb{R}.$$

By Lemma 8.5 and Proposition 2 of [11] we notice that as

$$n \to \infty, \quad \omega_1^{(\mathcal{F})}\left(f, \frac{1}{n}\right) \to 0$$

and we get $D^*(J_n f, f) \to 0$. *Based on Theorem 3.4 of [67] we trivially have that* J_n *is a fuzzy linear operator.*

By Theorem 8.8 we have

$$
\begin{aligned}
[(J_n(f))(x)]^r &= \left[(FR)\int_{-\pi}^{\pi} K_n(t) \odot f(x+t)dt\right]^r \\
&= \left[\int_{-\pi}^{\pi} K_n(t) f_{-}^{(r)}(x+t)dt, \int_{-\pi}^{\pi} K_n(t) f_{+}^{(r)}(x+t)dt\right],
\end{aligned}
$$

$\forall r \in [0,1], \forall x \in \mathbb{R}$. I.e. $(J_n f)_{\pm}^{(r)} = \tilde{J}_n(f_{\pm}^{(r)}), \forall r \in [0,1]$. Here $f_{\pm}^{(r)} \in C_{2\pi}(\mathbb{R})$, $\forall r \in [0,1]$. Let $f, g \in C_{2\pi}^{(\mathcal{F})}(\mathbb{R})$ such that $f \succsim g$ iff $f_{\pm}^{(r)} \geq g_{\pm}^{(r)}, \forall r \in [0,1]$, respectively. Then

$$\int_{-\pi}^{\pi} K_n(t) f_{\pm}^{(r)}(x+t)dt \geq \int_{-\pi}^{\pi} K_n(t) g_{\pm}^{(r)}(x+t)dt, \quad \forall r \in [0,1],$$

respectively, $\forall x \in \mathbb{R}$, i.e.

$$(J_n f)_{\pm}^{(r)} \geq (J_n g)_{\pm}^{(r)}, \quad \forall r \in [0,1],$$

respectively, iff

$$(J_n f) \succsim (J_n g), \quad n \in \mathbb{N}.$$

That is proving that J_n is a fuzzy positive operator. In fact almost all fuzzy operators defined via fuzzy summation or fuzzy integration are fuzzy positive linear operators.

We further need to use

Theorem 8.11 (Shisha and Mond (1968), [97]). *Let* $\tilde{L}_1, \tilde{L}_2, \ldots$, *be linear positive operators, whose common domain* K *consists of real functions with domain* $(-\infty, \infty)$. *Suppose* $1, \cos x, \sin x, f$ *belong to* K, *where* f *is an everywhere continuous,* 2π-*periodic function, with usual modulus of continuity* ω_1. *Let* $-\infty < a < b < \infty$, *and suppose that for* $n = 1, 2, \ldots, \tilde{L}_n(1)$ *is bounded in* n *over* $[a, b]$. *Then for* $n = 1, 2, \ldots,$

$$\|\tilde{L}_n f - f\|_\infty \leq \|f\|_\infty \|\tilde{L}_n(1) - 1\|_\infty + \|\tilde{L}_n 1 + 1\|_\infty \omega_1(f, \mu_n), \qquad (8.1)$$

where

$$\mu_n = \pi \left\|\tilde{L}_n\left(\sin^2\left(\frac{t-x}{2}\right)\right)(x)\right\|_\infty^{1/2}, \qquad (8.2)$$

where $\|\cdot\|_\infty$ *is the supremum norm over* $[a, b]$. *In particular, if* $\tilde{L}_n(1) = 1$, *as is often the case, the last inequality (8.1) reduces to*

$$\|\tilde{L}_n f - f\|_\infty \leq 2\omega_1(f, \mu_n). \qquad (8.3)$$

Remarks ([97]). (1) In forming $\tilde{L}_n\left(\sin^2\left(\frac{t-x}{2}\right)\right)$, t is the independent variable.

(2) We have that

$$
\mu_n^2 \leq \left(\frac{\pi^2}{2}\right)\left(\|\tilde{L}_n(1) - 1\|_\infty + \|\cos x\|_\infty\|(\tilde{L}_n(\cos t))(x) - \cos x\|_\infty \right.
$$
$$
\left. + \|\sin x\|_\infty\|(\tilde{L}_n(\sin t))(x) - \sin x\|_\infty\right). \tag{8.4}
$$

So if $\tilde{L}_n F$ converges uniformly to F in $[a, b]$ for $F(t) \equiv 1$, $\cos t$, $\sin t$, then as $n \to \infty$, we get by (8.4) that $\mu_n \to 0$ and $\omega_1(f, \mu_n) \to 0$, thus by (8.1) giving us $\tilde{L}_n f \to f$ uniformly on $[a, b]$, $\forall f$ continuous 2π-periodic function on $\mathbb{R}$. Hence we get with rates in an inequality form, quantitatively, the famous trigonometric Korovkin theorem, see [78].

8.2 Main Results

We present the fuzzy analog of Trigonometric Shisha–Mond inequality of Theorem 8.11.

Theorem 8.12. *Let $\{L_n\}_{n\in\mathbb{N}}$ be a sequence of fuzzy positive linear operators on $C_{2\pi}^{(\mathcal{F})}(\mathbb{R})$. We assume that there exists a corresponding sequence $\{\tilde{L}_n\}_{n\in\mathbb{N}}$ of positive linear operators on $C_{2\pi}(\mathbb{R})$ with the property*

$$
(L_n(f))_\pm^{(r)} = \tilde{L}_n(f_\pm^{(r)}), \tag{8.5}
$$

respectively, $\forall r \in [0, 1]$, $\forall f \in C_{2\pi}^{(\mathcal{F})}(\mathbb{R})$. We assume that $\{\tilde{L}_n(1)\}_{n\in\mathbb{N}}$ is bounded in n over $[a, b] \subseteq \mathbb{R}$. Then for $n \in \mathbb{N}$ we have

$$
D^*(L_n f, f) \leq \|\tilde{L}_n 1 - 1\|_\infty D^*(f, \tilde{o}) + \|\tilde{L}_n(1) + 1\|_\infty \omega_1^{(\mathcal{F})}(f, \mu_n), \tag{8.6}
$$

where

$$
\mu_n = \pi \left\|\tilde{L}_n\left(\sin^2\left(\frac{t-x}{2}\right)\right)(x)\right\|_\infty^{1/2}, \tag{8.7}
$$

where D^* and $\|\cdot\|_\infty$-sup norm are taken over $[a, b]$. In particular, if $\tilde{L}_n(1) = 1$, then we get

$$
D^*(L_n f, f) \leq 2\omega_1^{(\mathcal{F})}(f, \mu_n), \quad n \in \mathbb{N}. \tag{8.8}
$$

Here $\omega_1^{(\mathcal{F})}$ is the fuzzy modulus of continuity over $\mathbb{R}$.

Proof. We would like to estimate

$$
\begin{aligned}
D^*(L_n f, f) &= \sup_{x \in [a,b]} D\big((L_n f)(x), f(x)\big) \\
&= \sup_{x \in [a,b]} \sup_{r \in [0,1]} \max\{|(L_n f)_-^{(r)}(x) - (f)_-^{(r)}(x)|, \\
&\qquad |(L_n f)_+^{(r)}(x) - (f)_+^{(r)}(x)|\} \\
&= \sup_{x \in [a,b]} \sup_{r \in [0,1]} \max\{|(\tilde{L}_n(f_-^{(r)})(x) - (f)_-^{(r)}(x)|, \\
&\qquad |\tilde{L}_n(f_+^{(r)})(x) - (f)_+^{(r)}(x)|\} \\
&= \sup_{r \in [0,1]} \max\{\|\tilde{L}_n f_-^{(r)} - f_-^{(r)}\|_\infty, \|\tilde{L}_n f_+^{(r)} - f_+^{(r)}\|_\infty\}
\end{aligned}
$$

(by Theorem 8.11)

$$
\begin{aligned}
&\leq \sup_{r \in [0,1]} \max\{(\|f_-^{(r)}\|_\infty \|\tilde{L}_n 1 - 1\|_\infty \\
&\quad + \|\tilde{L}_n(1) + 1\|_\infty \omega_1(f_-^{(r)}, \mu_n)), (\|f_+^{(r)}\|_\infty \|\tilde{L}_n 1 - 1\|_\infty \\
&\quad + \|\tilde{L}_n(1) + 1\|_\infty \omega_1(f_+^{(r)}, \mu_n))\} \\
&\leq \|\tilde{L}_n 1 - 1\|_\infty \sup_{r \in [0,1]} \max(\|f_-^{(r)}\|_\infty, \|f_+^{(r)}\|_\infty) \\
&\quad + \|\tilde{L}_n(1) + 1\|_\infty \sup_{r \in [0,1]} \max\{\omega_1(f_-^{(r)}, \mu_n), \omega_1(f_+^{(r)}, \mu_n)\}
\end{aligned}
$$

(by Proposition 8.6)

$$
= \|\tilde{L}_n 1 - 1\|_\infty D^*(f, \tilde{o}) + \|\tilde{L}_n(1) + 1\|_\infty \omega_1^{(\mathcal{F})}(f, \mu_n),
$$

proving (8.6).   □

We present now the first Fuzzy Trigonometric Korovkin theorem.

Theorem 8.13. *Let $\{L_n\}_{n \in \mathbb{N}}$ be a sequence of fuzzy positive linear operators on $C_{2\pi}^{(\mathcal{F})}(\mathbb{R})$. We assume that there exists a corresponding sequence $\{\tilde{L}_n\}_{n \in \mathbb{N}}$ of positive linear operators on $C_{2\pi}(\mathbb{R})$ with the property*

$$
(L_n(f))_\pm^{(r)} = \tilde{L}_n(f_\pm^{(r)}), \tag{8.9}
$$

respectively, $\forall r \in [0,1]$, $\forall f \in C_{2\pi}^{(\mathcal{F})}(\mathbb{R})$.

Furthermore assume that $\tilde{L}_n(1) \overset{u}{\to} 1$, $\tilde{L}_n(\sin x) \overset{u}{\to} \sin x$, $\tilde{L}_n(\cos x) \overset{u}{\to} \cos x$, as $n \to \infty$, uniformly over $x \in [a,b] \subseteq \mathbb{R}$. Then $D^*(L_n f, f) \to 0$, as $n \to \infty$, over $[a,b]$, $\forall f \in C_{2\pi}^{(\mathcal{F})}(\mathbb{R})$. I.e. $L_n f \overset{D^*}{\to} f$, over $[a,b]$, that is $L_n \to I$ fuzzy unit operator, over $[a,b]$, as $n \to \infty$.

Proof. From (8.4) we get $\mu_n \to 0$, as $n \to \infty$. By Lemma 8.5 any $f \in C_{2\pi}^{(\mathcal{F})}(\mathbb{R})$ is uniformly continuous on $\mathbb{R}$ and thus $\omega_1^{(\mathcal{F})}(f, \mu_n) \to 0$. Also $\tilde{L}_n 1$ are bounded in n over $[a,b]$. Hence by (8.6) we get $D^*(L_n f, f) \to 0$ as $n \to \infty$, over $[a,b]$.   □

Application 8.14. As an example for Theorem 8.13 we mention the fuzzy Jackson operator J_n we discussed earlier, see Theorem 8.10, etc. Notice that $\tilde{J}_n(1) = 1$, $\forall n \in \mathbb{N}$ and by Theorem 2.2, p. 204, [56] of Jackson, we have that

$$\|\tilde{J}_n g - g\|_\infty \leq C\omega_1\left(g, \frac{1}{n}\right), \quad \forall g \in C_{2\pi}(\mathbb{R}),$$

$C > 0$ universal constant, giving us also that $\tilde{J}_n(\sin x) \xrightarrow{u} \sin x$, $\tilde{J}_n(\cos x) \xrightarrow{u} \cos x$ over $\mathbb{R}$, as $n \to \infty$. Furthermore, by [96] we get $\mu_n \leq \frac{C}{n}$, $C > 0$ constant independent of f and n. Then by Theorem 8.12, inequality (8.8) we obtain

$$D^*(J_n f, f) \leq C\omega_1^{(\mathcal{F})}\left(f, \frac{1}{n}\right), \quad n \in \mathbb{N},$$

where $C > 0$ universal constant. That is reconfirming Theorem 8.10.   □

9

FUZZY GLOBAL SMOOTHNESS PRESERVATION

Here we present the *property of global smoothness preservation* for fuzzy linear operators acting on spaces of fuzzy continuous functions. Basically we transfer the property of *real global smoothness preservation* into the fuzzy setting, via some *natural realization* condition fulfilled by almost all example-fuzzy linear operators. The derived inequalities involve fuzzy moduli of continuity and we give examples. This chapter relies on [21].

9.1 Basics

Motivation to this chapter are [26], [30]. In general and in applications, we prefer to have nice and fit approximations. That is the approximants, in this case operators, e.g. see [24], [32] *should not wiggle* more than the approximated functions. This feature is expressed with the property of *global smoothness preservation* by the approximating operators. In general let L be a fuzzy linear operator acting on a space of fuzzy continuous functions T defined on a metric space (X, d) and taking values in $\mathbb{R}_{\mathcal{F}}$, the set of fuzzy real numbers.

We say that L preserve the *property of global smoothness preservation*, iff

$$\omega_1^{(\mathcal{F})}(Lf, \delta) \leq c\omega_1^{(\mathcal{F})}(f, \delta), \quad \forall f \in T, \ \forall \delta > 0,$$

G.A. Anastassiou: Fuzzy Mathematics: Approximation The., STUDFUZZ 251, pp. 115–124.
springerlink.com © Springer-Verlag Berlin Heidelberg 2010

where $\omega_1^{(\mathcal{F})}$ is the fuzzy first modulus of continuity and $c > 0$ a universal constant possibly depended only on L.

In the real setting the analogous to the above inequality is true under certain conditions, see [26], [25], [30], [33]. So here we establish related inequalities in the fuzzy setting for various spaces T, by transferring from the real case under a minimal and natural *realization* assumption on L, see Theorems 9.10, 9.12, 9.14 and Proposition 9.15.

This assumption is fulfilled by almost all fuzzy operators defined via fuzzy summation or fuzzy integration. We provide also some interesting examples that motivate and fulfill our general theory. On the way to prove the main results we prove other important side results.

We need

Definition 9.1. Let $f: X \to \mathbb{R}_{\mathcal{F}}$, we call it *fuzzy bounded*, iff $D(f(x), \tilde{o}) \leq M$, $M > 0$, $\forall x \in X$ and we denote the related space by $B_{\mathcal{F}}(X)$. We call f *fuzzy continuous*, iff whenever $x_n, x \in X$ with $x_n \xrightarrow{d} x$ we have $D(f(x_n), f(x)) \to 0$, as $n \to \infty$, and this is true for all $x \in X$, and we denote the related space by $C_{\mathcal{F}}(X)$. We denote by $C_{\mathcal{F}}^U(X) = \{f: X \to \mathbb{R}_{\mathcal{F}} \mid \forall \varepsilon > 0, \exists \delta > 0$ so that whenever $x, y \in X$ with $d(x, y) \leq \delta$, then $D(f(x), f(y)) \leq \varepsilon\}$, the space of *fuzzy uniformly continuous functions on* X.

Set

$$\begin{aligned} C_{\mathcal{F}}^B(X) &= B_{\mathcal{F}}(X) \cap C_{\mathcal{F}}(X), \\ C_{\mathcal{F}}^{BU}(X) &= B_{\mathcal{F}}(X) \cap C_{\mathcal{F}}^U(X). \end{aligned}$$

If X is compact metric space then $C_{\mathcal{F}}^B(X) = C_{\mathcal{F}}(X) = C_{\mathcal{F}}^{BU}(X) = C_{\mathcal{F}}^U(X)$, see [2], pp. 52–53. Denote $[f] = [f_-^{(r)}, f_+^{(r)}]$, which means

$$[f(x)]^r = [f_-^{(r)}(x), f_+^{(r)}(x)], \quad \forall x \in X, \ \forall r \in [0, 1].$$

We use

Definition 9.2. Let $f: (X, d) \to \mathbb{R}_{\mathcal{F}}$. We define the (first) *fuzzy modulus of continuity* of f by

$$\omega_1^{(\mathcal{F})}(f, \delta) := \sup_{\substack{x, y \in X \\ d(x, y) \leq \delta}} D(f(x), f(y)), \quad \forall \delta > 0.$$

It holds

Proposition 9.3 (see [24]). *Let* $f: (X, d) \to \mathbb{R}_{\mathcal{F}}$. *Assume*

$$\omega_1^{(\mathcal{F})}(f, \delta), \quad \omega_1(f_-^{(r)}, \delta), \quad \omega_1(f_+^{(r)}, \delta), \quad r \in [0, 1]$$

that they are all finite $\forall \delta > 0$. *Here* ω_1 *is the usual real modulus of continuity. Then*

$$\omega_1^{(\mathcal{F})}(f, \delta) = \sup_{r \in [0, 1]} \max\{\omega_1(f_-^{(r)}, \delta), \omega_1(f_+^{(r)}, \delta)\}, \quad \forall \delta > 0. \tag{9.1}$$

We need

Definition 9.4. Let L be a map on $C_{\mathcal{F}}(X)$, we call it a *fuzzy linear operator*, iff

$$L(c_1 \odot f_1 \oplus c_2 \odot f_2) = c_1 \odot L(f_1) \oplus c_2 \odot L(f_2), \quad \forall c_1, c_2 \in \mathbb{R}, \ f_1, f_2 \in C_{\mathcal{F}}(X). \tag{9.2}$$

The linear operator L is called *fuzzy bounded*, iff

$$D^*(Lf, \tilde{o}) \le A D^*(f, \tilde{o}), \quad \forall f \in C_{\mathcal{F}}^B(X), \tag{9.3}$$

where $A > 0$ depending only on L. We denote $\|L\| = \inf A$, the norm of the operator. Similarly (9.3) is defined on $C_{\mathcal{F}}^{BU}(X)$.

We need

Proposition 9.5. *Let L be a fuzzy linear operator acting on $C_{\mathcal{F}}^B(X)$, or $C_{\mathcal{F}}^{BU}(X)$, respectively. We assume that to L there corresponds a bounded linear operator $\tilde{L} \ne 0$ acting on $C_B(X)$ (bounded continuous real valued functions on X), or $C_U^B(X)$ (bounded uniformly continuous real valued functions on X), respectively, with the property*

$$(Lf)_{\pm}^{(r)}(x) = \big(\tilde{L}(f_{\pm}^{(r)})\big)(x),$$

$\forall r \in [0, 1], \forall x \in X, \forall f \in C_{\mathcal{F}}^B(X)$ or $C_{\mathcal{F}}^{BU}(X)$, respectively. Then L is a fuzzy bounded linear operator on $C_{\mathcal{F}}^B(X)$, or $C_{\mathcal{F}}^{BU}(X)$, respectively, with $\|L\| \le \|\tilde{L}\|$.

Proof. Here $0 < \|\tilde{L}\| < \infty$ by assumption. We observe that

$$
\begin{aligned}
D^*(Lf, \tilde{o}) &= \sup_{x \in X} \sup_{r \in [0,1]} \max\{|(Lf)_{-}^{(r)}(x)|, |(Lf)_{+}^{(r)}(x)|\} \\
&= \sup_{x \in X} \sup_{r \in [0,1]} \max\{|\tilde{L}(f_{-}^{(r)})(x)|, |\tilde{L}(f_{+}^{(r)})(x)|\} \\
&\le \sup_{r \in [0,1]} \max\{\|\tilde{L}(f_{-}^{(r)})\|_{\infty}, \|\tilde{L}(f_{+}^{(r)})\|_{\infty}\} \\
&\le \|\tilde{L}\| \sup_{r \in [0,1]} \max\{\|f_{-}^{(r)}\|_{\infty}, \|f_{+}^{(r)}\|_{\infty}\} = \|\tilde{L}\| D^*(f, \tilde{o}).
\end{aligned}
$$

I.e. we got that $D^*(Lf, \tilde{o}) \le \|\tilde{L}\| D^*(f, \tilde{o})$. Hence L is a fuzzy bounded operator. Clearly then $\|L\|$ is finite and $\|L\| \le \|\tilde{L}\|$. $\square$

We require also

Proposition 9.6. *Let $[a, b] \subset \mathbb{R}$ and $t_0 \in [a, b]$ fixed. Let $\tilde{L}$ be a linear operator that maps $C([a, b])$ into $Z := \{g \in C([a, b]) \mid g(t_0) = 0\}$, i.e. $(\tilde{L}g)(t_0) = 0, \forall g \in C([a, b])$. Furthermore assume that*

$$\omega_1(\tilde{L}g, \delta) \le c\omega_1(g, \delta), \quad \forall \delta > 0, \ \forall g \in C([a, b]),$$

where c is universal constant possibly depended on $\tilde{L}$, i.e. $\tilde{L}$ has the property of preservation of global smoothness. Then $\tilde{L}$ is a bounded linear operator.

Proof. By $\omega_1(g, \delta) \leq 2\|g\|_\infty$ we have

$$\omega_1(\tilde{L}g, \delta) \leq 2c\|g\|_\infty, \quad \text{all } 0 < \delta \leq b - a.$$

Thus

$$\sup_{\substack{x \in [a,b] \\ |x-t_0| \leq \delta}} |(\tilde{L}g)(t_0) - (\tilde{L}g)(x)| \leq \sup_{\substack{x,y \in [a,b] \\ |x-y| \leq \delta}} |(\tilde{L}g)(x) - (\tilde{L}g)(y)| \leq 2c\|g\|_\infty.$$

Setting $\delta = b - a$ we have

$$\|\tilde{L}g\|_\infty \leq 2c\|g\|_\infty,$$

proving that $\tilde{L}$ is a bounded operator.   □

We present

Example 9.7. Let $f \in C_\mathcal{F}([0,1])$, we define the *fuzzy Bernstein operator*

$$(B_n^{(\mathcal{F})}f)(x) = \sum_{k=0}^{n}{}^* \binom{n}{k} x^k (1-x)^{n-k} \odot f\left(\frac{k}{n}\right), \quad \forall x \in [0,1], \ n \in \mathbb{N}. \quad (9.4)$$

Notice that $B_n^{(\mathcal{F})}$ is a fuzzy linear operator mapping $C_\mathcal{F}([0,1])$ into itself. Furthermore we easily see that

$$D^*(B_n^{(\mathcal{F})}f, \tilde{o}) \leq 1 \cdot D^*(f, \tilde{o}) < \infty,$$

i.e. (see also [24]) $B_n^{(\mathcal{F})}$ is a fuzzy bounded operator.
By p. 642, [66], we have that

$$D^*(B_n^{(\mathcal{F})}f, f) \leq \frac{3}{2}\omega_1^{(\mathcal{F})}\left(f, \frac{1}{\sqrt{n}}\right), \quad \forall n \in \mathbb{N}. \quad (9.5)$$

Let the real Bernstein operator

$$(B_n g)(x) = \sum_{k=0}^{n} \binom{n}{k} x^k (1-x)^{n-k} g\left(\frac{k}{n}\right), \quad \forall x \in [0,1], \ g \in C([0,1]), \quad (9.6)$$

which converges uniformly to g with rates. We notice that

$$(B_n^{(\mathcal{F})}f)_\pm^{(r)} = B_n(f_\pm^{(r)}), \quad \forall r \in [0,1], \quad (9.7)$$

respectively in $\pm$, $\forall f \in C_\mathcal{F}([0,1])$. That is an important property motivating our theory next. Also from [26] and [30], p. 244 we have that

$$\omega_1(B_n g, \delta) \leq 2\omega_1(g, t), \quad \forall g \in C([0,1]), \quad (9.8)$$

and number 2 is the best constant. I.e. real Bernstein operators possess the *global smoothness preservation property*.

We also present

Example 9.8. Let $f \in C_{2\pi}^{(\mathcal{F})}(\mathbb{R})$ (space of 2π-periodic fuzzy continuous functions on $\mathbb{R}$). We define (see [66], p. 646) the *fuzzy Jackson operator*

$$J_n(f)(x) = (FR) \int_{-\pi}^{\pi} K_n(t) \odot f(x+t)\, dt, \quad \forall n \in \mathbb{N}, \qquad (9.9)$$

where $((FR) \int)$ the Fuzzy–Riemann integral is as in [67], [66], p. 644. Here $K_n(t) = L_{n'}(t)$, $n' = \left[\frac{n}{2}\right] + 1$, and

$$L_m(t) = \lambda_m^{-1} \left[\frac{\sin(mt/2)}{\sin(t/2)}\right]^4, \qquad (9.10)$$

such that

$$\int_{-\pi}^{\pi} L_m(t)dt = 1, \quad \forall m \in \mathbb{N}.$$

That is $\int_{-\pi}^{\pi} K_n(t)dt = 1$. It is known that $K_n(t) \geq 0$ being an even trigonometric polynomial of order n. The fuzzy operator J_n is linear, by the linearity of Fuzzy–Riemann integral, see [67].

By Lemma 13.2, p. 644 of [66] we get

$$
\begin{aligned}
D((J_n f)(x), \tilde{o}) &= D\left((FR)\int_{-\pi}^{\pi} K_n(t) \odot f(x+t)dt, \tilde{o}\right) \\
&\leq \int_{-\pi}^{\pi} D(K_n(t) \odot f(x+t), \tilde{o})dt \\
&= \int_{-\pi}^{\pi} K_n(t) D(f(x+t), \tilde{o})dt \\
&\leq D^*(f, \tilde{o}) \int_{-\pi}^{\pi} K_n(t)dt = D^*(f, \tilde{o}) < \infty.
\end{aligned}
$$

I.e. we have
$$D^*(J_n f, \tilde{o}) \leq 1 \cdot D^*(f, \tilde{o}), \qquad (9.11)$$

proving J_n a fuzzy bounded operator, $\forall n \in \mathbb{N}$.

By Theorem 13.14, p. 647 of [66] we have that

$$D((J_n f)(x), f(x)) \leq C\omega_1^{(\mathcal{F})}\left(f, \frac{1}{n}\right), \quad \forall n \in \mathbb{N}, \ x \in \mathbb{R}, \qquad (9.12)$$

$C > 0$ universal constant, that is convergence with rates.

Next we mention the real Jackson operator

$$(\tilde{J}_n(g))(x) = \int_{-\pi}^{\pi} K_n(t)g(x+t)dt, \quad \forall n \in \mathbb{N}, \ g \in C_{2\pi}(\mathbb{R}). \qquad (9.13)$$

From Theorem 2.2, p. 204, [56] we have that

$$\|\tilde{J}_n g - g\|_\infty \leq C\omega_1\left(g, \frac{1}{n}\right), \quad \forall n \in \mathbb{N}, \ g \in C_{2\pi}(\mathbb{R}), \tag{9.14}$$

$C > 0$ universal constant, that is convergence with rates. By [67] we observe that

$$(J_n f(x))_\pm^{(r)} = \tilde{J}_n(f_\pm^{(r)})(x), \quad \forall x \in \mathbb{R}, \ r \in [0, 1], \tag{9.15}$$

respectively in $\pm$, $\forall f \in C_{2\pi}^{(\mathcal{F})}(\mathbb{R})$.

Next we see

$$
\begin{aligned}
|(\tilde{J}_n g)(x) - (\tilde{J}_n g)(y)| &= \left| \int_{-\pi}^{\pi} K_n(t)g(x+t)dt - \int_{-\pi}^{\pi} K_n(t)g(y+t)dt \right| \\
&\leq \int_{-\pi}^{\pi} K_n(t)|g(x+t) - g(y+t)|dt \\
&\leq \int_{-\pi}^{\pi} K_n(t)\omega_1(g, |x-y|)dt = \omega_1(g, |x-y|),
\end{aligned}
$$

i.e. giving us

$$\omega_1(\tilde{J}_n g, |x-y|) \leq \omega_1(g, |x-y|),$$

and

$$\omega_1(\tilde{J}_n g, \delta) \leq \omega_1(g, \delta), \quad \forall \delta > 0, \ \forall g \in C_{2\pi}(\mathbb{R}), \tag{9.16}$$

that is proving that property of preservation of global smoothness for the real Jackson operators. Similarly we see that

$$
\begin{aligned}
D((J_n f)(x), &(J_n f)(y)) \\
&= D\left((FR)\int_{-\pi}^{\pi} K_n(t) \odot f(x+t)dt, (FR)\int_{-\pi}^{\pi} K_n(t) \odot f(y+t)dt \right) \\
&\quad \text{(by Lemma 13.2, p. 644 of [66])} \\
&\leq \int_{-\pi}^{\pi} D(K_n(t) \odot f(x+t), K_n(t) \odot f(y+t))dt \\
&= \int_{-\pi}^{\pi} K_n(t)D(f(x+t), f(y+t))dt \\
&\leq \int_{-\pi}^{\pi} K_n(t)\omega_1^{(\mathcal{F})}(f, |x-y|)dt = \omega_1^{(\mathcal{F})}(f, |x-y|),
\end{aligned}
$$

i.e. giving us

$$\omega_1^{(\mathcal{F})}(J_n f, \delta) \leq \omega_1^{(\mathcal{F})}(f, \delta), \quad \forall \delta > 0, \ \forall f \in C_{2\pi}^{(\mathcal{F})}(\mathbb{R}). \tag{9.17}$$

That is proving the *property of fuzzy global smoothness preservation* for the fuzzy Jackson operators.

Next let $X := (X, d)$ be a compact metric space, and $C(X)$ the space of continuous real-valued functions on X endowed with $\| \cdot \|_\infty$. Let $f \in C(X)$, we define the first modulus of continuity of it by

$$\omega_1(f, \delta) := \sup_{\substack{x, y \\ d(x,y) \leq \delta}} |f(x) - f(y)|, \quad \forall \delta > 0. \tag{9.18}$$

Let $\mathrm{Lip}(X)$ the sub-space of all $g \in C(X)$ with finite semi-norm

$$|g|_{\mathrm{Lip}} := \sup_{d(x,y)>0} \frac{|g(x) - g(y)|}{d(x,y)}. \tag{9.19}$$

It is known $\mathrm{Lip}(X)$ is a dense subset of $C(X)$.

Let $\tilde{\omega}_1(f, \delta)$ the *least concave majorant* of $\omega_1(f, \delta)$, this also measures smoothness of f.

We would like to mention the following related fundamental result.

Theorem 9.9 (see [26] and [30], p. 234). *Let X be a compact metric space, and $L: C(X) \to C(X)$, $L \neq 0$, be a bounded linear operator mapping $\mathrm{Lip}(X)$ to $\mathrm{Lip}(X)$ such that for all $g \in \mathrm{Lip}(X)$,*

$$|Lg|_{\mathrm{Lip}} \leq c|g|_{\mathrm{Lip}}, \tag{9.20}$$

with constant c possibly depending on L, but independent of g. Then for all $f \in C(X)$ and $\delta > 0$ we have

$$\omega_1(Lf, \delta) \leq \|L\| \tilde{\omega}_1 \left(f, \frac{c\delta}{\|L\|} \right), \tag{9.21}$$

where $\|L\|$ denotes the operator norm of L on $C(X)$. It is known that over $[a, b] \subseteq \mathbb{R}$ we have $\omega_1 \leq \tilde{\omega}_1 \leq 2\omega_1$, etc.

9.2 Main Results

Examples 9.7, 9.8 and Theorem 9.9 motivate the main results next.

Theorem 9.10. *Let (X, d) be a compact metric space. Let L be a fuzzy linear operator acting on $C_{\mathcal{F}}(X)$. Assume for L there corresponds a real linear operator $\tilde{L}$ on $C(X)$ with the property*

$$(L(f))_\pm^{(r)}(x) = \tilde{L}(f_\pm^{(r)})(x), \quad \forall r \in [0, 1], \ \forall x \in X, \tag{9.22}$$

respectively in $\pm$, $\forall f \in C_{\mathcal{F}}(X)$. Furthermore it holds

$$\omega_1(\tilde{L}g, \delta) \leq c\omega_1(g, \delta), \quad \forall \delta > 0, \ \forall g \in C(X), \tag{9.23}$$

where $c > 0$ is a universal constant may be depended only on $\tilde{L}$. Then

$$\omega_1^{(\mathcal{F})}(Lf, \delta) \leq c\omega_1^{(\mathcal{F})}(f, \delta), \quad \forall \delta > 0, \ \forall f \in C_{\mathcal{F}}(X). \tag{9.24}$$

That is L possesses the property of preservation of fuzzy global smoothness.

Proof. Notice here that all $f_{\pm}^{(r)} \in C(X)$. Also $\omega_1^{(\mathcal{F})}(f, \delta)$; $\omega_1(f_{\pm}^{(r)}, \delta)$, $\forall r \in [0, 1]$, $\forall \delta > 0$, are all finite, easily observed since X is compact. Then by Proposition 9.3 we see that

$$
\begin{aligned}
\omega_1^{(\mathcal{F})}(Lf, \delta) \quad &\leq \quad \sup_{r \in [0,1]} \max\{\omega_1((Lf)_{-}^{(r)}, \delta), \omega_1((Lf)_{+}^{(r)}, \delta)\} \\
&\overset{(9.22)}{=} \quad \sup_{r \in [0,1]} \max\{\omega_1(\tilde{L}(f_{-}^{(r)}), \delta), \omega_1(\tilde{L}(f_{+}^{(r)}), \delta)\} \\
&\overset{(9.23)}{\leq} \quad \sup_{r \in [0,1]} \max\{c\omega_1(f_{-}^{(r)}, \delta), c\omega_1(f_{+}^{(r)}, \delta)\} \\
&= \quad c \sup_{r \in [0,1]} \max\{\omega_1(f_{-}^{(r)}, \delta), \omega_1(f_{+}^{(r)}, \delta)\} \overset{(9.1)}{=} c\omega_1^{(\mathcal{F})}(f, \delta).
\end{aligned}
$$

That is proving (9.24).   □

We give

Example 9.11. Using Theorem 9.10 and (9.8) we obtain

$$\omega_1^{(\mathcal{F})}(B_n^{(\mathcal{F})}(f), \delta) \leq 2\omega_1^{(\mathcal{F})}(f, \delta), \quad \forall \delta > 0, \ \forall f \in C_{\mathcal{F}}([0, 1]). \tag{9.25}$$

We present also

Theorem 9.12. *Let Θ open convex subset of a real normed vector space $(V, \|\cdot\|)$. Let L be a fuzzy linear operator acting on $C_{\mathcal{F}}^B(\Theta)$. Assume for L there corresponds a real linear operator $\tilde{L}$ on $C_B(\Theta)$ with the property*

$$(L(f))_{\pm}^{(r)}(x) = \tilde{L}(f_{\pm}^{(r)})(x), \quad \forall r \in [0, 1], \ \forall x \in \Theta, \tag{9.26}$$

respectively in $\pm$, $\forall f \in C_{\mathcal{F}}^B(\Theta)$. Furthermore it holds

$$\omega_1(\tilde{L}g, \delta) \leq c\omega_1(g, \delta), \quad \forall \delta > 0, \ \forall g \in C_B(\Theta), \tag{9.27}$$

where $c > 0$ is a universal constant may be depended only on $\tilde{L}$. Then

$$\omega_1^{(\mathcal{F})}(Lf, \delta) \leq c\omega_1^{(\mathcal{F})}(f, \delta), \quad \forall \delta > 0, \ \forall f \in C_{\mathcal{F}}^B(\Theta). \tag{9.28}$$

That is L possesses the property of preservation of fuzzy global smoothness.

Proof. Since $f \in C_{\mathcal{F}}^B(\Theta)$, this implies that $f_{\pm}^{(r)} \in C_B(\Theta)$, $\forall r \in [0, 1]$. Consequently we get that all $\omega_1^{(\mathcal{F})}(f, \delta)$, $\omega_1(f_{\pm}^{(r)}, \delta)$, $\forall r \in [0, 1]$ are finite, $\forall \delta > 0$. Then the proof goes the same way as the proof of Theorem 9.10.□

We give

Example 9.13. In [32] we proved that $C_{2\pi}^{(\mathcal{F})}(\mathbb{R}) = {}_{2\pi}C_{\mathcal{F}}^{U}(\mathbb{R})$, the space of fuzzy uniformly continuous 2π-periodic functions. Also $C_{2\pi}^{(\mathcal{F})}(\mathbb{R}) = {}_{2\pi}C_{B}^{(\mathcal{F})}(\mathbb{R})$ the space of fuzzy bounded continuous 2π-periodic functions on $\mathbb{R}$. Clearly, working as in Theorem 9.12 we can derive again (9.17).

We give also

Theorem 9.14. *Let Θ open convex subset of a real normed vector space $(V, \|\cdot\|)$. Let L a fuzzy linear operator acting on $C_{\mathcal{F}}^{U}(\Theta)$. Assume for L there corresponds a real linear operator $\tilde{L}$ on $C_{U}(\Theta)$ (the space of uniformly continuous real valued functions on Θ) with the property*

$$(L(f))_{\pm}^{(r)}(x) = \tilde{L}(f_{\pm}^{(r)})(x), \quad \forall r \in [0,1], \ \forall x \in \Theta, \qquad (9.29)$$

respectively in $\pm$, $\forall f \in C_{\mathcal{F}}^{U}(\Theta)$. Furthermore it holds

$$\omega_1(\tilde{L}g, \delta) \le c\omega_1(g, \delta), \quad \forall \delta > 0, \ \forall g \in C_{U}(\Theta), \qquad (9.30)$$

where $c > 0$ is a universal constant may be depended only on $\tilde{L}$. Then

$$\omega_1^{(\mathcal{F})}(Lf, \delta) \le c\omega_1^{(\mathcal{F})}(f, \delta), \quad \forall \delta > 0, \ \forall f \in C_{\mathcal{F}}^{U}(\Theta). \qquad (9.31)$$

That is L possesses the property of preservation of fuzzy global smoothness.

Proof. Since $f \in C_{\mathcal{F}}^{U}(\Theta)$, this implies that $f_{\pm}^{(r)} \in C_{U}(\Theta)$, $\forall r \in [0,1]$. Consequently, by [24] and as in [30], p. 281, 298, we get that all $\omega_1^{(\mathcal{F})}(f, \delta)$, $\omega_1(f_{\pm}^{(r)}, \delta)$, $\forall r \in [0,1]$ are finite, $\forall \delta > 0$. Then the proof goes the same way as the proof of Theorem 9.10. □

We finish with

Proposition 9.15. *Let $[a,b] \subset \mathbb{R}$, $t_0 \in [a,b]$ fixed. Let L be a fuzzy linear operator acting on $C_{\mathcal{F}}([a,b])$. Assume for L there corresponds a real linear operator $\tilde{L}$ from $C([a,b])$ into $Z = \{g \in C([a,b]) \mid g(t_0) = 0\}$, with the property*

$$(Lf)_{\pm}^{(r)}(x) = \tilde{L}(f_{\pm}^{(r)})(x), \quad \forall r \in [0,1], \ \forall x \in [a,b], \qquad (9.32)$$

respectively in $\pm$, $\forall f \in C_{\mathcal{F}}([a,b])$. Furthermore it holds

$$\omega_1(\tilde{L}g, \delta) \le c\omega_1(g, \delta), \quad \forall \delta > 0, \ \forall g \in C([a,b]), \qquad (9.33)$$

where $c > 0$ is a universal constant may be depended only on $\tilde{L}$. Then
(1)

$$\omega_1^{(\mathcal{F})}(Lf, \delta) \le c\omega_1^{(\mathcal{F})}(f, \delta), \quad \forall \delta > 0, \ \forall f \in C_{\mathcal{F}}([a,b]). \qquad (9.34)$$

I.e. L possesses the property of preservation of fuzzy global smoothness.

(2) L *is a fuzzy bounded linear operator with* $\|L\| \le \|\tilde{L}\| < \infty.$

Proof. (1) Same as in Theorem 9.10.

(2) By Proposition 9.6 we get that $\tilde{L}$ is a bounded real linear operator. And by Proposition 9.5, for $X = [a,b]$ and L, $\tilde{L}$ acting on $C_{\mathcal{F}}([a,b])$, $C([a,b])$, respectively, we establish the claim, that L is a fuzzy bounded linear operator. $\square$

10
FUZZY KOROVKIN THEORY AND INEQUALITIES

Here we study the fuzzy positive linear operators acting on fuzzy continuous functions. We prove the fuzzy Riesz representation theorem, the fuzzy Shisha–Mond type inequalities and fuzzy Korovkin type theorems regarding the fuzzy convergence of fuzzy positive linear operators to the fuzzy unit in various cases. Special attention is paid to the study of fuzzy weak convergence of finite positive measures to the unit Dirac measure. All convergences are with rates and are given via fuzzy inequalities involving the fuzzy modulus of continuity of the engaged fuzzy valued function. The assumptions for the Korovkin theorems are minimal and of natural realization, fulfilled by almost all example – fuzzy positive linear operators. The surprising fact is that the real Korovkin test functions assumptions carry over here in the fuzzy setting and they are the only enough to impose the conclusions of fuzzy Korovkin theorems. We give a lot of examples and applications to our theory, namely: to fuzzy Bernstein operators, to fuzzy Shepard operators, to fuzzy Szasz–Mirakjan and fuzzy Baskakov-type operators and to fuzzy convolution type operators.

We work in general, basically over real normed vector space domains that are compact and convex or just convex. On the way to prove the main theorems we establish a lot of other interesting and important side results This chapter relies on [24].

G.A. Anastassiou: Fuzzy Mathematics: Approximation The., STUDFUZZ 251, pp. 125–157.
springerlink.com   © Springer-Verlag Berlin Heidelberg 2010

10.1 Background

Motivation for this chapter are [102], [18]. In [102] the authors establish some Korovkin type theory for fuzzy random variables under some specialized assumptions. In [18] the author proves the first fuzzy Shisha–Mond type inequalities and the first basic fuzzy Korovkin theorem over $[a, b] \subseteq \mathbb{R}$.

The surprising fact is that the basic assumptions of real Korovkin theory for the test functions 1, id, id^2 carry over there and here and they are the only ones needed. Of course a natural realization condition is needed in the fuzzy setting to prove the fuzzy convergence. Here we continue the first study [18], see also Chapter 7, now over compact convex subsets of real normed vector spaces $(V, \| \cdot \|)$ or just convex subsets. On the way there we prove the interesting fuzzy Riesz representation theorem and its implications. Also we prove a lot of other needed interesting results which by themselves independently have their own merit, e.g. about fuzzy continuity and boundedness and about abstract fuzzy modulus of continuity properties, etc. So this chapter is essentially the study with rates and quantitatively of the fuzzy convergence of a sequence of fuzzy positive linear operators to the fuzzy unit operator.

In duality one can see it as the study with rates and quantitatively of the fuzzy weak convergence of a sequence of finite positive measures to the unit Dirac measure. The concept of positivity we use in our operators is the natural analog of the real case. The same thing with linearity, in our case is over $\mathbb{R}$, in [102] is only over $\mathbb{R}_+$.

Our fuzzy integral here is defined according to the excellent and great article [75]. Finally a lot of examples are given. The application of the fuzzy Korovkin theory to fuzzy Szasz–Mirakjan and fuzzy Baskakov-type operators is very elaborate. It takes a great deal of probabilistic and fuzzy real analysis tools and work to accomplish the results. Also it reveals a lot as these operators are defined on $\mathbb{R}_+$, they are infinite fuzzy sums, and still fulfill our basic realization condition defined over compact sets, see Assumption 10.20 and (10.46).

Remark 10.1. (1) Let $(u_k)_{k \in \mathbb{N}} \in \mathbb{R}_{\mathcal{F}}$. We denote the *fuzzy infinite series* by $\sum^{*}{}_{k=1}^{\infty} u_k$ and we say that it converges to $u \in \mathbb{R}_{\mathcal{F}}$ iff $\lim_{n \to \infty} D \left(\sum^{*}{}_{k=1}^{n} u_k, u \right)$ $= 0$. We denote the last by $\sum^{*}{}_{k=1}^{\infty} u_k = u$. Let $(u_k)_{k \in \mathbb{N}}, (v_k)_{k \in \mathbb{N}}, u, v \in \mathbb{R}_{\mathcal{F}}$ such that

$$\sum_{k=1}^{\infty}{}^{*} u_k = u, \sum_{k=1}^{\infty}{}^{*} v_k = v.$$

Then

$$\sum_{k=1}^{\infty}{}^{*} (u_k \oplus v_k) = u \oplus v = \sum_{k=1}^{\infty}{}^{*} u_k \oplus \sum_{k=1}^{\infty}{}^{*} v_k.$$

The last is true since

$$\lim_{n\to\infty} D\left(\sum_{k=1}^{n}{}^{*}(u_k \oplus v_k), u \oplus v\right) = \lim_{n\to\infty} D\left(\left(\sum_{k=1}^{n}{}^{*} u_k\right) \oplus \left(\sum_{k=1}^{n}{}^{*} v_k\right), u \oplus v\right)$$

$$\leq \lim_{n\to\infty} \left(D\left(\sum_{k=1}^{n}{}^{*} u_k, u\right) + D\left(\sum_{k=1}^{n}{}^{*} v_k, v\right)\right) = 0.$$

Let $\sum_{k=1}^{*\infty} u_k = u \in \mathbb{R}_{\mathcal{F}}$ then one has that

$$\sum_{k=1}^{\infty}(u_k)_{-}^{(r)} = u_{-}^{(r)} = \left(\sum_{k=1}^{\infty}{}^{*} u_k\right)_{-}^{(r)}$$

and

$$\sum_{k=1}^{\infty}(u_k)_{+}^{(r)} = u_{+}^{(r)} = \left(\sum_{k=1}^{\infty}{}^{*} u_k\right)_{+}^{(r)}, \quad \forall r \in [0,1].$$

We prove the last claim: We have that

$$0 = \lim_{n\to\infty} D\left(\sum_{k=1}^{n} u_k, u\right)$$

$$= \lim_{n\to\infty} \sup_{r\in[0,1]} \max\left\{\left|\sum_{k=1}^{n}(u_k)_{-}^{(r)} - u_{-}^{(r)}\right|, \left|\sum_{k=1}^{n}(u_k)_{+}^{(r)} - u_{+}^{(r)}\right|\right\}$$

$$\geq \lim_{n\to\infty} \left\{\left|\sum_{k=1}^{n}(u_k)_{-}^{(r)} - u_{-}^{(r)}\right|, \left|\sum_{k=1}^{n}(u_k)_{+}^{(r)} - u_{+}^{(r)}\right|\right\}, \quad \forall r \in [0,1],$$

proving the claim.

Also we need: let $(u_k)_{k\in\mathbb{N}} \in \mathbb{R}_{\mathcal{F}}$ with $\sum_{k=1}^{\infty}{}^{*} u_k = u \in \mathbb{R}_{\mathcal{F}}$, then clearly one

has for any $\lambda \in \mathbb{R}$ that $\sum_{k=1}^{\infty}{}^{*} \lambda u_k = \lambda u.$

2) From [75] we see: Let $u_n = \{(u_{n-}^{(r)}, u_{n+}^{(r)}) \mid 0 \leq r \leq 1\} \in \mathbb{R}_{\mathcal{F}}$ such that $\sum_{n=1}^{\infty} u_{n-}^{(r)} = u_{-}^{(r)}$ and $\sum_{n=1}^{\infty} u_{n+}^{(r)} = u_{+}^{(r)}$ converge uniformly in $r \in [0,1]$, then $u = \{(u_{-}^{(r)}, u_{+}^{(r)}) \mid 0 \leq r \leq 1\} \in \mathbb{R}_{\mathcal{F}}$ and $u = \sum_{n=1}^{\infty} u_n$. I.e. we have

$$\sum_{n=1}^{\infty}\{(u_{n-}^{(r)}, u_{n+}^{(r)}) \mid 0 \leq r \leq 1\} = \left\{\left(\sum_{n=1}^{\infty} u_{n-}^{(r)}, \sum_{n=1}^{\infty} u_{n+}^{(r)}\right) \mid 0 \leq r \leq 1\right\}.$$

We use the following

Definition 10.2. Let $U \subseteq (M, d)$ a metric space and let $f: U \to \mathbb{R}_{\mathcal{F}}$. We define the (*first*) *fuzzy modulus of continuity* of f by

$$\omega_1^{(\mathcal{F})}(f; \delta) := \sup_{\substack{x,y \in U \\ d(x,y) \leq \delta}} D(f(x), f(y))$$

for $0 < \delta \leq \text{diameter}(U)$. If $\delta > \text{diam}(U)$ then we define

$$\omega_1^{(\mathcal{F})}(f; \delta) := \omega_1^{(\mathcal{F})}(f; \text{diam}(U)).$$

Proposition 10.3. *Let* $U \subseteq (M, d)$ *metric space and* $f: U \to \mathbb{R}_{\mathcal{F}}$. *Assume that*

$$\omega_1^{(\mathcal{F})}(f, \delta), \quad \omega_1(f_-^{(r)}, \delta), \quad \omega_1(f_+^{(r)}, \delta)$$

are finite for any $\delta > 0$. *Here* ω_1 *is the usual real modulus of continuity. Then it holds*

$$\omega_1^{(\mathcal{F})}(f; \delta) = \sup_{r \in [0,1]} \max\{\omega_1(f_-^{(r)}, \delta), \omega_1(f_+^{(r)}, \delta)\}.$$

Proof. Let $x, y \in U: d(x, y) \leq \delta$. We have

$$D(f(x), f(y)) = \sup_{r \in [0,1]} \max\{|(f(x))_-^{(r)} - (f(y))_-^{(r)}|, |(f(x))_+^{(r)} - (f(y))_+^{(r)}|\}$$

$$\leq \sup_{r \in [0,1]} \max\{\omega_1(f_-^{(r)}, \delta), \omega_1(f_+^{(r)}, \delta)\}.$$

Hence

$$\omega_1^{(\mathcal{F})}(f, \delta) \leq \sup_{r \in [0,1]} \max\{\omega_1(f_-^{(r)}, \delta), \omega_1(f_+^{(r)}, \delta)\}.$$

For any $r \in [0, 1]$ and any $x, y \in U: d(x, y) \leq \delta$ we see that

$$\omega_1^{(\mathcal{F})}(f; \delta) \geq D(f(x), f(y)) \geq |(f(x))_-^{(r)} - (f(y))_-^{(r)}|, |(f(x))_+^{(r)} - (f(y))_+^{(r)}|.$$

Therefore

$$\omega_1(f_{\pm}^{(r)}; \delta) \leq \omega_1^{(\mathcal{F})}(f; \delta), \quad \forall r \in [0, 1].$$

Hence

$$\sup_{r \in [0,1]} \max\{\omega_1(f_-^{(r)}; \delta), \omega_1(f_+^{(r)}; \delta)\} \leq \omega_1^{(\mathcal{F})}(f; \delta)$$

proving the claim. $\qquad \qquad \square$

We need

Definition 10.4. Let U open or compact $\subseteq (M, d)$ metric space and $f: U \to \mathbb{R}_{\mathcal{F}}$. We say that f is *fuzzy continuous* at $x_0 \in U$ iff whenever $x_n \to x_0$, then $D(f(x_n), f(x_0)) \to 0$. If f is continuous for every $x_0 \in U$, we then call f a *fuzzy continuous real number valued function*. We denote the related space by $C_{\mathcal{F}}(U)$. Similarly one defines $C_{\mathcal{F}}([a,b])$, $[a,b] \subseteq \mathbb{R}$.

Definition 10.5. Let $f \colon K \to \mathbb{R}_{\mathcal{F}}$, K open or compact $\subseteq (M,d)$ metric space. We call f a *fuzzy uniformly continuous real number valued function*, iff $\forall \varepsilon > 0$, $\exists \delta > 0$: whenever $d(x,y) \leq \delta$, $x,y \in K$, implies that $D(f(x),f(y)) \leq \varepsilon$. We denote the related space by $C_{\mathcal{F}}^{U}(K)$.

Definition 10.6. Let $f \colon U \to \mathbb{R}_{\mathcal{F}}$, $U \subseteq (M,d)$ metric space. If $D(f(x),\tilde{o}) \leq M$, $\forall x \in U$, $M \geq 0$, we call f a *fuzzy bounded real number valued function*.

In particular if $f \in C_{\mathcal{F}}([a,b])$, $[a,b] \subseteq \mathbb{R}$, then f is a fuzzy bounded function, also $\omega_{1}^{(\mathcal{F})}(f;\delta) < \infty$ for any $0 < \delta \leq b - a$. Also notice that $C_{\mathcal{F}}^{U}(K) = C_{\mathcal{F}}(K)$, for K compact $\subseteq (V, \|\cdot\|)$ real normed vector space.

We use

Proposition 10.7. *Let $K \subseteq (V, \|\cdot\|)$ a real normed vector space and*

$$\omega_{1}^{(\mathcal{F})}(f;\delta) = \sup_{\substack{x,y \in K \\ \|x-y\| \leq \delta}} D(f(x),f(y)), \quad \delta > 0,$$

the fuzzy modulus of continuity for $f \colon K \to \mathbb{R}_{\mathcal{F}}$. Then

(1) *If $f \in C_{\mathcal{F}}^{U}(K)$, K open convex or compact convex $\subseteq (V, \|\cdot\|)$, then $\omega_{1}^{(\mathcal{F})}(f;\delta) < \infty$, $\forall \delta > 0$.*

(2) *Assume that K is open convex or compact convex $\subseteq (V, \|\cdot\|)$, then $\omega_{1}^{(\mathcal{F})}(f;\delta)$ is continuous on $\mathbb{R}_{+}$ in δ for $f \in C_{\mathcal{F}}^{U}(K)$.*

(3) *Assume that K is convex, then*

$$\omega_{1}^{(\mathcal{F})}(f,t_{1}+t_{2}) \leq \omega_{1}^{(\mathcal{F})}(f,t_{1}) + \omega_{1}^{(\mathcal{F})}(f,t_{2}), \quad t_{1},t_{2} \geq 0,$$

that is the subadditivity property is true. Also it holds

$$\omega_{1}^{(\mathcal{F})}(f,n\delta) \leq n\omega_{1}^{(\mathcal{F})}(f,\delta),$$

and

$$\omega_{1}^{(\mathcal{F})}(f,\lambda\delta) \leq \lceil\lambda\rceil \omega_{1}^{(\mathcal{F})}(f,\delta) \leq (\lambda+1)\omega_{1}^{(\mathcal{F})}(f,\delta),$$

where $n \in \mathbb{N}$, $\lambda > 0$, $\delta > 0$, $\lceil\cdot\rceil$ is the ceiling of the number.

(4) *Clearly in general $\omega_{1}^{(\mathcal{F})}(f;\delta) \geq 0$ and is increasing in $\delta > 0$ and $\omega_{1}^{(\mathcal{F})}(f;0) = 0$.*

(5) *Let K be open or compact $\subseteq (V, \|\cdot\|)$. Then $\omega_{1}^{(\mathcal{F})}(f;\delta) \to 0$ as $\delta \downarrow 0$ iff $f \in C_{\mathcal{F}}^{U}(K)$.*

(6) *It holds*

$$\omega_1^{(\mathcal{F})}(f \oplus g; \delta) \leq \omega_1^{(\mathcal{F})}(f; \delta) + \omega_1^{(\mathcal{F})}(g; \delta),$$

for $\delta > 0$, any $f, g \colon K \to \mathbb{R}_{\mathcal{F}}$, $K \subseteq (V, \|\cdot\|)$ is arbitrary.

Proof. (1) Here K is open convex. Let here $f \in C_{\mathcal{F}}^U(K)$, iff $\forall \varepsilon > 0$, $\exists \delta > 0 \colon \|x - y\| \leq \delta$ implies $D(f(x), f(y)) < \varepsilon$. Let $\varepsilon_0 > 0$ then $\exists \delta_0 > 0 \colon \|x - y\| \leq \delta_0$ with $D(f(x), f(y)) < \varepsilon_0$, hence $\omega_1(f, \delta_0) \leq \varepsilon_0 < \infty$.

Let $\delta > 0$ arbitrary and $x, y \in K$ such that $\|x - y\| \leq \delta$. Choose $n \in \mathbb{N} \colon n\delta_0 > \delta$, and set $x_i = x + \frac{i}{n}(y - x)$, $0 \leq i \leq n$. Notice that all $x_i \in K$. Then

$$
\begin{aligned}
D(f(x), f(y)) &= D\left(\sideset{}{^*}\sum_{i=0}^{n-1} f(x_i), \sideset{}{^*}\sum_{i=0}^{n-1} f(x_{i+1}) \right) \\
&\leq D(f(x), f(x_1)) + D(f(x_1), f(x_2)) + D(f(x_2), f(x_3)) \\
&\quad + \cdots + D(f(x_{n-1}), f(y)) \\
&\leq n\omega_1^{(\mathcal{F})}(f; \delta_0) \leq n\varepsilon_0 < \infty,
\end{aligned}
$$

Since $\|x_i - x_{i+1}\| = \frac{1}{n}\|x - y\| \leq \frac{1}{n}\delta < \delta_0$. Thus $\omega_1^{(\mathcal{F})}(f; \delta) \leq n\varepsilon_0 < \infty$, proving the claim. If K is compact, then claim is obvious.

(2) Let $x, y \in K$ and let $\|x - y\| \leq t_1 + t_2$, then there exists a point $z \in \overline{xy}$, $z \in K \colon \|x - z\| \leq t_1$ and $\|y - z\| \leq t_2$, where $t_1, t_2 > 0$.

Notice that

$$D(f(x), f(y)) \leq D(f(x), f(z)) + D(f(z), f(y)) \leq \omega_1^{(\mathcal{F})}(f, t_1) + \omega_1^{(\mathcal{F})}(f, t_2).$$

Hence

$$\omega_1^{(\mathcal{F})}(f; t_1 + t_2) \leq \omega_1^{(\mathcal{F})}(f, t_1) + \omega_1^{(\mathcal{F})}(f, t_2),$$

proving (3).

Then by the obvious (4), we get

$$0 \leq \omega_1^{(\mathcal{F})}(f; t_1 + t_2) - \omega_1^{(\mathcal{F})}(f; t_1) \leq \omega_1^{(\mathcal{F})}(f; t_2),$$

and

$$|\omega_1^{(\mathcal{F})}(f; t_1 + t_2) - \omega_1^{(\mathcal{F})}(f; t_1)| \leq \omega_1^{(\mathcal{F})}(f; t_2).$$

Let $f \in C_{\mathcal{F}}^U(K)$, then $\lim_{t_2 \downarrow 0} \omega_1^{(\mathcal{F})}(f, t_2) = 0$ by (5). Hence $\omega_1^{(\mathcal{F})}(f, \cdot)$ is continuous on $\mathbb{R}_+$.

(5) ($\Rightarrow$) Let $\omega_1^{(\mathcal{F})}(f; \delta) \to 0$ as $\delta \downarrow 0$. Then $\forall \varepsilon > 0$, $\exists \delta > 0$ with $\omega_1^{(\mathcal{F})}(f; \delta) \leq \varepsilon$. I.e. $\forall x, y \in K \colon \|x - y\| \leq \delta$ we get $D(f(x), f(y)) \leq \varepsilon$. That is $f \in C_{\mathcal{F}}^U(K)$.

($\Leftarrow$) Let $f \in C_{\mathcal{F}}^U(K)$. Then $\forall \varepsilon > 0$, $\exists \delta > 0 \colon$ whenever $\|x - y\| \leq \delta$, $x, y \in K$, it implies $D(f(x), f(y)) \leq \varepsilon$. I.e. $\forall \varepsilon > 0$, $\exists \delta > 0 \colon \omega_1^{(\mathcal{F})}(f; \delta) \leq \varepsilon$. That is $\omega_1^{(\mathcal{F})}(f; \delta) \to 0$ as $\delta \downarrow 0$.

(6) Notice that

$$D\big(f(x) \oplus g(x), f(y) \oplus g(y)\big) \le D(f(x), f(y)) + D(g(x), g(y)).$$

That is (6) now is clear.   □

We also need

Lemma 10.8. *Let K be a compact subset of the real normed vector space $(V, \|\cdot\|)$ and $f \in C_{\mathcal{F}}(K)$. Then f is a fuzzy bounded function.*

Proof. Let $x_n, x \in K$ such that $x_n \to x$, as $n \to \infty$, then $D(f(x_n), f(x)) \to 0$ by continuity of f. But

$$D(f(x_n), f(x)) = \sup_{r \in [0,1]} \max\{|(f(x_n))_{-}^{(r)} - (f(x))_{-}^{(r)}|, |(f(x_n))_{+}^{(r)} - (f(x))_{+}^{(r)}|\}.$$

Hence $|(f(x_n))_{\pm}^{(r)} - (f(x))_{\pm}^{(r)}| \to 0$, all $0 \le r \le 1$, as $n \to +\infty$. That is $(f(x_n))_{\pm}^{(r)} \to (f(x))_{\pm}^{(r)}$, all $0 \le r \le 1$, as $n \to +\infty$. Hence $(f)_{\pm}^{(r)} \in C(K)$, all $0 \le r \le 1$. Consequently, $(f)_{\pm}^{(r)}$ are bounded over K, all $0 \le r \le 1$. Here

$$D(f(x), \tilde{o}) = \sup_{r \in [0,1]} \max\{|(f(x))_{-}^{(r)}|, |(f(x))_{+}^{(r)}|\}.$$

From basic fuzzy theory we get that

$$(f(x))_{-}^{(0)} \le (f(x))_{-}^{(r)} \le (f(x))_{-}^{(1)},$$

and

$$(f(x))_{+}^{(1)} \le (f(x))_{+}^{(r)} \le (f(x))_{+}^{(0)},$$

for all $r \in [0,1]$ and for any $x \in K$. Thus

$$|(f(x))_{-}^{(r)}| \le \max\{|(f(x))_{-}^{(0)}|, |(f(x))_{-}^{(1)}|\},$$

and

$$|(f(x))_{+}^{(r)}| \le \max\{|(f(x))_{+}^{(0)}|, |(f(x))_{+}^{(1)}|\}, \quad \text{all } 0 \le r \le 1, \text{ for any } x \in K.$$

Therefore

$$D(f(x), \tilde{o}) \le \max\{|(f(x))_{\pm}^{(0)}|, |(f(x))_{\pm}^{(1)}|\} \le \max\{\|f_{\pm}^{(0)}\|_\infty, \|f_{\pm}^{(1)}\|_\infty\} =: M,$$

$\forall x \in K$, where $M \ge 0$, that is proving the claim. I.e. for all $0 \le r \le 1$, $-M \le (f(x))_{\pm}^{(r)} \le M, \forall x \in K \Rightarrow X_{\{-M\}} \le f(x) \le X_{\{M\}}$, where $f(x) \in \mathbb{R}_{\mathcal{F}}$.   □

10.2   Related Results

We start with

Proposition 10.9. *Let U open or compact $\subseteq (M,d)$ metric space, $f \in C_{\mathcal{F}}(U)$. Then $f_{\pm}^{(r)}$ are equicontinuous with respect to $r \in [0,1]$ over U, respectively in $\pm$.*

Proof. Easy. $\qquad\qquad\qquad\qquad\qquad\qquad\qquad\qquad\qquad\qquad\qquad\qquad\square$

For the reverse we have

Proposition 10.10. *Let $f_{-}^{(r)}$, $f_{+}^{(r)}$ be equicontinuous with respect to $r \in [0,1]$ on U-open or compact $\subseteq (M,d)$-metric space, respectively in $\pm$, then $f \in C_{\mathcal{F}}(U)$.*

Proof. We have $\forall \varepsilon > 0, \exists \delta_1 > 0 \colon |f_{-}^{(r)}(x) - f_{-}^{(r)}(x_0)| < \varepsilon, \forall x \in U \colon d(x, x_0) < \delta_1$, all $r \in [0,1]$, $x_0 \in U$.

Similarly, we have $\forall \varepsilon > 0, \exists \delta_2 > 0 \colon |f_{+}^{(r)}(x) - f_{+}^{(r)}(x_0)| < \varepsilon, \forall x \in U \colon d(x, x_0) < \delta_2$, all $r \in [0,1]$.

Taking $\delta := \min(\delta_1, \delta_2) > 0$ we get for

$$D(f(x), f(x_0)) = \sup_{r \in [0,1]} \max\{|f_{-}^{(r)}(x) - f_{-}^{(r)}(x_0)|, |f_{+}^{(r)}(x) - f_{+}^{(r)}(x_0)|\},$$

that if $x \in U$ with $d(x, x_0) < \delta$ then

$$|f_{-}^{(r)}(x) - f_{-}^{(r)}(x_0)|, |f_{+}^{(r)}(x) - f_{+}^{(r)}(x_0)| < \varepsilon,$$

for all $r \in [0,1]$, and so is the max of both over all $r \in [0,1]$. That is $D(f(x), f(x_0)) < \varepsilon$, proving the claim. $\qquad\qquad\qquad\qquad\qquad\square$

We need

Definition 10.11. Let $L \colon C_{\mathcal{F}}(U) \hookrightarrow C_{\mathcal{F}}(U)$, where U is open or compact $\subseteq (M,d)$ metric space, such that

$$L(c_1 f + c_2 g) = c_1 L(f) + c_2 L(g), \quad \forall c_1, c_2 \in \mathbb{R}.$$

We call L a *fuzzy linear operator*.

We give the following example of a fuzzy linear operator, etc.

Definition 10.12. Let $f \colon [0,1] \to \mathbb{R}_{\mathcal{F}}$ be a fuzzy real function. The fuzzy algebraic polynomial defined by

$$B_n^{(\mathcal{F})}(f)(x) = \sum_{k=0}^{n}{}^{*} \binom{n}{k} x^k (1-x)^{n-k} \odot f\left(\frac{k}{n}\right), \quad \forall x \in [0,1],$$

will be called the *fuzzy Bernstein operator*.

We do have

Theorem 10.13 (see p. 642, [66], S. Gal). *If* $f \in C_{\mathcal{F}}([0,1])$, *then*

$$D\big(B_n^{(\mathcal{F})}(f)(x), f(x)\big) \le \frac{3}{2}\omega_1^{(\mathcal{F})}\left(f; \frac{1}{\sqrt{n}}\right), \quad n \in \mathbb{N}, \ \forall x \in [0,1],$$

i.e.,

$$\lim_{n \to +\infty} D^*\big(B_n^{(\mathcal{F})}(f), f\big) = 0,$$

that is $B_n^{(\mathcal{F})} f \to_{n \to +\infty} f$, *fuzzy uniform convergence.*

We also need

Definition 10.14. Let $f, g: U \to \mathbb{R}_{\mathcal{F}}, U \subseteq (M, d)$ metric space. We denote $f \succsim g$, iff $f(x) \succsim g(x), \forall x \in U$, iff $f_+^{(r)}(x) \ge g_+^{(r)}(x)$ and $f_-^{(r)}(x) \ge g_-^{(r)}(x)$, $\forall x \in U, \forall r \in [0,1]$, iff $f_+^{(r)} \ge g_+^{(r)}$ and $f_-^{(r)} \ge g_-^{(r)}, \forall r \in [0,1]$.

We give

Definition 10.15. Let $L: C_{\mathcal{F}}(U) \hookrightarrow C_{\mathcal{F}}(U)$ be a fuzzy linear operator, U open or compact $\subseteq (M, d)$ metric space. We say that L is *positive*, iff whenever $f, g \in C_{\mathcal{F}}(U)$ are such that $f \succsim g$ then $L(f) \succsim L(g)$, iff

$$(L(f))_+^{(r)} \ge (L(g))_+^{(r)}$$

and

$$(L(f))_-^{(r)} \ge (L(g))_-^{(r)}, \quad \forall r \in [0,1].$$

Here we denote

$$[L(f)]^r = \big[(L(f))_-^{(r)}, (L(f))_+^{(r)}\big], \quad \forall r \in [0,1].$$

An example of a fuzzy positive linear operator is the fuzzy Bernstein operator on the domain $[0, 1]$, etc.

For the definition of general fuzzy integral we follow [75] next.

Definition 10.16. Let (Ω, Σ, μ) be a complete σ-finite measure space. We call $F: \Omega \to \mathbb{R}_{\mathcal{F}}$ *measurable* iff $\forall$ closed $B \subseteq \mathbb{R}$ the function $F^{-1}(B): \Omega \to [0, 1]$ defined by

$$F^{-1}(B)(\omega) := \sup_{x \in B} F(\omega)(x), \quad \text{all } \omega \in \Omega$$

is measurable, see [75].

Theorem 10.17 ([75]). *For* $F: \Omega \to \mathbb{R}_{\mathcal{F}}$, $F(\omega) = \{(F_-^{(r)}(\omega), F_+^{(r)}(\omega)) \mid 0 \le r \le 1\}$, *the following are equivalent.*

(1) F *is measurable,*

(2) $\forall r \in [0,1], F_-^{(r)}, F_+^{(r)}$ *are measurable.*

Following [75], given that for each $r \in [0,1]$, $F_-^{(r)}$, $F_+^{(r)}$ are integrable we have that the parametrized representation

$$\left\{ \left(\int_A F_-^{(r)} \, d\mu, \int_A F_+^{(r)} \, d\mu \right) \mid 0 \le r \le 1 \right\}$$

is a fuzzy real number for each $A \in \Sigma$.

The last fact leads to

Definition 10.18 ([75]). A measurable function $F: \Omega \to \mathbb{R}_{\mathcal{F}}$,

$$F(\omega) = \left\{ (F_-^{(r)}(\omega), F_+^{(r)}(\omega)) \mid 0 \le r \le 1 \right\}$$

is called *integrable* if for each $r \in [0,1]$, $F_{\pm}^{(r)}$ are integrable, or equivalently, if $F_{\pm}^{(0)}$ are integrable. In this case, the fuzzy integral of F over $A \in \Sigma$ is defined by

$$\int_A F \, d\mu := \left\{ \left(\int_A F_-^{(r)} \, d\mu, \int_A F_+^{(r)} \, d\mu \right) \mid 0 \le r \le 1 \right\}.$$

By [75], F is integrable iff $\omega \to \|F(\omega)\|_{\mathcal{F}}$ is real-valued integrable.

We need also

Theorem 10.19 ([75]). *Let $F, G: \Omega \to \mathbb{R}_{\mathcal{F}}$ be integrable. Then*

(1) *Let $a, b \in \mathbb{R}$, then $aF + bG$ is integrable and for each $A \in \Sigma$,*

$$\int_A (aF + bG) \, d\mu = a \int_A F \, d\mu + b \int_A G \, d\mu;$$

(2) *$D(F, G)$ is a real-valued integrable function and for each $A \in \Sigma$,*

$$D\left(\int_A F \, d\mu, \int_A G \, d\mu \right) \le \int_A D(F, G) \, d\mu.$$

In particular,

$$\left\| \int_A F \, d\mu \right\|_{\mathcal{F}} \le \int_A \|F\|_{\mathcal{F}} \, d\mu.$$

We need to state the following.

Assumption 10.20. Let L be a fuzzy positive linear operator from $C_{\mathcal{F}}(K)$, K compact $\subseteq (M, d)$ metric space, into itself. Here we *assume* that there exists a positive linear operator $\tilde{L}$ from $C(K)$ into itself with the property

$$(Lf)_{\pm}^{(r)} = \tilde{L}(f_{\pm}^{(r)}),$$

respectively, for all $r \in [0,1]$, $\forall f \in C_{\mathcal{F}}(K)$.

As an example again we mention the fuzzy Bernstein operator and the real Bernstein operator fulfilling the above assumption on $[0,1]$, etc.

Remark 10.21 (following Assumption 10.20). Then by Riesz representation theorem we have

$$\tilde{L}(f_{\pm}^{(r)})(x) = \int_K f_{\pm}^{(r)}(t)\mu_x(dt),$$

(the last is true for all real continuous functions on K) respectively, $\forall r \in [0,1]$, where μ_x is the unique positive finite Borel measure on K, $\forall x \in K$.

Furthermore in the last integral we can take as μ_x the unique positive finite completed Borel measure of the same mass on K as the initial one. Hence

$$\{(\tilde{L}(f_-^{(r)})(x), \tilde{L}(f_+^{(r)})(x)) \mid r \in [0,1]\}$$
$$= \left\{\left(\int_K f_-^{(r)}(t)\mu_x(dt), \int_K f_+^{(r)}(t)\mu_x(dt)\right) \mid r \in [0,1]\right\}$$
$$= \int_K f(t)\mu_x(dt),$$

the general fuzzy integral with respect to μ_x, $\forall x \in K$, see Definition 10.18. But we also have

$$(L(f))(x) = \{((Lf)_-^{(r)}(x), (Lf)_+^{(r)}(x)) \mid r \in [0,1]\}$$
$$= \{(\tilde{L}(f_-^{(r)})(x), \tilde{L}(f_+^{(r)})(x)) \mid r \in [0,1]\}.$$

Based on the above we have proved the following *Fuzzy Riesz Representation Theorem*.

Theorem 10.22. *Let L be a fuzzy positive linear operator from $C_{\mathcal{F}}(K)$ into itself as in Assumption 10.20, K compact $\subseteq (M,d)$ metric space. Then for each $x \in K$ there exists a unique positive finite completed Borel measure μ_x on K such that*

$$(Lf)(x) = \int_K f(t)\mu_x(dt), \quad \forall f \in C_{\mathcal{F}}(K).$$

The other way around follows.

Remark 10.23. Let $\tilde{L}$ be a positive linear operator from $C(K)$ into itself, K compact $\subseteq (M,d)$ metric space. Then by the basic Riesz representation theorem we have

$$(\tilde{L}(g))(x) = \int_K g(t)\mu_x(dt), \quad \forall g \in C(K),$$

where μ_x is a unique positive finite completed Borel measure on K.

Next consider any $f \in C_{\mathcal{F}}(K)$, thus

$$\tilde{L}(f_{\pm}^{(r)})(x) = \int_K f_{\pm}^{(r)}(t)\mu_x(dt), \quad \text{all } r \in [0,1].$$

Then

$$\left\{ (\tilde{L}(f_{-}^{(r)})(x), \tilde{L}(f_{+}^{(r)})(x)) \mid r \in [0,1] \right\}$$

$$= \left\{ \left(\int_K f_{-}^{(r)}(t)\mu_x(dt), \int_K f_{+}^{(r)}(t)\mu_x(dt) \right) \mid r \in [0,1] \right\}$$

$$= \int_K f(t)\mu_x(dt) =: L(f)(x),$$

by Definition 10.18. Clearly by Theorem 10.19, L is a linear operator.

Next let $f \succeq h$; $f, h \in C_{\mathcal{F}}(K)$, iff $f_{\pm}^{(r)} \geq h_{\pm}^{(r)}$, respectively, all $r \in [0,1]$. Thus

$$\int_K f_{\pm}^{(r)}(t)\mu_x(dt) \geq \int_K h_{\pm}^{(r)}(t)\mu_x(dt),$$

respectively, all $r \in [0,1]$. But

$$(Lf)(x) = \left\{ ((Lf)_{-}^{(r)}(x), (Lf)_{+}^{(r)}(x)) \mid r \in [0,1] \right\}.$$

That is by Definition 10.18 we have

$$(Lf)_{\pm}^{(r)}(x) = \int_K f_{\pm}^{(r)}(t)\mu_x(dt),$$

etc. Hence we observe $(Lf)_{\pm}^{(r)} \geq (Lh)_{\pm}^{(r)}$, respectively, all $r \in [0,1]$, i.e. $Lf \succeq Lh$. We have proved that L is a fuzzy positive linear operator.

We get the important conclusion by using also Proposition 10.10.

Theorem 10.24. *Any positive linear operator $\tilde{L}$ from $C(K)$ into itself, K compact $\subseteq (M,d)$ metric space, induces a unique fuzzy positive linear operator L on $C_{\mathcal{F}}(K)$. It holds*

$$(Lf)_{\pm}^{(r)} = \tilde{L}(f_{\pm}^{(r)}), \quad \forall f \in C_{\mathcal{F}}(K), \quad \text{all } r \in [0,1].$$

If additionally $\tilde{L}(f_{\pm}^{(r)})$ are equicontinuous with respect to $r \in [0,1]$, respectively in $\pm$, then $Lf \in C_{\mathcal{F}}(K)$ whenever $f \in C_{\mathcal{F}}(K)$.

10.3   Main Results

We need

Lemma 10.25. *Let $(V, \|\cdot\|)$ be a real normed vector space and Q is a subset of V which is star-shaped relative to its fixed point x_0. Let $f\colon Q \to \mathbb{R}_{\mathcal{F}}$ with $0 < \omega_1^{(\mathcal{F})}(f; \delta) < \infty$, any $\delta > 0$. Then*

$$D(f(t), f(x_0)) \le \left\lceil \frac{\|t - x_0\|}{\delta} \right\rceil \omega_1^{(\mathcal{F})}(f; \delta),$$

for all $\delta > 0$, $\lceil \cdot \rceil$ denotes the ceiling of the number, $\forall t \in Q$.

Clearly Lemma 10.25 is true when Q is convex and we have

Lemma 10.26. *Let $(V, \|\cdot\|)$ be a real normed vector space and $Q \subseteq V$, Q is convex. Let $f\colon Q \to \mathbb{R}_{\mathcal{F}}$ with $0 < \omega_1^{(\mathcal{F})}(f; \delta) < \infty$, any $\delta > 0$. Then*

$$D(f(x), f(y)) \le \left\lceil \frac{\|x - y\|}{\delta} \right\rceil \omega_1^{(\mathcal{F})}(f; \delta), \quad all\ \delta > 0,\ \forall x, y \in Q.$$

In particular we obtain

Lemma 10.27. *Let $f\colon [a, b] \to \mathbb{R}_{\mathcal{F}}$ with $0 < \omega_1^{(\mathcal{F})}(f; \delta) < \infty$, any $\delta > 0$. Then*

$$D(f(x), f(y)) \le \left\lceil \frac{|x - y|}{\delta} \right\rceil \omega_1^{(\mathcal{F})}(f; \delta),$$

all $\delta > 0$, $\forall x, y \in [a, b] \subseteq \mathbb{R}$.

Proof of Lemma 10.25. Let $t \ne x_0\colon \|t - x_0\| < \delta$, i.e. $\left\lceil \frac{\|t - x_0\|}{\delta} \right\rceil = 1$, then

$$D(f(t), f(x_0)) \le \omega_1^{(\mathcal{F})}(f; \delta) =: w = 1 \cdot w.$$

Now for another $t\colon \|t - x_0\| > \delta$ we have that $(n - 1)\delta < \|t - x_0\| \le n\delta$, for some $n \ge 2$ integer. That is

$$\left\lceil \frac{\|t - x_0\|}{\delta} \right\rceil = n.$$

Consider the points

$$t + \frac{j(x_0 - t)}{n} \in Q, \quad j = 0, 1, \ldots, n$$

and see that $\left\| \frac{(t - x_0)}{n} \right\| \le \delta.$

Then observe that

$$
D(f(t), f(x_0)) = D\left(f(t) \oplus \sum_{k=1}^{n-1}{}^* f\left(t + \frac{k(x_0 - t)}{n}\right), \sum_{k=1}^{n-1}{}^* f\left(t + \frac{k(x_0 - t)}{n}\right) \right.
$$

$$
\left. \oplus f(x_0) \right)
$$

$$
\leq D\left(f(t), f\left(t + \left(\frac{x_0 - t}{n}\right)\right) \right)
$$

$$
+ D\left(f\left(t + \frac{(x_0 - t)}{n}\right), f\left(t + \frac{2(x_0 - t)}{n}\right) \right)
$$

$$
+ D\left(f\left(t + \frac{2(x_0 - t)}{n}\right), f\left(t + \frac{3(x_0 - t)}{n}\right) \right)
$$

$$
+ \cdots + D\left(f\left(t + (n-1)\left(\frac{x_0 - t}{n}\right)\right), f(x_0) \right) \leq nw.
$$

Thus

$$
D(f(t), f(x_0)) \leq nw = \left\lceil \frac{\|t - x_0\|}{\delta} \right\rceil \omega_1^{(\mathcal{F})}(f; \delta),
$$

proving the claim.   □

Sometimes it is useful

Lemma 10.28. Let $f: Q \to \mathbb{R}_{\mathcal{F}}$, Q convex $\subseteq (V, \|\cdot\|)$ with $0 < \omega_1^{(\mathcal{F})}(f; \delta) < \infty$, any $\delta > 0$. Then

$$
D(f(x), f(y)) \leq \left(1 + \frac{\|x - y\|}{\delta}\right) \omega_1^{(\mathcal{F})}(f; \delta), \quad \text{all } \delta > 0, \ \forall x, y \in Q.
$$

It holds

Lemma 10.29. Let $f: Q \to \mathbb{R}_{\mathcal{F}}$, Q convex $\subseteq (V, \|\cdot\|)$ with $0 < \omega_1^{(\mathcal{F})}(f; \delta) < \infty$, any $\delta > 0$. Then

$$
D(f(t), f(x)) \leq \left(1 + \frac{\|t - x\|^2}{\delta^2}\right) \omega_1^{(\mathcal{F})}(f; \delta), \quad \forall t, x \in Q, \ \forall \delta > 0.
$$

In particular, let $f: [a, b] \to \mathbb{R}_{\mathcal{F}}$, $[a, b] \subseteq \mathbb{R}$, with $0 < \omega_1^{(\mathcal{F})}(f; \delta) < \infty$, any $\delta > 0$. Then

$$
D(f(t), f(x)) \leq \left(1 + \frac{(t - x)^2}{\delta^2}\right) \omega_1^{(\mathcal{F})}(f; \delta), \quad \forall t, x \in [a, b], \ \forall \delta > 0.
$$

Proof. If $\|t - x\| > \delta$ then $\frac{\|t - x\|}{\delta} > 1$. Thus by Lemma 10.28 we have

$$
D(f(t), f(x)) \leq \left(1 + \frac{\|t - x\|}{\delta}\right) \omega_1^{(\mathcal{F})}(f; \delta) \leq \left(1 + \frac{\|t - x\|^2}{\delta^2}\right) \omega_1^{(\mathcal{F})}(f; \delta).
$$

I.e. when $\|t - x\| > \delta$ we get

$$D(f(t), f(x)) \leq \left(1 + \frac{\|t - x\|^2}{\delta^2}\right) \omega_1^{(\mathcal{F})}(f; \delta).$$

But the last is obviously true when $\|t - x\| \leq \delta$. $\qquad\qquad\square$

Convention 10.30. From now on, unless otherwise stated, the σ-fields we consider will be the power sets of the spaces we are working on to the effect that every real-valued function there is measurable and the measures considered are complete.

We need

Definition 10.31. Let K be a compact convex $\subseteq (V, \|\cdot\|)$ real normed vector space. Let $(\mu_n)_{n\in\mathbb{N}}$ be a sequence of finite positive measures on K. We say that μ_n converges *fuzzy weakly* to Dirac measure δ_{x_0}, $x_0 \in K$, iff

$$\lim_{n\to\infty} D\left(\int_K f \, d\mu_n, f(x_0)\right) = 0, \quad \forall f \in C_{\mathcal{F}}(K).$$

We denote it by

$$\mu_n \overset{\mathcal{F}}{\rightrightarrows} \delta_{x_0} \quad \text{as } n \to \infty.$$

Clearly real weak convergence cannot imply fuzzy weak convergence.

We study here the degree of fuzzy weak convergence with rates.

We present the first main result.

Theorem 10.32. *Let K be a convex and compact subset of the real vector space $(V, \|\cdot\|)$. Let $x_0 \in K$ fixed and μ a finite measure on K with $\mu(K) = m > 0$. Let also $f \in C_{\mathcal{F}}(K)$. Then*

$$D\left(\int_K f(t)\mu(dt), f(x_0)\right) \tag{10.1}$$

$$\leq |m - 1| D(f(x_0), \tilde{o}) + \omega_1^{(\mathcal{F})}(f; \delta) \left(\int_K \left\lceil \frac{\|t - x_0\|}{\delta} \right\rceil \mu(dt)\right),$$

for all $0 < \delta \leq \mathrm{diam}(K)$. When $m = 1$, that is when μ is a probability measure on K we get

$$D\left(\int_K f(t)\mu(dt), f(x_0)\right) \leq \omega_1^{(\mathcal{F})}(f; \delta) \left(\int_K \left\lceil \frac{\|t - x_0\|}{\delta} \right\rceil \mu(dt)\right), \tag{10.2}$$

for all $0 < \delta \leq \mathrm{diam}(K)$.

Proof. By Lemma 10.8 we have that

$$D(f(x), \tilde{o}) \leq M, \quad \forall x \in K, \quad M \geq 0,$$

so that f is integrable. Also we see that

$$
\begin{aligned}
m \odot f(x_0) &= \left\{ (m f(x_0)_-^{(r)}, m f(x_0)_+^{(r)}) \mid 0 \leq r \leq 1 \right\} \\
&= \left\{ \left(\int_K (f(x_0))_-^{(r)} \, d\mu, \int_K (f(x_0))_+^{(r)} \, d\mu \right) \mid 0 \leq r \leq 1 \right\} \\
&= \int_K f(x_0) \, d\mu.
\end{aligned}
$$

We then observe

$$
D\left(\int_K f(t) \mu(dt), f(x_0) \right)
$$

$$
\leq D\left(\int_K f(t) \mu(dt), f(x_0) \odot m \right) + D\big(f(x_0) \odot m, f(x_0) \big)
$$

$$
= D\left(\int_K f(t) \mu(dt), \int_K f(x_0) \mu(dt) \right) + D\big(f(x_0) \odot m, f(x_0) \big)
$$

$$
\leq \int_K D\big(f(t), f(x_0) \big) \mu(dt) + |m - 1| D(f(x_0), \tilde{o})
$$

(the last comes by Theorem 10.19, and Lemma 1.2, respectively)

$$
\leq \left(\int_K \left\lceil \frac{\|t - x_0\|}{\delta} \right\rceil \mu(dt) \right) \omega_1^{(\mathcal{F})}(f; \delta) + |m - 1| D(f(x_0), \tilde{o}),
$$

$0 < \delta \leq \operatorname{diam}(K)$ (the last comes by Lemma 10.26). We have established
(10.1) and (10.2). □

Remark 10.33. 1) By the use of geometric moment theory methods
(method of optimal distance, see [74], [4]) we can find best upper bounds for
various interesting cases to $\int_K \left\lceil \frac{\|t - x_0\|}{\delta} \right\rceil \mu(dt)$ subject to the given moment
condition

$$
\left(\int_K \|t - x_0\|^r \mu(dt) \right)^{1/r} = D_r(x_0). \tag{10.3}
$$

Here $r > 0$, $D_r(x_0) > 0$ are given. For the existence of μ we assume
$D_r(x_0) \leq m^{1/r} \operatorname{diam}(K)$. I.e. to find the optimal quantity

$$
K(x_0) := \frac{1}{m} \sup_\mu \left(\int_K \left\lceil \frac{\|t - x_0\|}{\delta} \right\rceil \mu(dt) \right) w
$$

over all measures μ with $\mu(K) = m$ that fulfill (10.3). Here we assume that
for fixed $0 < \delta \leq \operatorname{diam}(K)$ we have that

$$
\omega_1^{(\mathcal{F})}(f; \delta) \leq w,
$$

where $w > 0$ is given. The precise calculation of quantity $K(x_0)$ is given in
Theorem 7.2.1, pp. 212–213 of [4] for all cases.

2) Consider $i_{\mathcal{F}} \in \mathbb{R}_{\mathcal{F}}$ such that

$$i_{\mathcal{F}} := \left\{ (i_-^{(r)}, i_+^{(r)}) \mid 0 \le r \le 1 \text{ with all } i_\pm^{(r)} = 1 \right\},$$

actually $i_{\mathcal{F}} = \mathcal{X}_{\{1\}}$. Then the fuzzy integral

$$\int_K \left\lceil \frac{\|t - x_0\|}{\delta} \right\rceil \odot i_{\mathcal{F}} d\mu$$

$$= \left\{ \left(\int_K \left\lceil \frac{\|t - x_0\|}{\delta} \right\rceil i_-^{(r)} d\mu, \int_K \left\lceil \frac{\|t - x_0\|}{\delta} \right\rceil i_+^{(r)} d\mu \right) \mid 0 \le r \le 1 \right\}$$

$$= \left\{ \left(\int_K \left\lceil \frac{\|t - x_0\|}{\delta} \right\rceil d\mu, \int_K \left\lceil \frac{\|t - x_0\|}{\delta} \right\rceil d\mu \right) \mid 0 \le r \le 1 \right\}$$

$$= \left(\int_K \left\lceil \frac{\|t - x_0\|}{\delta} \right\rceil d\mu \right) \odot \{(1,1) \mid 0 \le r \le 1\} = \left(\int_K \left\lceil \frac{\|t - x_0\|}{\delta} \right\rceil d\mu \right)$$

$$\odot i_{\mathcal{F}}.$$

I.e. we got that

$$\int_K \left\lceil \frac{\|t - x_0\|}{\delta} \right\rceil \odot i_{\mathcal{F}} d\mu = \left(\int_K \left\lceil \frac{\|t - x_0\|}{\delta} \right\rceil d\mu \right) \odot i_{\mathcal{F}}.$$

Also notice that $D(i_{\mathcal{F}}, \tilde{o}) = 1$. Then the equality sign in (10.1) cannot be attained but it can be arbitrarily closely approached by the close approximations to $\hat{f}(t) := w \left\lceil \frac{\|t - x_0\|}{\delta} \right\rceil \odot i_{\mathcal{F}}$, where $w > 0$ is a prescribed value of $\omega_1^{(\mathcal{F})}(\cdot; \delta)$.

We give

Corollary 10.34 (to Theorem 10.32). *We get that*

$$D \left(\int_K f(t)\mu(dt), f(x_0) \right) \le |m - 1| D(f(x_0), \tilde{o})$$

$$+ \omega_1^{(\mathcal{F})}(f; \delta) \left(m + \frac{1}{\delta} \int_K \|t - x_0\| \mu(dt) \right). \tag{10.4}$$

For $r \ge 1$ we obtain

$$D \left(\int_K f(t)\mu(dt), f(x_0) \right) \le |m - 1| D(f(x_0), \tilde{o})$$

$$+ \omega_1^{(\mathcal{F})}(f; \delta) \left(m + \frac{D_r(x_0)}{\delta} m^{1 - \left(\frac{1}{r} \right)} \right), \tag{10.5}$$

and

$$D \left(\int_K f(t)\mu(dt), f(x_0) \right) \le |m - 1| D(f(x_0), \tilde{o})$$

$$+ \omega_1^{(\mathcal{F})} \left(f; \left(\int_K \|t - x_0\|^r \mu(dt) \right)^{1/r} \right) m(1 + m^{-1/r}). \tag{10.6}$$

When $r = 2$ we get

$$D\left(\int_K f(t)\mu(dt), f(x_0)\right) \leq |m - 1|D(f(x_0), \tilde{o})$$

$$+ \omega_1^{(\mathcal{F})}\left(f; \left(\int_K \|t - x_0\|^2\mu(dt)\right)^{1/2}\right)(m + \sqrt{m}). \quad (10.7)$$

Proof. Obvious. $\square$

We further present

Theorem 10.35. *Same assumptions as in Theorem 10.32. Then*

$$D\left(\int_K f(t)\mu(dt), f(x_0)\right) \leq |m - 1|D(f(x_0), \tilde{o})$$

$$+ \omega_1^{(\mathcal{F})}(f; \delta)\left(m + \frac{\int_K \|t - x\|^2\mu(dt)}{\delta^2}\right), \quad (10.8)$$

for all $0 < \delta \leq diam(K)$.
  If

$$\delta = \left(\int_K \|t - x\|^2\mu(dt)\right)^{1/2}$$

then it holds

$$D\left(\int_K f(t)\mu(dt), f(x_0)\right) \leq |m - 1|D(f(x_0), \tilde{o})$$

$$+ \omega_1^{(\mathcal{F})}\left(f; \left(\int_K \|t - x\|^2\mu(dt)\right)^{1/2}\right)(m + 1). \quad (10.9)$$

For $m = 1$ we get

$$D\left(\int_K f(t)\mu(dt), f(x_0)\right) \leq 2\omega_1^{(\mathcal{F})}\left(f; \left(\int_K \|t - x\|^2\mu(dt)\right)^{1/2}\right). \quad (10.10)$$

Corollary 10.36 (to Theorems 10.32, 10.35). *It holds*

$$D\left(\int_K f(t)\mu(dt), f(x_0)\right) \leq |m - 1|D(f(x_0), \tilde{o})$$

$$+ \omega_1^{(\mathcal{F})}\left(f; \left(\int_K \|t - x_0\|^2\mu(dt)\right)^{1/2}\right)\min\{(m + \sqrt{m}),$$

$$(m + 1)\}. \quad (10.11)$$

In case of $K = [a, b] \subseteq \mathbb{R}$ *we have*

$$D\left(\int_{[a,b]} f(t)\mu(dt), f(x_0)\right) \leq |m - 1|D(f(x_0), \tilde{o})$$

$$+ \omega_1^{(\mathcal{F})}\left(f; \left(\int_{[a,b]}(t - x_0)^2\mu(dt)\right)^{1/2}\right) \min\{(m + \sqrt{m}),$$

$$(m + 1)\}. \tag{10.12}$$

Clearly by (10.11) or (10.12), as $m \to 1$ *and* $\int_K \|t - x_0\|^2\mu(dt) \to 0$ *or* $\int_{[a,b]}(t - x_0)^2\mu(dt) \to 0$, *we get the degree of approximation of fuzzy weak convergence* $\mu \overset{\mathcal{F}}{\rightrightarrows} \delta_{x_0}$ *with rates, respectively.*

Proof of Theorem 10.35. Most here are as in the proof of Theorem 10.32. Furthermore, we make use of Lemma 10.29 and we have

$$D\left(\int_K f(t)\mu(dt), f(x_0)\right)$$

$$\leq \int_K D(f(t), f(x_0))\mu(dt) + |m - 1|D(f(x_0), \tilde{o})$$

$$\leq |m - 1|D(f(x_0), \tilde{o}) + \left(\int_K \left(1 + \frac{\|t - x\|^2}{\delta^2}\right)\mu(dt)\right)\omega_1^{(\mathcal{F})}(f; \delta)$$

$$= |m - 1|D(f(x_0), \tilde{o}) + \omega_1^{(\mathcal{F})}(f; \delta)\left(m + \frac{\int_K \|t - x\|^2\mu(dt)}{\delta^2}\right).$$

$\square$

Remark 10.37. (1) Theorems 10.32, 10.35 and Corollaries 10.34, 10.36 are true also when μ is the completed Borel measure, same proofs.

(2) Theorems 10.32, 10.35 and Corollaries 10.34, 10.36 are true also when *f is integrable in the fuzzy sense*, see Definition 10.18. So we do not need always to assume that $f \in C_{\mathcal{F}}(K)$. Also we do not need to assume always that K is compact. Again same proofs.

Based on Theorem 10.22 and Remark 10.37(1) we present the following *fuzzy Shisha–Mond inequalities* result.

Theorem 10.38. *Let* K *be a convex and compact subset of the real normed vector space* $(V, \|\cdot\|)$. *Let* L *be a fuzzy positive linear operator from* $C_{\mathcal{F}}(K)$ *into itself with the property that there exists positive linear operator* $\tilde{L}$ *from* $C(K)$ *into itself with* $(Lf)_{\pm}^{(r)} = \tilde{L}(f_{\pm}^{(r)})$, *respectively for all* $r \in [0, 1]$, $\forall f \in C_{\mathcal{F}}(K)$. *Then*

$$D(L(f)(x), f(x)) \leq |\tilde{L}(1)(x) - 1|D(f(x), \tilde{o}) \tag{10.13}$$

$$+ \omega_1^{(\mathcal{F})}\left(f; ((\tilde{L}(\|\cdot - x\|^2))(x))^{1/2}\right) \min\{(\tilde{L}(1)(x) + \sqrt{\tilde{L}(1)(x)}),$$

$$(\tilde{L}(1)(x) + 1)\}.$$

Furthermore we get

$$D^*(Lf, f) \leq D^*(f, \tilde{o}) \|\tilde{L}1 - 1\|_\infty \qquad (10.14)$$
$$+ \min\{\|\tilde{L}(1) + \sqrt{\tilde{L}(1)}\|_\infty, \|\tilde{L}(1) + 1\|_\infty\} \omega_1^{(\mathcal{F})}(f; \|(\tilde{L}(\| \cdot - x\|^2))(x)\|_\infty^{1/2}),$$

and $\| \cdot \|_\infty$ stands for the sup-norm over K. In particular, if $\tilde{L}(1) = 1$ then (10.14) reduces to

$$D^*(Lf, f) \leq 2\omega_1^{(\mathcal{F})}(f; \|(\tilde{L}(\| \cdot - x\|^2))(x)\|_\infty^{1/2}). \qquad (10.15)$$

When $K = [a, b] \subseteq \mathbb{R}$ then we obtain

$$D^*(Lf, f) \leq D^*(f, \tilde{o}) \|\tilde{L}(1) - 1\|_\infty \qquad (10.16)$$
$$+ \min\{\|\tilde{L}(1) + \sqrt{\tilde{L}(1)}\|_\infty,$$
$$\|\tilde{L}(1) + 1\|_\infty\} \omega_1^{(\mathcal{F})}(f; \|(\tilde{L}((\cdot - x)^2))(x)\|_\infty^{1/2}).$$

Here one has that (see [96])

$$\|(\tilde{L}((\cdot - x)^2))(x)\|_\infty \leq \|\tilde{L}(t^2)(x) - x^2\|_\infty + 2c\|L(t)(x) - x\|_\infty$$
$$+ c^2 \|L(1)(x) - 1\|_\infty, \qquad (10.17)$$

where $c := \max(|a|, |b|)$.

Consequently we get the following *fuzzy Korovkin theorem*, see also [78].

Theorem 10.39. *Let $(L_n)_{n \in \mathbb{N}}$ be a sequence of fuzzy positive linear operators from $C_{\mathcal{F}}([a, b])$ into itself, $[a, b] \subseteq \mathbb{R}$, with the property that there exists a sequence $(\tilde{L}_n)_{n \in \mathbb{N}}$ of positive linear operators from $C([a, b])$ into itself with $(L_n f)_\pm^{(r)} = \tilde{L}_n(f_\pm^{(r)})$, respectively for all $r \in [0, 1]$, $\forall f \in C_{\mathcal{F}}([a, b])$, $\forall n \in \mathbb{N}$. Furthermore assume that*

$$\tilde{L}_n(1) \overset{u}{\to} 1, \quad \tilde{L}_n(id) \overset{u}{\to} id, \quad \tilde{L}_n(id^2) \overset{u}{\to} id^2, \quad as \; n \to +\infty,$$

where id is the identity map. Then $D^*(L_n f, f) \to 0$ as $n \to \infty$, $\forall f \in C_{\mathcal{F}}([a, b])$. Inequality (10.16) gives above convergence quantitatively and with rates.

Proof. By Theorem 10.38, $\tilde{L}_n(1)$, $n \in \mathbb{N}$ being bounded and the other assumptions. $\qquad \square$

Example 10.40. Applying (10.16) we obtain

$$D^*(B_n^{(\mathcal{F})} f, f) \leq 2\omega_1^{(\mathcal{F})}\left(f; \frac{1}{2\sqrt{n}}\right), \quad \forall f \in C_{\mathcal{F}}([0, 1]), \quad \forall n \in \mathbb{N}. \qquad (10.18)$$

Let now f be of Lipschitz type i.e.

$$D(f(x), f(y)) \leq M|x - y|, \quad M > 0, \quad \forall x, y \in [0, 1].$$

Then by (10.18) we get the improved

$$D^*(B_n^{(\mathcal{F})}f, f) \le \frac{M}{\sqrt{n}}, \tag{10.19}$$

while by Theorem 10.13 we only produce

$$D^*(B_n^{(\mathcal{F})}f, f) \le \frac{3M}{2\sqrt{n}}.$$

We present now the more general fuzzy Korovkin type result.

Theorem 10.41. *Let $(L_n)_{n\in\mathbb{N}}$ be a sequence of fuzzy positive linear operators from $C_{\mathcal{F}}(K)$ into itself, where K is a convex compact subset of the real normed vector space $(V, \|\cdot\|)$. It has the property that there exists a sequence $(\tilde{L}_n)_{n\in\mathbb{N}}$ of positive linear operators from $C(K)$ into itself with $(L_n f)_{\pm}^{(r)} = \tilde{L}_n(f_{\pm}^{(r)})$, respectively, for all $r \in [0,1], \forall f \in C_{\mathcal{F}}(K), \forall n \in \mathbb{N}$. Furthermore assume that*

$$\tilde{L}_n 1 \stackrel{u}{\to} 1 \quad and \quad \left\|(\tilde{L}_n(\|\cdot -x\|^2))(x)\right\|_\infty \to 0, \quad as\ n \to \infty.$$

Then $D^*(L_n f, f) \to 0$ as $n \to +\infty, \forall f \in C_{\mathcal{F}}(K)$. Inequality (10.14) gives above convergence quantitatively and with rates.

Proof. By Theorem 10.38 and that $w_1^{(\mathcal{F})}(f, \delta) \to 0$ as $\delta \to 0$. $\qquad\square$

Example 10.42. Let K convex compact $\subseteq (V, \|\cdot\|)$-Banach space. Let $x \in K$ and $\{x_1, \ldots, x_n\}$ $(n \in \mathbb{N})$ be a finite set of distinct points in K. The *fuzzy second Shepard metric interpolation operator* is defined by ($f \in C_{\mathcal{F}}(K)$)

$$S_n^2(f; x) := \quad S^2\{x_1, \ldots, x_n\}(f, x)$$

$$:= \begin{cases} \sum_{i=1}^{n}{}^* f(x_i) \odot \dfrac{\prod\limits_{j=1, j\neq i}^{n} \|x - x_j\|^2}{\sum\limits_{\ell=1}^{n} \prod\limits_{k=1, k\neq\ell}^{n} \|x - x_k\|^2}, & \text{if } x \notin \{x_1, \ldots, x_n\}, \\[6pt] & \hspace{3.5em}(10.20) \\ f(x), & \text{if } x \in \{x_1, \ldots, x_n\}. \end{cases}$$

Obviously S_n^2 is a fuzzy positive linear operator.

Notice $S_n^2(f; x_i) = f(x_i)$, all $i = 1, \ldots, n$. Actually we have that $S_n^2(f) \in C_{\mathcal{F}}(K)$. It also fulfills

$$((S_n^2(f))(x))_{\pm}^{(r)} = \tilde{S}_n^2(f_{\pm}^{(r)})(x), \quad \forall r \in [0,1],$$

where $\tilde{S}_n^2$ is the corresponding real valued positive linear operator, see [68], [95], [4]. We have $\tilde{S}_n^2(1)(x) = 1$ for all $x \in K$. Furthermore it holds

$$\left((\tilde{S}_n^2(\|t - x\|^2))(x)\right)^{1/2} = n^{1/2} \left(\sum_{i=1}^{n} \|x - x_i\|^{-2}\right)^{-1/2}. \tag{10.21}$$

We then apply (10.13) of Theorem 10.38: Let $x \in K$ then

$$D((S_n^2 f)(x), f(x)) \leq 2\omega_1^{(\mathcal{F})} \left(f; n^{1/2} \left(\sum_{i=1}^{n} \|x - x_i\|^{-2} \right)^{-1/2} \right). \quad (10.22)$$

Consequently we get finally:

$$D^*(S_n^2 f, f) \leq 2\omega_1^{(\mathcal{F})} \left(f; n^{1/2} \left\| \left(\sum_{i=1}^{n} \|x - x_i\|^{-2} \right)^{-1/2} \right\|_{\infty} \right), \quad \forall n \in \mathbb{N}. \quad (10.23)$$

We give

Theorem 10.43. *Let K be a convex subset of the real normed vector space $(V, \| \cdot \|)$. Let $x_0 \in K$ be fixed and μ a completed Borel finite measure on K with $\mu(K) = m > 0$. Let also $f \in C_{\mathcal{F}}^B(K)$, i.e. f is fuzzy continuous and bounded on K. Then*

$$D \left(\int_K f(t) \mu(dt), f(x_0) \right) \leq |m - 1| D(f(x_0), \tilde{o}) \quad (10.24)$$

$$+ \omega_1^{(\mathcal{F})} \left(f; \left(\int_K \|t - x_0\|^2 d\mu(t) \right)^{1/2} \right) \min\{m + 1,$$

$$m + \sqrt{m}\}.$$

If $K = \mathbb{R}$ or $\mathbb{R}_+$ we obtain

$$D \left(\int_{\mathbb{R} \text{ or } \mathbb{R}_+} f(t) \mu(dt), f(x_0) \right) \leq |m - 1| D(f(x_0), \tilde{o}) \quad (10.25)$$

$$+ \omega_1^{(\mathcal{F})} \left(f; \left(\int_{\mathbb{R} \text{ or } \mathbb{R}_+} (t - x_0)^2 d\mu(t) \right)^{1/2} \right) \min\{m + 1,$$

$$m + \sqrt{m}\},$$

for any $f \in C_{\mathcal{F}}^B(\mathbb{R})$ or $f \in C_{\mathcal{F}}^B(\mathbb{R}_+)$, respectively. Inequalities (10.24) and (10.25) are still valid if the related σ-field is the power set of K.

Proof. Clearly f is integrable. Again we have

$$m \odot f(x_0) = \int_K f(x_0) \, d\mu.$$

We then observe

$$D\left(\int_K f(t)\mu(dt), f(x_0)\right)$$

$$\leq D\left(\int_K f(t)\mu(dt), f(x_0) \odot m\right) + D(f(x_0) \odot m, f(x_0))$$

$$= D\left(\int_K f(t)\mu(dt), \int_K f(x_0)\mu(dt)\right) + D(f(x_0) \odot m, f(x_0))$$

(by Theorem 10.19 and Lemma 1.2)
$$\leq \int_K D(f(t), f(x_0))\mu(dt)$$

$$+|m-1|D(f(x_0), \tilde{o})$$

(by Lemmas 10.28, 10.29)
$$\leq \omega_1^{(\mathcal{F})}(f;\delta) \min\left\{\int_K \left(1 + \frac{\|t-x_0\|}{\delta}\right)d\mu(t),\right.$$

$$\left.\int_K \left(1 + \frac{\|t-x_0\|^2}{\delta^2}\right)d\mu(t)\right\}$$

$$+|m-1|D(f(x_0), \tilde{o})$$

$$= |m-1|D(f(x_0), \tilde{o}) + \min\left\{m + \frac{1}{\delta}\int_K \|t-x_0\|d\mu(t),\right.$$

$$\left. m + \frac{1}{\delta^2}\int_K \|t-x_0\|^2 d\mu(t)\right\}\omega_1^{(\mathcal{F})}(f;\delta)$$

$$\leq |m-1|D(f(x_0), \tilde{o}) + \omega_1^{(\mathcal{F})}(f;\delta) \min\left\{m + \frac{1}{\delta}\left(\int_K \|t-x_0\|^2 d\mu(t)\right)^{1/2}\right.$$

$$\left.\sqrt{m}, m + \frac{1}{\delta^2}\int_K \|t-x_0\|^2 d\mu(t)\right\}$$

$$\left(\text{setting } \delta := \left(\int_K \|t-x_0\|^2 d\mu(t)\right)^{1/2}\right)$$

$$= |m-1|D(f(x_0), \tilde{o}) + \omega_1^{(\mathcal{F})}(f;\delta) \min\{m+1, m+\sqrt{m}\},$$

proving the claim.   □

We present

Application 10.44. We consider here $f \in C_{\mathcal{F}}^B(\mathbb{R}_+)$. For $t \geq 0$ and $g \in C_B(\mathbb{R}_+)$ the real Szasz–Mirakjan operator is defined as

$$(\tilde{M}_n g)(t) := e^{-nt} \sum_{k=0}^{\infty} g\left(\frac{k}{n}\right) \frac{(nt)^k}{k!}, \tag{10.26}$$

and the real Baskakov-type operator is defined as

$$(\tilde{V}_n g)(t) := \sum_{k=0}^{\infty} g\left(\frac{k}{n}\right) \binom{n+k-1}{k} \frac{t^k}{(1+t)^{n+k}} .$$ (10.27)

Consider here X_j real independently and identically distributed random variables and put $S_n := \sum_{j=1}^{n} X_j$, $n \in \mathbb{N}$. Let E denote the expectation operator. Then $\tilde{M}_n$ and $\tilde{V}_n$ operators are of the form $E(g(s_n/n))$. In one case the random variable X has the distribution $P_X := e^{-t} \sum_{k=0}^{\infty} \frac{t^k}{k!} \delta_k$ and in the other

$$P_X := \sum_{k=0}^{\infty} \left(\frac{1}{1+t}\right) \left(\frac{t}{1+t}\right)^k \delta_k,$$

that is the Poisson and the geometric distribution, respectively, where δ_k denotes the Dirac measure at k. In both cases, $E(X) = t$, while the variance $\mathrm{Var}(X) = t$ and $\mathrm{Var}(X) = (t + t^2)$, respectively. Clearly the standard deviation $\sigma_{s_n/n}$ in Poisson case is $\sqrt{\frac{t}{n}}$ and in geometric case is $\sqrt{\frac{t+t^2}{n}}$.

Among others we will apply here Theorem 10.43, its inequality (10.25). We will see that the related measure μ here is the distribution $F_{s_n/n}$. Naturally, we define the *fuzzy Szasz–Mirakjan operator* $(f \in C_{\mathcal{F}}^B(\mathbb{R}_+))$,

$$(M_n f)(t) := e^{-nt} \odot \sum_{k=0}^{\infty}{}^* f\left(\frac{k}{n}\right) \odot \frac{(nt)^k}{k!} ,$$ (10.28)

and the *fuzzy Baskakov-type operator*

$$(V_n f)(t) := \sum_{k=0}^{\infty}{}^* f\left(\frac{k}{n}\right) \odot \binom{n+k-1}{k} \frac{t^k}{(1+t)^{n+k}} , \quad t \geq 0.$$ (10.29)

Clearly for $\tilde{M}_n$ operator the corresponding probability measure μ is

$$\mu_1 = e^{-nt} \sum_{k=0}^{\infty} \frac{(nt)^k}{k!} \delta_{k/n},$$ (10.30)

and for $\tilde{V}_n$ operator the probability measure μ is

$$\mu_2 = \sum_{k=0}^{\infty} \binom{n+k-1}{k} \frac{t^k}{(1+t)^{n+k}} \delta_{k/n}.$$ (10.31)

Hence for $f \in C_{\mathcal{F}}^B(\mathbb{R}_+)$ we have

$$(\tilde{M}_n f_{\pm}^{(r)})(t) = \int_{\mathbb{R}_+} f_{\pm}^{(r)} \mu_1(dt)$$ (10.32)

and

$$(\tilde{V}_n f_\pm^{(r)})(t) = \int_{\mathbb{R}_+} f_\pm^{(r)} \mu_2(dt), \quad \forall r \in [0,1], \tag{10.33}$$

respectively, and these real integrals exist, i.e. for every $r \in [0,1]$ the functions $f_\pm^{(r)}$ are integrable with respect to μ_i, $i = 1, 2$. Then it follows from [67] and the *Lebesgue-dominated convergence theorem* that the parametrized representation

$$\left\{ \left(\int_A f_-^{(r)} d\mu_i, \int_A f_+^{(r)} d\mu_i \right) \mid 0 \le r \le 1 \right\}$$

is a fuzzy real number for each $A \subseteq \mathbb{R}_+$, for $i = 1, 2$. Thus by Definition 10.18 we get that $\int_{\mathbb{R}_+} f \, d\mu_i$, $i = 1, 2$ exist as fuzzy real numbers.

From Theorem 10.45 next we obtain

$$|(\tilde{M}_n g)(t)| \le |g(t)| + 2\omega_1 \left(g; \sqrt{\frac{t}{n}} \right), \tag{10.34}$$

$$|(\tilde{V}_n g)(t)| \le |g(t)| + 2\omega_1 \left(g; \sqrt{\frac{t + t^2}{n}} \right), \tag{10.35}$$

$\forall g \in C_B(\mathbb{R}_+)$, $t > 0$, where ω_1 is the real modulus of continuity. I.e. $(\tilde{M}_n g)(t)$, $(\tilde{V}_n g)(t)$ as infinite series converge, $\forall n \in \mathbb{N}$. Hence easily we see that $\tilde{M}_n$, $\tilde{V}_n$ are positive linear operators. Here for $f \in C_\mathcal{F}^B(\mathbb{R}_+)$ we get that $f_\pm^{(r)}$ are equicontinuous, respectively, and uniformly bounded in $r \in [0,1]$ over $\mathbb{R}_+$.

We notice by Proposition 10.3 that

$$\begin{aligned} |(\tilde{M}_n f_\pm^{(r)})(t)| &\le |f_\pm^{(r)}(t)| + 2\omega_1 \left(f_\pm^{(r)}; \sqrt{\frac{t}{n}} \right) \\ &\le D^*(f, \tilde{o}) + 2\omega_1^{(\mathcal{F})} \left(f; \sqrt{\frac{t}{n}} \right) =: M_1(f) \end{aligned}$$

and

$$\begin{aligned} |(\tilde{V}_n f_\pm^{(r)})(t)| &\le |f_\pm^{(r)}(t)| + 2\omega_1 \left(f_\pm^{(r)}; \sqrt{\frac{t + t^2}{n}} \right) \\ &\le D^*(f, \tilde{o}) + 2\omega_1^{(\mathcal{F})} \left(f; \sqrt{\frac{t + t^2}{n}} \right) =: M_2(f), \quad \forall r \in [0,1]. \end{aligned}$$

I.e.

$$\begin{aligned} |(\tilde{M}_n f_\pm^{(r)})(t)| &\le M_1(f), \\ |(\tilde{V}_n f_\pm^{(r)})(t)| &\le M_2(f), \end{aligned} \tag{10.36}$$

where the constants $M_1(f)$, $M_2(f) \geq 0$, $\forall r \in [0, 1]$.

So as infinite series $(\tilde{M}_n f_{\pm}^{(r)})(t)$, $(\tilde{V}_n f_{\pm}^{(r)})(t)$ converge $\forall r \in [0, 1]$, for any $n \in \mathbb{N}$. For convenience call the weights $w_{kn}(t) := e^{-nt} \frac{(nt)^k}{k!} > 0$ and

$$\theta_{kn}(t) := \binom{n+k-1}{k} \frac{t^k}{(1+t)^{n+k}} > 0, \quad t > 0, \quad k, n \in \mathbb{N},$$

so that

$$(\tilde{M}_n g)(t) = \sum_{k=0}^{\infty} g\left(\frac{k}{n}\right) w_{kn}(t),$$

and

$$(\tilde{V}_n g)(t) = \sum_{k=0}^{\infty} g\left(\frac{k}{n}\right) \theta_{kn}(t), \quad \forall g \in C_B(\mathbb{R}_+).$$

We notice for $f \in C_{\mathcal{F}}^B(\mathbb{R}_+)$ the following by basic properties of fuzzy numbers, see [67]:

$$(f(x))_{-}^{(0)} \leq (f(x))_{-}^{(r)} \leq (f(x))_{-}^{(1)} \leq (f(x))_{+}^{(1)} \leq (f(x))_{+}^{(r)} \leq (f(x))_{+}^{(0)}, \quad \forall x \in \mathbb{R}_+.$$

We get that

$$
\begin{aligned}
|(f(x))_{-}^{(r)}| &\leq \max\{|(f(x))_{-}^{(0)}|, |(f(x))_{-}^{(1)}|\} \\
&= \frac{1}{2}\left(|(f(x))_{-}^{(0)}| + |(f(x))_{-}^{(1)}| - \left||(f(x))_{-}^{(0)}| - |(f(x))_{-}^{(1)}|\right|\right) \\
&=: A_{-}^{0,1}(x) \in C_B(\mathbb{R}_+).
\end{aligned}
$$

Also it holds

$$
\begin{aligned}
|(f(x))_{+}^{(r)}| &\leq \max\{|(f(x))_{+}^{(0)}|, |(f(x))_{+}^{(1)}|\} \\
&= \frac{1}{2}\left(|(f(x))_{+}^{(0)}| + |(f(x))_{+}^{(1)}| - \left||(f(x))_{+}^{(0)}| - |(f(x))_{+}^{(1)}|\right|\right) \\
&=: A_{+}^{0,1}(x) \in C_B(\mathbb{R}_+).
\end{aligned}
$$

I.e. we have obtained that

$$
\begin{aligned}
0 &\leq |(f(x))_{-}^{(r)}| \leq A_{-}^{0,1}(x), \\
0 &\leq |(f(x))_{+}^{(r)}| \leq A_{+}^{0,1}(x), \quad \forall r \in [0, 1], \forall x \in \mathbb{R}_+. \quad (10.37)
\end{aligned}
$$

Consequently by (10.36) we have

$$
\begin{aligned}
0 &\leq |(\tilde{M}_n(A_{\pm}^{0,1}))(t)| \leq M_1(A_{\pm}^{0,1}) < \infty, \\
0 &\leq |(\tilde{V}_n(A_{\pm}^{0,1}))(t)| \leq M_2(A_{\pm}^{0,1}) < \infty, \quad (10.38)
\end{aligned}
$$

respectively, i.e. $(\tilde{M}_n(A_{\pm}^{0,1}))(t)$, $(\tilde{V}_n(A_{\pm}^{0,1}))(t)$ converge as series, $\forall n \in \mathbb{N}$. Since $\tilde{M}_n$, $\tilde{V}_n$ are positive linear operators we have

$$0 \leq |(\tilde{M}_n(f_{\pm}^{(r)}))(t)| \leq (\tilde{M}_n(|f_{\pm}^{(r)}|))(t) \leq \tilde{M}_n(A_{\pm}^{0,1})(t)$$

and
$$0 \leq |(\tilde{V}_n(f_\pm^{(r)}))(t)| \leq (\tilde{V}_n(|f_\pm^{(r)}|))(t) \leq \tilde{V}_n(A_\pm^{(0,1)}(t), \tag{10.39}$$

respectively in $\pm$ and $\forall r \in [0,1]$.

In detail one has

$$\left| \left(f\left(\frac{k}{n}\right) \right)_\pm^{(r)} \right| w_{kn}(t) \leq A_\pm^{0,1}\left(\frac{k}{n}\right) w_{kn}(t),$$

$$\left| \left(f\left(\frac{k}{n}\right) \right)_\pm^{(r)} \right| \theta_{kn}(t) \leq A_\pm^{0,1}\left(\frac{k}{n}\right) \theta_{kn}(t), \tag{10.40}$$

$\forall r \in [0,1]$, $k \in \mathbb{N}$, $n \in \mathbb{N}$ fixed, $t > 0$ fixed, respectively in $\pm$.

Thus by Weierstrass M-test we obtain that $\tilde{M}_n(|f_\pm^{(r)}|)(t)$, $\tilde{V}_n(|f_\pm^{(r)}|)(t)$ as series *converge uniformly* in $r \in [0,1]$, respectively in $\pm$, $\forall n \in \mathbb{N}$. And easily we get by Cauchy criterion for series uniform convergence that

$$\tilde{M}_n(f_\pm^{(r)})(t), \quad \tilde{V}_n(f_\pm^{(r)})(t)$$

as series *converge uniformly* in $r \in [0,1]$, respectively in $\pm$, $\forall n \in \mathbb{N}$.

We then notice that

$$\mathbb{R}_{\mathcal{F}} \ni \int_{\mathbb{R}_+} f \, d\mu_1 = \left\{ \left(\int_{\mathbb{R}_+} f_-^{(r)} d\mu_1, \int_{\mathbb{R}_+} f_+^{(r)} d\mu_1 \right) \mid r \in [0,1] \right\}$$

$$= \left\{ \left(e^{-nt} \sum_{k=0}^{\infty} f_-^{(r)}\left(\frac{k}{n}\right) \frac{(nt)^k}{k!}, e^{-nt} \sum_{k=0}^{\infty} f_+^{(r)}\left(\frac{k}{n}\right) \frac{(nt)^k}{k!} \right) \mid r \in [0,1] \right\}$$

$$= e^{-nt} \odot \left\{ \left(\sum_{k=0}^{\infty} f_-^{(r)}\left(\frac{k}{n}\right) \frac{(nt)^k}{k!}, \sum_{k=0}^{\infty} f_+^{(r)}\left(\frac{k}{n}\right) \frac{(nt)^k}{k!} \right) \mid r \in [0,1] \right\}$$

$$= e^{-nt} \odot \left\{ \sum_{k=0}^{\infty} \frac{(nt)^k}{k!} \left(f_-^{(r)}\left(\frac{k}{n}\right), f_+^{(r)}\left(\frac{k}{n}\right) \right) \mid r \in [0,1] \right\}$$

(by using Remark 10.1(2) and earlier comments)

$$= e^{-nt} \sum_{k=0}^{\infty} \frac{(nt)^k}{k!} \odot \left\{ \left(f_-^{(r)}\left(\frac{k}{n}\right), f_+^{(r)}\left(\frac{k}{n}\right) \right) \mid r \in [0,1] \right\}$$

$$= e^{-nt} \sum_{k=0}^{\infty} \frac{(nt)^k}{k!} \odot f\left(\frac{k}{n}\right) = (M_n(f))(t).$$

I.e. we proved that

$$\int_{\mathbb{R}_+} f \, d\mu_1 = (M_n f)(t). \tag{10.41}$$

Similarly we can prove that

$$\int_{\mathbb{R}_+} f \, d\mu_2 = (V_n f)(t). \tag{10.42}$$

So we have that

$$(M_n f)(t), \quad (V_n f)(t) \in \mathbb{R}_{\mathcal{F}}, \quad \forall t \in \mathbb{R}_+.$$

We next observe for $t > 0$ that

$$
\begin{aligned}
D\big((M_n(f))(t), \tilde{o}\big) &= e^{-nt} D\left(\sideset{}{^*}\sum_{k=0}^{\infty} f\left(\frac{k}{n}\right) \odot \frac{(nt)^k}{k!}, \tilde{o}\right) \\
&= e^{-nt} \lim_{m \to \infty} D\left(\sideset{}{^*}\sum_{k=0}^{m} f\left(\frac{k}{n}\right) \odot \frac{(nt)^k}{k!}, \tilde{o}\right) \\
&\leq e^{-nt} \lim_{m \to \infty} \sum_{k=0}^{m} \frac{(nt)^k}{k!} D\left(f\left(\frac{k}{n}\right), \tilde{o}\right) \\
&\qquad \text{(here } D(f(x), \tilde{o}) \leq M, \; M \geq 0) \\
&\leq M e^{-nt} \lim_{m \to \infty} \sum_{k=0}^{m} \frac{(nt)^k}{k!} = M.
\end{aligned}
$$

Hence $D^*(M_n f, \tilde{o}) \leq M$. I.e. $M_n f(t)$ is fuzzy bounded over $\mathbb{R}^+$.

Similarly we see that for $t > 0$ that

$$
\begin{aligned}
D\big((V_n f)(t), \tilde{o}\big) &= D\left(\sideset{}{^*}\sum_{k=0}^{\infty} f\left(\frac{k}{n}\right) \odot \binom{n+k-1}{k} \frac{t^k}{(1+t)^{n+k}}, \tilde{o}\right) \\
&= \lim_{m \to \infty} D\left(\sideset{}{^*}\sum_{k=0}^{m} f\left(\frac{k}{n}\right) \odot \binom{n+k-1}{k} \frac{t^k}{(1+t)^{n+k}}, \tilde{o}\right) \\
&\leq \lim_{m \to \infty} \sum_{k=0}^{m} D\left(f\left(\frac{k}{n}\right) \odot \binom{n+k-1}{k} \frac{t^k}{(1+t)^{n+k}}, \tilde{o}\right) \\
&= \lim_{m \to \infty} \sum_{k=0}^{m} \binom{n+k-1}{k} \frac{t^k}{(1+t)^{n+k}} D\left(f\left(\frac{k}{n}\right), \tilde{o}\right) \\
&\leq M \left(\sum_{k=0}^{\infty} \binom{n+k-1}{k} \frac{t^k}{(1+t)^{n+k}}\right) = M.
\end{aligned}
$$

I.e. $D^*(V_n f, \tilde{o}) \leq M$, that is $V_n f$ is fuzzy bounded. Using Remark 10.1(1) and Lemma 1.3(iv) we get easily that M_n, V_n are fuzzy linear operators.

Next we prove their fuzzy positivity.

Indeed we have: let $f, g \in C_{\mathcal{F}}^B(\mathbb{R}_+)$ then $g_\pm^{(r)}$, $f_\pm^{(r)} \in C_B(\mathbb{R}_+)$, $\forall r \in [0, 1]$. So assume that $f \succsim g$ iff $f_-^{(r)} \geq g_-^{(r)}$ and $f_+^{(r)} \geq g_+^{(r)}$, $\forall r \in [0, 1]$. Then

$$f_-^{(r)}\left(\frac{k}{n}\right) \geq g_-^{(r)}\left(\frac{k}{n}\right)$$

and

$$f_+^{(r)}\left(\frac{k}{n}\right) \geq g_+^{(r)}\left(\frac{k}{n}\right), \quad \forall r \in [0, 1],$$

and

$$f_-^{(r)}\left(\frac{k}{n}\right)\frac{(nt)^k}{k!} \geq g_-^{(r)}\left(\frac{k}{n}\right)\frac{(nt)^k}{k!},$$

and

$$f_+^{(r)}\left(\frac{k}{n}\right)\frac{(nt)^k}{k!} \geq g_+^{(r)}\left(\frac{k}{n}\right)\frac{(nt)^k}{k!}, \qquad \forall r \in [0,1].$$

Consequently it holds

$$e^{-nt}\sum_{k=0}^{\infty}f_{\mp}^{(r)}\left(\frac{k}{n}\right)\frac{(nt)^k}{k!} \geq e^{-nt}\sum_{k=0}^{\infty}g_{\mp}^{(r)}\left(\frac{k}{n}\right)\frac{(nt)^k}{k!}, \qquad \forall r \in [0,1],$$

respectively. Therefore

$$\tilde{M}_n f_{\mp}^{(r)}(t) \geq \tilde{M}_n g_{\mp}^{(r)}(t), \qquad \forall r \in [0,1],$$

respectively in $\pm$.

Next we use Remark 10.1(1). We notice that

$$
\begin{aligned}
[(M_n f)(t)]^r &= \left[(M_n f)_-^{(r)}(t), (M_n f)_+^{(r)}(t)\right] \\
&= \left[e^{-nt}\left(\sum_{k=0}^{\infty}{}^* f\left(\frac{k}{n}\right)\odot\frac{(nt)^k}{k!}\right)_-^{(r)},\right. \\
&\qquad \left. e^{-nt}\left(\sum_{k=0}^{\infty}{}^* f\left(\frac{k}{n}\right)\odot\frac{(nt)^k}{k!}\right)_+^{(r)}\right] \\
&= \left[e^{-nt}\sum_{k=0}^{\infty}f_-^{(r)}\left(\frac{k}{n}\right)\frac{(nt)^k}{k!}, e^{-nt}\sum_{k=0}^{\infty}f_+^{(r)}\left(\frac{k}{n}\right)\frac{(nt)^k}{k!}\right] \\
&= \left[\tilde{M}_n(f_-^{(r)})(t), \tilde{M}_n(f_+^{(r)}(t)\right].
\end{aligned}
$$

I.e. $(M_n f)_{\pm}^{(r)} = \tilde{M}_n(f_{\pm}^{(r)})$, $\forall r \in [0,1]$, respectively, (that is fulfilling the basic condition of Assumption 10.20), $\forall n \in \mathbb{N}$. Therefore $(M_n(f))_{\pm}^{(r)} \geq (M_n g)_{\pm}^{(r)}$, respectively, iff $M_n f \succsim M_n g$. We have proved M_n's positivity. The V_n's positivity follows similarly.

We need to state

Theorem 10.45. Let $f \in C_B(\mathbb{R}_+)$. Then

$$|(\tilde{M}_n f)(t) - f(t)| \leq 2\omega_1\left(f, \sqrt{\frac{t}{n}}\right), \qquad \forall t > 0, \ n \in \mathbb{N}, \qquad (10.43)$$

and

$$|(\tilde{V}_n f)(t) - f(t)| \leq 2\omega_1\left(f, \sqrt{\frac{t+t^2}{n}}\right), \qquad \forall t > 0, \ n \in \mathbb{N}, \qquad (10.44)$$

where ω_1 is the real basic first modulus of continuity.

Proof. Let μ be a probability measure on $\mathbb{R}_+$. We notice that

$$\left| \int_{\mathbb{R}_+} f \, d\mu - f(t) \right| = \left| \int_{\mathbb{R}_+} (f - f(t)) \, d\mu \right|$$

$$\text{(by Corollary 7.1.1, p. 209 of [4])}$$

$$\leq \int_{\mathbb{R}_+} |f(x) - f(t)| \, d\mu(x)$$

$$\leq \omega_1(f;\delta) \int_{\mathbb{R}_+} \left\lceil \frac{|x-t|}{\delta} \right\rceil d\mu(x) \ (\delta > 0)$$

$$\leq \omega_1(f;\delta) \left(\int_{\mathbb{R}_+} \left(1 + \frac{|x-t|}{\delta} \right) d\mu(x) \right)$$

$$= \omega_1(f;\delta) \left(1 + \frac{1}{\delta} \int_{\mathbb{R}_+} |x-t| d\mu(x) \right)$$

$$\text{(by Cauchy–Schwarz inequality)}$$

$$\leq \omega_1(f;\delta) \left(1 + \frac{1}{\delta} \left(\int_{\mathbb{R}_+} (x-t)^2 d\mu(x) \right)^{1/2} \right)$$

$$\left(\text{by choosing } \delta := \left(\int_{\mathbb{R}_+} (x-t)^2 d\mu(x) \right)^{1/2} \right)$$

$$= 2\omega_1 \left(f; \left(\int_{\mathbb{R}_+} (x-t)^2 d\mu(x) \right)^{1/2} \right).$$

So we have proved that

$$\left| \int_{\mathbb{R}_+} f \, d\mu - f(t) \right| \leq 2\omega_1 \left(f \left(\int_{\mathbb{R}_+} (x-t)^2 d\mu(x) \right)^{1/2} \right). \tag{10.45}$$

Using now the content of Application 10.20 and setting as $\mu = F_{S_n/n}$ in (10.45) we get that the operator values $(\tilde{M}_n f)(t)$, $(\tilde{V}_n f)(t)$ are of the form

$$E \left(f \left(\frac{S_n}{n} \right) \right) = \int_{\mathbb{R}_+} f \, dF_{S_n/n}.$$

In this case $\left(\int_{\mathbb{R}_+} (x-t)^2 d\mu(x) \right)^{1/2}$ is the standard deviation $\sigma_{S_n/n}$, which in Poisson case is $\sqrt{\frac{t}{n}}$ and in geometric case is $\sqrt{\frac{t+t^2}{n}}$.   □

Using now all of the above and (10.25) of Theorem 10.43 we obtain

Theorem 10.46. *Operators M_n, V_n are well-defined on $C_{\mathcal{F}}^B(\mathbb{R}_+)$, $n \in \mathbb{N}$ and they are fuzzy positive linear operators there. Also for each $f \in C_{\mathcal{F}}^B(\mathbb{R}_+)$ we have that $M_n f$, $V_n f$ are uniformly in n fuzzy bounded. Furthermore it holds*

$$
\begin{aligned}
(M_n f)_\pm^{(r)} &= \tilde{M}(f_\pm^{(r)}), \\
(V_n f)_\pm^{(r)} &= \tilde{V}_n(f_\pm^{(r)}),
\end{aligned}
\tag{10.46}
$$

$\forall r \in [0,1]$, *respectively, any $f \in C_{\mathcal{F}}^B(\mathbb{R}_+)$. It holds*

$$
D((M_n f)(t), f(t)) \le 2\omega_1^{(\mathcal{F})}\left(f; \sqrt{\frac{t}{n}}\right),
\tag{10.47}
$$

and

$$
D((V_n f)(t), f(t)) \le 2\omega_1^{(\mathcal{F})}\left(f; \sqrt{\frac{t + t^2}{n}}\right), \quad \forall t > 0,\ n \in \mathbb{N},\ f \in C_{\mathcal{F}}^B(\mathbb{R}_+).
\tag{10.48}
$$

Application 10.47. Let a real normed vector space $(V, \|\cdot\|)$, let $x_0 \in V$ and a sequence $(\mu_n)_{n\in\mathbb{N}}$ of positive finite measures on V with $\mu_n(V) = m_n > 0$, $n \in \mathbb{N}$, where $m_n \le \gamma$, $\gamma > 0$. Let $f \colon V \to \mathbb{R}_{\mathcal{F}}$ be fuzzy bounded function i.e. $D(f(x), \tilde{o}) \le M$, $M \ge 0$, $\forall x \in V$. Thus $D(f(x + x_0), \tilde{o}) \le M$, $\forall x \in V$, so that $f(\cdot + x_0)$ is integrable by Theorem 10.17 and Definition 10.18.

We define the *fuzzy convolution operator*

$$
(L_n f)(x_0) = \int_V f(t + x_0)\mu_n(dt),
\tag{10.49}
$$

for any $x_0 \in V$, $\forall n \in \mathbb{N}$. See that $(L_n f)(x_0) \in \mathbb{R}_{\mathcal{F}}$ and L_n is a fuzzy linear operator by Theorem 10.19.

By the representation

$$
f(t + x_0) = \left\{ (f_-^{(r)}(t + x_0), f_+^{(r)}(t + x_0)) \mid r \in [0,1] \right\}, \quad \forall t \in V
$$

we get that: Let $f, g \colon V \to \mathbb{R}_{\mathcal{F}}$ fuzzy bounded functions such that $f \succsim g$ iff $f_-^{(r)} \ge g_-^{(r)}$ and $f_+^{(r)} \ge g_+^{(r)}$, $\forall r \in [0,1]$ iff $f_\pm^{(r)}(x) \ge g_\pm^{(r)}(x)$, $\forall x \in V$, $\forall r \in [0,1]$, respectively, then $f_\pm^{(r)}(t + x_0) \ge g_\pm^{(r)}(t + x_0)$, $\forall t \in V$, $\forall r \in [0,1]$, respectively, then

$$
\int_V f_\pm^{(r)}(t + x_0)d\mu(t) \ge \int_V g_\pm^{(r)}(t + x_0)d\mu(t), \quad \forall r \in [0,1],
$$

respectively. Hence by Definition 10.18 we get that

$$
\left(\int_V f(t + x_0)d\mu(t)\right)_\pm^{(r)} \ge \left(\int_V g(t + x_0)d\mu(t)\right)_\pm^{(r)}, \quad \forall r \in [0,1],
$$

respectively, iff $((L_nf)(x_0))_\pm^{(r)} \geq ((L_ng)(x_0))_\pm^{(r)}$, $\forall r \in [0,1]$, respectively, iff $(L_nf)(x_0) \succsim (L_ng)(x_0)$, any $x_0 \in V$ iff $(L_nf) \succsim (L_ng)$, $n \in \mathbb{N}$. That is $(L_n)_{n \in \mathbb{N}}$ is a sequence of *fuzzy positive operators*.

As before we obtain

$$
\begin{aligned}
&D\big((L_nf)(x_0), f(x_0)\big) \\
&\quad = D\left(\int_V f(t+x_0)\mu_n(dt), f(x_0)\right) \\
&\quad \leq D\left(\int_V f(t+x_0)\mu_n(dt), f(x_0) \odot m_n\right) + D(f(x_0) \odot m_n, f(x_0)) \\
&\quad = D\left(\int_V f(t+x_0)\mu_n(dt), \int_V f(x_0)\mu_n(dt)\right) + D(f(x_0) \odot m_n, f(x_0)) \\
&\quad \leq \int_V D(f(t+x_0), f(x_0))\mu_n(dt) + |m_n - 1|D(f(x_0), \tilde{o}) \quad (\delta > 0) \\
&\quad \leq \int_V \omega_1^{(\mathcal{F})}(f; \delta)\left\lceil \frac{\|t\|}{\delta}\right\rceil \mu_n(dt) + |m_n - 1|D(f(x_0), \tilde{o}) \\
&\quad \leq |m_n - 1|M + \omega_1^{(\mathcal{F})}(f; \delta)\left(\int_V \left(1 + \frac{\|t\|}{\delta}\right)\mu_n(dt)\right) \\
&\quad = |m_n - 1|M + \omega_1^{(\mathcal{F})}(f; \delta)\left(m_n + \frac{1}{\delta}\int_V \|t\|\mu_n(dt)\right) \\
&\quad \leq |m_n - 1|M + \omega_1^{(\mathcal{F})}(f; \delta)\left(m_n + \frac{1}{\delta}\left(\int_V \|t\|^2\mu_n(dt)\right)^{1/2}\sqrt{m_n}\right) \\
&\quad \left(\text{choosing } \delta := \left(\int_V \|t\|^2\mu_n(dt)\right)^{1/2}\right) \\
&\quad = M|m_n - 1| + \omega_1^{(\mathcal{F})}\left(f; \left(\int_V \|t\|^2\mu_n(dt)\right)^{1/2}\right)(m_n + \sqrt{m_n}).
\end{aligned}
$$

Finally we have established the following *fuzzy convolution result*.

Theorem 10.48. Let $(V, \|\cdot\|)$ *real normed vector space and* $(\mu_n)_{n \in \mathbb{N}}$ *positive finite measures on* V: $\mu_n(V) = m_n > 0$, $n \in \mathbb{N}$, *where* $m_n \leq \gamma$, $\gamma > 0$. *Let* $f: V \to \mathbb{R}_\mathcal{F}$: $D^*(f, \tilde{o}) \leq M$, $M \geq 0$. *Define*

$$(L_nf)(x) = \int_V f(t+x)\mu_n(dt), \quad \forall x \in V. \tag{10.50}$$

Then $(L_n)_{n \in \mathbb{N}}$ *is a well-defined sequence of fuzzy positive linear operators. It holds*

$$D^*(L_nf, f) \leq M|m_n - 1| + \omega_1^{(\mathcal{F})}\left(f; \left(\int_V \|t\|^2\mu_n(dt)\right)^{1/2}\right)(m_n + \sqrt{m_n}), \quad n \in \mathbb{N}. \tag{10.51}$$

If $m_n = 1$ then

$$D^*(L_n f, f) \leq 2\omega_1^{(\mathcal{F})}\left(f; \left(\int_V \|t\|^2 \mu_n(dt)\right)^{1/2}\right). \tag{10.52}$$

11

HIGHER ORDER FUZZY KOROVKIN THEORY USING INEQUALITIES

Here is studied with rates the fuzzy uniform and L_p, $p \geq 1$, convergence of a sequence of fuzzy positive linear operators to the fuzzy unit operator acting on spaces of fuzzy differentiable functions. This is done quantitatively via fuzzy Korovkin type inequalities involving the fuzzy modulus of continuity of a fuzzy derivative of the engaged function. From there we deduce general fuzzy Korovkin type theorems with high rate of convergence. The surprising fact is that basic real positive linear operator simple assumptions enforce here the fuzzy convergences. At the end we give applications. The results are univariate and multivariate. The assumptions are minimal and natural fulfilled by almost all example—fuzzy positive linear operators. This chapter follows [20].

11.1 Introduction

Motivation for this chapter are [4], [18], [24], [32], [78], [96]. In fact this is continuation of [18], [24]. Here first we translate the necessary measure theory approximation results from [4] into the language of real positive linear operators, then by combining the facts, e.g. use of Proposition 11.2, we transfer results at the fuzzy level.

G.A. Anastassiou: Fuzzy Mathematics: Approximation The., STUDFUZZ 251, pp. 159–190.
springerlink.com   © Springer-Verlag Berlin Heidelberg 2010

Applications are on univariate and multivariate Bernstein operators. At the beginning we provide all necessary fuzzy terminology, definitions and theorems we use here. In that background section we prove some results that they stand by themselves, such as in positivity. The basic ingredient to establish the results is the bridge between real operators to fuzzy ones. It is the natural realization condition: Assumption 11.21, see (11.17). This is fulfilled by almost all example positive operators, in fact by all summation and integration operators: real and fuzzy. The concept of fuzzy positivity we use is the natural analog of the real positivity, the same thing with linearity.

We need

Definition 11.1. Let $U \subseteq (M, d)$ a metric space and let $f: U \to \mathbb{R}_{\mathcal{F}}$. We define the (*first*) *fuzzy modulus of continuity* of f by

$$\omega_1^{(\mathcal{F})}(f; \delta) := \sup_{\substack{x, y \in U \\ d(x, y) \leq \delta}} D(f(x), f(y))$$

for $0 < \delta \leq$ diameter(U). If $\delta >$ diam(U) then we define

$$\omega_1^{(\mathcal{F})}(f; \delta) := \omega_1^{(\mathcal{F})}(f; \text{diam}(U)).$$

Proposition 11.2 (see [24]). *Let $U \subseteq (M, d)$ metric space and $f: U \to \mathbb{R}_{\mathcal{F}}$. Assume that*

$$\omega_1^{(\mathcal{F})}(f, \delta), \quad \omega_1(f_-^{(r)}, \delta), \quad \omega_1(f_+^{(r)}, \delta)$$

are finite for any $\delta > 0$. Here ω_1 is the usual real modulus of continuity, i.e. for $g: U \to \mathbb{R}$ we define

$$\omega_1(g; \delta) := \sup_{\substack{x, y \in U \\ d(x, y) \leq \delta}} |g(x) - g(y)|,$$

etc. Then

$$\omega_1^{(\mathcal{F})}(f; \delta) = \sup_{r \in [0,1]} \max\{\omega_1(f_-^{(r)}, \delta), \omega_1(f_+^{(r)}, \delta)\}.$$

We need

Definition 11.3. Let U open or compact $\subseteq (M, d)$ metric space and $f: U \to \mathbb{R}_{\mathcal{F}}$. We say that f is *fuzzy continuous* at $x_0 \in U$ iff whenever $x_n \to x_0$, then $D(f(x_n), f(x_0)) \to 0$. If f is continuous for every $x_0 \in U$, we then call f a *fuzzy continuous real number valued function*. We denote the related space by $C_{\mathcal{F}}(U)$. Similarly one defines $C_{\mathcal{F}}([a, b])$, $[a, b] \subseteq \mathbb{R}$, etc.

Definition 11.4. Let $f: K \to \mathbb{R}_{\mathcal{F}}$, K open or compact $\subseteq (M, d)$ metric space. We call f a *fuzzy uniformly continuous real number valued function*, iff $\forall \varepsilon > 0$, $\exists \delta > 0$: whenever $d(x, y) \leq \delta$, $x, y \in K$, implies that $D(f(x), f(y)) \leq \varepsilon$. We denote the related space by $C_{\mathcal{F}}^U(K)$.

Definition 11.5. Let $f: U \to \mathbb{R}_{\mathcal{F}}$, $U \subseteq (M, d)$ metric space. If $D(f(x), \tilde{o}) \leq M$, $\forall x \in U$, $M \geq 0$, we call f a *fuzzy bounded real number valued function*.

In particular if $f \in C_{\mathcal{F}}([a, b])$, $[a, b] \subseteq \mathbb{R}$, then f is a fuzzy bounded function, also $\omega_1^{(\mathcal{F})}(f; \delta) < \infty$ for any $0 < \delta \leq b - a$, etc. Also notice that $C_{\mathcal{F}}^U(K) = C_{\mathcal{F}}(K)$, for K compact $\subseteq (V, \| \cdot \|)$ real normed vector space.

We use

Proposition 11.6 (see [24]) *Let $K \subseteq (V, \| \cdot \|)$ a real normed vector space and*

$$\omega_1^{(\mathcal{F})}(f; \delta) = \sup_{\substack{x, y \in K \\ \|x-y\| \leq \delta}} D(f(x), f(y)), \quad \delta > 0,$$

the fuzzy modulus of continuity for $f: K \to \mathbb{R}_{\mathcal{F}}$. Then

(1) *If $f \in C_{\mathcal{F}}^U(K)$, K open convex or compact convex $\subseteq (V, \| \cdot \|)$, then $\omega_1^{(\mathcal{F})}(f; \delta) < \infty$, $\forall \delta > 0$.*

(2) *Assume that K is open convex or compact convex $\subseteq (V, \| \cdot \|)$, then $\omega_1^{(\mathcal{F})}(f; \delta)$ is continuous on $\mathbb{R}_+$ in δ for $f \in C_{\mathcal{F}}^U(K)$.*

(3) *Assume that K is convex, then*

$$\omega_1^{(\mathcal{F})}(f, t_1 + t_2) \leq \omega_1^{(\mathcal{F})}(f, t_1) + \omega_1^{(\mathcal{F})}(f, t_2), \quad t_1, t_2 \geq 0,$$

that is the subadditivity property is true. Also it holds

$$\omega_1^{(\mathcal{F})}(f, n\delta) \leq n\omega_1^{(\mathcal{F})}(f, \delta),$$

and

$$\omega_1^{(\mathcal{F})}(f, \lambda\delta) \leq \lceil \lambda \rceil \omega_1^{(\mathcal{F})}(f, \delta) \leq (\lambda + 1)\omega_1^{(\mathcal{F})}(f, \delta),$$

where $n \in \mathbb{N}$, $\lambda > 0$, $\delta > 0$, $\lceil \cdot \rceil$ is the ceiling of the number.

(4) *Clearly in general $\omega_1^{(\mathcal{F})}(f; \delta) \geq 0$ and is increasing in $\delta > 0$ and $\omega_1^{(\mathcal{F})}(f; 0) = 0$.*

(5) *Let K be open or compact $\subseteq (V, \| \cdot \|)$. Then $\omega_1^{(\mathcal{F})}(f; \delta) \to 0$ as $\delta \downarrow 0$ iff $f \in C_{\mathcal{F}}^U(K)$.*

(6) *It holds*

$$\omega_1^{(\mathcal{F})}(f \oplus g; \delta) \leq \omega_1^{(\mathcal{F})}(f; \delta) + \omega_1^{(\mathcal{F})}(g; \delta),$$

for $\delta > 0$, any $f, g: K \to \mathbb{R}_{\mathcal{F}}$, $K \subseteq (V, \| \cdot \|)$ is arbitrary.

We also need

Lemma 11.7 (see [24]) *Let K be a compact subset of the real normed vector space $(V, \| \cdot \|)$ and $f \in C_{\mathcal{F}}(K)$. Then f is a fuzzy bounded function.*

We use

Proposition 11.8 (see [24]). *Let U open or compact $\subseteq (M,d)$ metric space, $f \in C_{\mathcal{F}}(U)$. Then $f_{\pm}^{(r)}$ are equicontinuous with respect to $r \in [0,1]$ over U, respectively in $\pm$.*

For the reverse we have

Proposition 11.9 (see [24]). *Let $f_{-}^{(r)}$, $f_{+}^{(r)}$ be equicontinuous with respect to $r \in [0,1]$ on U-open or compact $\subseteq (M,d)$-metric space, respectively in $\pm$, then $f \in C_{\mathcal{F}}(U)$.*

We mention

Definition 11.10 (see [53]). *Let $x,y \in \mathbb{R}_{\mathcal{F}}$. If there exists a $z \in \mathbb{R}_{\mathcal{F}}$, such that $x = y + z$, then, we call z the H-difference of x and y, denoted by $z := x - y$.*

Definition 11.11 (see [53]). *Let $T := [x_0, x_0 + \beta] \subset \mathbb{R}$, with $\beta > 0$. A function $f \colon T \to \mathbb{R}_{\mathcal{F}}$ is differentiable at $x \in T$, if there exists a $f'(x) \in \mathbb{R}_{\mathcal{F}}$, such that the limits*

$$\lim_{h \to 0^+} \frac{f(x+h) - f(x)}{h}, \quad \lim_{h \to 0^+} \frac{f(x) - f(x-h)}{h},$$

exist and are equal to $f'(x)$. We call f' the derivative of f at x. If f is differentiable at any $x \in T$, we call f differentiable and it has derivative over T, the function f'. Here is assumed that $f(x+h)-f(x), f(x)-f(x-h)$ exist for small h.

Similarly we define higher order fuzzy derivatives. Regarding functions of several variables one can define the same way partial derivatives in the fuzzy sense.

Let Q be a compact convex subset of $\mathbb{R}^k$, $k > 1$ and $n \in \mathbb{N}$. By $C_{\mathcal{F}}^n(Q)$ we mean all the functions from Q into $\mathbb{R}_{\mathcal{F}}$ that are n-times continuously differentiable in the fuzzy sense.

We use

Theorem 11.12 (see [71]). *Let $f \colon [a,b] \subseteq \mathbb{R} \to \mathbb{R}_{\mathcal{F}}$ be fuzzy differentiable. Let $t \in [a,b]$, $0 \leq r \leq 1$. Clearly*

$$[f(t)]^r = \left[(f(t))_{-}^{(r)}, (f(t))_{+}^{(r)} \right] \subseteq \mathbb{R}. \tag{11.1}$$

Then $(f(t))_{\pm}^{(r)}$ are differentiable and

$$[f'^r = \left[((f(t))_{-}^{(r)})_{+}^{'(r)})' \right], \tag{11.2}$$

i.e.

$$(f')_{\pm}^{(r)} = (f_{\pm}^{(r)})', \quad \text{for any } r \in [0,1]. \tag{11.3}$$

We make

Remark 11.13. 1) Let $f \in C_{\mathcal{F}}^n([a,b])$. Then by Theorem 11.12 and Proposition 11.8 we obtain $f_{\pm}^{(r)} \in C^n([a,b])$ and

$$[f^{(i)}(t)]^r = [((f(t))_-^{(r)})^{(i)}, ((f(t))_+^{(r)})^{(i)}], \tag{11.4}$$

for $i = 0, 1, 2, \ldots, n$, and, in particular, we have that

$$(f^{(i)})_{\pm}^{(r)} = (f_{\pm}^{(r)})^{(i)}, \tag{11.5}$$

for any $r \in [0,1]$.

2) Let $f \in C_{\mathcal{F}}^n(Q)$, denote $f_\alpha := \frac{\partial^\alpha f}{\partial x^\alpha}$, where $\alpha := (\alpha_1, \ldots, \alpha_k)$, $\alpha_i \in \mathbb{Z}^+$, $i = 1, \ldots, k$ and $0 < |\alpha| := \sum_{i=1}^{k} \alpha_i \leq n$, $n > 1$. Then we get by Theorem 11.12 that

$$(f_{\pm}^{(r)})_\alpha = (f_\alpha)_{\pm}^{(r)}, \tag{11.6}$$

for any $r \in [0,1]$ and any $\alpha \colon |\alpha| \leq n$. Here $f_{\pm}^{(r)} \in C^n(Q)$.

We also make

Remark 11.14 (see [15], Remark 3). Let $r \in [0,1]$, $x_i^{(r)}$, $y_i^{(r)} \in \mathbb{R}$, $i = 1, \ldots, m \in \mathbb{N}$. Assume

$$\sup_{r \in [0,1]} \max(x_i^{(r)}, y_i^{(r)}) \in \mathbb{R}, \quad i = 1, \ldots, m.$$

Then

$$\sup_{r \in [0,1]} \max\left(\sum_{i=1}^m x_i^{(r)}, \sum_{i=1}^m y_i^{(r)} \right) \leq \sum_{i=1}^m \sup_{r \in [0,1]} \max(x_i^{(r)}, y_i^{(r)}). \tag{11.7}$$

We use

Lemma 11.15. *Let* $r \in [0,1]$, $\alpha \in I$, I *a finite index set, and* $x_\alpha^{(r)}, y_\alpha^{(r)} \in \mathbb{R}$. *Assume*

$$A := \max_{\alpha \in I} \sup_{r \in [0,1]} \max\{x_\alpha^{(r)}, y_\alpha^{(r)}\} \in \mathbb{R}$$

and

$$B := \sup_{r \in [0,1]} \max\left\{ \max_{\alpha \in I} x_\alpha^{(r)}, \max_{\alpha \in I} y_\alpha^{(r)} \right\} \in \mathbb{R}. \tag{11.8}$$

It holds

$$A = B. \tag{11.9}$$

Proof. 1) First we prove $B \leq A$. We see that

$$x_\alpha^{(r)}, y_\alpha^{(r)} \leq A, \quad \forall r \in [0,1] \text{ and } \forall \alpha \in I.$$

Then

$$A_1 := \max_{\alpha \in I} x_\alpha^{(r)} \leq A, \quad A_2 := \max_{\alpha \in I} y_\alpha^{(r)} \leq A$$

and
$$\max(A_1, A_2) \le A.$$

Thus
$$\sup_{r \in [0,1]} \max(A_1, A_2) \le A.$$

2) We last prove $A \le B$. We notice
$$x_\alpha^{(r)} \le \max_{\alpha \in I} x_\alpha^{(r)}, \quad \forall r \in [0,1], \quad \forall \alpha \in I$$

and
$$y_\alpha^{(r)} \le \max_{\alpha \in I} y_\alpha^{(r)}, \quad \forall r \in [0,1], \quad \forall \alpha \in I.$$

Then
$$\max\{x_\alpha^{(r)}, y_\alpha^{(r)}\} \le \max\left\{\max_{\alpha \in I} x_\alpha^{(r)}, \max_{\alpha \in I} y_\alpha^{(r)}\right\}, \quad \forall r \in [0,1], \forall \alpha \in I.$$

Furthermore
$$\sup_{r \in [0,1]} \max\{x_\alpha^{(r)}, y_\alpha^{(r)}\} \le \sup_{r \in [0,1]} \max\left\{\max_{\alpha \in I} x_\alpha^{(r)}, \max_{\alpha \in I} y_\alpha^{(r)}\right\} = B, \quad \forall \alpha \in I,$$

and finally
$$\max_{\alpha \in I} \sup_{r \in [0,1]} \max\{x_\alpha^{(r)}, y_\alpha^{(r)}\} \le B.$$

$\square$

We mention

Definition 11.16. Let $L\colon C_{\mathcal{F}}(U) \hookrightarrow C_{\mathcal{F}}(U)$, where U is open or compact $\subseteq (M,d)$ metric space, such that
$$L(c_1 f + c_2 g) = c_1 L(f) + c_2 L(g), \quad \forall c_1, c_2 \in \mathbb{R}. \tag{11.10}$$

We call L a *fuzzy linear operator*.

We give the following example of a fuzzy linear operator, etc.

Definition 11.17. Let $f\colon [0,1] \to \mathbb{R}_{\mathcal{F}}$ be a fuzzy real function. The fuzzy algebraic polynomial defined by
$$B_N^{(\mathcal{F})}(f)(x) = \sum_{k=0}^{N}{}^* \binom{N}{k} x^k (1-x)^{N-k} \odot f\left(\frac{k}{N}\right), \quad \forall x \in [0,1], \ N \in \mathbb{N}, \tag{11.11}$$

will be called the *fuzzy Bernstein operator*.

We do have

Theorem 11.18 (see p. 642, [66], S. Gal). *If $f \in C_{\mathcal{F}}([0,1])$, then*
$$D\big(B_N^{(\mathcal{F})}(f)(x), f(x)\big) \le \frac{3}{2} \omega_1^{(\mathcal{F})}\left(f; \frac{1}{\sqrt{N}}\right), \quad N \in \mathbb{N}, \ \forall x \in [0,1], \tag{11.12}$$

i.e.,

$$\lim_{N \to +\infty} D^* \left(B_N^{(\mathcal{F})}(f), f \right) = 0, \tag{11.13}$$

that is $B_N^{(\mathcal{F})} f \to_{n \to +\infty} f$, *fuzzy uniform convergence.*

We also need

Definition 11.19. Let $f, g \colon U \to \mathbb{R}_{\mathcal{F}}, U \subseteq (M, d)$ metric space. We denote $f \succsim g$, iff $f(x) \succsim g(x)$, $\forall x \in U$, iff $f_+^{(r)}(x) \ge g_+^{(r)}(x)$ and $f_-^{(r)}(x) \ge g_-^{(r)}(x)$, $\forall x \in U$, $\forall r \in [0, 1]$, iff $f_+^{(r)} \ge g_+^{(r)}$ and $f_-^{(r)} \ge g_-^{(r)}$, $\forall r \in [0, 1]$.

We give

Definition 11.20. Let $L \colon C_{\mathcal{F}}(U) \hookrightarrow C_{\mathcal{F}}(U)$ be a fuzzy linear operator, U open or compact $\subseteq (M, d)$ metric space. We say that L is *positive*, iff whenever $f, g \in C_{\mathcal{F}}(U)$ are such that $f \succsim g$ then $L(f) \succsim L(g)$, iff

$$(L(f))_+^{(r)} \ge (L(g))_+^{(r)} \tag{11.14}$$

and

$$(L(f))_-^{(r)} \ge (L(g))_-^{(r)}, \quad \forall r \in [0, 1]. \tag{11.15}$$

Here we denote

$$[L(f)]^r = \left[(L(f))_-^{(r)}, (L(f))_+^{(r)} \right], \quad \forall r \in [0, 1]. \tag{11.16}$$

An example of a fuzzy positive linear operator is the fuzzy Bernstein operator on the domain $[0, 1]$, etc.

We will use this type of supposition.

Assumption 11.21. Let L be a fuzzy positive linear operator from $C_{\mathcal{F}}(K)$, K compact $\subseteq (M, d)$ metric space, into itself. Here we *assume* that there exists a positive linear operator $\tilde{L}$ from $C(K)$ into itself with the property

$$(Lf)_\pm^{(r)} = \tilde{L}(f_\pm^{(r)}), \tag{11.17}$$

respectively, for all $r \in [0, 1]$, $\forall f \in C_{\mathcal{F}}(K)$.

As an example again we mention the fuzzy Bernstein operator and the real Bernstein operator fulfilling the above assumption on $[0, 1]$. etc.

We will use

Theorem 11.22. *Let (X, O) be a linearly ordered vector space, (O denotes the order) and G a majorant subspace of X (i.e. for any x in X there exist z and y in G such that $zOxOy$). If $f \colon G \to \mathbb{R}$ is a positive linear map, then there exists (a not necessarily unique) $F \colon X \to \mathbb{R}$ positive linear map, such that $F(x) = f(x)$, $\forall x \in G$.*

Above Theorem 11.22 is a special case of the famous Kantorovich Extension Theorem, due to L.V. Kantorovich [73]. For a proof see [1], Theorem 2.8, p. 26. See also [54], p. 72, Proposition 1.

We need

Lemma 11.23. *Let* $\tilde{L}\colon C^n(Q) \to C(Q)$ *be a positive linear operator,* $n \in \mathbb{N}$, Q *compact* $\subseteq \mathbb{R}^k$, $k \geq 1$. *Then there exists a unique finite Borel measure* μ_x, $x \in Q$, *such that*

$$(\tilde{L}(f))(x) = \int_Q f(t)d\mu_x(t), \quad \forall f \in C^n(Q).$$

Notice $(\tilde{L}(1))(x) = \mu_x(Q) < \infty$.

Proof. Clearly $(\tilde{L}(\cdot))(x)$ is a positive linear functional on $C^n(Q)$. Also since $1 \in C^n(Q)$, $C^n(Q)$ is a majorant subspace of $C(Q)$. Thus, by Theorem 11.22 there exists a positive linear functional $M\colon C(Q) \to \mathbb{R}$ extending $\tilde{L}$, i.e. $\tilde{L} = M\big|_{C^n(Q)}$. Consequently, by Riesz Representation theorem there exists a unique finite Borel measure μ_x for this M, such that

$$M(f) = \int_Q f(t)d\mu_x(t), \quad \forall f \in C(Q).$$

Therefore

$$\tilde{L}(f)(x) = \int_Q f(t)d\mu_x(t), \quad \forall f \in C^n(Q).$$

We prove uniqueness μ_x regarding $\tilde{L}(\cdot)(x)$. Let us assume that there exists another finite Borel measure ν such that

$$\int_Q f(t)d\mu_x(t) = \int_Q f(t)d\nu(t), \quad \forall f \in C^n(Q).$$

In particular, we have

$$\int_Q p(t)d\mu_x(t) = \int_Q p(t)d\nu(t),$$

for all polynomials p. Since by the Stone–Weierstrass approximation theorem the polynomials are uniformly dense in $C(Q)$, it follows for any $f \in C(Q)$ that there exists a sequence $\{p_n\}_{n\in\mathbb{N}}$ of polynomials that converges uniformly to f on Q. In particular $\{p_n\}$ is uniformly bounded. Now from

$$\int_Q p_n(t)d\mu_x(t) = \int_Q p_n(t)d\nu(t)$$

and the Lebesgue Dominated Convergence Theorem, by taking the limits in the last equation, we find

$$\int_Q f(t)d\mu_x(t) = \int_Q f(t)d\nu(t).$$

Hence, by the uniqueness of the measure in the Riesz representation theorem, we get indeed that $\mu_x = \nu$. $\qquad\square$

We give

Remark 11.24. We recall from [4], p. 210–211 the function

$$\phi_n(x) := \int_0^{|x|} \left\lceil \frac{t}{h} \right\rceil \frac{(|x| - t)^{n-1}}{(n-1)!} dt, \tag{11.18}$$

$x \in \mathbb{R}$, $n \in \mathbb{N}$, where $\lceil \cdot \rceil$ is the ceiling of the number.
We have

$$\phi_n(x) = \int_0^{|x|} \int_0^{x_1} \cdots \left(\int_0^{x_{n-1}} \left\lceil \frac{x_n}{h} \right\rceil dx_n \right) \cdots dx_1, \tag{11.19}$$

and

$$\phi_n(x) = \frac{1}{n!} \left(\sum_{j=0}^{\infty} (|x| - jh)_+^n \right), \quad x \in \mathbb{R}. \tag{11.20}$$

Also it holds

$$\phi_n(x) \le \left(\frac{|x|^{n+1}}{(n+1)!h} + \frac{|x|^n}{2n!} + \frac{h|x|^{n-1}}{8(n-1)!} \right), \tag{11.21}$$

and

$$\phi_n(x) = \int_0^x \phi_{n-1}(t)dt, \quad x \in \mathbb{R}_+, \quad n \in \mathbb{N}. \tag{11.22}$$

And from [4], p. 217 we have

$$\phi_n(x) \le \frac{|x|^n}{n!} \left(1 + \frac{|x|}{(n+1)h} \right), \quad x \in \mathbb{R}, \quad n \in \mathbb{N}. \tag{11.23}$$

11.2   Univariate Results

We need to mention

Theorem 11.25 (by Corollary 7.2.2, p. 219–220 in [4] and Geometric Moment theory [13]). *Consider the positive linear operator*

$$\tilde{L} \colon C([a, b]) \to C([a, b]). \tag{11.24}$$

Let

$$\begin{aligned}
c_k(x) &:= & \tilde{L}((t - x)^k, x), \quad k = 0, 1, \ldots, n, \ n \in \mathbb{N}; \\
d_n(x) &:= & \left[\tilde{L}(|t - x|^n, x) \right]^{1/n}; \\
c(x) &:= & \max(x - a, b - x) \quad (c(x) \ge (b - a)/2).
\end{aligned} \tag{11.25}$$

Let $g \in C^n([a,b])$ such that $\omega_1(g^{(n)}, h) \leq w$, where $w, h > 0$, $0 < h \leq b-a$.
Then

$$|\tilde{L}(g,x) - g(x)| \leq |g(x)| |c_0(x) - 1| + \sum_{k=1}^{n} \frac{|g^{(k)}(x)|}{k!} |c_k(x)|$$

$$+ w\phi_n(c(x)) \left(\frac{d_n(x)}{c(x)}\right)^n, \quad \forall x \in [a,b]. \ (11.26)$$

Inequality (11.26) is sharp. It is attained in a certain sense by $w\phi_n((t-x)_+)$ and a measure $\tilde{\mu}_x$ supported by $\{x, b\}$ when $x-a \leq b-x$, also attained by $w\phi_n((x-t)_+)$ and a measure $\tilde{\mu}_x$ supported by $\{x, a\}$ when $x-a \geq b-x$: in each case with masses $c_0(x) - \left(\frac{d_n(x)}{c(x)}\right)^n$ and $\left(\frac{d_n(x)}{c(x)}\right)^n$, respectively.

Note 11.26. Assuming $\omega_1(g^{(n)}, h) > 0$ for $h > 0$ and all as in the context of Theorem 11.25. We prefer to write

$$|\tilde{L}(g,x) - g(x)| \leq |g(x)| |c_0(x) - 1| + \sum_{k=1}^{n} \frac{|g^{(k)}(x)|}{k!} |c_k(x)|$$

$$+ \phi_n(c(x)) \left(\frac{d_n(x)}{c(x)}\right)^n \omega_1(g^{(n)}, h), \quad (11.27)$$

$\forall x \in [a,b]$.
    We need also

Theorem 11.27 (By Theorem 7.3.5, p. 231–232 in [4] and Lemma 11.23).
Consider the positive linear operator

$$\tilde{L} \colon C^n([a,b]) \to C([a,b]), \quad n \in \mathbb{N}.$$

Assume that
$$\tilde{L}(1,x) > 0,$$

and
$$\tilde{L}(|t-x|^{n+1}, x) > 0, \quad x \in [a,b], \quad (11.28)$$

Consider also $\rho > 0$. Consider $g \in C^n([a,b])$, $n \geq 1$, with $\omega_1(g^{(n)}, \delta) > 0$ for any $\delta > 0$. Then

$$|\tilde{L}(g,x) - g(x)| \leq |g(x)| |\tilde{L}(1,x) - 1| + \sum_{k=1}^{n} \frac{|g^{(k)}(x)|}{k!} |\tilde{L}((t-x)^k, x)| +$$

$$\left[\frac{n\rho^2}{8} + \frac{\rho}{2} + \frac{1}{(n+1)}\right] \frac{(\tilde{L}(1,x))^{1/(n+1)}}{\rho n!} (\tilde{L}(|t-x|^{n+1}, x))^{\frac{n}{(n+1)}}$$

$$\omega_1\left(g^{(n)}, \rho\left(\frac{\tilde{L}(|t-x|^{n+1}, x)}{\tilde{L}(1,x)}\right)^{\frac{1}{(n+1)}}\right), \quad (11.29)$$

$x \in [a, b]$, $n \in \mathbb{N}$.
   We will use also

Theorem 11.28 (by Theorem 7.2.2, p. 216–217, in [4] and Lemma 11.23).
Consider the positive linear operator

$$\tilde{L} \colon C^n([a, b]) \to C([a, b]), \quad n \in \mathbb{N}.$$

Assume that

$$\tilde{L}(1, x) > 0, \tag{11.30}$$

and

$$\tilde{L}(|t - x|^{n+1}, x) > 0, \quad x \in [a, b].$$

Consider $g \in C^n([a, b])$, $n \geq 1$, *with* $\omega_1(g^{(n)}, \delta) > 0$ *for any* $\delta > 0$. *Then*

$$|\tilde{L}(g, x) - g(x)| \leq |g(x)| \, |\tilde{L}(1, x) - 1| + \sum_{k=1}^{n} \frac{|g^{(k)}(x)|}{k!} |\tilde{L}((t - x)^k, x)|$$

$$+ \frac{\left(\tilde{L}(|t - x|^{n+1}, x)\right)^{n/(n+1)}}{n!} \left((\tilde{L}(1)(x))^{\frac{1}{(n+1)}}\right)$$

$$+ \frac{1}{(n+1)}\right) \omega_1\left(g^{(n)}, (\tilde{L}(|t - x|^{n+1}, x))^{\frac{1}{n+1}}\right),$$

$$x \in [a, b], \ n \in \mathbb{N}. \tag{11.31}$$

Remark 11.29. 1) If $\|\tilde{L}(1)\|_\infty = 0$, then $\tilde{L} = 0$ the trivial operator, therefore without loss of generality we may assume $\tilde{L} \not\equiv 0$, and as a result we have $\|\tilde{L}(1)\|_\infty > 0$.
   2) By using Hölder's inequality and Riesz Representation theorem, for $\tilde{L}$ as in (11.24), we easily derive that

$$|\tilde{L}((t - x)^k, x)| \leq \left(\tilde{L}(1, x)\right)^{1 - \frac{k}{n}} \left(\tilde{L}(|t - x|^n, x)\right)^{\frac{k}{n}}, \tag{11.32}$$

under the assumption $\tilde{L}(1, x) > 0$, $0 < k \leq n$. And it holds

$$\|\tilde{L}((t - x)^k, x)\|_\infty \leq \|\tilde{L}(1)\|_\infty^{1 - \frac{k}{n}} \|\tilde{L}(|t - x|^n, x)\|_\infty^{\frac{k}{n}}, \quad 0 < k \leq n, \ \tilde{L} \not\equiv 0. \tag{11.33}$$

Similarly, by Lemma 11.23, for $\tilde{L} \colon C^n([a, b]) \to C([a, b])$ positive linear operator, it holds

$$|\tilde{L}((t - x)^k, x)| \leq \left(\tilde{L}(1, x)\right)^{1 - \frac{k}{(n+1)}} \left(\tilde{L}(|t - x|^{n+1}, x)\right)^{\frac{k}{(n+1)}}, \ 0 < k < n + 1. \tag{11.34}$$

And also it holds

$$\|\tilde{L}((t - x)^k, x)\|_\infty \leq \|\tilde{L}(1)\|_\infty^{1 - \frac{k}{(n+1)}} \|\tilde{L}(|t - x|^{n+1}, x)\|_\infty^{\frac{k}{(n+1)}}, \ 0 < k < n + 1. \tag{11.35}$$

3) If $\tilde{L}(1, x) = 0$, $x \in [a, b]$, then easily we get that $\tilde{L}(g, x) = 0$, $\forall g \in C^n([a, b])$. So that inequalities (11.26), (11.27) and (11.31) hold trivially.

4) If $d_n(x) = 0$, $x \in [a, b]$, then we get that $c_k(x) = 0$, $k = 0, 1, \ldots, n$, since the associated measure μ_x is concentrated at $\{x\}$ only, etc., and inequalities (11.26) and (11.27) hold again, in fact as equalities.

5) Similarly, if $\tilde{L}(|t - x|^{n+1}, x) = 0$, $x \in [a, b]$, then we get again $\tilde{L}((t - x)^k, x) = 0$, $k = 1, \ldots, n$, since the associated measure μ_x is concentrated at $\{x\}$ only, etc., and inequalities (11.29) and (11.31) hold again, in fact as equalities.

6) If $\omega_1(g^{(n)}, h) = 0$ for some $h > 0$ then $g^{(n)}$ is the constant function, furthermore inequalities (11.26), (11.27), (11.29) and (11.31) are again valid.

We give the first main fuzzy result.

Theorem 11.30. *Consider the fuzzy positive linear operator*

$$L \colon C_{\mathcal{F}}^n([a, b]) \to C_{\mathcal{F}}([a, b]), \quad n \in \mathbb{N}, \tag{11.36}$$

with the property

$$(Lf)_{\pm}^{(r)} = \tilde{L}(f_{\pm}^{(r)}), \tag{11.37}$$

respectively, for all $r \in [0, 1]$, $\forall f \in C_{\mathcal{F}}^n([a, b])$. Here $\tilde{L}$ is a positive linear operator such that

$$\tilde{L} \colon C([a, b]) \to C([a, b]). \tag{11.38}$$

Let

$$
\begin{aligned}
c_k(x) &:= \tilde{L}\big((t - x)^k, x\big), \quad k = 0, 1, \ldots, n; \\
d_n(x) &:= \big(\tilde{L}(|t - x|^n, x)\big)^{1/n}; \\
c(x) &:= \max(x - a, b - x).
\end{aligned}
\tag{11.39}
$$

Let $f \in C_{\mathcal{F}}^n([a, b])$ such that $\omega_1^{(\mathcal{F})}(f^{(n)}, h) > 0$ for any $h > 0$. Then
1)

$$
\begin{aligned}
D\big((Lf)(x), f(x)\big) &\leq |c_0(x) - 1| D(f(x), \tilde{o}) + \sum_{k=1}^{n} \frac{|c_k(x)|}{k!} D(f^{(k)}(x), \tilde{o}) \\
&\quad + \phi_n(c(x)) \left(\frac{d_n(x)}{c(x)}\right)^n \omega_1^{(\mathcal{F})}(f^{(n)}, h),
\end{aligned}
\tag{11.40}
$$

$x \in [a, b]$, and
2)

$$
\begin{aligned}
D^*(Lf, f) &\leq \|\tilde{L}1 - 1\|_\infty D^*(f, \tilde{o}) + \sum_{k=1}^{n} \frac{\|\tilde{L}((t - x)^k, x)\|_\infty}{k!} D^*(f^{(k)}, \tilde{o}) \\
&\quad + \|\phi_n(c(x))\|_\infty \left\| \frac{\tilde{L}(|t - x|^n, x)}{c(x)^n} \right\|_\infty \omega_1^{(\mathcal{F})}(f^{(n)}, h).
\end{aligned}
\tag{11.41}
$$

Proof. We have the following

$$D((Lf)(x), f(x)) = \sup_{r \in [0,1]} \max\{|(Lf)_-^{(r)}(x) - f_-^{(r)}(x)|,$$

$$|(Lf)_+^{(r)}(x) - f_+^{(r)}(x)|\}$$

$$= \sup_{r \in [0,1]} \max\{|(\tilde{L}(f_-^{(r)}))(x) - f_-^{(r)}(x)|, |(\tilde{L}(f_+^{(r)}))(x) - f_+^{(r)}(x)|\}$$

(by Remark 11.13 (11.1) and (11.27))

$$\leq \sup_{r \in [0,1]} \max\left\{ |f_-^{(r)}(x)| \, |c_0(x) - 1| + \sum_{k=1}^{n} \frac{|(f^{(k)})_-^{(r)}(x)|}{k!} |c_k(x)| \right.$$

$$+ \phi_n(c(x)) \left(\frac{d_n(x)}{c(x)} \right)^n \omega_1\left((f^{(n)})_-^{(r)}, h\right), |f_+^{(r)}(x)| \, |c_0(x) - 1|$$

$$+ \sum_{k=1}^{n} \frac{|(f^{(k)})_+^{(r)}(x)|}{k!} |c_k(x)| + \phi_n(c(x)) \left(\frac{d_n(x)}{c(x)} \right)^n \omega_1\left((f^{(n)})_+^{(r)}, h\right) \right\}$$

$$\overset{(11.7)}{\leq} |c_0(x) - 1| \sup_{r \in [0,1]} \max\{|f_-^{(r)}(x)|, |f_+^{(r)}(x)|\}$$

$$+ \sum_{k=1}^{n} \frac{|c_k(x)|}{k!} \sup_{r \in [0,1]} \max\{|(f^{(k)})_-^{(r)}(x)|, |(f^{(k)})_+^{(r)}(x)|\}$$

$$+ \phi_n(c(x)) \left(\frac{d_n(x)}{c(x)} \right)^n \sup_{r \in [0,1]} \max\{\omega_1((f^{(n)})_-^{(r)}, h), \omega_1((f^{(n)})_+^{(r)}, h)\}$$

(by Proposition 11.2)

$$= |c_0(x) - 1| D(f(x), \tilde{o})$$

$$+ \sum_{k=1}^{n} \frac{|c_k(x)|}{k!} D(f^{(k)}(x), \tilde{o}) + \phi_n(c(x)) \left(\frac{d_n(x)}{c(x)} \right)^n \omega_1^{(\mathcal{F})}(f^{(n)}, h).$$

That is proving (11.40). $\qquad\qquad\qquad\qquad\qquad\qquad\qquad\qquad\qquad\qquad\qquad$ □

We proceed with the next main result.

Theorem 11.31. *Consider the fuzzy positive linear operator*

$$L: C_{\mathcal{F}}^n([a,b]) \to C_{\mathcal{F}}([a,b]), \quad n \in \mathbb{N},$$

with the property

$$(Lf)_\pm^{(r)} = \tilde{L}(f_\pm^{(r)}),$$

respectively, for all $r \in [0,1]$, $\forall f \in C_{\mathcal{F}}^n([a,b])$. Here $\tilde{L}$ is a positive linear operator from $C^n([a,b])$ into $C([a,b])$. Additionally assume that $\tilde{L}(1,x) > 0$, $x \in [a,b]$ and consider $\rho > 0$. Let $f \in C_{\mathcal{F}}^n([a,b])$ such that $\omega_1^{(\mathcal{F})}(f^{(n)}, h) > 0$ for any $h > 0$. Then

1)

$$
\begin{aligned}
D((Lf)(x), f(x)) \;\le\; & |\tilde{L}(1,x) - 1| D(f(x), \tilde{o}) \\
& + \sum_{k=1}^{n} \frac{|\tilde{L}((t-x)^k, x)|}{k!} D(f^{(k)}(x), \tilde{o}) \\
& + \left[\frac{n\rho^2}{8} + \frac{\rho}{2} + \frac{1}{(n+1)} \right] \qquad (11.42) \\
& \times \frac{(\tilde{L}(1,x))^{\frac{1}{(n+1)}}}{\rho n!} \left(\tilde{L}(|t-x|^{n+1}, x) \right)^{\frac{n}{(n+1)}} \\
& \times \omega_1^{(\mathcal{F})} \left(f^{(n)}, \rho \left(\frac{\tilde{L}(|t-x|^{n+1}, x)}{\tilde{L}(1,x)} \right)^{\frac{1}{(n+1)}} \right),
\end{aligned}
$$

$n \in \mathbb{N}$, $x \in [a, b]$.

2) *And by assuming* $\tilde{L}(1, x) > 0$, $\forall x \in [a, b]$, *it holds*

$$
\begin{aligned}
D^*(Lf, f) \;\le\; & \|\tilde{L}1 - 1\|_\infty D^*(f, \tilde{o}) + \sum_{k=1}^{n} \frac{\|\tilde{L}((t-x)^k, x)\|_\infty}{k!} D^*(f^{(k)}, \tilde{o}) \\
& + \left[\frac{n\rho^2}{8} + \frac{\rho}{2} + \frac{1}{(n+1)} \right] \frac{\|\tilde{L}(1,x)\|_\infty^{\frac{1}{(n+1)}}}{\rho n!} \|\tilde{L}(|t-x|^{n+1}, x)\|_\infty^{\frac{n}{(n+1)}} \\
& \times \omega_1^{(\mathcal{F})} \left(f^{(n)}, \rho \left\| \frac{\tilde{L}(|t-x|^{n+1}, x)}{\tilde{L}(1,x)} \right\|_\infty^{\frac{1}{(n+1)}} \right). \qquad (11.43)
\end{aligned}
$$

Proof. Here we are using (11.29), see also Remark 11.29(11.5). The proof is similar to the proof of Theorem 11.30 and is omitted.   □

We present

Theorem 11.32. *Consider the fuzzy positive linear operator*

$$
L \colon C_{\mathcal{F}}^n([a, b]) \to C_{\mathcal{F}}([a, b]), \quad n \in \mathbb{N}, \qquad (11.44)
$$

with the property

$$
(Lf)_\pm^{(r)} = \tilde{L}(f_\pm^{(r)}), \qquad (11.45)
$$

respectively, for all $r \in [0, 1]$, $\forall f \in C_{\mathcal{F}}^n([a, b])$. *Here* $\tilde{L}$ *is a positive linear operator such that*

$$
\tilde{L} \colon C^n([a, b]) \to C([a, b]). \qquad (11.46)
$$

Let $f \in C_{\mathcal{F}}^n([a, b])$ *such that* $\omega_1^{(\mathcal{F})}(f^{(n)}, h) > 0$ *for any* $h > 0$. *Then*

1)

$$D((Lf)(x), f(x)) \leq |\tilde{L}(1,x) - 1| D(f(x), \tilde{o})$$
$$+ \sum_{k=1}^{n} \frac{|(\tilde{L}((\cdot - x)^k))(x)|}{k!} D(f^{(k)}, (x), \tilde{o})$$
$$+ \frac{(\tilde{L}(|\cdot - x|^{n+1})(x))^{\frac{n}{(n+1)}}}{n!} \left((\tilde{L}(1,x))^{\frac{1}{(n+1)}} + \frac{1}{(n+1)} \right)$$
$$\times \omega_1^{(\mathcal{F})} \left(f^{(n)}, (\tilde{L}(|\cdot - x|^{n+1})(x))^{\frac{1}{(n+1)}} \right), \quad (11.47)$$

$\forall x \in [a, b]$.

And also it holds

2)

$$D^*(Lf, f) \leq \|\tilde{L}1 - 1\|_\infty D^*(f, \tilde{o}) + \sum_{k=1}^{n} \frac{\|(\tilde{L}((\cdot - x)^k))(x)\|_\infty}{k!} D^*(f^{(k)}, \tilde{o})$$
$$+ \frac{\|(\tilde{L}(|\cdot - x|^{n+1}))(x)\|_\infty^{\frac{n}{(n+1)}}}{n!} \left\| (\tilde{L}(1))^{\frac{1}{(n+1)}} + \frac{1}{(n+1)} \right\|_\infty$$
$$\times \omega_1^{(\mathcal{F})} \left(f^{(n)}, \|(\tilde{L}(|\cdot - x|^{n+1}))(x)\|_\infty^{\frac{1}{n+1}} \right). \quad (11.48)$$

Proof. Here we use (11.31) and see also Remark 11.29 (11.3) & (11.5). We do have

$$D((Lf)(x), f(x)) = \sup_{r \in [0,1]} \max\{|(Lf)_-^{(r)}(x) - f_-^{(r)}(x)|, |(Lf)_+^{(r)}(x) - f_+^{(r)}(x)|\}$$

$$= \sup_{r \in [0,1]} \max\{|\tilde{L}(f_-^{(r)})(x) - f_-^{(r)}(x)|, |\tilde{L}(f_+^{(r)})(x) - f_+^{(r)}(x)|\}$$

$$\leq \sup_{r \in [0,1]} \max\left\{ |f_-^{(r)}(x)| |\tilde{L}(1,x) - 1| + \sum_{k=1}^{n} \frac{|(f^{(k)})_-^{(r)}(x)|}{k!} |(\tilde{L}((\cdot - x)^k))(x)| \right.$$
$$+ \frac{((\tilde{L}(|\cdot - x|^{n+1}))(x))^{\frac{n}{(n+1)}}}{n!} \left((\tilde{L}(1,x))^{\frac{1}{(n+1)}} + \frac{1}{(n+1)} \right)$$
$$\times \omega_1 \left((f^{(n)})_-^{(r)}, ((\tilde{L}(|\cdot - x|^{n+1}))(x))^{\frac{1}{(n+1)}} \right), |f_+^{(r)}(x)| |\tilde{L}(1,x) - 1|$$
$$+ \sum_{k=1}^{n} \frac{|(f^{(k)})_+^{(r)}(x)|}{k!} |(\tilde{L}((\cdot - x)^k))(x)|$$
$$+ \frac{((\tilde{L}(|\cdot - x|^{n+1}))(x))^{\frac{n}{(n+1)}}}{n!} \left((\tilde{L}(1,x))^{\frac{1}{(n+1)}} + \frac{1}{(n+1)} \right)$$
$$\left. \times \omega_1 \left((f^{(n)})_+^{(r)}, ((\tilde{L}(|\cdot - x|^{n+1}))(x))^{\frac{1}{(n+1)}} \right) \right\}$$

$$\leq |\tilde{L}(1,x) - 1| \sup_{r \in [0,1]} \max\{|f_-^{(r)}(x)|, |f_+^{(r)}(x)|\}$$

$$+ \sum_{k=1}^{n} \frac{|(\tilde{L}((\cdot - x)^k))(x)|}{k!} \sup_{r\in[0,1]} \max\{|(f^{(k)})_{-}^{(r)}(x)|, |(f^{(k)})_{+}^{(r)}(x)|\}$$

$$+ \frac{((\tilde{L}(|\cdot - x|^{n+1}))(x))^{\frac{n}{(n+1)}}}{n!} \left((\tilde{L}(1,x))^{\frac{1}{(n+1)}} + \frac{1}{(n+1)} \right)$$

$$\times \sup_{r\in[0,1]} \max\{\omega_1((f^{(n)})_{-}^{(r)}, ((\tilde{L}(|\cdot - x|^{n+1}))(x))^{\frac{1}{(n+1)}}),$$

$$\omega_1((f^{(n)})_{+}^{(r)}, ((\tilde{L}(|\cdot - x|^{n+1}))(x))^{\frac{1}{(n+1)}})\}$$

$$= |\tilde{L}(1,x) - 1| D(f(x), \tilde{o}) + \sum_{k=1}^{n} \frac{|(\tilde{L}((\cdot - x)^k))(x)|}{k!} D(f^{(k)}(x), \tilde{o})$$

$$+ \frac{(\tilde{L}(|\cdot - x|^{n+1})(x))^{\frac{n}{(n+1)}}}{n!} \left((\tilde{L}(1,x))^{\frac{1}{(n+1)}} + \frac{1}{(n+1)} \right)$$

$$\times \omega_1^{(\mathcal{F})}(f^{(n)}, ((\tilde{L}(|\cdot - x|^{n+1}))(x))^{\frac{1}{(n+1)}}).$$

$\square$

Note 11.33. If $\omega_1^{(\mathcal{F})}(f^{(n)}, h) = 0$ for some $h > 0$, then by Proposition 11.2 we get $\omega_1((f^{(n)})_{-}^{(r)}, h), \omega_1((f^{(n)})_{+}^{(r)}, h) = 0, \forall r \in [0,1]$, i.e. $(f_{\pm}^{(r)})^{(n)} = (f^{(n)})_{\pm}^{(r)}$ are constant real valued functions, $\forall r \in [0,1]$. Consequently (see Remark 11.29(11.6) and repeat proofs) inequalities (11.40), (11.41), (11.42), (11.43), (11.47) and (11.48) are again valid.

We give the following fuzzy Korovkin type theorem ([78]).

Theorem 11.34. *Consider the sequence of fuzzy positive linear operators*

$$L_N : C_{\mathcal{F}}^n([a,b]) \to C_{\mathcal{F}}([a,b]), \quad n \in \mathbb{N}, \forall N \in \mathbb{N}, \tag{11.49}$$

with the property

$$(L_N f)_{\pm}^{(r)} = \tilde{L}_N(f_{\pm}^{(r)}), \tag{11.50}$$

respectively, for all $r \in [0,1]$, $\forall f \in C_{\mathcal{F}}^n([a,b])$, $\forall N \in \mathbb{N}$. Here $\{\tilde{L}_N\}_{N\in\mathbb{N}}$ is a sequence of positive linear operators such that

$$\tilde{L}_N : C^n([a,b]) \to C([a,b]).$$

Assume that $\|\tilde{L}_N(1)\|_\infty \le \gamma, \forall N \in \mathbb{N}$, for some $\gamma > 0$. Furthermore assume that $\tilde{L}_N 1 \overset{u}{\to} 1$ and $\|(\tilde{L}_N(|\cdot - x|^{n+1}))(x)\|_\infty \to 0$, as $N \to \infty$. Then $D^*(L_N f, f) \to 0$ as $N \to \infty$, $\forall f \in C_{\mathcal{F}}^n([a,b])$. I.e. $L_N \to I$, as $N \to \infty$, fuzzy and uniformly, where I is the fuzzy unit operator.

Proof. From (11.48) we get

$$D^*(L_N f, f) \leq \|\tilde{L}_N 1 - 1\|_\infty D^*(f, \tilde{o}) \tag{11.51}$$
$$+ \sum_{k=1}^{n} \frac{\|(\tilde{L}_N((\cdot - x)^k))(x)\|_\infty}{k!} D^*(f^{(k)}, \tilde{o})$$
$$+ \frac{\|(\tilde{L}_N(|\cdot - x|^{n+1}))(x)\|_\infty^{\frac{n}{(n+1)}}}{n!} \left\| (\tilde{L}_N(1))^{\frac{1}{(n+1)}} + \frac{1}{(n+1)} \right\|_\infty$$
$$\times \omega_1^{(\mathcal{F})} \left(f^{(n)}, \|(\tilde{L}_N(|\cdot - x|^{n+1}))(x)\|_\infty^{\frac{1}{(n+1)}} \right), \quad \forall N \in \mathbb{N}.$$

Also by (11.35) we have

$$\|\tilde{L}_N((t-x)^k, x)\|_\infty \leq \|\tilde{L}_N(1)\|_\infty^{1 - \frac{k}{(n+1)}} \|\tilde{L}_N(|t-x|^{n+1}, x)\|_\infty^{\frac{k}{(n+1)}}, \tag{11.52}$$

any $0 < k < n+1$, $\forall N \in \mathbb{N}$.

Now by using (11.52), (11.51) and the assumptions of the theorem we conclude that $D^*(L_N f, f) \to 0$, as $N \to \infty$. $\qquad\square$

Comment. Inequality (11.51), proof of Theorem 11.34, gives the convergence of $L_N \to I$, quantitatively, and at higher rate, reflecting the higher order fuzzy differentiability of f.

11.3 Multidimensional Results

We need

Theorem 11.35 (By Theorem 7.4.1, p. 236 of [4] and the Riesz Representation Theorem). *Take $Q := \{x \in \mathbb{R}^k : \|x\|_{\ell_1} \leq 1\}$, $k \geq 1$, $x \in Q$. Let $\tilde{L}: C(Q) \to C(Q)$ positive linear operator with $\tilde{L}(1, x) = 1$ and $f \in C^n(Q)$, $n \in \mathbb{N}$. Here $f_\alpha = \frac{\partial^\alpha f}{\partial x^\alpha}$, where $\alpha := (\alpha_1, \dots, \alpha_k)$, $\alpha_i \in \mathbb{Z}^+$, $i = 1, \dots, k$, and $0 < |\alpha| := \sum_{i=1}^{k} \alpha_i \leq n$. Assume for $h > 0$ we have $w := \max_{|\alpha|=n} \omega_1(f_\alpha, h) > 0$, where ω_1 is the usual modulus of continuity with respect to $\|\cdot\|_{\ell_1}$ relative to Q. Then*

$$|\tilde{L}(f, x) - f(x)| \leq \sum_{j=1}^{n} \left(\sum_{|\alpha|=j} \left(\frac{|f_\alpha(x)|}{\alpha_1! \cdots \alpha_k!} \left| \tilde{L}\left(\left(\prod_{i=1}^{k} (z_i - x_i)^{\alpha_i} \right), x \right) \right| \right) \right)$$
$$+ (\tilde{L}(\|z - x\|_{\ell_1}, x)) \frac{\phi_n(1 + \|x\|_{\ell_1})}{(1 + \|x\|_{\ell_1})} w. \tag{11.53}$$

Proof. We add the following, in order to transfer here Theorem 7.4.1, p. 236 of [4]. Let $g_z(t) := f(x + t(z - x))$, $x, z \in Q$, all $0 \leq t \leq 1$. For

$j = 1, \ldots, n$ we have

$$g_z^{(j)}(t) = \left[\left(\sum_{i=1}^{k}(z_i - x_i)\frac{\partial}{\partial x_i}\right)^j f\right](x + t(z - x))$$

with

$$g_z^{(j)}(0) = \left[\left(\sum_{i=1}^{k}(z_i - x_i)\frac{\partial}{\partial x_i}\right)^j f\right](x), \quad \forall x, z \in Q.$$

More precisely, we get

$$\frac{g_z^{(j)}(0)}{j!} = \sum_{|\alpha|=j} \frac{\left(\prod_{i=1}^{k}(z_i - x_i)^{\alpha_i}\right)}{\prod_{i=1}^{k}\alpha_i!} f_\alpha(x). \tag{11.54}$$

$\square$

We will use

Theorem 11.36 (By Theorem 7.4.2, p. 237 of [4]). *Let μ be a measure of mass $m > 0$ on $Q \subseteq \mathbb{R}^k$, $k \geq 1$ compact and convex. Assume*

$$\frac{1}{(n+1)}\left(\frac{1}{m}\int_Q \|x - x_0\|_{\ell_1}^{n+1}\mu(dx)\right)^{\frac{1}{(n+1)}} =: h > 0, \tag{11.55}$$

where x_0 is a fixed point of Q. Also, let $f \in C^n(Q)$: $\max_{|\alpha|=n}\omega_1(f_\alpha, h) \leq w$, where $w > 0$. Then

$$\left|\int_Q f\,d\mu - f(x_0)\right| \leq |m - 1||f(x_0)| + \left|\sum_{j=1}^{n}\frac{1}{j!}\int_Q g_x^{(j)}(0)\mu(dx)\right|$$
$$+ mwh^n\left[\frac{3}{2}\frac{(n+1)^n}{n!} + \frac{(n+1)^{n-1}}{8(n-1)!}\right], \tag{11.56}$$

where $g_x(t) := f(x_0 + t(x - x_0))$, $t \geq 0$. Above inequality (11.56) is trivially true if $h = 0$, or if $w = 0$ with $h > 0$.

Translating last Theorem 11.36 into the terminology of positive linear operators and by expanding we have

Theorem 11.37. *Let $Q \subseteq \mathbb{R}^k$, $k \geq 1$ compact and convex. Let $\tilde{L}$ be a positive linear operator from $C(Q)$ into $C(Q)$. Assume $\tilde{L}(1, x) > 0$, $\forall x \in Q$.*

Consider $f \in C^n(Q)$, $n \in \mathbb{N}$. *Then*

$$|\tilde{L}(f)(x) - f(x)| \leq |\tilde{L}(1)(x) - 1| \, |f(x)|$$

$$+ \sum_{j=1}^{n} \left\{ \sum_{|\alpha|=j} \left[\frac{|f_\alpha(x)|}{\prod\limits_{i=1}^{k} \alpha_i!} \left| \left(\tilde{L}\left(\prod_{i=1}^{k} (z_i - x_i)^{\alpha_i} \right) \right)(x) \right| \right] \right\}$$

$$+ \left(\tilde{L}(1)(x) \right)^{\frac{1}{(n+1)}} \left(\tilde{L}(\|z - x\|_{\ell_1}^{n+1})(x) \right)^{\frac{n}{(n+1)}} \left[\frac{3}{2n!} + \frac{1}{8(n-1)!(n+1)} \right]$$

$$\times \max_{|\alpha|=n} \omega_1 \left(f_\alpha, \frac{1}{(n+1)} \left(\frac{(\tilde{L}(\|z - x\|_{\ell_1}^{n+1}))(x)}{\tilde{L}(1)(x)} \right)^{\frac{1}{(n+1)}} \right), \qquad (11.57)$$

$\forall x \in Q$.

We have

Corollary 11.38 (to Theorem 11.37). *All as in Theorem 11.37 with* $\tilde{L}(1,x) = 1$, $\forall x \in Q$. *Then*

$$|\tilde{L}(f)(x) - f(x)| \leq \sum_{j=1}^{n} \left\{ \sum_{|\alpha|=j} \left[\frac{|f_\alpha(x)|}{\prod\limits_{i=1}^{k} \alpha_i!} \left| \left(\tilde{L}\left(\prod_{i=1}^{k} (z_i - x_i)^{\alpha_i} \right) \right)(x) \right| \right] \right\}$$

$$+ \left(\tilde{L}(\|z - x\|_{\ell_1}^{n+1})(x) \right)^{\frac{n}{(n+1)}} \left[\frac{3}{2n!} + \frac{1}{8(n-1)!(n+1)} \right] \qquad (11.58)$$

$$\times \max_{|\alpha|=n} \omega_1 \left(f_\alpha, \frac{1}{(n+1)} \left((\tilde{L}(\|z - x\|_{\ell_1}^{n+1}))(x) \right)^{\frac{1}{(n+1)}} \right),$$

$\forall x \in Q$, $\forall f \in C^n(Q)$.

We further give

Theorem 11.39. *Let* Q *be a compact and convex subset of* $\mathbb{R}^k$, $k \geq 1$, *and let* $x_0 := (x_{01}, \ldots, x_{0k}) \in Q$ *be fixed and let* μ *be a measure on* Q *of mass* $m \geq 0$. *Consider* $f \in C^n(Q)$, $n \in \mathbb{N}$, *and suppose that each* nth *order partial derivative* $f_\alpha := \frac{\partial^\alpha f}{\partial x^\alpha}$, *where* $\alpha := (\alpha_1, \ldots, \alpha_k)$, $\alpha_i \in \mathbb{Z}^+$, $i = 1, \ldots, k$ *and* $|\alpha| := \sum\limits_{i=1}^{n} \alpha_i = n$, *has relative to* Q *and the* $\|\cdot\|_{\ell_1}$, *a modulus of continuity* $\omega_1(f_\alpha, h) \leq w$. *Here we take*

$$h := \left(\int_Q \|z - x_0\|_{\ell_1}^{n+1} d\mu(z) \right)^{\frac{1}{(n+1)}}. \qquad (11.59)$$

Then

$$\left| \int_Q f \, d\mu - f(x_0) \right| \leq |m - 1| \, |f(x_0)| + \left| \sum_{j=1}^{n} \frac{1}{j!} \int_Q g_x^{(j)}(0) \mu(dx) \right|$$

$$+ \frac{wh^n}{n!} \left(m^{\frac{1}{(n+1)}} + \frac{1}{(n+1)} \right),$$

where $g_x(t) := f(x_0 + t(x - x_0))$, $t \geq 0$, $x \in Q$.

Proof. Here we have

$$f(z_1, \ldots, z_k) = g_z(1) = \sum_{j=0}^{n} \frac{g_z^{(j)}(0)}{j!} + R_n(z, 0),$$

where

$$R_n(z, 0) := \int_0^1 \left(\int_0^{t_1} \cdots \left(\int_0^{t_{n-1}} (g_z^{(n)}(t_n) - g_z^{(n)}(0))dt_n \right) \cdots \right) dt_1, \quad z \in Q.$$

In [4], p. 236 we got that (see 7.4.4 there)

$$|R_n(z, 0)| \leq w\phi_n(\|z - x_0\|_{\ell_1}), \quad \forall z \in Q.$$

But by (11.23) we have

$$\begin{aligned}
w\phi_n(\|z - x_0\|_{\ell_1}) &\leq w\frac{\|z - x_0\|_{\ell_1}^n}{n!}\left(1 + \frac{\|z - x_0\|_{\ell_1}}{(n+1)h}\right) \\
&= \frac{w}{n!}\left(\|z - x_0\|_{\ell_1}^n + \frac{\|z - x_0\|_{\ell_1}^{n+1}}{(n+1)h}\right).
\end{aligned}$$

Hence

$$\int_Q |R_n(z, 0)| \, d\mu(z) \leq w \int_Q \phi_n\|z - x_0\|_{\ell_1}) d\mu(z)$$

$$\leq \frac{w}{n!}\left[\int_Q \|z - x_0\|_{\ell_1}^n \, d\mu(z) + \frac{1}{(n+1)h}\int_Q \|z - x_0\|_{\ell_1}^{n+1} d\mu(z)\right]$$

(not to have a trivial case we take $\mu(Q) = m > 0$)

$$\leq \frac{w}{n!}\left[m^{\frac{1}{(n+1)}}\left(\int_Q \|z - x_0\|_{\ell_1}^{n+1} d\mu(z)\right)^{\frac{n}{(n+1)}}\right.$$

$$\left. + \frac{1}{h}\left(\frac{1}{(n+1)}\int_Q \|z - x_0\|_{\ell_1}^{n+1} d\mu(z)\right)\right]$$

(we choose

$$h := \left(\int_Q \|z - x_0\|_{\ell_1}^{n+1} d\mu(z)\right)^{\frac{1}{(n+1)}} > 0,$$

the case $h = 0$ is trivial and not discussed here).

$$= \frac{wh^n}{n!}\left(m^{\frac{1}{(n+1)}} + \frac{1}{(n+1)}\right).$$

That is, we got that

$$\int_Q |R_n(z, 0)| d\mu(z) \leq \frac{wh^n}{n!}\left(m^{\frac{1}{(n+1)}} + \frac{1}{(n+1)}\right).$$

The validity of (11.59) is now clear.   □

By using Riesz Representation theorem, Theorem 11.39 and expanding we obtain

Theorem 11.40. *Let* Q *be a compact and convex subset of* $\mathbb{R}^k$, $k \geq 1$. *Let* $\tilde{L}$ *be a positive linear operator from* $C(Q)$ *into itself. Consider* $f \in C^n(Q)$, $n \in \mathbb{N}$. *Then*

$$|\tilde{L}(f)(x) - f(x)| \leq |\tilde{L}(1)(x) - 1| |f(x)|$$

$$+ \sum_{j=1}^{n} \left\{ \sum_{|\alpha|=j} \left[\frac{|f_\alpha(x)|}{\prod_{i=1}^{k} \alpha_i!} \left| \left(\tilde{L}\left(\prod_{i=1}^{k} (z_i - x_i)^{\alpha_i} \right) \right)(x) \right| \right] \right\}$$

$$+ \frac{((\tilde{L}(\|z - x\|_{\ell_1}^{n+1}))(x))^{\frac{n}{(n+1)}}}{n!} \left[((\tilde{L}(1))(x))^{\frac{1}{(n+1)}} + \frac{1}{(n+1)} \right]$$

$$\times \max_{|\alpha|=n} \omega_1\left(f_\alpha, ((\tilde{L}(\|z - x\|_{\ell_1}^{n+1}))(x))^{\frac{1}{(n+1)}} \right), \quad \forall x \in Q. \quad (11.60)$$

Next we give the fuzzy multidimensional related results.

Theorem 11.41. *Let* $Q := \{x \in \mathbb{R}^k : \|x\|_{\ell_1} \leq 1\}$, $k \geq 1$. *Consider the fuzzy positive linear operator*

$$L : C_{\mathcal{F}}^n(Q) \to C(Q), \quad n \in \mathbb{N}, \quad (11.61)$$

with the property

$$(Lf)_{\pm}^{(r)} = \tilde{L}(f_{\pm}^{(r)}), \quad (11.62)$$

respectively, for all $r \in [0,1]$, $\forall f \in C_{\mathcal{F}}^n(Q)$. *Here* $\tilde{L}$ *is a positive linear operator such that*

$$\tilde{L} : C(Q) \to C(Q), \quad (11.63)$$

with $\tilde{L}(1,x) = 1$. *Also* $f_\alpha = \frac{\partial^\alpha f}{\partial x^\alpha}$ *is the fuzzy partial derivative of* f, *where* $\alpha := (\alpha_1, \ldots, \alpha_k)$, $\alpha_i \in \mathbb{Z}^+$, $i = 1, \ldots, k$, *and* $0 < |\alpha| := \sum_{i=1}^{k} \alpha_i \leq n$. *Assume for* $h > 0$ *we have that* $w^{(\mathcal{F})} := \max_{|\alpha|=n} \omega_1^{(\mathcal{F})}(f_\alpha, h) > 0$, *where* $\omega_1^{(\mathcal{F})}$ *is the fuzzy modulus of continuity with respect to* $\|\cdot\|_{\ell_1}$ *relative to* Q. *Then*

1) $$D((Lf)(x), f(x))$$

$$\leq \sum_{j=1}^{n} \left\{ \sum_{|\alpha|=j} \left[\frac{\left| (\tilde{L}((\prod_{i=1}^{k} (z_i - x_i)^{\alpha_i}))(x) \right|}{\prod_{i=1}^{k} \alpha_i!} D(f_\alpha(x), \tilde{o}) \right] \right\}$$

$$+ ((\tilde{L}(\|z - x\|_{\ell_1}))(x)) \frac{\phi_n(1 + \|x\|_{\ell_1})}{(1 + \|x\|_{\ell_1})} w^{(\mathcal{F})}, \forall x \in Q. \quad (11.64)$$

$$2) \quad D^*(Lf, f) \leq \sum_{j=1}^{n} \left\{ \sum_{|\alpha|=j} \left[\frac{\left\| (\tilde{L}((\prod_{i=1}^{k}(z_i - x_i)^{\alpha_i}))(x) \right\|_{\infty}}{\prod_{i=1}^{k} \alpha_i!} D^*(f_\alpha, \tilde{o}) \right] \right\}$$

$$+ \left\| (\tilde{L}(\|z - x\|_{\ell_1}))(x) \right\|_{\infty} \left\| \frac{\phi_n(1 + \|x\|_{\ell_1})}{(1 + \|x\|_{\ell_1})} \right\|_{\infty} w^{(\mathcal{F})}. \qquad (11.65)$$

Proof. We have the following

$$D((Lf)(x), f(x)) = \sup_{r \in [0,1]} \max\{|(Lf)_-^{(r)} - f_-^{(r)}(x)|,$$

$$|(Lf)_+^{(r)} - f_+^{(r)}(x)|\}$$

$$\overset{(11.62)}{=} \sup_{r \in [0,1]} \max\{|\tilde{L}(f_-^{(r)}) - f_-^{(r)}(x)|, |\tilde{L}(f_+^{(r)}) - f_+^{(r)}(x)|\}$$

(by Remark 11.13 (11.2), (11.6) and (11.53))

$$\leq \sup_{r \in [0,1]} \max\left\{ \sum_{j=1}^{n} \left[\sum_{|\alpha|=j} \left(\frac{|(f_\alpha)_-^{(r)}(x)|}{\prod_{i=1}^{k} \alpha_i!} \left| \left(\tilde{L}\left(\prod_{i=1}^{k}(z_i - x_i)^{\alpha_i} \right) \right)(x) \right| \right] \right.$$

$$+ (\tilde{L}(\|z - x\|_{\ell_1})(x)) \frac{\phi_n(1 + \|x\|_{\ell_1})}{(1 + \|x\|_{\ell_1})} \max_{|\alpha|=n} \omega_1((f_\alpha)_-^{(r)}, h),$$

$$\sum_{j=1}^{n} \left[\sum_{|\alpha|=j} \left(\frac{|(f_\alpha)_+^{(r)}(x)|}{\prod_{i=1}^{k} \alpha_i} \left| \left(\tilde{L}\left(\prod_{i=1}^{k}(z_i - x_i)^{\alpha_i} \right) \right)(x) \right| \right) \right]$$

$$+ (\tilde{L}(\|z - x\|_{\ell_1})(x)) \frac{\phi_n(1 + \|x\|_{\ell_1})}{(1 + \|x\|_{\ell_1})} \max_{|\alpha|=n} \omega_1((f_\alpha)_+^{(r)}, h) \right\}$$

$$\overset{(11.7)}{\leq} \sum_{j=1}^{n} \left[\sum_{|\alpha|=j} \left(\frac{D(f_\alpha(x), \tilde{o})}{\prod_{i=1}^{k} \alpha_i!} \left| \left(\tilde{L}\left(\prod_{i=1}^{k}(z_i - x_i)^{\alpha_i} \right) \right)(x) \right| \right) \right]$$

$$+ (\tilde{L}(\|z - x\|_{\ell_1})(x)) \frac{\phi_n(1 + \|x\|_{\ell_1})}{(1 + \|x\|_{\ell_1})} \sup_{r \in [0,1]}$$

$$\times \max\left\{ \max_{|\alpha|=n} \omega_1((f_\alpha)_-^{(r)}, h), \max_{|\alpha|=n} \omega_1((f_\alpha)_+^{(r)}, h) \right\}$$

(by Lemma 11.15)

$$= \sum_{j=1}^{n} \left[\sum_{|\alpha|=j} \left(\frac{D(f_\alpha(x), \tilde{o})}{\prod_{i=1}^{k} \alpha_i!} \left| \left(\tilde{L}\left(\prod_{i=1}^{k}(z_i - x_i)^{\alpha_i} \right) \right)(x) \right| \right) \right]$$

$$+ \, (\tilde{L}(\|z - x\|_{\ell_1})(x)) \frac{\phi_n(1 + \|x\|_{\ell_1})}{(1 + \|x\|_{\ell_1})} \max_{|\alpha|=n} \sup_{r \in [0,1]}$$

$$\times \, \max\{\omega_1((f_\alpha)_-^{(r)}, h), \omega_1((f_\alpha)_+^{(r)}, h)\}$$

$$\overset{\text{(by Proposition 11.2)}}{=} \sum_{j=1}^{n} \left[\sum_{|\alpha|=j} \left(\frac{D(f_\alpha(x), \tilde{o})}{\prod\limits_{i=1}^{k} \alpha_i!} \right. \right.$$

$$\left. \left. \left| \left(\tilde{L} \left(\prod_{i=1}^{k} (z_i - x_i)^{\alpha_i} \right) \right)(x) \right| \right) \right]$$

$$+ \, (\tilde{L}(\|z - x\|_{\ell_1}(x)) \frac{\phi_n(1 + \|x\|_{\ell_1})}{(1 + \|x\|_{\ell_1})} \max_{|\alpha|=n} \omega_1^{(\mathcal{F})}(f_\alpha, h).$$

Inequality (11.64) is established. $\qquad \square$

The next result follows.

Theorem 11.42. *Let $Q \subseteq \mathbb{R}^k$, $k \geq 1$ be a compact convex subset. Consider the fuzzy positive linear operator*

$$L \colon C_{\mathcal{F}}^n(Q) \to C(Q), \quad n \in \mathbb{N},$$

with the property

$$(Lf)_\pm^{(r)} = \tilde{L}(f_\pm^{(r)}),$$

respectively, for all $r \in [0,1]$, $\forall f \in C_{\mathcal{F}}^n(Q)$. Here $\tilde{L}$ is a positive linear operator such that $\tilde{L} \colon C(Q) \to C(Q)$, with $\tilde{L}(1, x) > 0$, $\forall x \in Q$. Consider $f \in C_{\mathcal{F}}^n(Q)$. Then
1)

$$D((Lf)(x), f(x)) \leq |\tilde{L}(1)(x) - 1| D(f(x), \tilde{o})$$

$$+ \sum_{j=1}^{n} \left[\sum_{|\alpha|=j} \left(\frac{\left| (\tilde{L}(\prod\limits_{i=1}^{k} (z_i - x_i)^{\alpha_i}))(x) \right|}{\prod\limits_{i=1}^{k} \alpha_i!} D(f_\alpha(x), \tilde{o}) \right) \right]$$

$$+ \, (\tilde{L}(1)(x))^{\frac{1}{(n+1)}} (\tilde{L}(\|z - x\|_{\ell_1}^{n+1})(x))^{\frac{n}{(n+1)}} \left(\frac{3}{2n!} + \frac{1}{8(n-1)!(n+1)} \right)$$

$$\times \max_{|\alpha|=n} \omega_1^{(\mathcal{F})} \left(f_\alpha, \frac{1}{(n+1)} \left(\frac{(\tilde{L}(\|z - x\|_{\ell_1}^{n+1}))(x)}{\tilde{L}(1)(x)} \right)^{\frac{1}{(n+1)}} \right), \quad (11.66)$$

$\forall x \in Q$.
Also it holds

2)

$$D^*(Lf, f) \leq \|\tilde{L}(1) - 1\|_\infty D^*(f, \tilde{o})$$

$$+ \sum_{j=1}^{n} \left[\sum_{|\alpha|=j} \left(\frac{\|(\tilde{L}(\prod_{i=1}^{k}(z_i - x_i)^{\alpha_i}))(x)\|_\infty}{\prod_{i=1}^{k} \alpha_i!} D^*(f_\alpha, \tilde{o}) \right) \right]$$

$$+ \|\tilde{L}(1)\|_\infty^{\frac{1}{(n+1)}} \left(\|(\tilde{L}(\|z - x\|_{\ell_1}^{n+1}))(x)\|_\infty^{\frac{n}{(n+1)}} \right.$$

$$\times \left(\frac{3}{2n!} + \frac{1}{8(n-1)!(n+1)} \right) \tag{11.67}$$

$$\times \max_{|\alpha|=n} \omega_1^{(\mathcal{F})} \left(f_\alpha, \frac{1}{(n+1)} \left\| \frac{(\tilde{L}(\|z - x\|_{\ell_1}^{n+1}))(x)}{\tilde{L}(1)(x)} \right\|_\infty^{\frac{1}{(n+1)}} \right).$$

Proof. We observe the following:

$$D((Lf)(x), f(x)) = \sup_{r \in [0,1]} \max\{|(Lf)_-^{(r)}(x) - f_-^{(r)}(x)|, |(Lf)_+^{(r)}(x)$$

$$- f_+^{(r)}(x)|\}$$

$$= \sup_{r \in [0,1]} \max\{|\tilde{L}(f_-^{(r)})(x) - f_-^{(r)}(x)|, |\tilde{L}(f_+^{(r)})(x) - f_+^{(r)}(x)|\}$$

$$\overset{(11.57)}{\leq} \sup_{r \in [0,1]} \max\left\{ |\tilde{L}(1)(x) - 1| |f_-^{(r)}(x)| \right.$$

$$+ \sum_{j=1}^{n} \left[\sum_{|\alpha|=j} \left(\frac{|(f_\alpha)_-^{(r)}(x)|}{\prod_{i=1}^{k} \alpha_i!} \left| \left(\tilde{L}\left(\prod_{i=1}^{k}(z_i - x_i)^{\alpha_i} \right) \right)(x) \right| \right) \right]$$

$$+ (\tilde{L}(1)(x))^{\frac{1}{(n+1)}} \left(\tilde{L}(\|z - x\|_{\ell_1}^{n+1})(x) \right)^{\frac{n}{(n+1)}} \left[\frac{3}{2n!} + \frac{1}{8(n-1)!(n+1)} \right]$$

$$\times \max_{|\alpha|=n} \omega_1 \left((f_\alpha)_-^{(r)}, \frac{1}{(n+1)} \left(\frac{\tilde{L}(\|z - x\|_{\ell_1}^{n+1})(x)}{\tilde{L}(1)(x)} \right)^{\frac{1}{(n+1)}} \right), |\tilde{L}(1)(x) - 1|$$

$$|f_+^{(r)}(x)| + \sum_{j=1}^{n} \left[\sum_{|\alpha|=j} \left(\frac{|(f_\alpha)_+^{(r)}(x)|}{\prod_{i=1}^{k} \alpha_i!} \left(\tilde{L}\left(\prod_{i=1}^{k}(z_i - x_i)^{\alpha_i} \right) \right)(x) \right| \right) \right]$$

$$+ (\tilde{L}(1)(x))^{\frac{1}{(n+1)}} \left(\tilde{L}(\|z - x\|_{\ell_1}^{n+1})(x) \right)^{\frac{n}{(n+1)}} \left[\frac{3}{2n!} + \frac{1}{8(n-1)!(n+1)} \right]$$

$$\times \max_{|\alpha|=n} \omega_1 \left((f_\alpha)_+^{(r)}, \frac{1}{(n+1)} \left(\frac{\tilde{L}(\|z - x\|_{\ell_1}^{n+1})(x)}{\tilde{L}(1)(x)} \right)^{\frac{1}{(n+1)}} \right) \right\}$$

$$\leq |\tilde{L}(1)(x) - 1|D(f(x), \tilde{o}) + \sum_{j=1}^{n}\left[\sum_{|\alpha|=j}\left(\frac{\left|\left(\tilde{L}\left(\prod_{i=1}^{k}(z_i - x_i)^{\alpha_i}\right)(x)\right)\right|}{\prod_{i=1}^{k}\alpha_i!}\right.\right.$$

$$\left.\left.D(f_\alpha(x), \tilde{o})\right)\right] + (\tilde{L}(1)(x))^{\frac{1}{(n+1)}}\left(\tilde{L}(\|z - x\|_{\ell_1}^{n+1})(x)\right)^{\frac{n}{(n+1)}}\left[\frac{3}{2n!}\right.$$

$$+ \frac{1}{8(n-1)!(n+1)}\Bigg]\sup_{r\in[0,1]}\max\left\{\max_{|\alpha|=n}\omega_1\left((f_\alpha)_-^{(r)},\right.\right.$$

$$\frac{1}{(n+1)}\left(\frac{\tilde{L}(\|z-x\|_{\ell_1}^{n+1})(x)}{\tilde{L}(1)(x)}\right)^{\frac{1}{(n+1)}}\right),$$

$$\max_{|\alpha|=n}\omega_1\left((f_\alpha)_+^{(r)}, \frac{1}{(n+1)}\left(\frac{\tilde{L}(\|z-x\|_{\ell_1}^{n+1})(x)}{\tilde{L}(1)(x)}\right)^{\frac{1}{(n+1)}}\right)\right\}$$

(by Lemma 11.15 and Prop. 11.2)
$$= |\tilde{L}(1)(x) - 1|D(f(x), \tilde{o})$$

$$+ \sum_{j=1}^{n}\left[\sum_{|\alpha|=j}\left(\frac{\left|\left(\tilde{L}\left(\prod_{i=1}^{k}(z_i - x_i)^{\alpha_i}\right)(x)\right)\right|}{\prod_{i=1}^{k}\alpha_i!}\right)D(f_\alpha(x), \tilde{o})\right]$$

$$+ (\tilde{L}(1)(x))^{\frac{1}{(n+1)}}\left(\tilde{L}(\|z - x\|_{\ell_1}^{n+1})(x)\right)^{\frac{n}{(n+1)}}\left[\frac{3}{2n!} + \frac{1}{8(n-1)!(n+1)}\right]$$

$$\times \max_{|\alpha|=n}\omega_1^{(\mathcal{F})}\left(f_\alpha, \frac{1}{(n+1)}\left(\frac{(\tilde{L}(\|z-x\|_{\ell_1}^{n+1}))(x)}{\tilde{L}(1)(x)}\right)^{\frac{1}{(n+1)}}\right).$$

Inequality (11.66) is proved.   □

Using Theorem 11.40 and working similarly as in the proof of Theorem 11.42 we obtain the very important

Theorem 11.43. *Let $Q \subseteq \mathbb{R}^k$, $k \geq 1$ be a compact convex subset. Consider the fuzzy positive linear operator*

$$L: C_{\mathcal{F}}^n(Q) \to C(Q), \quad n \in \mathbb{N},$$

with the property

$$(Lf)_\pm^{(r)} = \tilde{L}(f_\pm^{(r)}),$$

respectively, for all $r \in [0, 1]$, $\forall f \in C_{\mathcal{F}}^n(Q)$. Here $\tilde{L}$ is a positive linear operator from $C(Q)$ into itself. Consider $f \in C_{\mathcal{F}}^n(Q)$. Then

1)

$$D((Lf)(x), f(x)) \leq |\tilde{L}(1)(x) - 1| D(f(x), \tilde{o})$$

$$+ \sum_{j=1}^{n} \left\{ \sum_{|\alpha|=j} \left[\frac{|(\tilde{L}(\prod_{i=1}^{k} (z_i - x_i)^{\alpha_i}))(x)|}{\prod_{i=1}^{k} \alpha_i!} D(f_\alpha(x), \tilde{o}) \right] \right\}$$

$$+ \frac{(\tilde{L}(\|z - x\|_{\ell_1}^{n+1})(x))^{\frac{n}{(n+1)}}}{n!} \left[((\tilde{L}(1))(x))^{\frac{1}{(n+1)}} + \frac{1}{(n+1)} \right]$$

$$\times \max_{|\alpha|=n} \omega_1^{(\mathcal{F})} \left(f_\alpha, ((\tilde{L}(\|z - x\|_{\ell_1}^{n+1}))(x))^{\frac{1}{(n+1)}} \right), \qquad (11.68)$$

$\forall x \in Q$.

And

2)

$$D^*(Lf, f) \leq \|\tilde{L}1 - 1\|_\infty D^*(f, \tilde{o})$$

$$+ \sum_{j=1}^{n} \left[\sum_{|\alpha|=j} \left(\frac{\|(\tilde{L}(\prod_{i=1}^{k} (z_i - x_i)^{\alpha_i}))(x)\|_\infty}{\prod_{i=1}^{k} \alpha_i!} D^*(f_\alpha, \tilde{o}) \right) \right]$$

$$+ \frac{\|(\tilde{L}(\|z - x\|_{\ell_1}^{n+1}))(x)\|_\infty^{\frac{n}{(n+1)}}}{n!} \left\| (\tilde{L}(1))^{\frac{1}{(n+1)}} + \frac{1}{(n+1)} \right\|_\infty$$

$$\times \max_{|\alpha|=n} \omega_1^{(\mathcal{F})} \left(f_\alpha, \|(\tilde{L}(\|z - x\|_{\ell_1}^{n+1}))(x)\|_\infty^{\frac{1}{n+1}} \right). \qquad (11.69)$$

Next we give a fuzzy multivariate Korovkin type result.

Theorem 11.44. *Let $Q \subseteq \mathbb{R}^k$, $k \geq 1$ be a compact convex subset. Consider the sequence of fuzzy positive linear operators*

$$L_N \colon C_{\mathcal{F}}^n(Q) \to C(Q), \quad n \geq 1, \quad \forall N \in \mathbb{N}$$

with the property

$$(L_N f)_\pm^{(r)} = \tilde{L}_N(f_\pm^{(r)}),$$

respectively, for all $r \in [0, 1]$, $\forall f \in C_{\mathcal{F}}^n(Q)$. Here $\tilde{L}_N$ is a sequence of positive linear operators from $C(Q)$ into itself, $\forall N \in \mathbb{N}$. Assume $\|\tilde{L}_N(1)\| \leq \gamma$, $\gamma > 0$, $\forall N \in \mathbb{N}$, and $\tilde{L}_N(1) \xrightarrow{u} 1$,

$$\|(\tilde{L}_N(\|z - x\|^{n+1}))(x)\|_\infty \to 0, \quad as \ N \to \infty.$$

Then $D^*(L_N f, f) \to 0$, as $N \to \infty$, $\forall f \in C_{\mathcal{F}}^n(Q)$ at higher rate. I.e. $L_N \to I$, as $N \to \infty$, fuzzy and uniformly.

Proof. We use (11.69) and the following.

By Hölder's inequality and Riesz Representation Theorem we obtain that

$$\left\|(\tilde{L}_N(\|z - x\|_{\ell_1}^j))(x)\right\|_\infty \leq \gamma^{1 - \frac{j}{n+1}} \left\|(\tilde{L}_N(\|z - x\|_{\ell_1}^{n+1}))(x)\right\|_\infty^{\frac{j}{(n+1)}}$$

for $j = 1, \ldots, n$. Therefore we get

$$\left\|(\tilde{L}_N(\|z - x\|_{\ell_1}^j))(x)\right\|_\infty \to 0, \quad 1 \leq j \leq n,$$

as $N \to \infty$. Notice that

$$(\tilde{L}_N(\|z - x\|_{\ell_1}^j))(x) = \sum_{|\alpha| = j} \frac{j!}{\prod\limits_{i=1}^{k} \alpha_i!} \left(\tilde{L}_N\left(\prod_{i=1}^{k} |z_i - x_i|^{\alpha_i}\right)\right)(x).$$

Hence, since in the last equality all parts are nonnegative, we have

$$\frac{j!}{\prod\limits_{i=1}^{k} \alpha_i!} \left(\tilde{L}_N\left(\prod_{i=1}^{k} |z_i - x_i|^{\alpha_i}\right)\right)(x) \leq (\tilde{L}_N(\|z - x\|_{\ell_1}^j))(x).$$

Consequently we find that

$$\left\|\left(\tilde{L}_N\left(\prod_{i=1}^{k}(z_i - x_i)^{\alpha_i}\right)\right)(x)\right\|_\infty \leq \frac{\prod\limits_{i=1}^{k} \alpha_i!}{j!} \left\|(\tilde{L}_N(\|z - x\|_{\ell_1}^j))(x)\right\|_\infty.$$

Thus

$$\left\|\left(\tilde{L}_N\left(\prod_{i=1}^{k}(z_i - x_i)^{\alpha_i}\right)\right)(x)\right\|_\infty \to 0,$$

as $N \to \infty$, for all α: $|\alpha| = j$, $j = 1, \ldots, n$. The claim is now established.
$\square$

11.4 L_p-estimates, $p \geq 1$

From [24] we have

Theorem 11.45. *Let K be a convex and compact subset of the real normed vector space $(V, \|\cdot\|)$. Let L be a fuzzy positive linear operator from $C_{\mathcal{F}}(K)$ into itself with the property that there exists positive linear operator $\tilde{L}$ from $C(K)$ into itself with $(Lf)_{\pm}^{(r)} = \tilde{L}(f_{\pm}^{(r)})$, respectively for all $r \in [0, 1]$, $\forall f \in C_{\mathcal{F}}(K)$. Then*

$$D(L(f)(x), f(x)) \leq \|\tilde{L}(1)(x) - 1\| D(f(x), \tilde{o})$$
$$+ \omega_1^{(\mathcal{F})}\left(f, ((\tilde{L}(\|\cdot - x\|^2))(x))^{1/2}\right) \min\{(\tilde{L}(1)(x)$$
$$+ \sqrt{\tilde{L}(1)(x)}), (\tilde{L}(1)(x) + 1)\}, \quad \forall x \in K. \tag{11.70}$$

Furthermore we get

$$D^*(Lf, f) \leq D^*(f, \tilde{o})\|\tilde{L}1 - 1\|_\infty \tag{11.71}$$

$$+ \min\{\|\tilde{L}(1) + \sqrt{\tilde{L}(1)}\|_\infty, \|\tilde{L}(1) + 1\|_\infty\}\omega_1^{(\mathcal{F})}(f, \|(\tilde{L}(\|\cdot - x\|^2))(x)\|_\infty^{1/2}),$$

and $\|\cdot\|_\infty$ stands for the sup-norm over K. In particular, if $\tilde{L}(1) = 1$ then (11.71) reduces to

$$D^*(Lf, f) \leq 2\omega_1^{(\mathcal{F})}(f, \|(\tilde{L}(\|\cdot - x\|^2))(x)\|_\infty^{1/2}). \tag{11.72}$$

We give

Theorem 11.46. *All as in the assumptions of Theorem 11.45, plus $(K, \mathcal{A}, \mu)$ is a Borel measure space with $\mu(K) < \infty$, $p \geq 1$. Then*

$$\left(\int_K D^p(L(f)(x), f(x))d\mu(x)\right)^{1/p}$$

$$\leq \|\tilde{L}1 - 1\|_\infty \left(\int_K D^p(f(x), \tilde{o})d\mu(x)\right)^{1/p}$$

$$+ \min\{\|\tilde{L}(1) + \sqrt{\tilde{L}(1)}\|_\infty, \|\tilde{L}(1) + 1\|_\infty\}$$

$$\times \omega_1^{(\mathcal{F})}(f, \|(\tilde{L}(\|\cdot - x\|^2))(x)\|_\infty^{1/2})(\mu(K))^{1/p}. \tag{11.73}$$

Proof. Let $f, g \in C_\mathcal{F}(K)$ and let $x_n \xrightarrow{\|\cdot\|} x_0$, $n \to \infty$, where $\{x_n\}_{n\in\mathbb{N}}$, $x_0 \in K$. We have

$$D(f(x_n), g(x_n)) \leq D(f(x_n), f(x_0)) + D(f(x_0), g(x_0)) + D(g(x_0), g(x_n))$$

and

$$D(f(x_0), g(x_0)) \leq D(f(x_0), f(x_n) + D(f(x_n), g(x_n)) + D(g(x_n), g(x_0)).$$

Letting $n \to +\infty$, from the continuity of f and g we find

$$\lim_{n\to\infty} D(f(x_n), g(x_n)) = D(f(x_0), g(x_0)).$$

Therefore the function $F(x) = D(f(x), g(x))$, $x \in K$ is a continuous real valued function. Thus $D(L(f)(x), f(x))$ is continuous, hence Borel measurable. Finally using (11.70) and by integrating we obtain (11.73). $\qquad\square$

From now on the measure of integration will be the Lebesgue measure λ, and $\|\cdot\|_p$, $p \geq 1$ will be the L_p-norm.
We present

Theorem 11.47. *Assume $\tilde{L}(1,x) > 0$, $\forall x \in [a,b]$. All the rest as in Theorem 11.31. Then*

$$
\|D((Lf)(x), f(x))\|_p \leq \|\tilde{L}1 - 1\|_\infty \|D(f(x), \tilde{o})\|_p
$$
$$
+ \sum_{k=1}^{n} \frac{\|\tilde{L}((t-x)^k, x)\|_\infty}{k!} \|D(f^{(k)}(x), \tilde{o})\|_p \tag{11.74}
$$
$$
+ (b-a)^{1/p} \left[\frac{n\rho^2}{8} + \frac{\rho}{2} + \frac{1}{(n+1)} \right] \frac{(\|\tilde{L}(1)\|_\infty)^{\frac{1}{(n+1)}}}{\rho n!}
$$
$$
\|(\tilde{L}(|t-x|^{n+1}, x))\|_\infty^{\frac{n}{(n+1)}} \omega_1^{(\mathcal{F})} \left(f^{(n)}, \rho \left\| \left(\frac{\tilde{L}(|t-x|^{n+1}, x)}{\tilde{L}(1,x)} \right)^{\frac{1}{(n+1)}} \right\|_\infty \right),
$$

$n \in \mathbb{N}$.
We also have

Theorem 11.48. *Assume all as in Theorem 11.32. Then*

$$
\|D((Lf)(x), f(x))\|_p \leq \|\tilde{L}1 - 1\|_\infty \|D(f(x), \tilde{o})\|_p
$$
$$
+ \sum_{k=1}^{n} \frac{\|(\tilde{L}((\cdot - x)^k))(x)\|_\infty}{k!} \|D(f^{(k)}(x), \tilde{o})\|_p
$$
$$
+ \frac{\|(\tilde{L}(|\cdot - x|^{n+1})(x))\|_\infty^{\frac{n}{(n+1)}}}{n!} \left\| (\tilde{L}(1))^{\frac{1}{(n+1)}} + \frac{1}{(n+1)} \right\|_\infty
$$
$$
\times \omega_1^{(\mathcal{F})} \left(f^{(n)}, \|(\tilde{L}(|\cdot - x|^{n+1})(x))\|_\infty^{\frac{1}{(n+1)}} \right) (b-a)^{1/p}. \tag{11.75}
$$

Furthermore we list the following multivariate fuzzy L_p results, $p \geq 1$.

Theorem 11.49. *Assume all as in Theorem 11.41. Then*

$$
\|D((Lf)(x), f(x))\|_p \tag{11.76}
$$
$$
\leq \sum_{j=1}^{n} \left\{ \sum_{|\alpha|=j} \left[\frac{\|(\tilde{L}((\prod_{i=1}^{k}(z_i - x_i)^{\alpha_i}))(x)\|_\infty}{\prod_{i=1}^{k} \alpha_i!} \|D(f_\alpha(x), \tilde{o})\|_p) \right] \right\}
$$
$$
+ \|(\tilde{L}(\|z - x\|_{\ell_1}))(x)\|_\infty \left\| \frac{\phi_n(1 + \|x\|_{\ell_1})}{(1 + \|x\|_{\ell_1})} \right\|_\infty w^{(\mathcal{F})}(\lambda(Q))^{1/p}.
$$

We continue with

Theorem 11.50. *Assume all as in Theorem 11.42. Then*

$$\|D((Lf)(x), f(x))\|_p \le \|\tilde{L}(1) - 1\|_\infty \|D(f(x), \tilde{o})\|_p \tag{11.77}$$

$$+ \sum_{j=1}^{n} \left[\sum_{|\alpha|=j} \left(\frac{\|(\tilde{L}(\prod_{i=1}^{k}(z_i - x_i)^{\alpha_i}))(x)\|_\infty}{\prod_{i=1}^{k} \alpha_i!} \|D(f_\alpha(x), \tilde{o})\| \right) \right]$$

$$+ (\|\tilde{L}(1)\|_\infty)^{\frac{1}{(n+1)}} \|(\tilde{L}(\|z - x\|_{\ell_1}^{n+1})(x))\|_\infty^{\frac{n}{(n+1)}} \left(\frac{3}{2n!} + \frac{1}{8(n-1)!(n+1)} \right)$$

$$\times \max_{|\alpha|=n} \omega_1^{(\mathcal{F})} \left(f_\alpha, \frac{1}{(n+1)} \left\| \left(\frac{(\tilde{L}(\|z - x\|_{\ell_1}^{n+1}))(x)}{\tilde{L}(1)(x)} \right) \right\|_\infty^{\frac{1}{(n+1)}} \right) \lambda(Q)^{1/p}.$$

We finish the main results with

Theorem 11.51. *Assume all as in Theorem 11.43. Then*

$$\|D((Lf)(x), f(x))\|_p \le \|\tilde{L}(1) - 1\|_\infty \|D(f(x), \tilde{o})\|_p \tag{11.78}$$

$$+ \sum_{j=1}^{n} \left\{ \sum_{|\alpha|=j} \left[\frac{\|(\tilde{L}(\prod_{i=1}^{k}(z_i - x_i)^{\alpha_i}))(x)\|_\infty}{\prod_{i=1}^{k} \alpha_i!} \|D(f_\alpha(x), \tilde{o})\|_p \right] \right\}$$

$$+ \frac{\|(\tilde{L}(\|z - x\|_{\ell_1}^{n+1})(x))\|_\infty^{\frac{n}{(n+1)}}}{n!} \left\| (\tilde{L}(1))^{\frac{1}{(n+1)}} + \frac{1}{(n+1)} \right\|_\infty$$

$$\times \max_{|\alpha|=n} \omega_1^{(\mathcal{F})} \left(f_\alpha, \|((\tilde{L}(\|z - x\|_{\ell_1}^{n+1}))(x))\|_\infty^{\frac{1}{(n+1)}} \right) \lambda(Q)^{1/p}.$$

11.5 Applications

Let $f \in C_{\mathcal{F}}^1([0,1])$, and the real Bernstein operators

$$(B_N(g))(x) := \sum_{k=0}^{N} g\left(\frac{k}{N} \right) \binom{N}{k} x^k (1-x)^{N-k}, \ \forall x \in [0,1], \ \forall g \in C^1([0,1]).$$

We have $B_N 1 = 1$, $(B_N(id))(x) = x$, also $(B_N(\cdot - x))(x) = 0$, $\forall x \in [0,1]$.
Furthermore we get

$$(B_N((\cdot - x)^2))(x) = \frac{x(1-x)}{N} \le \frac{1}{4N},$$

with equality at $x = \frac{1}{2}$. By (11.11) and (11.17) we obtain

$$D(B_N^{(\mathcal{F})}(f)(x), f(x)) \leq \frac{3}{2}\sqrt{\frac{x(1-x)}{N}}\omega_1^{(\mathcal{F})}\left(f', \sqrt{\frac{x(1-x)}{N}}\right)$$

$$\leq \frac{3}{4\sqrt{N}}\omega_1^{(\mathcal{F})}\left(f', \frac{1}{2\sqrt{N}}\right), \qquad (11.79)$$

$$\forall N \in \mathbb{N}, \ \forall x \in [0,1], \ \forall f \in C_{\mathcal{F}}^1([0,1]).$$

Clearly then $\lim_{N \to +\infty} D^*(B_N^{(\mathcal{F})}(f), f) = 0$, fuzzy and uniformly, $\forall f \in C_{\mathcal{F}}^1([0,1])$, at higher speed than (11.13).

2) Inequality (11.68) for $n = 1$ becomes

$$D((Lf)(x), f(x)) \leq |\tilde{L}(1)(x) - 1|D(f(x), \tilde{o}) \qquad (11.80)$$

$$+ \sum_{i=1}^k \left(|(\tilde{L}(z_i - x_i))(x)|D\left(\frac{\partial f}{\partial x_i}(x), \tilde{o}\right)\right)$$

$$+ \sqrt{((\tilde{L}(\|z - x\|_{\ell_1}^2))(x))}\left[\sqrt{(\tilde{L}(1))(x)} + \frac{1}{2}\right]$$

$$\times \max_{i=\{1,\ldots,k\}} \omega_1^{(\mathcal{F})}\left(\frac{\partial f}{\partial x_i}, \sqrt{((\tilde{L}(\|z - x\|_{\ell_1}^2))(x))}\right), \quad \forall x \in Q.$$

If $\tilde{L}(1)(x) = 1$, $(\tilde{L}(z_i - x_i))(x) = 0$, $\forall x \in Q$, all $i = 1, \ldots, k$, $n = 1$, then (11.68) reduces to

$$D((Lf)(x), f(x)) \leq \frac{3}{2}\sqrt{((\tilde{L}(\|z - x\|_{\ell_1}^2))(x))} \qquad (11.81)$$

$$\times \max_{i \in \{1,\ldots,k\}} \omega_1^{(\mathcal{F})}\left(\frac{\partial f}{\partial x_i}, \sqrt{((\tilde{L}(\|z - x\|_{\ell_1}^2))(x))}\right),$$

$\forall x \in Q$.

Let $g \in C([0,1])^2$, the two-dimensional Bernstein polynomials of g are defined by

$$(B_{m,\bar{n}}(g))(t_1, t_2) := \sum_{k=0}^m \sum_{\ell=0}^{\bar{n}} g\left(\frac{k}{m}, \frac{\ell}{\bar{n}}\right)\binom{m}{k}\binom{\bar{n}}{\ell}t_1^k(1-t_1)^{m-k}t_2^\ell(1-t_2)^{\bar{n}-\ell},$$

$$(11.82)$$

for all $t := (t_1, t_2) \in [0,1]^2$, all $(m, \bar{n}) \in \mathbb{N}^2$. It is known that $B_{m,\bar{n}}(g) \to g$ uniformly on $[0,1]^2$. Clearly $(B_{m,\bar{n}}(1))(t_1 t_2) = 1$, $\forall (t_1, t_2) \in [0,1]^2$, $\forall (m, \bar{n}) \in \mathbb{N}^2$. Using Schwarz's inequality we get

$$\sqrt{((B_{m,\bar{n}}(\|\cdot - t\|_{\ell_1}^2))(t))} \leq \left(\sqrt{\frac{t_1(1-t_1)}{m}} + \sqrt{\frac{t_2(1-t_2)}{\bar{n}}}\right)$$

$$\leq \frac{1}{2}\left(\frac{1}{\sqrt{m}} + \frac{1}{\sqrt{n}}\right), \qquad (11.83)$$

$\forall (m, \bar{n}) \in \mathbb{N}^2, \ \forall t \in [0, 1]^2.$

We have easily that

$$(B_{m,\bar{n}}(z_i - t_i))(t_1, t_2) = 0, \quad i = 1, 2. \tag{11.84}$$

Next we define the fuzzy two-dimensional Bernstein operators as follows

$$(B_{m,\bar{n}}^{(\mathcal{F})}(f))(t_1, t_2) := \sum_{k=1}^{m}{}^{*} \sum_{\ell=0}^{\bar{n}}{}^{*} f\left(\frac{k}{m}, \frac{\ell}{\bar{n}}\right)$$

$$\odot \binom{m}{k} \binom{\bar{n}}{\ell} t_1^k (1 - t_1)^{m-k} t_2^\ell (1 - t_2)^{\bar{n}-\ell}, \tag{11.85}$$

$$\forall (t_1, t_2) \in [0, 1]^2, \ \forall (m, \bar{n}) \in \mathbb{N}^2, \ \forall f \in C_{\mathcal{F}}([0, 1]^2).$$

We observe as valid the following

$$(B_{m,\bar{n}}^{(\mathcal{F})}(f))_{\pm}^{(r)} = B_{m,\bar{n}}(f_{\pm}^{(r)}), \tag{11.86}$$

respectively, for all $r \in [0, 1]$, $\forall f \in C_{\mathcal{F}}([0, 1]^2)$. Finally, by (11.81) and (11.83) we derive that

$$D\big((B_{m,\bar{n}}^{(\mathcal{F})}(f))(t_1, t_2), f(t_1, t_2)\big)$$

$$\leq \frac{3}{2} \left(\sqrt{\frac{t_1(1 - t_1)}{m}} + \sqrt{\frac{t_2(1 - t_2)}{\bar{n}}} \right)$$

$$\times \max_{i \in \{1, 2\}} \left\{ \omega_1^{(\mathcal{F})} \left(\frac{\partial f}{\partial x_i}, \left(\sqrt{\frac{t_1(1 - t_1)}{m}} + \sqrt{\frac{t_2(1 - t_2)}{\bar{n}}} \right) \right) \right\}$$

$$\leq \frac{3}{4} \left(\frac{1}{\sqrt{m}} + \frac{1}{\sqrt{\bar{n}}} \right) \max \left\{ \omega_1^{(\mathcal{F})} \left(\frac{\partial f}{\partial x_1}, \frac{1}{2} \left(\frac{1}{\sqrt{m}} + \frac{1}{\sqrt{\bar{n}}} \right) \right), \right.$$

$$\left. \omega_1^{(\mathcal{F})} \left(\frac{\partial f}{\partial x_2}, \frac{1}{2} \left(\frac{1}{\sqrt{m}} + \frac{1}{\sqrt{\bar{n}}} \right) \right) \right\}, \tag{11.87}$$

$$\forall (m, \bar{n}) \in \mathbb{N}^2, \ \forall (t_1, t_2) \in [0, 1]^2, \ \forall f \in C_{\mathcal{F}}^1([0, 1]^2).$$

Clearly then $\lim\limits_{m,\bar{n} \to \infty} D^*(B_{m,\bar{n}}^{(\mathcal{F})}(f), f) = 0$, fuzzy and uniformly,

$\forall f \in C_{\mathcal{F}}^1([0, 1]^2)$, at a higher rate.

One can give many similar other applications of the produced theorems.

12

FUZZY WAVELET LIKE OPERATORS

The basic wavelet type operators A_k, B_k, C_k, D_k, $k \in \mathbb{Z}$ were studied extensively in the real case, e.g., see [9]. Here they are extended to the fuzzy setting and are defined similarly via a real valued scaling function. Their pointwise and uniform convergence with rates to the fuzzy unit operator I is presented. The produced Jackson type inequalities involve the fuzzy first modulus of continuity and usually are proved to be sharp, in fact attained. Furthermore all fuzzy wavelet like operators A_k, B_k, C_k, D_k preserve monotonicity in the fuzzy sense. Here we do not suppose any kind of orthogonality condition on the scaling function φ, and the operators act on fuzzy valued continuous functions over $\mathbb{R}$. This chapter follows [14].

12.1 Background

We use the following

Definition 12.1 ([11]). Let $f \colon \mathbb{R} \to \mathbb{R}_{\mathcal{F}}$ be a fuzzy real number valued function.

We define the (*first*) *fuzzy modulus of continuity of* f by

$$\omega_1^{(\mathcal{F})}(f, \delta) := \sup_{\substack{x, y \in \mathbb{R} \\ |x-y| \le \delta}} D(f(x), f(y)), \quad \delta > 0.$$

G.A. Anastassiou: Fuzzy Mathematics: Approximation The., STUDFUZZ 251, pp. 191–207.
springerlink.com   © Springer-Verlag Berlin Heidelberg 2010

Denote by $C(\mathbb{R}, \mathbb{R}_{\mathcal{F}})$ the space of fuzzy continuous functions and by $C_b(\mathbb{R}, \mathbb{R}_{\mathcal{F}})$ the space of bounded fuzzy continuous functions on $\mathbb{R}$ with respect to the metric D.

Definition 12.2 ([11]). Let $f \colon \mathbb{R} \to \mathbb{R}_{\mathcal{F}}$. We call f a *uniformly continuous fuzzy real number valued function*, iff for any $\varepsilon > 0$ there exists $\delta > 0$: whenever $|x - y| \le \delta$; $x, y \in \mathbb{R}$, implies that $D(f(x), f(y)) \le \varepsilon$. We denote it as $f \in C_{\mathcal{F}}^{U}(\mathbb{R})$.

Proposition 12.3 ([11]). *Let* $f \in C_{\mathcal{F}}^{U}(\mathbb{R})$. *Then* $\omega_1^{(\mathcal{F})}(f, \delta) < +\infty$, *any* $\delta > 0$.

Proposition 12.4 ([11]). *It holds*

(i) $\omega_1^{(\mathcal{F})}(f, \delta)$ *is nonnegative and nondecreasing in* $\delta > 0$, *any* $f \colon \mathbb{R} \to \mathbb{R}_{\mathcal{F}}$.

(ii) $\lim_{\delta \downarrow 0} \omega_1^{(\mathcal{F})}(f, \delta) = \omega_1^{(\mathcal{F})}(f, 0) = 0$, *iff* $f \in C_{\mathcal{F}}^{U}(\mathbb{R})$.

(iii) $\omega_1^{(\mathcal{F})}(f, \delta_1 + \delta_2) \le \omega_1^{(\mathcal{F})}(f, \delta_1) + \omega_1^{(\mathcal{F})}(f, \delta_2)$, $\delta_1, \delta_2 > 0$, *any* $f \colon \mathbb{R} \to \mathbb{R}_{\mathcal{F}}$.

(iv) $\omega_1^{(\mathcal{F})}(f, n\delta) \le n\omega_1^{(\mathcal{F})}(f, \delta)$, $\delta > 0$, $n \in \mathbb{N}$, *any* $f \colon \mathbb{R} \to \mathbb{R}_{\mathcal{F}}$.

(v) $\omega_1^{(\mathcal{F})}(f, \lambda\delta) \le \lceil \lambda \rceil \omega_1^{(\mathcal{F})}(f, \delta) \le (\lambda + 1)\omega_1^{(\mathcal{F})}(f, \delta)$, $\lambda > 0$, $\delta > 0$, *where* $\lceil \cdot \rceil$ *is the ceiling of the number, any* $f \colon \mathbb{R} \to \mathbb{R}_{\mathcal{F}}$.

(vi) $\omega_1^{(\mathcal{F})}(f \oplus g, \delta) \le \omega_1^{(\mathcal{F})}(f, \delta) + \omega_1^{(\mathcal{F})}(g, \delta)$, $\delta > 0$, *any* $f, g \colon \mathbb{R} \to \mathbb{R}_{\mathcal{F}}$.

(vii) $\omega_1^{(\mathcal{F})}(f, \cdot)$ *is continuous on* $\mathbb{R}_+$, *for* $f \in C_{\mathcal{F}}^{U}(\mathbb{R})$.

12.2   Results

We present the first main result.

Theorem 12.5. *Let* $f \in C(\mathbb{R}, \mathbb{R}_{\mathcal{F}})$ *and the scaling function* $\varphi(x)$ *a real valued bounded function with* $\operatorname{supp} \varphi(x) \subseteq [-a, a]$, $0 < a < +\infty$, $\varphi(x) \ge 0$, *such that* $\sum\limits_{j=-\infty}^{\infty} \varphi(x - j) \equiv 1$ *on* $\mathbb{R}$. *For* $k \in \mathbb{Z}$, $x \in \mathbb{R}$ *put*

$$(B_k f)(x) := \sum_{j=-\infty}^{\infty}{}^{*} f\left(\frac{j}{2^k}\right) \odot \varphi(2^k x - j), \tag{12.1}$$

which is a fuzzy wavelet like operator. Then

$$D(B_k f)(x), f(x)) \le \omega_1^{(\mathcal{F})}\left(f, \frac{a}{2^k}\right), \tag{12.2}$$

and

$$D^*(B_k f, f) \leq \omega_1^{(\mathcal{F})}\left(f, \frac{a}{2^k}\right), \qquad (12.3)$$

all $x \in \mathbb{R}$, and $k \in \mathbb{Z}$. If $f \in C_{\mathcal{F}}^U(\mathbb{R})$, then as $k \to +\infty$ we get $\omega_1^{(\mathcal{F})}\left(f, \frac{a}{2^k}\right) \to 0$ and $\lim_{k \to +\infty} B_k f = f$, pointwise and uniformly with rates.

Proof. Notice that

$$(B_k f)(x) = \sideset{}{^*}\sum_{\substack{j \\ 2^k x - j \in [a,a]}} f\left(\frac{j}{2^k}\right) \odot \varphi(2^k x - j).$$

We would like to estimate

$$
\begin{aligned}
D((B_k f)(x), f(x)) &= D\left(\sideset{}{^*}\sum_{\substack{j \\ 2^k x - j \in [-a,a]}} f\left(\frac{j}{2^k}\right) \odot \varphi(2^k x - j), f(x) \odot 1 \right) \\
&= D\left(\sideset{}{^*}\sum_{\substack{j \\ 2^k x - j \in [-a,a]}} f\left(\frac{j}{2^k}\right) \odot \varphi(2^k x - j), \right. \\
&\qquad \left. f(x) \odot \sum_{j=-\infty}^{\infty} \varphi(2^k x - j) \right) \\
&= D\left(\sideset{}{^*}\sum_{\substack{j \\ 2^k x - j \in [-a,a]}} f\left(\frac{j}{2^k}\right) \odot \varphi(2^k x - j), \right. \\
&\qquad \left. \sideset{}{^*}\sum_{\substack{j \\ 2^k x - j \in [-a,a]}} f(x) \odot \varphi(2^k x - j) \right) \\
&\leq \sum_{\substack{j \\ 2^k x - j \in [-a,a]}} \varphi(2^k x - j) D\left(f\left(\frac{j}{2^k}\right), f(x) \right)
\end{aligned}
$$

$$\leq \sum_{\substack{j \\ 2^k x - j \in [-a,a]}} \varphi(2^k x - j)\omega_1^{(\mathcal{F})}\left(f, \left|\frac{j}{2^k} - x\right|\right)$$

$$\left(\text{here } x - \frac{j}{2^k} \in \left[-\frac{a}{2^k}, \frac{a}{2^k}\right]\right)$$

$$\leq \left(\sum_{\substack{j \\ 2^k x - j \in [-a,a]}} \varphi(2^k x - j)\right)\omega_1^{(\mathcal{F})}\left(f, \frac{a}{2^k}\right)$$

$$= 1 \cdot \omega_1^{(\mathcal{F})}\left(f, \frac{a}{2^k}\right). \quad \square$$

It follows the next important result.

Theorem 12.6. *Let $f \in C_b(\mathbb{R}, \mathbb{R}_{\mathcal{F}})$ and the scaling function $\varphi(x)$ a real valued function with supp $\varphi(x) \subseteq [-a,a]$, $0 < a < +\infty$, φ is continuous on $[-a,a]$, $\varphi(x) \geq 0$, such that $\sum_{j=-\infty}^{\infty} \varphi(x - j) = 1$ on $\mathbb{R}$ (then $\int_{-\infty}^{\infty} \varphi(x)dx = 1$). Define*

$$\varphi_{kj}(t) := \quad 2^{k/2}\varphi(2^k t - j), \quad \text{for } k, j \in \mathbb{Z}, \quad t \in \mathbb{R}, \quad (12.4)$$

$$\langle f, \varphi_{kj}\rangle := \quad (FR)\int_{\frac{j-a}{2^k}}^{\frac{j+a}{2^k}} f(t) \odot \varphi_{kj}(t)dt, \quad (12.5)$$

and set

$$(A_k f)(x) := \sum_{j=-\infty}^{\infty}{}^{*} \langle f, \varphi_{kj}\rangle \odot \varphi_{kj}(x), \quad x \in \mathbb{R}, \quad (12.6)$$

which a fuzzy wavelet like operator. Then

$$D((A_k f)(x), f(x)) \leq \omega_1^{(\mathcal{F})}\left(f, \frac{a}{2^{k-1}}\right), \quad x \in \mathbb{R}, \ k \in \mathbb{Z}, \quad (12.7)$$

and

$$D^*((A_k f), f) \leq \omega_1^{(\mathcal{F})}\left(f, \frac{a}{2^{k-1}}\right). \quad (12.8)$$

If $f \in C_{\mathcal{F}}^U(\mathbb{R})$ and bounded, then again we get $A_k \to$ unit operator I with rates as $k \to +\infty$.

Proof. Since φ is compactly supported we have

$$\varphi_{kj}(t) \neq 0 \text{ iff } -a \leq 2^k t - j \leq a, \text{ iff } \frac{j-a}{2^k} \leq t \leq \frac{j+a}{2^k}.$$

Also it holds that

$$(A_k f)(x) := \sum_{\substack{j \\ 2^k x - j \in [-a,a]}}{}^{*} \langle f, \varphi_{kj}\rangle \odot \varphi_{kj}(x), \quad k \in \mathbb{Z}.$$

We would like to estimate

$$D((A_k f)(x), f(x)) = D\left(\sum_{\substack{j \\ 2^k x - j \in [-a,a]}}^{*} \langle f, \varphi_{kj} \rangle \odot \varphi_{kj}(x), f(x)\right)$$

$$= D\left(\sum_{\substack{j \\ 2^k x - j \in [-a,a]}}^{*} \langle f, \varphi_{kj} \rangle \odot \varphi_{kj}(x),\right.$$

$$\left. f(x) \odot \sum_{\substack{j \\ 2^k x - j \in [-a,a]}} \varphi(2^k x - j)\right)$$

$$= D\left(\sum_{\substack{j \\ 2^k x - j \in [-a,a]}}^{*} \langle f, \varphi_{kj} \rangle \odot \varphi_{kj}(x),\right.$$

$$\left. \sum_{\substack{j \\ 2^k x - j \in [-a,a]}}^{*} f(x) \odot 2^{-k/2} \varphi_{kj}(x)\right)$$

$$\leq \sum_{\substack{j \\ 2^k x - j \in [-a,a]}} \varphi_{kj}(x) D(\langle f, \varphi_{kj} \rangle, 2^{-k/2} \odot f(x)) =: (*).$$

Next we estimate separately

$$D(\langle f, \varphi_{kj} \rangle, 2^{-k/2} \odot f(x))$$

$$= D\left((FR) \int_{\frac{j-a}{2^k}}^{\frac{j+a}{2^k}} f(t) \odot \varphi_{kj}(t) dt, 2^{-k/2} \odot f(x)\right)$$

$$= D\left(2^{k/2} \odot (FR) \int_{\frac{j-a}{2^k}}^{\frac{j+a}{2^k}} f(t) \odot \varphi(2^k t - j) dt, 2^{-k/2} \odot f(x)\right)$$

(in Fuzzy-Riemann integral we can have linear change of variables)

$$= D\left(2^{k/2} \odot (FR) \int_{j-a}^{j+a} f\left(\frac{u}{2^k}\right) \odot \varphi(u-j) \frac{du}{2^k}, 2^{-k/2} \odot f(x)\right)$$

$$= D\left(2^{-k/2} \odot (FR) \int_{j-a}^{j+a} f\left(\frac{u}{2^k}\right) \odot \varphi(u-j) du, 2^{-k/2} \odot f(x)\right)$$

$$= 2^{-k/2} D\left((FR) \int_{j-a}^{j+a} f\left(\frac{u}{2^k}\right) \odot \varphi(u-j) du, f(x) \odot 1\right) =: (**).$$

Notice that $\int_{-\infty}^{\infty} \varphi(u-j)du = 1$, $j \in \mathbb{Z}$ and by compact support of φ we have

$$\int_{j-a}^{j+a} \varphi(u-j)du = 1.$$

Hence

$$(**) = 2^{-k/2} D\left((FR)\int_{j-a}^{j+a} f\left(\frac{u}{2^k}\right) \odot \varphi(u-j)du,\right.$$

$$f(x) \odot \int_{j-a}^{j+a} \varphi(u-j)du\Bigg)$$

$$= 2^{-k/2} D\left((FR)\int_{j-a}^{j+a} f\left(\frac{u}{2^k}\right) \odot \varphi(u-j)du,\right.$$

$$(FR)\int_{j-a}^{j+a} f(x) \odot \varphi(u-j)du\Bigg)$$

$$\begin{array}{c} \text{(Lemma 1, [11]) and} \\ \leq \\ \text{(Lemma 12.7 next)} \end{array} \quad 2^{-k/2} \int_{j-a}^{j+a} D\left(f\left(\frac{u}{2^k}\right) \odot \varphi(u-j),\right.$$

$$f(x) \odot \varphi(u-j))\, du$$

$$= 2^{-k/2} \int_{j-a}^{j+a} \varphi(u-j) D\left(f\left(\frac{u}{2^k}\right), f(x)\right) du$$

$$\leq 2^{-k/2} \int_{j-a}^{j+a} \varphi(u-j)\omega_1^{(\mathcal{F})}\left(f, \left|\frac{u}{2^k} - x\right|\right) du$$

$$\left(\text{notice that} \quad -\frac{a}{2^{k-1}} \leq \frac{u}{2^k} - x \leq \frac{a}{2^{k-1}}\right)$$

$$\leq 2^{-k/2}\left(\int_{j-a}^{j+a} \varphi(u-j)du\right)\omega_1^{(\mathcal{F})}\left(f, \frac{a}{2^{k-1}}\right)$$

$$\leq 2^{-k/2}\omega_1^{(\mathcal{F})}\left(f, \frac{a}{2^{k-1}}\right).$$

I.e. we prove that

$$D(\langle f, \varphi_{kj}\rangle, 2^{-k/2} \odot f(x)) \leq 2^{-k/2}\omega_1^{(\mathcal{F})}\left(f, \frac{a}{2^{k-1}}\right).$$

Going back to $(*)$ we get

$$
(*) \quad \leq \quad \sum_{\substack{j \\ 2^k x - j \in [-a,a]}} \varphi_{kj}(x) 2^{-k/2} \omega_1^{(\mathcal{F})} \left(f, \frac{a}{2^{k-1}} \right)
$$

$$
= \quad \left(\sum_{\substack{j \\ 2^k x - j \in [-a,a]}} \varphi(2^k x - j) \right) \omega_1^{(\mathcal{F})} \left(f, \frac{a}{2^{k-1}} \right)
$$

$$
= \quad \left(\sum_{j=-\infty}^{\infty} \varphi(2^k x - j) \right) \omega_1^{(\mathcal{F})} \left(f, \frac{a}{2^{k-1}} \right) = 1 \cdot \omega_1^{(\mathcal{F})} \left(f, \frac{a}{2^{k-1}} \right), \quad x \in \mathbb{R}. \quad \Box
$$

Here we use

Lemma 12.7. Let $f \colon \mathbb{R} \to \mathbb{R}_{\mathcal{F}}$ fuzzy continuous and bounded, i.e. $\exists M_1 > 0 \colon D(f(x), \tilde{o}) \leq M_1, \forall x \in \mathbb{R}$. Let also $g \colon J \subseteq \mathbb{R} \to \mathbb{R}_+$ continuous and bounded, i.e. $\exists M_2 > 0 \colon g(x) \leq M_2, \forall x \in J$, where J is an interval. Then $f(x) \odot g(x)$ is fuzzy continuous function $\forall x \in J$.

Proof. Let $x_n, x_0 \in J, n = 1, 2, \ldots$, such that $x_n \to x_0$. Thus $D(f(x_n), f(x_0)) \to 0$, as $n \to +\infty$ and $|g(x_n) - g(x_0)| \to 0$. We need to establish that

$$
\Delta_n := D(f(x_n) \odot g(x_n), f(x_0) \odot g(x_0)) \to 0,
$$

as $n \to +\infty$. We have

$$
2\Delta_n = D(2 \odot (f(x_n) \odot g(x_n)), 2 \odot (f(x_0) \odot g(x_0)))
$$
$$
\text{(notice for } u \in \mathbb{R}_{\mathcal{F}} \text{ that } u \oplus u = 2 \odot u)
$$
$$
D(f(x_n) \odot g(x_n) \oplus f(x_n) \odot g(x_n) \oplus f(x_0) \odot g(x_n)
$$
$$
\oplus f(x_n) \odot g(x_0), f(x_0) \odot g(x_n) \oplus f(x_n) \odot g(x_0) \oplus f(x_0)
$$
$$
\odot g(x_0) \oplus f(x_0) \odot g(x_0))
$$
$$
\leq \quad D(f(x_n) \odot g(x_n), f(x_0) \odot g(x_n)) + D(f(x_n) \odot g(x_n), f(x_n) \odot g(x_0))
$$
$$
+ D(f(x_0) \odot g(x_n), f(x_0) \odot g(x_0)) + D(f(x_n) \odot g(x_0),
$$
$$
f(x_0) \odot g(x_0))
$$
$$
\text{(by Lemma 2.2, [31])}
$$
$$
\leq \quad g(x_n) D(f(x_n), f(x_0))
$$
$$
+ |g(x_n) - g(x_0)| D(f(x_n), \tilde{o})
$$
$$
+ |g(x_n) - g(x_0)| D(f(x_0), \tilde{o}) + g(x_0) D(f(x_n), f(x_0))
$$
$$
\leq \quad 2M_2 D(f(x_n), f(x_0)) + 2M_1 |g(x_n) - g(x_0)| \to 0, \quad \text{as } n \to +\infty. \quad \Box
$$

We proceed with the following related result.

Theorem 12.8. *All assumptions here are as in Theorem 12.5. Define for* $k \in \mathbb{Z}$, $x \in \mathbb{R}$ *the fuzzy wavelet like operator*

$$(C_k f)(x) := \sum_{j=-\infty}^{\infty}{}^{*} \left(2^k \odot (FR) \int_0^{2^{-k}} f\left(t + \frac{j}{2^k}\right) dt \right) \odot \varphi(2^k x - j). \quad (12.9)$$

Then

$$D((C_k f)(x), f(x)) \le \omega_1^{(\mathcal{F})} \left(f, \frac{a+1}{2^k} \right), \quad (12.10)$$

and

$$D^*((C_k f), f) \le \omega_1^{(\mathcal{F})} \left(f, \frac{a+1}{2^k} \right), \quad \text{all } k \in \mathbb{Z}, \ x \in \mathbb{R}. \quad (12.11)$$

When $f \in C_{\mathcal{F}}^U(\mathbb{R})$ *then as* $k \to +\infty$ *we get* $C_k \to I$ *with rates.*

Proof. We need to estimate

$$D((C_k f)(x), f(x))$$

$$= D \left(\sum_{\substack{j \\ 2^k x - j \in [-a,a]}}^{*} \left(2^k \odot (FR) \int_0^{2^{-k}} f\left(t + \frac{j}{2^k}\right) dt \right) \right.$$

$$\odot \varphi(2^k x - j), f(x) \odot 1 \Big)$$

$$= D \left(\sum_{\substack{j \\ 2^k x - j \in [-a,a]}}^{*} \left(2^k \odot (FR) \int_0^{2^{-k}} f\left(t + \frac{j}{2^k}\right) dt \right) \odot \varphi(2^k x - j), \right.$$

$$\sum_{\substack{j \\ 2^k x - j \in [-a,a]}}^{*} \left(2^k \odot (FR) \int_0^{2^{-k}} (f(x) \odot 1) dt \right) \odot \varphi(2^k x - j) \Big)$$

$$\le \sum_{\substack{j \\ 2^k x - j \in [-a,a]}} D \left(\left(2^k \odot (FR) \int_0^{2^{-k}} f\left(t + \frac{j}{2^k}\right) dt \right) \odot \varphi(2^k x - j), \right.$$

$$\left(2^k \odot (FR) \int_0^{2^{-k}} (f(x) \odot 1) dt \right) \odot \varphi(2^k x - j) \Big)$$

$$\le 2^k \sum_{\substack{j \\ 2^k x - j \in [-a,a]}} \varphi(2^k x - j) D \left((FR) \int_0^{2^{-k}} f\left(t + \frac{j}{2^k}\right) dt, \right.$$

$$(FR) \int_0^{2^{-k}} f(x) dt \Bigg)$$

(by Lemma 1, [11])
$$\leq \quad 2^k \sum_{\substack{j \\ 2^k x - j \in [-a,a]}} \varphi(2^k x - j)$$

$$\int_0^{2^{-k}} D\left(f\left(t + \frac{j}{2^k}\right), \; f(x)\right) dt =: (*)$$

(here $0 \leq t \leq \frac{1}{2^k}$ and $\left|x - \frac{j}{2^k}\right| \leq \frac{a}{2^k}$, thus $\left|t + \frac{j}{2^k} - x\right| \leq \frac{a+1}{2^k}$). Hence

$$(*) \leq 2^k \sum_{\substack{j \\ 2^k x - j \in [-a,a]}} \varphi(2^k x - j) \omega_1^{(\mathcal{F})}\left(f, \frac{a+1}{2^k}\right) 2^{-k} = \omega_1^{(\mathcal{F})}\left(f, \frac{a+1}{2^k}\right).$$

□

Next we give the corresponding result for the last fuzzy wavelet type operator we are dealing with here.

Theorem 12.9. *All assumptions here are as in Theorem 12.5. Define for* $k \in \mathbb{Z}$, $x \in \mathbb{R}$ *the fuzzy wavelet like operator*

$$(D_k f)(x) := \sum_{j=-\infty}^{\infty}{}^* \delta_{kj}(f) \odot \varphi(2^k x - j), \tag{12.12}$$

where

$$\delta_{kj}(f) := \sum_{\tilde{r}=0}^{n}{}^* w_{\tilde{r}} \odot f\left(\frac{j}{2^k} + \frac{\tilde{r}}{2^k n}\right), \; n \in \mathbb{N}, \; w_{\tilde{r}} \geq 0, \; \sum_{\tilde{r}=0}^{n} w_{\tilde{r}} = 1. \tag{12.13}$$

Then

$$D((D_k f)(x), f(x)) \leq \omega_1^{(\mathcal{F})}\left(f, \frac{a+1}{2^k}\right), \tag{12.14}$$

and

$$D^*(D_k f, f) \leq \omega_1^{(\mathcal{F})}\left(f, \frac{a+1}{2^k}\right), \quad \text{all } k \in \mathbb{Z}, \; x \in \mathbb{R}. \tag{12.15}$$

When $f \in C_{\mathcal{F}}^U(\mathbb{R})$ *then as* $k \to +\infty$ *we get* $D_k \to I$ *with rates.*

Proof. We need to upper bound

$$D((D_k f)(x), f(x))$$

$$= D\left(\sum_{\substack{j \\ 2^k x - j \in [-a,a]}}^{*} \left(\sum_{\tilde{r}=0}^{n}{}^{*} w_{\tilde{r}} \odot f\left(\frac{j}{2^k} + \frac{\tilde{r}}{2^k n} \right) \right) \cdot \varphi(2^k x - j), \right.$$

$$\left. \sum_{\substack{j \\ 2^k x - j \in [-a,a]}}^{*} f(x) \odot \varphi(2^k x - j) \right)$$

$$\leq \sum_{\substack{j \\ 2^k x - j \in [-a,a]}} \varphi(2^k x - j) D\left(\sum_{\tilde{r}=0}^{n}{}^{*} \left(w_{\tilde{r}} \odot f\left(\frac{j}{2^k} + \frac{\tilde{r}}{2^k n} \right) \right), \right.$$

$$\left. \sum_{\tilde{r}=0}^{n}{}^{*} (w_{\tilde{r}} \odot f(x)) \right)$$

$$\leq \sum_{\substack{j \\ 2^k x - j \in [-a,a]}} \varphi(2^k x - j) \sum_{\tilde{r}=0}^{n} w_{\tilde{r}} D\left(f\left(\frac{j}{2^k} + \frac{\tilde{r}}{2^k n} \right), f(x) \right)$$

$$\left(\text{notice that } \left| \frac{j}{2^k} + \frac{\tilde{r}}{2^k n} - x \right| \leq \frac{a+1}{2^k} \right)$$

$$\leq \sum_{\substack{j \\ 2^k x - j \in [-a,a]}} \varphi(2^k x - j) \sum_{\tilde{r}=0}^{n} w_{\tilde{r}} \omega_1^{(\mathcal{F})}\left(f, \frac{a+1}{2^k} \right) = \omega_1^{(\mathcal{F})}\left(f, \frac{a+1}{2^k} \right). \quad \square$$

Next we prove optimality for three of the above main results.

Proposition 12.10. *Inequality (12.2) is attained, that is sharp.*

Proof. Take $\varphi(x) = \chi_{[-\frac{1}{2}, \frac{1}{2})}(x)$, the characteristic function on $[-\frac{1}{2}, \frac{1}{2})$. Fix $u \in \mathbb{R}_{\mathcal{F}}$ and take $f(x) = q(x) \odot u$, where

$$q(x) := \begin{cases} 0, & x \leq -2^{-k-1} \\ 1, & x \geq 0, \\ 2^{k+1} x + 1, & -2^{-k-1} < x < 0, \end{cases}$$

$k \in \mathbb{Z}$ fixed, $x \in \mathbb{R}$. Clearly $q(x) \geq 0$. We observe that

$$
\begin{aligned}
(B_k f)(x) &= \sum_{j=-\infty}^{\infty}{}^{*} q\left(\frac{j}{2^k}\right) \odot u \odot \varphi(2^k x - j) \\
&= \left(\sum_{j=-\infty}^{\infty} q\left(\frac{j}{2^k}\right) \varphi(2^k x - j)\right) \odot u \\
&= \left(\sum_{j=0}^{\infty} \varphi(2^k x - j)\right) \odot u.
\end{aligned}
$$

Hence

$$
D((B_k f)(-2^{-k-1}), f(-2^{-k-1}) = D\left(\left(\sum_{j=0}^{\infty} \varphi\left(-\frac{1}{2} - j\right)\right) \odot u, \tilde{o}\right) = D(u, \tilde{o}).
$$

Furthermore we see that

$$
\begin{aligned}
\omega_1^{(\mathcal{F})}(f, 2^{-k-1}) &= \sup_{\substack{x,y \in \mathbb{R} \\ |x-y| \leq 2^{-k-1}}} D(f(x), f(y)) \\
&= \sup_{\substack{x,y \in \mathbb{R} \\ |x-y| \leq 2^{-k-1}}} D(q(x) \odot u, q(y) \odot u) \\
&\overset{\text{(by Lemma 2.2, [31])}}{\leq} \left(\sup_{\substack{x,y \in \mathbb{R} \\ |x-y| \leq 2^{-k-1}}} |q(x) - q(y)|\right) D(u, \tilde{o}) \\
&= 1 \cdot D(u, \tilde{o}).
\end{aligned}
$$

I.e. we got that

$$
\omega_1^{(\mathcal{F})}(f, 2^{-k-1}) \leq D(u, \tilde{o}).
$$

So that by (12.2) and the above we find

$$
D((B_k f)(-2^{-k-1}), f(-2^{-k-1})) = \omega_1^{(\mathcal{F})}(f, 2^{-k-1}),
$$

proving the sharpness of (12.2).   □

Proposition 12.11. *Inequalities (12.10) and (12.14) are attained, i.e. they are sharp.*

Proof. (I) Consider as optimal elements φ, q, u, and f, exactly as in the proof of Proposition 12.10. Here $a = \frac{1}{2}$. We observe that

$$\omega_1^{(\mathcal{F})}\left(f, \frac{a+1}{2^k}\right) = \omega_1^{(\mathcal{F})}\left(f, \frac{3}{2^{k+1}}\right) = \sup_{\substack{x,y \\ |x-y| \leq \frac{3}{2^{k+1}}}} D(f(x), f(y))$$

$$= \sup_{\substack{x,y \\ |x-y| \leq \frac{3}{2^{k+1}}}} D(q(x) \odot u, q(y) \odot u)$$

$$\begin{array}{c} \text{(by Lemma 2.2, [31])} \\ \leq \end{array} \left(\sup_{\substack{x,y \\ |x-y| \leq \frac{3}{2^{k+1}}}} |q(x) - q(y)| \right) D(u, \tilde{o})$$

$$= \left(\sup_{\substack{x,y \\ |x-y| \leq \frac{1}{2^{k+1}}}} |q(x) - q(y)| \right) D(u, \tilde{o}) = 1 \cdot D(u, \tilde{o}).$$

That is

$$\omega_1^{(\mathcal{F})}\left(f, \frac{a+1}{2^k}\right) \leq D(u, \tilde{o}).$$

Call

$$\gamma_{kj}(f) := 2^k \odot (FR) \int_0^{2^{-k}} f\left(t + \frac{j}{2^k}\right) dt.$$

We obtain

$$\begin{aligned} \gamma_{k(-1)}(f) &= 2^k \odot (FR) \int_0^{2^{-k}} \left(q\left(t - \frac{1}{2^k}\right) \odot u \right) dt \\ &= \left(2^k \int_0^{2^{-k}} q\left(t - \frac{1}{2^k}\right) dt \right) \odot u \\ &= \left(2^k \int_{-\frac{1}{2^k}}^{0} q(t) dt \right) \odot u \\ &= \left(2^k \int_{-\frac{1}{2^{k+1}}}^{0} q(t) dt \right) \odot u = \frac{1}{4} \odot u. \end{aligned}$$

I.e.

$$\gamma_{k(-1)}(f) = \frac{1}{4} \odot u.$$

Moreover $\gamma_{k(-2)}(f) = \tilde{o}$, and $\gamma_{kj}(f) = \tilde{o}$, all $j \leq -2$, and $\gamma_{kj}(f) = u$, all $j \geq 0$. Hence

$$(C_k f)(x) = \left[\frac{1}{4}\varphi(2^k x + 1) + \sum_{j=0}^{+\infty} \varphi(2^k x - j) \right] \odot u.$$

We easily see then that

$$(C_k f)\left(-\frac{1}{2^{k+1}}\right) = u, \quad \text{also } f\left(-\frac{1}{2^{k+1}}\right) = \tilde{o}.$$

Therefore

$$D\left((C_k f)\left(-\frac{1}{2^{k+1}}\right), f\left(-\frac{1}{2^{k+1}}\right)\right) = D(u, \tilde{o}).$$

From the above and (12.10) we conclude that

$$D\left((C_k f)\left(-\frac{1}{2^{k+1}}\right), f\left(-\frac{1}{2^{k+1}}\right)\right) = w_1^{(\mathcal{F})}\left(f, \frac{a+1}{2^k}\right), \quad k \in \mathbb{Z},$$

proving the sharpness of (12.10).

(II) The sharpness of (12.14) is treated similarly to (I). Notice that $\delta_{kj}(f) = u$, all $j \geq 0$, and $\delta_{kj}(f) = \tilde{o}$, all $j \leq -2$. We observe that

$$\varphi\left(2^k\left(-\frac{1}{2^{k+1}}\right) - (-1)\right) = \varphi\left(\frac{1}{2}\right) = 0.$$

Furthermore

$$D\left((D_k f)\left(-\frac{1}{2^{k+1}}\right), f\left(-\frac{1}{2^{k+1}}\right)\right)$$

$$= D\left(\sum_{j=-\infty}^{\infty}{}^* \delta_{kj}(f) \odot \varphi\left(2^k\left(-\frac{1}{2^{k+1}}\right) - j\right), \tilde{o}\right)$$

$$= D\left(\left(\sum_{j=0}^{\infty} 1\varphi\left(-\frac{1}{2} - j\right)\right) \odot u, \tilde{o}\right) = D(1 \odot u, \tilde{o}) = D(u, \tilde{o}).$$

So that by (12.14) and the above

$$D\left((D_k f)\left(-\frac{1}{2^{k+1}}\right), f\left(-\frac{1}{2^{k+1}}\right)\right) = w_1^{(\mathcal{F})}\left(f, \frac{a+1}{2^k}\right),$$

proving sharpness of (12.14).   □

Remark 12.12. We notice that

$$(L_k f)(x) = L_0(f(2^{-k} \cdot))(2^k x), \quad \text{all } x \in \mathbb{R}, \ k \in \mathbb{Z},$$

where $L_k = B_k, A_k, C_k, D_k$. Clearly L_k's are linear over $\mathbb{R}$ operators.

In the following we present a monotonicity result for the fuzzy wavelet like operators B_k and D_k. For that we need

Definition 12.13. Let $f\colon \mathbb{R} \to \mathbb{R}_{\mathcal{F}}$, then f is called a *nondecreasing* function iff whenever $x_1 \le x_2$, $x_1, x_2 \in \mathbb{R}$, we have that $f(x_1) \le f(x_2)$, i.e. $(f(x_1))_-^{(r)} \le (f(x_2))_-^{(r)}$ and $(f(x_1))_+^{(r)} \le (f(x_2))_+^{(r)}$, $\forall r \in [0,1]$.

Theorem 12.14. *Let $f \in C(\mathbb{R}, \mathbb{R}_{\mathcal{F}})$, and the scaling function $\varphi(x)$ a real valued bounded function with $\operatorname{supp} \varphi \subseteq [-a,a]$, $0 < a < +\infty$, such that*

(i) $\displaystyle\sum_{j=-\infty}^{\infty} \varphi(x-j) \equiv 1$ *on* $\mathbb{R}$,

(ii) *there exists a $b \in \mathbb{R}$ such that φ is nondecreasing for $x \le b$ and φ is nonincreasing for $x \ge b$,*

(the above imply $\varphi \ge 0$). Let $f(x)$ be nondecreasing fuzzy function, then $(B_k f)(x)$, $(D_k f)(x)$ are nondecreasing fuzzy valued functions for any $k \in \mathbb{Z}$.

Remark 12.15. We give two examples of φ's as in Theorem 12.14.

(i)
$$\varphi(x) = \begin{cases} 1, & -\tfrac{1}{2} \le x < \tfrac{1}{2}, \\ 0, & \text{elsewhere.} \end{cases}$$

(ii)
$$\varphi(x) = \begin{cases} x+1, & -1 \le x \le 0, \\ 1-x, & 0 < x \le 1, \\ 0, & \text{elsewhere.} \end{cases}$$

Proof of Theorem 12.14. Let $x_n, x \in \mathbb{R}$ such that $x_n \to x$, as $n \to +\infty$, then $D(f(x_n), f(x)) \to 0$ by fuzzy continuity of f. But we have

$$D(f(x_n), f(x)) = \sup_{r\in[0,1]} \max\{|(f(x_n))_-^{(r)}-(f(x))_-^{(r)}|, |(f(x_n))_+^{(r)}-(f(x))_+^{(r)}|\}.$$

That is, $|(f(x_n))_\pm^{(r)}-(f(x))_\pm^{(r)}| \to 0$, all $0 \le r \le 1$, as $n \to +\infty$, respectively. Therefore $(f)_\pm^{(r)} \in C(\mathbb{R},\mathbb{R})$, all $0 \le r \le 1$, i.e. real valued continuous functions on $\mathbb{R}$. Since f is fuzzy nondecreasing by Definition 12.13 we get that $(f)_\pm^{(r)}$ are nondecreasing, $\forall r \in [0,1]$, respectively. Then by Theorem 6.3, p. 156, [9], see also [36], we get that the corresponding real wavelet type operators map to the functions $(B_k(f)_\pm^{(r)})(x)$ that are nondecreasing on $\mathbb{R}$ for all $r \in [0,1]$, any $k \in \mathbb{Z}$. Also by Lemma 8.2, p. 186, [9], see also [7], we get that the corresponding real wavelet type operators map to the functions $(D_k(f)_\pm^{(r)})(x)$ that are nondecreasing on $\mathbb{R}$ for all $r \in [0,1]$, any $k \in \mathbb{Z}$. We notice for any $r \in [0,1]$ that

$$[(B_k f)(x)]^r = \sum_{j=-\infty}^{+\infty} \left[f\left(\frac{j}{2^k}\right)\right]^r \varphi(2^k x - j).$$

That is

$$\left[((B_k f)(x))_-^{(r)}, ((B_k f)(x))_+^{(r)}\right] = \sum_{j=-\infty}^{+\infty} \left[\left(f\left(\frac{j}{2^k}\right)\right)_-^{(r)}, \left(f\left(\frac{j}{2^k}\right)\right)_+^{(r)}\right]$$
$$\varphi(2^k x - j)$$

$$= \left[\sum_{j=-\infty}^{+\infty} \left(f\left(\frac{j}{2^k}\right)\right)_-^{(r)} \varphi(2^k x - j), \sum_{j=-\infty}^{+\infty} \left(f\left(\frac{j}{2^k}\right)\right)_+^{(r)} \varphi(2^k x - j)\right]$$

$$= \left[(B_k(f)_-^{(r)})(x), (B_k(f)_+^{(r)})(x)\right].$$

So whenever $x_1 \le x_2$ we get $(f)_\pm^{(r)}(x_1) \le (f)_\pm^{(r)}(x_2)$, respectively, and

$$(B_k(f)_\pm^{(r)})(x_1) \le (B_k(f)_\pm^{(r)})(x_2), \quad \forall r \in [0, 1].$$

Therefore $(B_k f)(x_1) \le (B_k f)(x_2)$, that is $(B_k f)$ is nondecreasing.
   Next we observe that

$$[(D_k f)(x)]^r = \sum_{j=-\infty}^{+\infty} \left(\sum_{\tilde{r}=0}^{n} w_{\tilde{r}} \left[f\left(\frac{j}{2^k} + \frac{\tilde{r}}{2^k n}\right)\right]^r\right) \varphi(2^k x - j).$$

That is

$$\left[((D_k f)(x))_-^{(r)}, ((D_k f)(x))_+^{(r)}\right]$$
$$= \sum_{j=-\infty}^{+\infty} \left(\sum_{\tilde{r}=0}^{n} w_{\tilde{r}} \left[\left(f\left(\frac{j}{2^k} + \frac{\tilde{r}}{2^k n}\right)\right)_-^{(r)}, \left(f\left(\frac{j}{2^k} + \frac{\tilde{r}}{2^k n}\right)\right)_+^{(r)}\right]\right)$$
$$\varphi(2^k x - j)$$

$$= \left[\sum_{j=-\infty}^{+\infty} \left(\sum_{\tilde{r}=0}^{n} w_{\tilde{r}} \left(f\left(\frac{j}{2^k} + \frac{\tilde{r}}{2^k n}\right)\right)_-^{(r)}\right) \varphi(2^k x - j),\right.$$

$$\left. \sum_{j=-\infty}^{+\infty} \left(\sum_{\tilde{r}=0}^{n} w_{\tilde{r}} \left(f\left(\frac{j}{2^k} + \frac{\tilde{r}}{2^k n}\right)\right)_+^{(r)}\right) \varphi(2^k x - j)\right]$$

$$= \left[(D_k(f)_-^{(r)})(x), (D_k(f)_+^{(r)})(x)\right].$$

So whenever $x_1 \le x_2$ we get

$$(D_k(f)_\pm^{(r)})(x_1) \le (D_k(f)_\pm^{(r)})(x_2), \quad \forall r \in [0, 1].$$

Therefore $(D_k f)(x_1) \le (D_k f)(x_2)$, so that $(D_k f)$ is nondecreasing.   □
   Finally we present the corresponding monotonicity results for the fuzzy wavelet like operators A_k, C_k.

Theorem 12.16. *Let $f \in C_b(\mathbb{R}, \mathbb{R}_{\mathcal{F}})$ and φ as in Theorem 12.14 which is continuous on $[-a, a]$. Let $f(x)$ be nondecreasing fuzzy function, then $(A_k f)(x)$ is a nondecreasing fuzzy valued function for any $k \in \mathbb{Z}$.*

Proof. Since f is fuzzy nondecreasing we get again that $(f)_{\pm}^{(r)}$ are nondecreasing, $\forall r \in [0, 1]$, respectively. Then by Theorem 6.1, p. 149, [9], see also [36], we get that the corresponding real wavelet type operators map to the functions $(A_k(f)_{\pm}^{(r)})(x)$ that are nondecreasing on $\mathbb{R}$ for all $r \in [0, 1]$, any $k \in \mathbb{Z}$.

Using Theorem 3.2 ([67]), for any $r \in [0, 1]$ we notice that

$$
\begin{aligned}
[\langle f, \varphi_{kj} \rangle]^r &= \left[\int_{\frac{j-a}{2^k}}^{\frac{j+a}{2^k}} (f(t) \odot \varphi_{kj}(t))_{-}^{(r)} dt, \int_{\frac{j-a}{2^k}}^{\frac{j+a}{2^k}} (f(t) \odot \varphi_{kj}(t))_{+}^{(r)} dt \right] \\
&= \left[\int_{\frac{j-a}{2^k}}^{\frac{j+a}{2^k}} (f(t))_{-}^{(r)} \varphi_{kj}(t) dt, \int_{\frac{j-a}{2^k}}^{\frac{j+a}{2^k}} (f(t))_{+}^{(r)} \varphi_{kj}(t) dt \right].
\end{aligned}
$$

We observe for any $r \in [0, 1]$ that

$$
[(A_k f)(x)]^r = \sum_{j=-\infty}^{+\infty} [\langle f, \varphi_{kj} \rangle]^r \varphi_{kj}(x).
$$

That is

$$
\begin{aligned}
&\left[((A_k f)(x))_{-}^{(r)}, ((A_k f)(x))_{+}^{(r)} \right] \\
&= \sum_{j=-\infty}^{+\infty} \left[\int_{\frac{j-a}{2^k}}^{\frac{j+a}{2^k}} (f(t))_{-}^{(r)} \varphi_{kj}(dt) dt, \int_{\frac{j-a}{2^k}}^{\frac{j+a}{2^k}} (f(t))_{+}^{(r)} \varphi_{kj}(t) dt \right] \varphi_{kj}(x) \\
&= \left[\sum_{j=-\infty}^{+\infty} \left(\int_{\frac{j-a}{2^k}}^{\frac{j+a}{2^k}} (f(t))_{-}^{(r)} \varphi_{kj}(t) dt \right) \varphi_{kj}(x), \right. \\
&\quad \left. \sum_{j=-\infty}^{+\infty} \left(\int_{\frac{j-a}{2^k}}^{\frac{j+a}{2^k}} (f(t))_{+}^{(r)} \varphi_{kj}(t) dt \right) \varphi_{kj}(x) \right] \\
&= \left[(A_k(f)_{-}^{(r)})(x), (A_k(f)_{+}^{(r)})(x) \right].
\end{aligned}
$$

So whenever $x_1 \leq x_2$ we have that $(f)_{\pm}^{(r)}(x_1) \leq (f)_{\pm}^{(r)}(x_2)$, respectively, and

$$
(A_k(f)_{\pm}^{(r)})(x_1) \leq (A_k(f)_{\pm}^{(r)})(x_2), \quad \forall r \in [0, 1].
$$

Hence $(A_k f)(x_1) \leq (A_k f)(x_2)$, that is $(A_k f)$ is nondecreasing.   □

Theorem 12.17. *Let f and φ as in Theorem 12.14. Let $f(x)$ be nondecreasing fuzzy function, then $(C_k f)(x)$ is a nondecreasing fuzzy valued function for any $k \in \mathbb{Z}$.*

Proof. By Lemma 8.2, p. 186, [9], see also [7], we get that the corresponding real wavelet type operators map to the functions $(C_k(f)_\pm^{(r)})(x)$ that are nondecreasing on $\mathbb{R}$ for all $r \in [0, 1]$, any $k \in \mathbb{Z}$. Using Theorem 3.2 ([67]), for any $r \in [0, 1]$ we notice that

$$
\begin{aligned}
[(C_k f)(x)]^r &= \sum_{j=-\infty}^{+\infty} \left[2^k \odot (FR) \int_0^{2^{-k}} f\left(t + \frac{j}{2^k}\right) dt \right]^r \varphi(2^k x - j) \\
&= \sum_{j=-\infty}^{+\infty} \left[2^k \odot (FR) \int_{2^{-k}j}^{2^{-k}(j+1)} f(t) dt \right]^r \varphi(2^k x - j) \\
&= \sum_{j=-\infty}^{+\infty} \left[2^k \int_{2^{-k}j}^{2^{-k}(j+1)} (f)_-^{(r)}(t) dt, 2^k \int_{2^{-k}j}^{2^{-k}(j+1)} (f)_+^{(r)}(t) dt \right] \\
&\quad \varphi(2^k x - j) \\
&= \left[\sum_{j=-\infty}^{+\infty} \left(2^k \int_{2^{-k}j}^{2^{-k}(j+1)} (f)_-^{(r)}(t) dt \right) \varphi(2^k x - j), \right. \\
&\quad \left. \sum_{j=-\infty}^{+\infty} \left(2^k \int_{2^{-k}j}^{2^{-k}(j+1)} (f)_+^{(r)}(t) dt \right) \varphi(2^k x - j) \right] \\
&= [(C_k(f)_-^{(r)})(x), (C_k(f)_+^{(r)})(x)].
\end{aligned}
$$

That is, for any $r \in [0, 1]$ we found

$$
[((C_k f)(x))_-^{(r)}, ((C_k f)(x))_+^{(r)}] = [(C_k(f)_-^{(r)})(x), (C_k(f)_+^{(r)})(x)].
$$

So whenever $x_1 \le x_2$ we have $(f)_\pm^{(r)}(x_1) \le (f)_\pm^{(r)}(x_2)$ and

$$
(C_k(f)_\pm^{(r)})(x_1) \le (C_k(f)_\pm^{(r)})(x_2), \quad \forall r \in [0, 1],
$$

respectively. Hence $(C_k f)(x_1) \le (C_k f)(x_2)$, that is $(C_k f)$ is nondecreasing.

13

ESTIMATES TO DISTANCES BETWEEN FUZZY WAVELET LIKE OPERATORS

The basic fuzzy wavelet like operators A_k, B_k, C_k, D_k, $k \in \mathbb{Z}$ were first introduced in [14], see also Chapter 12, where they were studied among others for their pointwise/uniform convergence with rates to the fuzzy unit operator I. Here we continue this study by estimating the fuzzy distances between these operators. We give the pointwise convergence with rates of these distances to zero. The related approximation is of higher order since we involve these higher order fuzzy derivatives of the engaged fuzzy continuous function f. The derived Jackson type inequalities involve the fuzzy (first) modulus of continuity. Some comparison inequalities are also given so we get better upper bounds to the distances we study. The defining of these operators scaling function φ is of compact support in $[-a, a]$, $a > 0$ and is not assumed to be orthogonal. This chapter is based on [23].

13.1 Background

This Chapter is motivated by [14], especially from the following result:

Theorem 13.1 ([14]). *Let f be a function from $\mathbb{R}$ into the fuzzy real numbers $\mathbb{R}_\mathcal{F}$ which is fuzzy continuous. Let $\varphi(x)$ be a real valued bounded*

G.A. Anastassiou: Fuzzy Mathematics: Approximation The., STUDFUZZ 251, pp. 209–236.
springerlink.com   © Springer-Verlag Berlin Heidelberg 2010

scaling function with supp $\varphi(x) \subseteq [-a, a]$, $a > 0$, $\varphi(x) \geq 0$, *such that* $\sum\limits_{j=-\infty}^{\infty} \varphi(x-j) \equiv 1$ *on* $\mathbb{R}$. *For* $k \in \mathbb{Z}$, $x \in \mathbb{R}$ *put*

$$(B_k f)(x) := \text{fuzzy sum} \sum_{j=-\infty}^{\infty} f\left(\frac{j}{2^k}\right) \odot \varphi(2^k x - j)$$

($\odot$ *denotes fuzzy multiplication*). *Clearly* B_k *is a fuzzy wavelet type operator.*

Then the fuzzy distance

$$D((B_k f)(x), f(x)) \leq \omega_1^{(\mathcal{F})}\left(f, \frac{a}{2^k}\right),$$

any $x \in \mathbb{R}$, *and* $k \in \mathbb{Z}$, *and*

$$\sup_{x \in \mathbb{R}} D((B_k f)(x), f(x)) \leq \omega_1^{(\mathcal{F})}\left(f, \frac{a}{2^k}\right).$$

Here $\omega_1^{(\mathcal{F})}$ *stands for the fuzzy (first) modulus of continuity. If* f *is fuzzy uniformly continuous then* $\lim\limits_{k \to +\infty} B_k f = f$ *uniformly with rates.*

All fuzzy wavelet like operators A_k, B_k, C_k, D_k, $k \in \mathbb{Z}$ are reintroduced here and we find upper bounds to their distances $D((B_k f)(x), (C_k f)(x))$, $D((D_k f)(x), (C_k f)(x))$, $D((B_k f)(x), (D_k f)(x))$, $D((A_k f)(x), (B_k f)(x))$, $D((A_k f)(x), (C_k f)(x))$, and $D((A_k f)(x), (D_k f)(x))$. Their proofs rely a lot on the found here upper bounds for $D\left((B_k f)(x), f\left(x - \frac{a}{2^k}\right)\right)$, $D\left((C_k f)(x), f\left(x - \frac{a}{2^k}\right)\right)$ and $D\left((D_k f)(x), f\left(x - \frac{a}{2^k}\right)\right)$, $k \in \mathbb{Z}$, $x \in \mathbb{R}$. The produced associated inequalities involve fuzzy (first) moduli of continuity of the engaged function and its fuzzy derivatives.

For fuzzy uniformly continuous functions f and its likewise derivatives we obtain pointwise convergence with rates to zero of all the above mentioned distances among the stated sequences of fuzzy wavelet like operators. For these see Section 13.2.

Our methods of proving here are based only on fuzzy theoretic concepts. We use the following

Definition 13.2 ([11]). Let $f: \mathbb{R} \to \mathbb{R}_{\mathcal{F}}$ be a fuzzy real number valued function. We define the *(first) fuzzy modulus of continuity* of f by

$$\omega_1^{(\mathcal{F})}(f, \delta) := \sup_{\substack{x, y \in \mathbb{R} \\ |x-y| \leq \delta}} D(f(x), f(y)), \quad \delta > 0.$$

Definition 13.3 ([11]). Let $f: \mathbb{R} \to \mathbb{R}_{\mathcal{F}}$. If $D(f(x), \tilde{o}) \leq M_1$, $\forall x \in \mathbb{R}$, $M_1 > 0$, we call f a *bounded fuzzy real number valued function.*

Definition 13.4 ([11]). Let $f: \mathbb{R} \to \mathbb{R}_{\mathcal{F}}$. We say that f is *fuzzy continuous* at $a \in \mathbb{R}$ if whenever $x_n \to a$, then $D(f(x_n), f(a)) \to 0$. If f is continuous for every $a \in \mathbb{R}$, then we call f a *fuzzy continuous real number valued function*. We denote it as $f \in C(\mathbb{R}, \mathbb{R}_\mathcal{F})$.

Definition 13.5 ([11]). Let $f \colon \mathbb{R} \to \mathbb{R}_\mathcal{F}$. We call f a *fuzzy uniformly continuous real number valued function*, iff for any $\varepsilon > 0$ there exists $\delta > 0$: whenever $|x - y| \le \delta$, $x, y \in \mathbb{R}$, implies that $D(f(x), f(y)) \le \varepsilon$. We denote it as $f \in C_\mathcal{F}^U(\mathbb{R})$.

Proposition 13.6 ([11]). *Let* $f \in C_\mathcal{F}^U(\mathbb{R})$. *Then* $\omega_1^{(\mathcal{F})}(f, \delta) < +\infty$, *any* $\delta > 0$.

Denote $f \colon \mathbb{R} \to \mathbb{R}_\mathcal{F}$ which is bounded and fuzzy continuous, as $f \in C_b(\mathbb{R}, \mathbb{R}_\mathcal{F})$.

Proposition 13.7 ([11]). *It holds*

(i) $\omega_1^{(\mathcal{F})}(f, \delta)$ *is nonnegative and nondecreasing in* $\delta > 0$, *any* $f \colon \mathbb{R} \to \mathbb{R}_\mathcal{F}$.

(ii) $\lim_{\delta \downarrow 0} \omega_1^{(\mathcal{F})}(f, \delta) = \omega_1^{(\mathcal{F})}(f, 0) = 0$, *iff* $f \in C_\mathcal{F}^U(\mathbb{R})$.

(iii) $\omega_1^{(\mathcal{F})}(f, \delta_1 + \delta_2) \le \omega_1^{(\mathcal{F})}(f, \delta_1) + \omega_1^{(\mathcal{F})}(f, \delta_2)$, $\delta_1, \delta_2 > 0$, *any* $f \colon \mathbb{R} \to \mathbb{R}_\mathcal{F}$.

(iv) $\omega_1^{(\mathcal{F})}(f, n\delta) \le n\omega_1^{(\mathcal{F})}(f, \delta)$, $\delta > 0$, $n \in \mathbb{N}$, *any* $f \colon \mathbb{R} \to \mathbb{R}_\mathcal{F}$.

(v) $\omega_1^{(\mathcal{F})}(f, \lambda\delta) \le \lceil \lambda \rceil \omega_1^{(\mathcal{F})}(f, \delta) \le (\lambda + 1)\omega_1^{(\mathcal{F})}(f, \delta)$, $\lambda > 0$, $\delta > 0$, *where* $\lceil \cdot \rceil$ *is the ceiling of the number, any* $f \colon \mathbb{R} \to \mathbb{R}_\mathcal{F}$.

(vi) $\omega_1^{(\mathcal{F})}(f \oplus g, \delta) \le \omega_1^{(\mathcal{F})}(f, \delta) + \omega_1^{(\mathcal{F})}(g, \delta)$, $\delta > 0$, *any* $f, g \colon \mathbb{R} \to \mathbb{R}_\mathcal{F}$.

(vii) $\omega_1^{(\mathcal{F})}(f, \cdot)$ *is continuous on* $\mathbb{R}_+$, *for* $f \in C_\mathcal{F}^U(\mathbb{R})$.

Lemma 13.8 ([11]). *Let* $f \in C(\mathbb{R}, \mathbb{R}_\mathcal{F})$, $r \in \mathbb{N}$. *Then the following integrals*

$$(FR) \int_a^{s_{r-1}} f(s_r)ds_r, (FR) \int_a^{s_{r-2}} \left(\int_a^{s_{r-1}} f(s_r)ds_r \right) ds_{r-1}, \dots,$$

$$(FR) \left(\int_a^s \int_a^{s_1} \cdots \left(\int_a^{s_{r-2}} \left(\int_a^{s_{r-1}} f(s_r)ds_r \right) ds_{r-1} \right) \cdots \right) ds_1,$$

are fuzzy continuous functions in $s_{r-1}, s_{r-2}, \dots, s$, *respectively. Here* s_{r-1}, $s_{r-2}, \dots, s \ge a$ *and all are real numbers.*

Here we use a lot the following fuzzy Taylor's formula.

Theorem 13.9 ([11]). *Let* $T := [x_0, x_0 + \beta] \subset \mathbb{R}$, *with* $\beta > 0$. *We assume that* $f^{(i)} \colon T \to \mathbb{R}_\mathcal{F}$ *are fuzzy differentiable for all* $i = 0, 1, \dots, n - 1$, *for any* $x \in T$ *(i.e., there exist in* $\mathbb{R}_\mathcal{F}$ *the H-differences* $f^{(i)}(x + h) - f^{(i)}(x)$,

$f^{(i)}(x) - f^{(i)}(x-h)$, $i = 0, 1, \ldots, n-1$ for all small $0 < h < \beta$. Furthermore there exist $f^{(i+1)}(x) \in \mathbb{R}_{\mathcal{F}}$ such that the limits in D-distance exist and

$$f^{(i+1)}(x) = \lim_{h \to 0+} \frac{f^{(i)}(x+h) - f^{(i)}(x)}{h} = \lim_{h \to 0+} \frac{f^{(i)}(x) - f^{(i)}(x-h)}{h},$$

for all $i = 0, 1, \ldots, n-1$). Also we assume that $f^{(i)}$, $i = 0, 1, \ldots, n$ are fuzzy continuous on T. Then for $s \geq a$, $s, a \in T$ we obtain

$$\begin{aligned} f(s) &= f(a) \oplus f'(a) \odot (s-a) \oplus f''(a) \odot \frac{(s-a)^2}{2!} \oplus \cdots \oplus f^{(n-1)}(a) \\ &\odot \frac{(s-a)^{n-1}}{(n-1)!} \oplus R_n(a,s) \end{aligned}$$

where

$$R_n(a,s) := (FR) \int_a^s \left(\int_a^{s_1} \cdots \left(\int_a^{s_{n-1}} f^{(n)} ds_n \right) ds_{n-1} \right) \cdots \right) ds_1.$$

Here $R_n(a,s)$ is fuzzy continuous over T as a function of s.

Note 13.10. This formula is invalid when $s < a$, as it is based on Theorem 3.6 of [53].

We denote by $C^N(\mathbb{R}, \mathbb{R}_{\mathcal{F}})$ the space of N-times continuously differentiable in the fuzzy sense functions from $\mathbb{R}$ into $\mathbb{R}_{\mathcal{F}}$, $N \in \mathbb{N}$. We also denote by $C_{\mathcal{F}}^{NU}(\mathbb{R})$, $N \in \mathbb{N}$, the space of functions $f \colon \mathbb{R} \to \mathbb{R}_{\mathcal{F}}$, such that the fuzzy derivatives exist up to order N and all f, $f'^{(N)}$ are fuzzy uniformly continuous.

Finally we make use of

Lemma 13.11 ([14]). *Let $f \colon \mathbb{R} \to \mathbb{R}_{\mathcal{F}}$ be fuzzy continuous and bounded function, i.e. $\exists M_1 > 0 \colon D(f(x), \tilde{o}) \leq M_1, \forall x \in \mathbb{R}$. Let also $g \colon$ from interval $J \subseteq \mathbb{R} \to \mathbb{R}_+$ continuous and bounded function. Then $f(x) \odot g(x)$ is a fuzzy continuous function on J.*

13.2   Results

We present the first main result.

Theorem 13.12. *Let $f \in C^N(\mathbb{R}, \mathbb{R}_{\mathcal{F}})$, $N \in \mathbb{N}$ and the scaling function $\varphi(x)$ a real valued bounded function with supp $\varphi(x) \subseteq [-a, a]$, $0 < a < +\infty$, $\varphi(x) \geq 0$, such that $\sum_{j=-\infty}^{\infty} \varphi(x-j) \equiv 1$ on $\mathbb{R}$. For $k \in \mathbb{Z}$, $x \in \mathbb{R}$ put*

$$(B_k f)(x) := \sum_{j=-\infty}^{\infty}{}^* f\left(\frac{j}{2^k}\right) \odot \varphi(2^k x - j), \tag{13.1}$$

which is a fuzzy wavelet like operator. Then it holds

$$D\left((B_k f)(x), f\left(x - \frac{a}{2^k}\right)\right) \leq$$

$$\sum_{i=1}^{N-1} \frac{a^i}{2^{i(k-1)} i!} \left(D(f^{(i)}(x), \tilde{o}) + \omega_1^{(\mathcal{F})}\left(f^{(i)}, \frac{a}{2^k}\right)\right)$$

$$+ \frac{a^N}{2^{N(k-1)} N!} \left(D(f^{(N)}(x), \tilde{o}) + 3\omega_1^{(\mathcal{F})}\left(f^{(N)}, \frac{a}{2^k}\right)\right)$$

$$=: \beta_k\left(\frac{a}{2^k}\right), \quad \text{for any } x \in \mathbb{R}, \ k \in \mathbb{Z}. \tag{13.2}$$

If $f \in C_{\mathcal{F}}^{NU}(\mathbb{R})$ and as $k \to +\infty$ we obtain with rates that

$$\lim_{k \to +\infty} D\left((B_k f)(x), f\left(x - \frac{a}{2^k}\right)\right) = 0.$$

Corollary 13.13 (to Theorem 13.12, for $N = 1$). *It holds*

$$D\left(B_k f)(x), f\left(x - \frac{a}{2^k}\right)\right) \leq \frac{a}{2^{k-1}} \left(D(f'(x), \tilde{o}) + 3\omega_1^{(\mathcal{F})}\left(f', \frac{a}{2^k}\right)\right),$$

for any $x \in \mathbb{R}$, $k \in \mathbb{Z}$.

Corollary 13.14 (to Theorem 13.12). *The following improvement of (13.2) holds*

$$D\left((B_k f)(x), f\left(x - \frac{a}{2^k}\right)\right) \leq \min\left(2\omega_1^{(\mathcal{F})}\left(f, \frac{a}{2^k}\right), \beta_k\left(\frac{a}{2^k}\right)\right),$$

for any $x \in \mathbb{R}$, $k \in \mathbb{Z}$.

Proof. From Theorem 13.1 we get

$$D((B_k f)(x), f(x)) \leq \omega_1^{(\mathcal{F})}\left(f, \frac{a}{2^k}\right).$$

Then we observe

$$D\left((B_k f)(x), f\left(x - \frac{a}{2^k}\right)\right) \leq D((B_k f)(x), f(x))$$

$$+ D\left(f(x), f\left(x - \frac{a}{2^k}\right)\right) \leq 2\omega_1^{(\mathcal{F})}\left(f, \frac{a}{2^k}\right). \qquad \square$$

Proof of Theorem 13.12. Because φ is of compact support in $[-a, a]$ we see that

$$(B_k f)(x) = \sum_{\substack{j \\ 2^k x - j \in [-a,a]}}^{*} f\left(\frac{j}{2^k}\right) \odot \varphi(2^k x - j). \tag{13.3}$$

That is for specific $x \in \mathbb{R}$, $k \in \mathbb{Z}$ we have the j's in (13.3) to fulfill

$$x - \frac{a}{2^k} \leq \frac{j}{2^k} \leq x + \frac{a}{2^k}.$$

Using the fuzzy Taylor formula (Theorem 1, [11]) we get

$$f\left(\frac{j}{2^k}\right) = \sum_{i=0}^{N-1}{}^{*} f^{(i)}\left(x - \frac{a}{2^k}\right) \odot \frac{\left(\frac{j+a}{2^k} - x\right)^i}{i!} \oplus \mathcal{R}_N\left(x - \frac{a}{2^k}, \frac{j}{2^k}\right), \quad (13.4)$$

where

$$\mathcal{R}_N\left(x - \frac{a}{2^k}, \frac{j}{2^k}\right) := (FR)\int_{x-\frac{a}{2^k}}^{\frac{j}{2^k}}\left(\int_{x-\frac{a}{2^k}}^{s_1}\left(\right. \right.$$

$$\left. \left. \cdots \left(\int_{x-\frac{a}{2^k}}^{s_{N-1}} f^{(N)}(s_N) ds_N\right) ds_{N-1} \cdots\right) ds_1. \quad (13.5)$$

Then

$$f\left(\frac{j}{2^k}\right) \odot \varphi(2^k x - j) = \sum_{i=0}^{N-1}{}^{*} f^{(i)}\left(x - \frac{a}{2^k}\right) \odot \frac{\left(\frac{j+a}{2^k} - x\right)^i}{i!} \odot \varphi(2^k x - j)$$

$$\oplus \mathcal{R}_N\left(x - \frac{a}{2^k}, \frac{j}{2^k}\right) \odot \varphi(2^k x - j)$$

and

$$\sum_{\substack{j \\ 2^k x - j \in [-a,a]}}^{*} f\left(\frac{j}{2^k}\right) \odot \varphi(2^k x - j) = \sum_{\substack{j \\ 2^k x - j \in [-a,a]}}^{*} \sum_{i=0}^{N-1}{}^{*} f^{(i)}\left(x - \frac{a}{2^k}\right)$$

$$\odot \frac{\left(\frac{j+a}{2^k} - x\right)^i}{i!} \odot \varphi(2^k x - j)$$

$$\oplus \sum_{\substack{j \\ 2^k x - j \in [-a,a]}}^{*} \mathcal{R}_N\left(x - \frac{a}{2^k}, \frac{j}{2^k}\right) \odot \varphi(2^k x - j).$$

That is we have

$$(B_k f)(x) = \sum_{i=0}^{N-1}{}^{*} \sum_{\substack{j \\ 2^k x - j \in [-a,a]}}^{*} f^{(i)}\left(x - \frac{a}{2^k}\right) \odot \frac{\left(\frac{j+a}{2^k} - x\right)^i}{i!} \odot \varphi(2^k x - j)$$

$$\oplus \sum_{\substack{j \\ 2^k x - j \in [-a,a]}}^{*} \mathcal{R}_N\left(x - \frac{a}{2^k}, \frac{j}{2^k}\right) \odot \varphi(2^k x - j). \quad (13.6)$$

Next we estimate

$$D\left((B_kf)(x), f\left(x - \frac{a}{2^k}\right)\right)$$

$$= D\left((B_kf)(x), \sum_{\substack{j \\ 2^kx-j\in[-a,a]}}^{*} \varphi(2^kx - j) \odot f\left(x - \frac{a}{2^k}\right)\right)$$

$$= D\left(\sum_{i=1}^{N-1}{}^{*} \sum_{\substack{j \\ 2^kx-j\in[-a,a]}}^{*} f^{(i)}\left(x - \frac{a}{2^k}\right) \odot \frac{\left(\frac{j+a}{2^k} - x\right)^i}{i!} \odot \varphi(2^kx - j)\right.$$

$$\left.\oplus \sum_{\substack{j \\ 2^kx-j\in[-a,a]}}^{*} \mathcal{R}_N\left(x - \frac{a}{2^k}, \frac{j}{2^k}\right) \odot \varphi(2^kx - j), \tilde{o}\right)$$

$$\le \sum_{i=1}^{N-1} \sum_{\substack{j \\ 2^kx-j\in[-a,a]}} \frac{\left(\frac{j+a}{2^k} - x\right)^i}{i!} \varphi(2^kx - j)D\left(f^{(i)}\left(x - \frac{a}{2^k}\right), \tilde{o}\right)$$

$$+ \sum_{\substack{j \\ 2^kx-j\in[-a,a]}} \varphi(2^kx - j)D\left(\mathcal{R}_N\left(x - \frac{a}{2^k}, \frac{j}{2^k}\right), \tilde{o}\right)$$

$$\le \sum_{i=1}^{N-1} \sum_{\substack{j \\ 2^kx-j\in[-a,a]}} \frac{a^i}{2^{i(k-1)}i!} \varphi(2^kx - j)\left(D(f^{(i)}(x), \tilde{o}) + \right.$$

$$\omega_1^{(\mathcal{F})}\left(f^{(i)}, \frac{a}{2^k}\right)\right)$$

$$+ \sum_{\substack{j \\ 2^kx-j\in[-a,a]}} \varphi(2^kx - j)D\left(\mathcal{R}_N\left(x - \frac{a}{2^k}, \frac{j}{2^k}\right), \tilde{o}\right) =: (*).$$

Next we work on

$$D\left(\mathcal{R}_N\left(x - \frac{a}{2^k}, \frac{j}{2^k}\right), \tilde{o}\right)$$

$$= D\left(\mathcal{R}_N\left(x - \frac{a}{2^k}, \frac{j}{2^k}\right) \oplus f^{(N)}\left(x - \frac{a}{2^k}\right)\right.$$

$$\left.\odot \frac{\left(\frac{j+a}{2^k} - x\right)^N}{N!}, f^{(N)}\left(x - \frac{a}{2^k}\right) \odot \frac{\left(\frac{j+a}{2^k} - x\right)^N}{N!}\right)$$

$$\leq D\left(\mathcal{R}_N\left(x - \frac{a}{2^k}, \frac{j}{2^k}\right), f^{(N)}\left(x - \frac{a}{2^k}\right) \odot \frac{\left(\frac{j+a}{2^k} - x\right)^N}{N!}\right)$$

$$+ D\left(f^{(N)}\left(x - \frac{a}{2^k}\right) \odot \frac{\left(\frac{j+a}{2^k} - x\right)^N}{N!}, \tilde{o}\right).$$

So that

$$D\left(\mathcal{R}_N\left(x - \frac{a}{2^k}, \frac{j}{2^k}\right), \tilde{o}\right) \leq D\left(\mathcal{R}_N\left(x - \frac{a}{2^k}, \frac{j}{2^k}\right), f^{(N)}\left(x - \frac{a}{2^k}\right)\right.$$

$$\left.\odot \frac{\left(\frac{j+a}{2^k} - x\right)^N}{N!}\right) \tag{13.7}$$

$$+ \frac{a^N}{2^{N(k-1)}N!}\left(D(f^{(N)}(x), \tilde{o}) + \omega_1^{(\mathcal{F})}\left(f^{(N)}, \frac{a}{2^k}\right)\right).$$

At the end we estimate

$$D\left(\mathcal{R}_N\left(x - \frac{a}{2^k}, \frac{j}{2^k}\right), f^{(N)}\left(x - \frac{a}{2^k}\right) \odot \frac{\left(\frac{j+a}{2^k} - x\right)^N}{N!}\right)$$

$$= D\left((FR)\int_{x-\frac{a}{2^k}}^{\frac{j}{2^k}}\left(\int_{x-\frac{a}{2^k}}^{s_1}\left(\int_{x-\frac{a}{2^k}}^{s_{N-1}} f^{(N)}(s_N)ds_N\right)ds_{N-1}\cdots\right)ds_1,$$

$$f^{(N)}\left(x - \frac{a}{2^k}\right)$$

$$\odot \int_{x-\frac{a}{2^k}}^{\frac{j}{2^k}}\left(\int_{x-\frac{a}{2^k}}^{s_1}\cdots\left(\int_{x-\frac{a}{2^k}}^{s_{N-1}} 1\, ds_N\right)ds_{N-1}\cdots\right)ds_1\right)$$

$$= D\left((FR)\int_{x-\frac{a}{2^k}}^{\frac{j}{2^k}}\left(\int_{x-\frac{a}{2^k}}^{s_1}\cdots\left(\int_{x-\frac{a}{2^k}}^{s_{N-1}} f^{(N)}(s_N)ds_N\right)ds_{N-1}\cdots\right)ds_1,$$

$$(FR)\int_{x-\frac{a}{2^k}}^{\frac{j}{2^k}}\left(\int_{x-\frac{a}{2^k}}^{s_1}\cdots\left(\int_{x-\frac{a}{2^k}}^{s_{N-1}} f^{(N)}\left(x - \frac{a}{2^k}\right)ds_N\right)ds_{N-1}\cdots\right)ds_1\right)$$

$$\underset{\text{(by Lemmas 1,4 of [11])}}{\leq} \int_{x-\frac{a}{2^k}}^{\frac{j}{2^k}}\left(\int_{x-\frac{a}{2^k}}^{s_1}\right.$$

$$\cdots \left(\int_{x-\frac{a}{2^k}}^{s_{N-1}} D\left(f^{(N)}(s_N), f^{(N)}\left(x - \frac{a}{2^k} \right) \right) ds_N \right) ds_{N-1} \cdots \right) ds_1$$

$$\leq \int_{x-\frac{a}{2^k}}^{\frac{j}{2^k}} \left(\int_{x-\frac{a}{2^k}}^{s_1} \cdots \left(\int_{x-\frac{a}{2^k}}^{s_{N-1}} \omega_1^{(\mathcal{F})}\left(f^{(N)}, \left| s_N + \frac{a}{2^k} - x \right| ds_N \right) ds_{N-1} \right.$$

$$\cdots \right) ds_1$$

$$\leq \omega_1^{(\mathcal{F})}\left(f^{(N)}, \left| \frac{j}{2^k} + \frac{a}{2^k} - x \right| \right) \int_{x-\frac{a}{2^k}}^{\frac{j}{2^k}} \left(\int_{x-\frac{a}{2^k}}^{s_1} \cdots \left(\int_{x-\frac{a}{2^k}}^{s_{N-1}} 1 \, ds_N \right) ds_{N-1} \right.$$

$$\cdots \right) ds_1$$

$$= \omega_1^{(\mathcal{F})}\left(f^{(N)}, \left| \frac{j}{2^k} + \frac{a}{2^k} - x \right| \right) \frac{\left(\frac{j}{2^k} - \left(x - \frac{a}{2^k} \right) \right)^N}{N!}$$

$$\leq \frac{a^N}{2^{N(k-1)} N!} \omega_1^{(\mathcal{F})}\left(f^{(N)}, \frac{a}{2^{k-1}} \right).$$

That is, we have found that

$$D\left(\mathcal{R}_N\left(x - \frac{a}{2^k}, \frac{j}{2^k} \right), f^{(N)}\left(x - \frac{a}{2^k} \right) \odot \frac{\left(\frac{j+a}{2^k} - x \right)^N}{N!} \right)$$

$$\leq \frac{a^N}{2^{N(k-1)} N!} \omega_1^{(\mathcal{F})}\left(f^{(N)}, \frac{a}{2^{k-1}} \right). \tag{13.8}$$

Therefore by (13.7) and (13.8) we obtain

$$D\left(\mathcal{R}_N\left(x - \frac{a}{2^k}, \frac{j}{2^k} \right), \tilde{o} \right) \leq \frac{a^N}{2^{N(k-1)} N!} \left(D(f^{(N)}, (x), \tilde{o}) + 3\omega_1^{(\mathcal{F})}\left(f^{(N)}, \frac{a}{2^k} \right) \right). \tag{13.9}$$

Using (13.9) into (∗) we get

$$(\ast) \quad \leq \quad \sum_{i=1}^{N-1} \frac{a^i}{2^{i(k-1)} i!} \left(D(f^{(i)}(x), \tilde{o}) + \omega_1^{(\mathcal{F})}\left(f^{(i)}, \frac{a}{2^k} \right) \right)$$

$$+ \frac{a^N}{2^{N(k-1)} N!} \left(D(f^{(N)}(x), \tilde{o}) + 3\omega_1^{(\mathcal{F})}\left(f^{(N)}, \frac{a}{2^k} \right) \right).$$

Inequality (13.2) has been established. □

The next related main result is given.

Theorem 13.15. *All assumptions are as in Theorem 13.12. Define for* $k \in \mathbb{Z}$, $x \in \mathbb{R}$ *the fuzzy wavelet like operator*

$$(D_k f)(x) := \sum_{j=-\infty}^{\infty}{}^{*} \delta_{kj}(f) \odot \varphi(2^k x - j), \tag{13.10}$$

where

$$\delta_{kj}(f) := \sum_{\tilde{r}=0}^{n}{}^{*} w_{\tilde{r}} \odot f\left(\frac{j}{2^k} + \frac{\tilde{r}}{2^k n}\right), \quad n \in \mathbb{N}, \; w_{\tilde{r}} \geq 0, \; \sum_{\tilde{r}=0}^{n} w_{\tilde{r}} = 1. \quad (13.11)$$

Then

$$D\left((D_k f)(x), f\left(x - \frac{a}{2^k}\right)\right)$$

$$\leq \sum_{i=1}^{N-1} \frac{(2a+1)^i}{2^{ki} i!} \left(D(f^{(i)}(x), \tilde{o}) + \omega_1^{(\mathcal{F})}\left(f^{(i)}, \frac{a}{2^k}\right)\right)$$

$$+ \frac{(2a+1)^N}{2^{kN} N!} \left(D(f^{(N)}(x), \tilde{o})\right.$$

$$\left. + 3\omega_1^{(\mathcal{F})}\left(f^{(N)}, \frac{a}{2^k}\right) + \omega_1^{(\mathcal{F})}\left(f^{(N)}, \frac{1}{2^k}\right)\right)$$

$$=: \Delta_k\left(\frac{a}{2^k}\right), \quad \text{for any } x \in \mathbb{R}, \; k \in \mathbb{Z}. \quad (13.12)$$

If $f \in C_{\mathcal{F}}^{NU}(\mathbb{R})$, as $k \to \infty$, we obtain with rates that

$$\lim_{k \to +\infty} D\left((D_k f)(x), f\left(x - \frac{a}{2^k}\right)\right) = 0.$$

Corollary 13.16 (to Theorem 13.15, for $N = 1$). *It holds*

$$D\left((D_k f)(x), f\left(x - \frac{a}{2^k}\right)\right) \leq \left(\frac{a}{2^{k-1}} + \frac{1}{2^k}\right)(D(f'(x), \tilde{o})$$

$$+ 3\omega_1^{(\mathcal{F})}\left(f', \frac{a}{2^k}\right) + \omega_1^{(\mathcal{F})}\left(f', \frac{1}{2^k}\right)\right),$$

for any $x \in \mathbb{R}, \; k \in \mathbb{Z}$.

Corollary 13.17 (to Theorem 13.15). *The following improvement of (13.12) holds*

$$D\left((D_k f)(x), f\left(x - \frac{a}{2^k}\right)\right) \leq \min\left(\left(\omega_1^{(\mathcal{F})}\left(f, \frac{a+1}{2^k}\right) + \omega_1^{(\mathcal{F})}\left(f, \frac{a}{2^k}\right)\right),\right.$$

$$\left. \Delta_k\left(\frac{a}{2^k}\right)\right),$$

for any $x \in \mathbb{R}, \; k \in \mathbb{Z}$.

Proof. From Theorem 4, inequality (14) of [14] we obtain

$$D((D_k f)(x), f(x)) \leq \omega_1^{(\mathcal{F})}\left(f, \frac{a+1}{2^k}\right).$$

Then we see that

$$D\left((D_k f)(x), f\left(x - \frac{a}{2^k}\right)\right) \leq D((D_k f)(x), f(x)) + D\left(f(x), f\left(x - \frac{a}{2^k}\right)\right)$$

$$\leq \omega_1^{(\mathcal{F})}\left(f, \frac{a+1}{2^k}\right) + \omega_1^{(\mathcal{F})}\left(f, \frac{a}{2^k}\right). \qquad \square$$

Proof of Theorem 13.15. Because φ is of compact support in $[-a, a]$ we observe that

$$(D_k f)(x) = \sum_{\substack{j \\ 2^k x - j \in [-a,a]}}^* \left(\sum_{\tilde{r}=0}^n{}^* w_{\tilde{r}} \odot f\left(\frac{j}{2^k} + \frac{\tilde{r}}{2^k n}\right)\right) \odot \varphi(2^k x - j).$$

(13.13)

Again for specific $x \in \mathbb{R}$, $k \in \mathbb{Z}$ we have that the j's in (13.12) satisfy

$$x - \frac{a}{2^k} \leq \frac{j}{2^k} \leq x + \frac{a}{2^k}.$$

Using the fuzzy Taylor formula (Theorem 1, [11]) we get

$$f\left(\frac{j}{2^k} + \frac{\tilde{r}}{2^k n}\right) = \sum_{i=0}^{N-1}{}^* f^{(i)}\left(x - \frac{a}{2^k}\right) \odot \frac{\left(\frac{(j+a)}{2^k} + \frac{\tilde{r}}{2^k n} - x\right)^i}{i!}$$

$$\oplus \mathcal{R}_N\left(x - \frac{a}{2^k}, \frac{j}{2^k} + \frac{\tilde{r}}{2^k n}\right),$$

(13.14)

where

$$\mathcal{R}_N\left(x - \frac{a}{2^k}, \frac{j}{2^k} + \frac{\tilde{r}}{2^k n}\right)$$

(13.15)

$$:= (FR)\int_{x-\frac{a}{2^k}}^{\left(\frac{j}{2^k}+\frac{\tilde{r}}{2^k n}\right)}\left(\int_{x-\frac{a}{2^k}}^{s_1} \cdots \left(\int_{x-\frac{a}{2^k}}^{s_{N-1}} f^{(N)}(s_N) ds_N\right) ds_{N-1}\right.$$

$$\left. \cdots \right) ds_1.$$

Then

$$\delta_{kj}(f) = \sum_{\tilde{r}=0}^n{}^* w_{\tilde{r}} \odot f\left(\frac{j}{2^k} + \frac{\tilde{r}}{2^k n}\right) = \sum_{i=0}^{N-1}{}^* f^{(i)}\left(x - \frac{a}{2^k}\right)$$

$$\odot \sum_{\tilde{r}=0}^n w_{\tilde{r}} \frac{\left(\frac{(j+a)}{2^k} + \frac{\tilde{r}}{2^k n} - x\right)^i}{i!}$$

$$\oplus \sum_{\tilde{r}=0}^n{}^* w_{\tilde{r}} \odot \mathcal{R}_N\left(x - \frac{a}{2^k}, \frac{j}{2^k} + \frac{\tilde{r}}{2^k n}\right),$$

(13.16)

and

$$(D_k f)(x) = \sum_{i=0}^{N-1}{}^* f^{(i)}\left(x - \frac{a}{2^k}\right) \tag{13.17}$$

$$\odot \sum_{\substack{j \\ 2^k x - j \in [-a,a]}} \varphi(2^k x - j)\left(\sum_{\tilde{r}=0}^{n} w_{\tilde{r}} \frac{\left(\frac{(j+a)}{2^k} + \frac{\tilde{r}}{2^k n} - x\right)^i}{i!}\right)$$

$$\oplus \sum_{\substack{j \\ 2^k x - j \in [-a,a]}} \varphi(2^k x - j)\left(\sum_{\tilde{r}=0}^{n}{}^* w_{\tilde{r}} \odot \right.$$

$$\left. \mathcal{R}_N\left(x - \frac{a}{2^k}, \frac{j}{2^k} + \frac{\tilde{r}}{2^k n}\right)\right).$$

Next we estimate

$$D\left((D_k f)(x), f\left(x - \frac{a}{2^k}\right)\right)$$

$$= D\left((D_k f)(x), \sum_{\substack{j \\ 2^k x - j \in [-a,a]}}{}^* \varphi(2^k x - j) \odot f\left(x - \frac{a}{2^k}\right)\right)$$

$$= D\left(\sum_{i=1}^{N-1}{}^* f^{(i)}\left(x - \frac{a}{2^k}\right)\right.$$

$$\odot \sum_{\substack{j \\ 2^k x - j \in [-a,a]}} \varphi(2^k x - j)\left(\sum_{\tilde{r}=0}^{n} w_{\tilde{r}} \frac{\left(\frac{(j+a)}{2^k} + \frac{\tilde{r}}{2^k n} - x\right)^i}{i!}\right)$$

$$\oplus \sum_{\substack{j \\ 2^k x - j \in [-a,a]}} \varphi(2^k x - j)\left(\sum_{\tilde{r}=0}^{n}{}^* w_{\tilde{r}} \odot \mathcal{R}_N\left(x - \frac{a}{2^k}, \frac{j}{2^k} + \frac{\tilde{r}}{2^k n}\right)\right), \tilde{o}\right)$$

$$\leq \sum_{i=1}^{N-1} \sum_{\substack{j \\ 2^k x - j \in [-a,a]}} \varphi(2^k x - j)\left(\sum_{\tilde{r}=0}^{n} w_{\tilde{r}} \frac{\left(\frac{(j+a)}{2^k} + \frac{\tilde{r}}{2^k n} - x\right)^i}{i!}\right)$$

$$D\left(f^{(i)}\left(x - \frac{a}{2^k}\right), \tilde{o}\right)$$

$$+ \sum_{\substack{j \\ 2^k x - j \in [-a,a]}} \varphi(2^k x - j)\left(\sum_{\tilde{r}=0}^{n} w_{\tilde{r}} D\left(\mathcal{R}_N\left(x - \frac{a}{2^k}, \frac{j}{2^k} + \frac{\tilde{r}}{2^k n}\right), \tilde{o}\right)\right)$$

$$\leq \sum_{i=1}^{N-1} \frac{(2a+1)^i}{2^{ki}i!} \left(D(f^{(i)}(x), \tilde{o}) + \omega_1^{(\mathcal{F})}\left(f^{(i)}, \frac{a}{2^k}\right) \right)$$

$$+ \sum_{\substack{j \\ 2^k x - j \in [-a,a]}} \varphi(2^k x - j) \left(\sum_{\tilde{r}=0}^{n} w_{\tilde{r}} D\left(\mathcal{R}_N\left(x - \frac{a}{2^k}, \frac{j}{2^k} + \frac{\tilde{r}}{2^k n}\right), \tilde{o}\right) \right) =: (*).$$

Next we work on

$$D\left(\mathcal{R}_N\left(x - \frac{a}{2^k}, \frac{j}{2^k} + \frac{\tilde{r}}{2^k n}\right), \tilde{o}\right)$$

$$\leq D\left(\mathcal{R}_N\left(x - \frac{a}{2^k}, \frac{j}{2^k} + \frac{\tilde{r}}{2^k n}\right), f^{(N)}\left(x - \frac{a}{2^k}\right) \odot\right.$$

$$\left.\frac{\left(\frac{(j+a)}{2^k} + \frac{\tilde{r}}{2^k n} - x\right)^N}{N!}\right)$$

$$+ D\left(f^{(N)}\left(x - \frac{a}{2^k}\right) \odot \frac{\left(\frac{(j+a)}{2^k} + \frac{\tilde{r}}{2^k n} - x\right)^N}{N!}, \tilde{o}\right).$$

So that

$$D\left(\mathcal{R}_N\left(x - \frac{a}{2^k}, \frac{j}{2^k} + \frac{\tilde{r}}{2^k n}\right), \tilde{o}\right)$$

$$\leq D\left(\mathcal{R}_N\left(x - \frac{a}{2^k}, \frac{j}{2^k} + \frac{\tilde{r}}{2^k n}\right), f^{(N)}\left(x - \frac{a}{2^k}\right) \odot\right.$$

$$\left.\frac{\left(\frac{(j+a)}{2^k} + \frac{\tilde{r}}{2^k n} - x\right)^N}{N!}\right)$$

$$+ \frac{(2a+1)^N}{2^{kN} N!} \left(D(f^{(N)}(x), \tilde{o}) + \omega_1^{(\mathcal{F})}\left(f^{(N)}, \frac{a}{2^k}\right) \right). \quad (13.18)$$

Next we observe, as in the proof of Theorem 13.12, that

$$D\left(\mathcal{R}_N\left(x - \frac{a}{2^k}, \frac{j}{2^k} + \frac{\tilde{r}}{2^k n}\right), f^{(N)}\left(x - \frac{a}{2^k}\right) \odot \frac{\left(\frac{j+a}{2^k} + \frac{\tilde{r}}{2^k n} - x\right)^N}{N!}\right)$$

$$\leq \frac{(2a+1)^N}{2^{kN} N!} \omega_1^{(\mathcal{F})}\left(f^{(N)}, \frac{2a+1}{2^k}\right). \quad (13.19)$$

Hence the previous inequality (13.19) implies

$$D\left(\mathcal{R}_N\left(x - \frac{a}{2^k}, \frac{j}{2^k} + \frac{\tilde{r}}{2^k n}\right), \tilde{o}\right) \quad (13.20)$$

$$\leq \frac{(2a+1)^N}{2^{kN} N!} \left(D(f^{(N)}(x), \tilde{o}) + 3\omega_1^{(\mathcal{F})}\left(f^{(N)}, \frac{a}{2^k}\right) + \omega_1^{(\mathcal{F})}\left(f^{(N)}, \frac{1}{2^k}\right) \right).$$

Using the last inequality (13.20) into $(*)$ we obtain

$$
(*) \;\leq\; \sum_{i=1}^{N-1} \frac{(2a+1)^i}{2^{ki} i!} \left(D(f^{(i)}(x), \tilde{o}) + \omega_1^{(\mathcal{F})} \left(f^{(i)}, \frac{a}{2^k} \right) \right) \quad (13.21)
$$

$$
+ \frac{(2a+1)^N}{2^{kN} N!} \left(D(f^{(N)}(x), \tilde{o}) + 3\omega_1^{(\mathcal{F})} \left(f^{(N)}, \frac{a}{2^k} \right) \right.
$$

$$
\left. + \omega_1^{(\mathcal{F})} \left(f^{(N)}, \frac{1}{2^k} \right) \right).
$$

By (13.21) the proof of the theorem is now finished. $\square$

We need also the following main result.

Theorem 13.18. *All assumptions are as in Theorem 13.12. Define for $k \in \mathbb{Z}$, $x \in \mathbb{R}$ the fuzzy wavelet like operator*

$$
(C_k f)(x) := \sum_{j=-\infty}^{\infty}{}^{*} \left(2^k \odot (FR) \int_0^{2^{-k}} f \left(t + \frac{j}{2^k} \right) dt \right) \odot \varphi(2^k x - j).
$$

$$
(13.22)
$$

Then

$$
D\left((C_k f)(x), f\left(x - \frac{a}{2^k} \right) \right) \leq \quad (13.23)
$$

$$
\sum_{i=1}^{N-1} \frac{(2a+1)^i}{2^{ki} i!} \left(D(f^{(i)}(x), \tilde{o}) + \omega_1^{(\mathcal{F})} \left(f^{(i)}, \frac{a}{2^k} \right) \right)
$$

$$
+ \frac{(2a+1)^N}{2^{kN} N!} \left(D(f^{(N)}(x), \tilde{o}) + 3\omega_1^{(\mathcal{F})} \left(f^{(N)}, \frac{a}{2^k} \right) \right.
$$

$$
\left. + \omega_1^{(\mathcal{F})} \left(f^{(N)}, \frac{1}{2^k} \right) \right)
$$

$$
= \Delta_k \left(\frac{a}{2^k} \right),
$$

all $k \in \mathbb{Z}$, $x \in \mathbb{R}$.

If $f \in C_{\mathcal{F}}^{NU}(\mathbb{R})$ and as $k \to +\infty$ we obtain with rates that

$$
\lim_{k \to +\infty} D\left((C_k f)(x), f\left(x - \frac{a}{2^k} \right) \right) = 0.
$$

Corollary 13.19 (to Theorem 13.18, for $N = 1$). *It holds*

$$
D\left((C_k f)(x), f\left(x - \frac{a}{2^k} \right) \right) \leq \left(\frac{a}{2^{k-1}} + \frac{1}{2^k} \right) \left(D(f'(x), \tilde{o}) \right.
$$

$$
\left. + 3\omega_1^{(\mathcal{F})} \left(f', \frac{a}{2^k} \right) + \omega_1^{(\mathcal{F})} \left(f', \frac{1}{2^k} \right) \right),
$$

for any $x \in \mathbb{R}$, $k \in \mathbb{Z}$.

Corollary 13.20 (to Theorem 13.18). *The following improvement of (13.23) holds*

$$D\left((C_k f)(x), f\left(x - \frac{a}{2^k}\right)\right) \leq$$

$$\min\left(\left(\omega_1^{(\mathcal{F})}\left(f, \frac{a+1}{2^k}\right) + \omega_1^{(\mathcal{F})}\left(f, \frac{a}{2^k}\right)\right), \Delta_k\left(\frac{a}{2^k}\right)\right),$$

for any $x \in \mathbb{R}$, $k \in \mathbb{Z}$.

Proof. From Theorem 3, inequality (10), of [14] we get

$$D((C_k f)(x), f(x)) \leq \omega_1^{(\mathcal{F})}\left(f, \frac{a+1}{2^k}\right).$$

Then we observe that

$$D\left((C_k f)(x), f\left(x - \frac{a}{2^k}\right)\right) \leq D((C_k f)(x), f(x)) + D\left(f(x), f\left(x - \frac{a}{2^k}\right)\right)$$

$$\leq \omega_1^{(\mathcal{F})}\left(f, \frac{a+1}{2^k}\right) + \omega_1^{(\mathcal{F})}\left(f, \frac{a}{2^k}\right). \qquad \square$$

Proof of Theorem 13.18. Because φ is of compact support in $[-a, a]$ we see that

$$(C_k f)(x) = \sum_{\substack{j \\ 2^k x - j \in [-a,a]}}^{*} \left(2^k \odot (FR) \int_0^{2^{-k}} f\left(t + \frac{j}{2^k}\right) dt\right) \odot \varphi(2^k x - j).$$

$$(13.24)$$

Hence for specific $x \in \mathbb{R}$, $k \in \mathbb{Z}$ we have the j's in (13.24) to satisfy

$$x - \frac{a}{2^k} \leq \frac{j}{2^k} \leq x + \frac{a}{2^k}.$$

Using the fuzzy Taylor formula (Theorem 1, [11]) we find

$$f\left(t + \frac{j}{2^k}\right) = \sum_{i=0}^{N-1}{}^{*} f^{(i)}\left(x - \frac{a}{2^k}\right) \odot \frac{\left(t + \frac{(j+a)}{2^k} - x\right)^i}{i!} \oplus R_N\left(x - \frac{a}{2^k}, t + \frac{j}{2^k}\right),$$

$$(13.25)$$

where

$$R_N\left(x - \frac{a}{2^k}, t + \frac{j}{2^k}\right) := (FR) \int_{x - \frac{a}{2^k}}^{t + \frac{j}{2^k}} \left(\int_{x - \frac{a}{2^k}}^{s_1}\right.$$

$$(13.26)$$

$$\cdots \left(\left(\int_{x - \frac{a}{2^k}}^{s_{N-1}} f^{(N)}(s_N) ds_N\right) ds_{N-1}\cdots\right) ds_1.$$

Then by Theorem 3.4, [67] we get

$$
2^k \odot (FR) \int_0^{2^{-k}} f\left(t + \frac{j}{2^k}\right) dt = \sum_{i=0}^{N-1}{}^* f^{(i)}\left(x - \frac{a}{2^k}\right)
$$

$$
\odot \frac{2^k}{i!} \int_0^{2^{-k}} \left(t + \frac{(j+a)}{2^k} - x\right)^i dt \qquad (13.27)
$$

$$
\oplus\ 2^k \odot (FR) \int_0^{2^{-k}} \mathcal{R}_N\left(x - \frac{a}{2^k}, t + \frac{j}{2^k}\right) dt.
$$

Thus we have

$$
(C_k f)(x) = \sum_{i=0}^{N-1}{}^* f^{(i)}\left(x - \frac{a}{2^k}\right)
$$

$$
\odot \sum_{\substack{j \\ 2^k x - j \in [-a,a]}} \varphi(2^k x - j) \left(\frac{2^k}{i!} \int_0^{2^{-k}} \left(t + \frac{(j+a)}{2^k} - x\right)^i dt\right) \qquad (13.28)
$$

$$
\oplus \sum_{\substack{j \\ 2^k x - j \in [-a,a]}} \varphi(2^k x - j) \left(2^k \odot (FR) \int_0^{2^{-k}} \mathcal{R}_N\left(x - \frac{a}{2^k}, t + \frac{j}{2^k}\right) dt\right).
$$

Next we estimate

$$
D\left((C_k f)(x), f\left(x - \frac{a}{2^k}\right)\right)
$$

$$
= D\left((C_k f)(x), \sum_{\substack{j \\ 2^k x - j \in [-a,a]}}{}^* \varphi(2^k x - j) \odot f\left(x - \frac{a}{2^k}\right)\right)
$$

$$
= D\left(\left(\sum_{i=1}^{N-1}{}^* f^{(i)}\left(x - \frac{a}{2^k}\right)\right.\right.
$$

$$
\odot \sum_{\substack{j \\ 2^k x - j \in [-a,a]}} \varphi(2^k x - j) \left(\frac{2^k}{i!} \int_0^{2^{-k}} \left(t + \frac{(j+a)}{2^k} - x\right)^i dt\right)
$$

$$
\oplus \sum_{\substack{j \\ 2^k x - j \in [-a,a]}} \varphi(2^k x - j)
$$

$$
\left.\left(2^k \odot (FR) \int_0^{2^{-k}} \mathcal{R}_N\left(x - \frac{a}{2^k}, t + \frac{j}{2^k}\right) dt\right)\right), \tilde{o}\right)
$$

$$\leq \sum_{i=1}^{N-1} \sum_{\substack{j \\ 2^k x - j \in [-a,a]}} \varphi(2^k x - j) \left(\frac{2^k}{i!} \int_0^{2^{-k}} \left(t + \frac{(j+a)}{2^k} - x \right)^i dt \right)$$

$$D \left(f^{(i)} \left(x - \frac{a}{2^k} \right), \tilde{o} \right)$$

$$+ \sum_{\substack{j \\ 2^k x - j \in [-a,a]}} \varphi(2^k x - j) \left(2^k D \right)$$

$$\left((FR) \int_0^{2^{-k}} \mathcal{R}_N \left(x - \frac{a}{2^k}, t + \frac{j}{2^k} \right) dt, \tilde{o} \right)$$

$$\overset{\text{(by Lemma 1, [11])}}{\leq} \sum_{i=1}^{N-1} \frac{(2a+1)^i}{2^{ki} i!} \left(D(f^{(i)}(x), \tilde{o}) + \omega_1^{(\mathcal{F})} \left(f^{(i)}, \frac{a}{2^k} \right) \right)$$

$$+ \sum_{\substack{j \\ 2^k x - j \in [-a,a]}} \varphi(2^k x - j)$$

$$\left(2^k \int_0^{2^{-k}} D \left(\mathcal{R}_N \left(x - \frac{a}{2^k}, t + \frac{j}{2^k} \right), \tilde{o} \right) dt \right)$$

$$= \quad : (*).$$

Next we work on

$$D \left(\mathcal{R}_N \left(x - \frac{a}{2^k}, t + \frac{j}{2^k} \right), \tilde{o} \right)$$

$$\leq D \left(\mathcal{R}_N \left(x - \frac{a}{2^k}, t + \frac{j}{2^k} \right), f^{(N)} \left(x - \frac{a}{2^k} \right) \odot \frac{\left(t + \frac{(j+a)}{2^k} - x \right)^N}{N!} \right)$$

$$+ D \left(f^{(N)} \left(x - \frac{a}{2^k} \right) \odot \frac{\left(t + \frac{(j+a)}{2^k} - x \right)^N}{N!}, \tilde{o} \right).$$

So that

$$D \left(\mathcal{R}_N \left(x - \frac{a}{2^k}, t + \frac{j}{2^k} \right), \tilde{o} \right)$$

$$\leq D \left(\mathcal{R}_N \left(x - \frac{a}{2^k}, t + \frac{j}{2^k} \right), f^{(N)} \left(x - \frac{a}{2^k} \right) \odot \frac{\left(t + \frac{(j+a)}{2^k} - x \right)^N}{N!} \right)$$

$$+ \frac{(2a+1)^N}{2^{kN} N!} \left(D(f^{(N)}(x), \tilde{o}) + \omega_1^{(\mathcal{F})} \left(f^{(N)}, \frac{a}{2^k} \right) \right), \qquad (13.29)$$

for $0 \leq t \leq 2^{-k}$.

Next, we act as in the proof of Theorem 13.12, we observe that

$$D\left(\mathcal{R}_N\left(x-\frac{a}{2^k},t+\frac{j}{2^k}\right),f^{(N)}\left(x-\frac{a}{2^k}\right)\odot\frac{\left(t+\frac{(j+a)}{2^k}-x\right)^N}{N!}\right)$$

$$\leq\frac{(2a+1)^N}{2^{kN}N!}\omega_1^{(\mathcal{F})}\left(f^{(N)},\frac{2a+1}{2^k}\right),\qquad(13.30)$$

for $0\leq t\leq 2^{-k}$.

Therefore we obtain

$$D\left(\mathcal{R}_N\left(x-\frac{a}{2^k},t+\frac{j}{2^k}\right),\tilde{o}\right)\leq\qquad(13.31)$$

$$\frac{(2a+1)^N}{2^{kN}N!}\left(D(f^{(N)}(x),\tilde{o})+3\omega_1^{(\mathcal{F})}\left(f^{(N)},\frac{a}{2^k}\right)\right.$$

$$\left.+\omega_1^{(\mathcal{F})}\left(f^{(N)},\frac{1}{2^k}\right)\right),$$

for $0\leq t\leq 2^{-k}$. Using the last inequality (13.31) into $(*)$ we get

$$(*)\quad\leq\quad\sum_{i=1}^{N-1}\frac{(2a+1)^i}{2^{ki}i!}\left(D(f^{(i)}(x),\tilde{o})+\omega_1^{(\mathcal{F})}\left(f^{(i)},\frac{a}{2^k}\right)\right)$$

$$+\frac{(2a+1)^N}{2^{kN}N!}\left(D(f^{(N)}(x),\tilde{o})+3\omega_1^{(\mathcal{F})}\left(f^{(N)},\frac{a}{2^k}\right)\right.$$

$$\left.+\omega_1^{(\mathcal{F})}\left(f^{(N)},\frac{1}{2^k}\right)\right).\qquad(13.32)$$

By (13.32) we have the validity of the theorem.   □

The previous results lead to the following important theorems.

Theorem 13.21. *All assumptions are as in Theorem 13.12. It holds*

$$D((B_k f)(x),(D_k f)(x))\leq\beta_k\left(\frac{a}{2^k}\right)+\Delta_k\left(\frac{a}{2^k}\right),\qquad(13.33)$$

for any $x\in\mathbb{R}$, $k\in\mathbb{Z}$. If $f\in C_{\mathcal{F}}^{NU}(\mathbb{R})$, as $k\to+\infty$, we obtain with rates that

$$\lim_{k\to+\infty}D((B_k f)(x),(D_k f)(x))=0.$$

Proof. By

$$D((B_k f)(x),(D_k f)(x))\leq D\left((B_k f)(x),f\left(x-\frac{a}{2^k}\right)\right)$$

$$+D\left((D_k f)(x),f\left(x-\frac{a}{2^k}\right)\right)$$

(by (13.2) and (13.12))

$$\leq\quad\beta_k\left(\frac{a}{2^k}\right)+\Delta_k\left(\frac{a}{2^k}\right).\qquad□$$

Theorem 13.22. *All assumptions are as in Theorem 13.12. It holds*

$$D((D_kf)(x),(C_kf)(x)) \le 2\Delta_k\left(\frac{a}{2^k}\right), \tag{13.34}$$

for any $x \in \mathbb{R}$, $k \in \mathbb{Z}$. If $f \in C_{\mathcal{F}}^{NU}(\mathbb{R})$, as $k \to +\infty$, we get with rates that

$$\lim_{k \to +\infty} D((D_kf)(x),(C_kf)(x)) = 0.$$

Proof. By

$$D((D_kf)(x),(C_kf)(x)) \le D\left((D_kf)(x), f\left(x - \frac{a}{2^k}\right)\right)$$
$$+ D\left((C_kf)(x), f\left(x - \frac{a}{2^k}\right)\right)$$

(by (13.12) and (13.23))
$$\le \quad \Delta_k\left(\frac{a}{2^k}\right) + \Delta_k\left(\frac{a}{2^k}\right) = 2\Delta_k\left(\frac{a}{2^k}\right). \quad \square$$

Theorem 13.23. *All assumptions are as in Theorem 13.12. It holds*

$$D((B_kf)(x),(C_kf)(x)) \le \beta_k\left(\frac{a}{2^k}\right) + \Delta_k\left(\frac{a}{2^k}\right), \tag{13.35}$$

all $x \in \mathbb{R}$, $k \in \mathbb{Z}$. If $f \in C_{\mathcal{F}}^{NU}(\mathbb{R})$, as $k \to +\infty$, we find with rates that

$$\lim_{k \to +\infty} D((B_kf)(x),(C_kf)(x)) = 0.$$

Proof. By

$$D((B_kf)(x),(C_kf)(x)) \le D\left((B_kf)(x), f\left(x - \frac{a}{2^k}\right)\right)$$
$$+ D\left((C_kf)(x), f\left(x - \frac{a}{2^k}\right)\right)$$

(by (13.2) and (13.23))
$$\le \quad \beta_k\left(\frac{a}{2^k}\right) + \Delta_k\left(\frac{a}{2^k}\right). \quad \square$$

It follows another family of basic interesting related results.

Theorem 13.24. *Here φ is as in Theorem 13.12 and $f \in C(\mathbb{R}, \mathbb{R}_{\mathcal{F}})$. Fuzzy wavelet like operators B_k defined by (13.1), and C_k defined by (13.22). Then*

$$D((B_kf)(x),(C_kf)(x)) \le \omega_1^{(\mathcal{F})}\left(f, \frac{1}{2^k}\right), \tag{13.36}$$

for any $x \in \mathbb{R}$, $k \in \mathbb{Z}$. If $f \in C_{\mathcal{F}}^U(\mathbb{R})$, as $k \to +\infty$, we obtain with rates that

$$\lim_{k \to +\infty} D((B_kf)(x),(C_kf)(x)) = 0.$$

Proof. We have that

$$D((B_k f)(x), (C_k f)(x)) = D\left(\sum_{j=-\infty}^{\infty}{}^{*} f\left(\frac{j}{2^k}\right) \odot \varphi(2^k x - j),\right.$$

$$\left.\sum_{j=-\infty}^{\infty}{}^{*} \left(2^k \odot (FR)\int_0^{2^{-k}} f\left(t + \frac{j}{2^k}\right) dt\right) \odot \varphi(2^k x - j)\right)$$

$$\leq \sum_{\substack{j \\ 2^k x - j \in [-a,a]}} \varphi(2^k x - j) D\left(f\left(\frac{j}{2^k}\right), 2^k \odot (FR)\int_0^{2^{-k}} f\left(t + \frac{j}{2^k}\right) dt\right)$$

$$= 2^k \sum_{\substack{j \\ 2^k x - j \in [-a,a]}} \varphi(2^k x - j) D\left((FR)\int_0^{2^{-k}} f\left(\frac{j}{2^k}\right) dt,\right.$$

$$\left.(FR)\int_0^{2^{-k}} f\left(t + \frac{j}{2^k}\right) dt\right)$$

$$\underset{\leq}{(\text{by Lemma 1, [11]})} \quad 2^k \sum_{\substack{j \\ 2^k x - j \in [-a,a]}} \varphi(2^k x - j) \int_0^{2^{-k}} D\left(f\left(\frac{j}{2^k}\right),\right.$$

$$\left.f\left(t + \frac{j}{2^k}\right)\right) dt$$

$$\leq 2^k \sum_{\substack{j \\ 2^k x - j \in [-a,a]}} \varphi(2^k x - j) \int_0^{2^{-k}} \omega_1^{(\mathcal{F})}(f, t) dt \leq \omega_1^{(\mathcal{F})}\left(f, \frac{1}{2^k}\right). \qquad \Box$$

Theorem 13.25. *Here φ is as in Theorem 13.12 and $f \in C(\mathbb{R}, \mathbb{R}_{\mathcal{F}})$. Fuzzy wavelet like operators D_k defined by (13.10) and (13.11), and C_k defined by (13.22). Then*

$$D((D_k f)(x), (C_k f)(x)) \leq \omega_1^{(\mathcal{F})}\left(f, \frac{1}{2^{k-1}}\right), \qquad (13.37)$$

for any $x \in \mathbb{R}$, $k \in \mathbb{Z}$. If $f \in C_{\mathcal{F}}^U(\mathbb{R})$, as $k \to +\infty$, we get with rates that

$$\lim_{k \to +\infty} D((D_k f)(x), (C_k f)(x)) = 0.$$

Proof. We have that

$$D((D_k f)(x), (C_k f)(x))$$

$$= D\left(\sum_{j=-\infty}^{\infty}{}^{*} \delta_{kj}(f) \odot \varphi(2^k x - j), \right.$$

$$\left. \sum_{j=-\infty}^{\infty}{}^{*} \left(2^k \odot (FR) \int_0^{2^{-k}} f\left(t + \frac{j}{2^k} \right) dt \right) \odot \varphi(2^k x - j) \right)$$

$$\leq \sum_{\substack{j \\ 2^k x - j \in [-a,a]}} \varphi(2^k x - j) D\left(\delta_{kj}(f), 2^k \odot (FR) \int_0^{2^{-k}} f\left(t + \frac{j}{2^k} \right) dt \right)$$

$$\leq \sum_{\substack{j \\ 2^k x - j \in [-a,a]}} \varphi(2^k x - j) \sum_{\tilde{r}=0}^{n} w_{\tilde{r}} D\left(f\left(\frac{j}{2^k} + \frac{\tilde{r}}{2^k n} \right), 2^k \odot \right.$$

$$\left. (FR) \int_0^{2^{-k}} f\left(t + \frac{j}{2^k} \right) dt \right)$$

$$= 2^k \sum_{\substack{j \\ 2^k x - j \in [-a,a]}} \varphi(2^k x - j) \sum_{\tilde{r}=0}^{n} w_{\tilde{r}} D\left((FR) \int_0^{2^{-k}} f\left(\frac{j}{2^k} + \frac{\tilde{r}}{2^k n} \right) dt, \right.$$

$$\left. (FR) \int_0^{2^{-k}} f\left(t + \frac{j}{2^k} \right) dt \right)$$

$$\begin{array}{c} \text{(by Lemma 1, [11])} \\ \leq \end{array} \quad 2^k \sum_{\substack{j \\ 2^k x - j \in [-a,a]}} \varphi(2^k x - j) \sum_{\tilde{r}=0}^{n} w_{\tilde{r}} \cdot$$

$$\int_0^{2^{-k}} D\left(f\left(\frac{j}{2^k} + \frac{\tilde{r}}{2^k n} \right), f\left(t + \frac{j}{2^k} \right) \right) dt$$

$$\leq 2^k \sum_{\substack{j \\ 2^k x - j \in [-a,a]}} \varphi(2^k x - j) \sum_{\tilde{r}=0}^{n} w_{\tilde{r}} \int_0^{2^{-k}} \omega_1^{(\mathcal{F})}\left(f, \left| \frac{\tilde{r}}{2^k n} - t \right| \right) dt$$

$$\leq 2^k \sum_{\substack{j \\ 2^k x - j \in [-a,a]}} \varphi(2^k x - j) \sum_{\tilde{r}=0}^{n} w_{\tilde{r}} \int_0^{2^{-k}} \omega_1^{(\mathcal{F})}\left(f, \frac{1}{2^{k-1}} \right) dt$$

$$= \omega_1^{(\mathcal{F})}\left(f, \frac{1}{2^{k-1}} \right). \quad \square$$

Theorem 13.26. *Here φ is as in Theorem 13.12 and $f \in C(\mathbb{R}, \mathbb{R}_\mathcal{F})$. Fuzzy wavelet like operators B_k defined by (13.1), and D_k defined by (13.10) and (13.11). Then*

$$D((B_k f)(x), (D_k f)(x)) \leq \omega_1^{(\mathcal{F})}\left(f, \frac{1}{2^k}\right),\tag{13.38}$$

for any $x \in \mathbb{R}$, $k \in \mathbb{Z}$. If $f \in C_\mathcal{F}^U(\mathbb{R})$, as $k \to +\infty$, we find with rates that

$$\lim_{k \to +\infty} D((B_k f)(x), (D_k f)(x)) = 0.$$

Proof. We see that

$$D((B_k f)(x), (D_k f)(x))$$

$$= D\left(\sum_{j=-\infty}^{\infty}{}^* f\left(\frac{j}{2^k}\right) \odot \varphi(2^k x - j), \sum_{j=-\infty}^{\infty}{}^* \delta_{kj}(f) \odot \varphi(2^k x - j)\right)$$

$$\leq \sum_{\substack{j \\ 2^k x - j \in [-a,a]}} \varphi(2^k x - j) D\left(f\left(\frac{j}{2^k}\right), \delta_{kj}(f)\right)$$

$$\leq \sum_{\substack{j \\ 2^k x - j \in [-a,a]}} \varphi(2^k x - j) \sum_{\tilde{r}=0}^{n} w_{\tilde{r}} D\left(f\left(\frac{j}{2^k}\right), f\left(\frac{j}{2^k} + \frac{\tilde{r}}{2^k n}\right)\right)$$

$$\leq \sum_{\substack{j \\ 2^k x - j \in [-a,a]}} \varphi(2^k x - j) \sum_{\tilde{r}=0}^{n} w_{\tilde{r}} \omega_1^{(\mathcal{F})}\left(f, \frac{\tilde{r}}{2^k n}\right) \leq \omega_1^{(\mathcal{F})}\left(f, \frac{1}{2^k}\right). \qquad \square$$

Next we present the corresponding comparison results based on the previously given theorems.

Corollary 13.27 (to Theorem 13.21, and Theorem 13.26). *All assumptions here are as in Theorem 13.12. It holds*

$$D((B_k f)(x), (D_k f)(x)) \leq \min\left(\omega_1^{(\mathcal{F})}\left(f, \frac{1}{2^k}\right), \beta_k\left(\frac{a}{2^k}\right) + \Delta_k\left(\frac{a}{2^k}\right)\right),\tag{13.39}$$

for any $x \in \mathbb{R}$, $k \in \mathbb{Z}$.

Corollary 13.28 (to Theorem 13.22 and Theorem 13.25). *All assumptions here are as in Theorem 13.12. It holds*

$$D((D_k f)(x), (C_k f)(x)) \leq \min\left(\omega_1^{(\mathcal{F})}\left(f, \frac{1}{2^{k-1}}\right), 2\Delta_k\left(\frac{a}{2^k}\right)\right),\tag{13.40}$$

for any $x \in \mathbb{R}$, $k \in \mathbb{Z}$.

Corollary 13.29 (to Theorem 13.23 and Theorem 13.24). *All assumptions here are as in Theorem 13.12. It holds*

$$D((B_k f)(x), (C_k f)(x)) \leq \min\left(\omega_1^{(\mathcal{F})}\left(f, \frac{1}{2^k}\right), \beta_k\left(\frac{a}{2^k}\right) + \Delta_k\left(\frac{a}{2^k}\right)\right),$$

$$(13.41)$$

all $x \in \mathbb{R}$, $k \in \mathbb{Z}$.

Next we study similarly the Fuzzy wavelet operator A_k.

Theorem 13.30. *Let* $f \in C_b(\mathbb{R}, \mathbb{R}_{\mathcal{F}})$ *and the scaling function* $\varphi(x)$ *a real valued function with* supp $\varphi(x) \subseteq [-a, a]$, $0 < a < +\infty$, φ *is continuous on* $[-a, a]$, $\varphi(x) \geq 0$, *such that* $\sum\limits_{j=-\infty}^{\infty} \varphi(x - 1) = 1$ *on* $\mathbb{R}$ *(then* $\int_{-\infty}^{\infty} \varphi(x)dx = 1$*). Define*

$$\varphi_{kj}(t) := \quad 2^{k/2}\varphi(2^k t - j), \quad \text{for } k, j \in \mathbb{Z}, \ t \in \mathbb{R}, \quad (13.42)$$

$$\langle f, \varphi_{kj} \rangle := \quad (FR) \int_{\frac{j-a}{2^k}}^{\frac{j+a}{2^k}} f(t) \odot \varphi_{kj}(t)dt, \quad (13.43)$$

and set

$$(A_k f)(x) := \sum_{j=-\infty}^{\infty}{}^* \langle f, \varphi_{kj} \rangle \odot \varphi_{kj}(x), \quad \text{any } x \in \mathbb{R}. \quad (13.44)$$

The fuzzy wavelet like operator $(B_k f)(x)$ *is defined by (13.1). Then it holds*

$$D((A_k f)(x), (B_k f)(x)) \leq \omega_1^{(\mathcal{F})}\left(f, \frac{a}{2^k}\right), \quad (13.45)$$

for any $x \in \mathbb{R}$, $k \in \mathbb{Z}$. *If* $f \in C_{\mathcal{F}}^U(\mathbb{R})$ *and bounded, then*

$$\lim_{k \to +\infty} D((A_k f)(x), (B_k f)(x)) = 0 \quad \text{with rates.}$$

Proof. We see easily that

$$(A_k f)(x) = \sum_{j=-\infty}^{\infty}{}^* \left((FR) \int_{j-a}^{j+a} f\left(\frac{u}{2^k}\right) \odot \varphi(u - j)du\right) \odot \varphi(2^k x - j).$$

$$(13.46)$$

Also it holds

$$\int_{j-a}^{j+a} \varphi(u - j)du = 1. \quad (13.47)$$

So we observe

$$D((A_k f)(x), (B_k f)(x))$$

$$= D\left(\sum_{j=-\infty}^{\infty}{}^* \left((FR)\int_{j-a}^{j+a} f\left(\frac{u}{2^k}\right) \odot \varphi(u-j)du\right)\right.$$

$$\left.\odot\varphi(2^k x - j), \sum_{j=-\infty}^{\infty}{}^* f\left(\frac{j}{2^k}\right) \odot \varphi(2^k x - j)\right)$$

$$\leq \sum_{\substack{j \\ 2^k x - j \in [-a,a]}} \varphi(2^k x - j)D\left((FR)\int_{j-a}^{j+a} f\left(\frac{u}{2^k}\right) \odot \varphi(u-j)du,\right.$$

$$\left.(FR)\int_{j-a}^{j+a} f\left(\frac{j}{2^k}\right) \odot \varphi(u-j)du\right)$$

$$\substack{\text{(by Lemma 1, [11]} \\ \leq \\ \text{and Lemma 2, [14])}} \sum_{\substack{j \\ 2^k x - j \in [-a,a]}} \varphi(2^k x - j)\int_{j-a}^{j+a} \varphi(u-j)$$

$$D\left(f\left(\frac{u}{2^k}\right), f\left(\frac{j}{2^k}\right)\right) du$$

$$\leq \sum_{\substack{j \\ 2^k x - j \in [-a,a]}} \varphi(2^k x - j)\int_{j-a}^{j+a} \varphi(u-j)\omega_1^{(\mathcal{F})}\left(f, \frac{a}{2^k}\right) du$$

$$= \omega_1^{(\mathcal{F})}\left(f, \frac{a}{2^k}\right). \qquad \square$$

Theorem 13.31. *Let φ, f, A_k as in Theorem 13.30. Let fuzzy wavelet like operator D_k as in (13.10) and (13.11). Then*

$$D((A_k f)(x), (D_k f)(x)) \leq \omega_1^{(\mathcal{F})}\left(f, \frac{a+1}{2^k}\right), \qquad (13.48)$$

for any $x \in \mathbb{R}$, $k \in \mathbb{Z}$. If $f \in C_{\mathcal{F}}^U(\mathbb{R})$ and bounded then

$$\lim_{k \to +\infty} D((A_k f)(x), (D_k f)(x)) = 0 \quad \text{with rates.}$$

Proof. We notice that

$$D((A_k f)(x), (D_k f)(x))$$

$$\stackrel{\text{(by (13.46))}}{=} D\left(\sum_{j=-\infty}^{\infty}{}^* \left((FR)\int_{j-a}^{j+a} f\left(\frac{u}{2^k}\right)\odot\varphi(u-j)du\right)\right.$$

$$\left.\cdot\,\varphi(2^k x - j), \sum_{j=-\infty}^{\infty}{}^* \delta_{kj}(f)\odot\varphi(2^k x - j)\right)$$

$$\leq \sum_{\substack{j\\2^k x-j\in[-a,a]}} \varphi(2^k x - j)D\left((FR)\int_{j-a}^{j+a} f\left(\frac{u}{2^k}\right)\right.$$

$$\left.\odot\,\varphi(u-j)du, \sum_{\tilde{r}=0}^{n}{}^* w_{\tilde{r}}\odot f\left(\frac{j}{2^k}+\frac{\tilde{r}}{2^k n}\right)\right)$$

$$\stackrel{\text{(by (13.47))}}{\leq} \sum_{\substack{j\\2^k x-j\in[-a,a]}} \varphi(2^k x - j)\sum_{\tilde{r}=0}^{n} w_{\tilde{r}}D\left((FR)\int_{j-a}^{j+a} f\left(\frac{u}{2^k}\right)\right.$$

$$\left.\odot\varphi(u-j)du, (FR)\int_{j-a}^{j+a} f\left(\frac{j}{2^k}+\frac{\tilde{r}}{2^k n}\right)\odot\varphi(u-j)du\right)$$

$$\stackrel{\substack{\text{(by Lemma 1, [11]}\\\leq\\\text{and Lemma 2, [14])}}}{} \sum_{\substack{j\\2^k x-j\in[-a,a]}} \varphi(2^k x - j)\sum_{\tilde{r}=0}^{n} w_{\tilde{r}}$$

$$\times\left(\int_{j-a}^{j+a}\varphi(u-j)D\left(f\left(\frac{u}{2^k}\right), f\left(\frac{j}{2^k}+\frac{\tilde{r}}{2^k n}\right)\right)du\right)$$

$$\leq \sum_{\substack{j\\2^k x-j\in[-a,a]}} \varphi(2^k x - j)\sum_{\tilde{r}=0}^{n} w_{\tilde{r}}\left(\int_{j-a}^{j+a}\varphi(u-j)\omega_1^{(\mathcal{F})}\left(f,\frac{a+1}{2^k}\right)du\right)$$

$$= \omega_1^{(\mathcal{F})}\left(f,\frac{a+1}{2^k}\right). \qquad \square$$

Theorem 13.32. *Let φ, f, A_k as in Theorem 13.30. Let fuzzy wavelet like operator C_k as in (13.22). Then*

$$D((A_k f)(x), (C_k f)(x)) \leq \omega_1^{(\mathcal{F})}\left(f,\frac{a+1}{2^k}\right), \qquad (13.49)$$

for any $x \in \mathbb{R}$, $k \in \mathbb{Z}$. If $f \in C_{\mathcal{F}}^U(\mathbb{R})$ and bounded, then

$$\lim_{k\to+\infty} D((A_k f)(x), (C_k f)(x)) = 0 \quad \text{with rates.}$$

Proof. We observe that

$$D((A_k f)(x), (C_k f)(x))$$

$$\stackrel{\text{(by (13.46) and (13.22))}}{=} D\left(\sum_{j=-\infty}^{\infty *}\left((FR)\int_{j-a}^{j+a} f\left(\frac{u}{2^k}\right)\odot\varphi(u-j)du\right.\right.$$

$$\left.\odot\varphi(2^k x - j), \sum_{j=-\infty}^{\infty *}\left(2^k\odot(FR)\int_0^{2^{-k}} f\left(t+\frac{j}{2^k}\right)dt\right)\odot\varphi(2^k x - j)\right)$$

$$\leq \sum_{\substack{j \\ 2^k x - j\in[-a,a]}} \varphi(2^k x - j)D\left((FR)\int_{j-a}^{j+a} f\left(\frac{u}{2^k}\right)\right.$$

$$\left.\odot\varphi(u-j)du, 2^k\odot(FR)\int_0^{2^{-k}} f\left(t+\frac{j}{2^k}\right)dt\right)$$

$$= \sum_{\substack{j \\ 2^k x - j\in[-a,a]}} \varphi(2^k x - j)D\left((FR)\int_{j-a}^{j+a} f\left(\frac{u}{2^k}\right)\odot\varphi(u-j)du,\right.$$

$$\left.(FR)\int_{j-a}^{j+a}\varphi(u-j)\odot\left[2^k\odot(FR)\int_0^{2^{-k}} f\left(t+\frac{j}{2^k}\right)dt\right]du\right)$$

$$\stackrel{\substack{\text{(by Lemma 1, [4]} \\ \text{and Lemma 2, [14])}}}{\leq} \sum_{\substack{j \\ 2^k x - j\in[-a,a]}} \varphi(2^k x - j)$$

$$\times\left(\int_{j-a}^{j+a}\varphi(u-j)D\left(f\left(\frac{u}{2^k}\right), 2^k\odot(FR)\int_0^{2^{-k}} f\left(t+\frac{j}{2^k}\right)dt\right)du\right)$$

$$= \sum_{\substack{j \\ 2^k x - j\in[-a,a]}} \varphi(2^k x - j)\left(\int_{j-a}^{j+a}\varphi(u-j)2^k\right.$$

$$\left.D\left((FR)\int_0^{2^{-k}} f\left(\frac{u}{2^k}\right)dt, (FR)\int_0^{2^{-k}} f\left(t+\frac{j}{2^k}\right)dt\right)du\right)$$

$$\stackrel{\text{(by Lemma 1, [11])}}{\leq} \sum_{\substack{j \\ 2^k x - j\in[-a,a]}} \varphi(2^k x - j)\left(\int_{j-a}^{j+a}\varphi(u-j)2^k\right.$$

$$\left.\times\left(\int_0^{2^{-k}} D\left(f\left(\frac{u}{2^k}\right), f\left(t+\frac{j}{2^k}\right)\right)dt\right)du\right)$$

$$\leq \sum_{\substack{j \\ 2^k x - j\in[-a,a]}} \varphi(2^k x - j)\left(\int_{j-a}^{j+a}\varphi(u-j)2^k\right.$$

$$\left.\left(\int_0^{2^{-k}} \omega_1^{(\mathcal{F})}\left(f, \left|\frac{u}{2^k} - \frac{j}{2^k} - t\right|\right)dt\right)du\right)$$

$$\leq \sum_{\substack{j \\ 2^k x - j \in [-a,a]}} \varphi(2^k x - j) \left(\int_{j-a}^{j+a} \varphi(u-j) 2^k \right.$$

$$\left. \left(\int_0^{2^{-k}} \omega_1^{(\mathcal{F})} \left(f, \frac{a+1}{2^k} \right) dt \right) du \right)$$

$$= \omega_1^{(\mathcal{F})} \left(f, \frac{a+1}{2^k} \right). \qquad \square$$

Example 13.33. The following scaling function φ fulfills the assumptions of the presented theorems

$$\varphi(x) = \begin{cases} x + 1, & -1 \leq x \leq 0, \\ 1 - x, & 0 < x \leq 1, \\ 0, & \text{elsewhere.} \end{cases}$$

Note 13.34. A_k, B_k, C_k, D_k are linear operators over $\mathbb{R}$.

Remark 13.35. On Theorems 13.24, 13.25, 13.26, 13.30, 13.31 and 13.32. It is enough to comment Theorem 13.24, similar conclusions can be derived from the rest of them. Assume that $f \in C(\mathbb{R}, \mathbb{R}_{\mathcal{F}})$ fulfills the following Lipschitz condition

$$D(f(x), f(y)) \leq M|x-y|^\rho, \quad 0 < \rho \leq 1, \ M > 0. \tag{13.50}$$

Then clearly it holds

$$\omega_1^{(\mathcal{F})} \left(f, \frac{1}{2^k} \right) \leq \frac{M}{2^{k\rho}}, \quad k \in \mathbb{Z}. \tag{13.51}$$

So that from (13.36) we obtain

$$D((B_k f)(x), (C_k f)(x)) \leq \frac{M}{2^{k\rho}}, \tag{13.52}$$

and $k \in \mathbb{Z}$, $x \in \mathbb{R}$. And consequently

$$D^*((B_k f), (C_k f)) \leq \frac{M}{2^{k\rho}}, \tag{13.53}$$

for any $k \in \mathbb{Z}$.

Finally we get the global error estimate

$$\sup_{\substack{f \in C(\mathbb{R}, \mathbb{R}_{\mathcal{F}}): \\ f \text{ as in } (13.50)}} D^*((B_k f), (C_k f)) \leq \frac{M}{2^{k\rho}}. \tag{13.54}$$

Etc.

Next we give two independent but related and useful results.

Proposition 13.36. *Let φ, f be both even functions as in Theorem 13.24. Let $(B_k f)$ defined by (13.1). Then $(B_k f)(x)$ is an even function.*

Proof. We observe for $x \in \mathbb{R}$ that

$$
(B_k f)(-x) = \sum_{j=-\infty}^{\infty}{}^* f\left(\frac{j}{2^k}\right) \odot \varphi(-2^k x - j)
$$

$$
= \sum_{j=-\infty}^{\infty}{}^* f\left(-\frac{j}{2^k}\right) \odot \varphi(2^k x + j) \quad \text{(a finite sum)}
$$

$$
= \sum_{j^*:=-j=\infty}^{-\infty}{}^* f\left(\frac{j^*}{2^k}\right) \odot \varphi(2^k x - j^*)
$$

$$
= \sum_{j^*=-\infty}^{\infty}{}^* f\left(\frac{j^*}{2^k}\right) \odot \varphi(2^k x - j^*) = (B_k f)(x). \qquad \square
$$

Proposition 13.37. *Let φ, f be both even functions as in Theorem 13.30. Let $(A_k f)$ defined by (13.43) and (13.44). Then $(A_k f)(x)$ is an even function.*

Proof. We observe for $x \in \mathbb{R}$ that

$$
(A_k f)(-x) \overset{(13.46)}{=} \sum_{j=-\infty}^{\infty}{}^* \left((FR) \int_{j-a}^{j+a} f\left(\frac{u}{2^k}\right) \odot \varphi(u - j) du \right)
$$

$$
\odot \varphi(-2^k x - j)
$$

$$
\text{(a finite sum)}
$$

$$
= \sum_{j=-\infty}^{\infty}{}^* \left((FR) \int_{j-a}^{j+a} f\left(-\frac{u}{2^k}\right) \odot \varphi(-u + j) du \right)
$$

$$
\odot \varphi(2^k x + j)
$$

$$
\text{(linear change of variables is valid in } (FR)\text{-integrals)}
$$

$$
= \sum_{j^*:=-j=\infty}^{-\infty}{}^* \left((FR) \int_{j^*-a}^{j^*+a} f\left(\frac{w}{2^k}\right) \odot \varphi(w - j^*) dw \right)
$$

$$
\odot \varphi(2^k x - j^*) \overset{(13.46)}{=} (A_k f)(x). \qquad \square
$$

Note. In this chapter we use only fuzzy methods in our proofs. However in Chapter 16 we use real and fuzzy analysis together in our proofs. Consequently, we get some improvements of some of our results here, namely improvements for Theorems 13.21, 13.22 and Theorem 13.23, and for Corollaries 13.27, 13.28 and Corollary 13.29. See there Theorem 16.16

14

FUZZY APPROXIMATION BY FUZZY CONVOLUTION OPERATORS

Here we study four sequences of naturally arising fuzzy integral operators of convolution type that are integral analogs of known fuzzy wavelet like operators, defined via a scaling function. Their fuzzy convergence with rates to the fuzzy unit operator is established through fuzzy inequalities involving the fuzzy modulus of continuity. Also their high order fuzzy approximation is given similarly by involving the fuzzy modulus of continuity of the Nth order ($N \geq 1$) H-fuzzy derivative of the engaged fuzzy number valued function. The *fuzzy global smoothness preservation property* of these operators is presented too. This chapter relies on [15].

14.1   Introduction

Here $C(\mathbb{R})$ is the space of continuous functions from $\mathbb{R}$ into $\mathbb{R}$, and $C^N(\mathbb{R})$, $N \geq 1$ the space of N-times continuously differentiable functions on $\mathbb{R}$. We denote by $C_B(\mathbb{R})$ the space of continuous and bounded functions from $\mathbb{R}$ into $\mathbb{R}$. For $f \colon \mathbb{R} \to \mathbb{R}$ we define its *first modulus of continuity* by

$$\omega_1(f, \delta) := \sup_{x,y \in \mathbb{R} \colon |x-y| \leq \delta} |f(x) - f(y)|, \quad \delta > 0.$$

G.A. Anastassiou: Fuzzy Mathematics: Approximation The., STUDFUZZ 251, pp. 237–261.
springerlink.com

Let $C_U(\mathbb{R})$ be the space of *uniformly continuous functions* from $\mathbb{R}$ into $\mathbb{R}$. For $f \in C_U(\mathbb{R})$ we have $\omega_1(f, \delta) < +\infty$, $\delta > 0$, see [30], p. 298 and [34]. Also it is a well-known fact, see [56], p. 40, that

$$\lim_{\delta \downarrow 0} \omega_1(f, \delta) = \omega_1(f, 0) = 0, \quad \text{iff } f \in C_U(\mathbb{R}).$$

Let φ be a real valued function of compact support $\subseteq [-a, a]$, $a > 0$, $\varphi \geq 0$, φ is Lebesgue measurable and such that

$$\int_{-\infty}^{\infty} \varphi(x - u)\, du = 1, \quad \forall x \in \mathbb{R}, \tag{14.1}$$

equivalently

$$\int_{-\infty}^{\infty} \varphi(u)\, du = 1. \tag{14.2}$$

Examples (i)

$$\varphi(x) := \chi_{\left[-\frac{1}{2}, \frac{1}{2}\right)}(x) = \begin{cases} 1, & x \in \left[-\frac{1}{2}, \frac{1}{2}\right), \\ 0, & \text{elsewhere}, \end{cases}$$

the characteristic function;
(ii)

$$\varphi(x) := \begin{cases} 1 - x, & 0 \leq x \leq 1 \\ 1 + x, & -1 \leq x \leq 0 \\ 0, & \text{elsewhere}, \end{cases}$$

the hat function. Here we follow [30], pp. 287–293, and [35].

We introduce the following positive linear operators for $k \in \mathbb{Z}$ and $f \in C_U(\mathbb{R})$.

(i) Case of A_k operators: here additionally we assume that φ is an even continuous function. Define

$$(A_k f)(x) := \int_{-\infty}^{\infty} r_k^f(u) \varphi(2^k x - u)\, du,$$

where

$$r_k^f(u) := 2^k \int_{-\infty}^{\infty} f(t) \varphi(2^k t - u)\, dt \tag{14.3}$$

is continuous in u.
(ii)

$$(B_k f)(x) := \int_{-\infty}^{\infty} f\left(\frac{u}{2^k}\right) \varphi(2^k x - u)\, du. \tag{14.4}$$

(iii)

$$(L_k f)(x) := \int_{-\infty}^{\infty} c_k^f(u) \varphi(2^k x - u)\, du, \tag{14.5}$$

where

$$c_k^f(u) := 2^k \int_{2^{-k}u}^{2^{-k}(u+1)} f(t)\, dt = 2^k \int_0^{2^{-k}} f\left(t + \frac{u}{2^k}\right) dt$$

is continuous in u.

(iv)

$$(\Gamma_k f)(x) := \int_{-\infty}^{\infty} \gamma_k^f(u)\varphi(2^k x - u)\, du, \qquad (14.6)$$

where

$$\gamma_k^f(u) := \sum_{j=0}^{n} w_j f\left(\frac{u}{2^k} + \frac{j}{2^k n}\right), \qquad n \in \mathbb{N} \text{ fixed}, \ w_j \geq 0, \ \sum_{j=0}^{n} w_j = 1,$$

is continuous in u. We have proved in [30], p. 289, p. 295, respectively, and [35] that A_k, B_k, L_k, Γ_k are shift invariant operators, and map continuous probabilistic distribution functions into continuous probabilistic distribution functions. More importantly we proved the following theorems [30], pp. 290–293 and [35]. The operators A_k, B_k, L_k, Γ_k, $k \in \mathbb{Z}$ converge to the unit operator I with rates as given by next

Theorem 14.1. *For $k \in \mathbb{Z}$,*

$$|(A_k f)(x) - f(x)| \leq \omega_1\left(f, \frac{a}{2^{k-1}}\right), \qquad (14.7)$$

$$|(B_k f)(x) - f(x)| \leq \omega_1\left(f, \frac{a}{2^k}\right), \qquad (14.8)$$

$$|(L_k f)(x) - f(x)| \leq \omega_1\left(f, \frac{a+1}{2^k}\right), \qquad (14.9)$$

and

$$|(\Gamma_k f)(x) - f(x)| \leq \omega_1\left(f, \frac{a+1}{2^k}\right). \qquad (14.10)$$

The operators A_k, B_k, L_k, Γ_k fulfill the property of *Global Smoothness Preservation*.

Theorem 14.2. *For all $f \in C_U(\mathbb{R})$ and any $\delta > 0$ we have*

$$\omega_1(A_k f, \delta) \leq \omega_1(f, \delta), \quad \omega_1(B_k f, \delta) \leq \omega_1(f, \delta),$$
$$\omega_1(L_k f, \delta) \leq \omega_1(f, \delta), \quad \text{and } \omega_1(\Gamma_k f, \delta) \leq \omega_1(f, \delta). \quad (14.11)$$

Inequalities (14.11) are sharp, attained by $f(x) = x \in C_U(\mathbb{R})$.

In [12] we continued the above study by establishing the high order of approximation of operators A_k, B_k, L_k, Γ_k to the unit operator I.

Theorem 14.3. *Let $f \in C^N(\mathbb{R})$, $N \geq 1$. Then one has*

$$|(A_k f)(x) - f(x)| \leq \sum_{i=1}^{N} \frac{|f^{(i)}(x)|}{i!} \frac{a^i}{2^{i(k-1)}} \tag{14.12}$$

$$+ \frac{a^N}{N! 2^{N(k-1)}} \omega_1 \left(f^{(N)}, \frac{a}{2^{k-1}} \right),$$

$$|(B_k f)(x) - f(x)| \leq \sum_{i=1}^{N} \frac{|f^{(i)}(x)|}{i!} \frac{a^i}{2^{ki}} \tag{14.13}$$

$$+ \frac{a^N}{N! 2^{kN}} \omega_1 \left(f^{(N)}, \frac{a}{2^k} \right),$$

$$\left\{ \begin{array}{l} |(L_k f)(x) - f(x)| \\ |(\Gamma_k f)(x) - f(x)| \end{array} \right. \leq \sum_{i=1}^{N} \frac{|f^{(i)}(x)|}{i!} \frac{(a+1)^i}{2^{ki}} \tag{14.14}$$

$$+ \frac{(a+1)^N}{N! 2^{kN}} \omega_1 \left(f^{(N)}, \frac{a+1}{2^k} \right),$$

for any $k \in \mathbb{Z}$, and any $x \in \mathbb{R}$. If $f^{(N)}$ is uniformly continuous function, then as $k \to +\infty$ we obtain that

$$(\mathcal{L}_k f)(x) \to f(x),$$

for $\mathcal{L}_k := A_k$, B_k, L_k, Γ_k, pointwise with rates.

In this chapter we generalize the above mentioned motivating results to the Fuzzy Theory setting. We study the same operators A_k, B_k, L_k, Γ_k, $k \in \mathbb{Z}$, when they act on *fuzzy real number valued functions*. So according to the context one can understand clearly, when these operators are applied to real valued functions, and when they are applied to fuzzy real valued functions.

The proofs of the presented results rely a lot on this introduction. For that we also need.

14.2 Background

We mention

Lemma 14.4. *Let $g: [a, b] \to \mathbb{R}_+$ has existing ordinary Riemann integral $\int_a^b g(x)dx$. Let $u \in \mathbb{R}_{\mathcal{F}}$ be fixed. Then $(FR) \int_a^b (u \odot g(x))dx$ exists in $\mathbb{R}_{\mathcal{F}}$, and*

$$u \odot \int_a^b g(x)\, dx = (FR) \int_a^b (u \odot g(x))\, dx. \tag{14.15}$$

Proof. Since $\int_a^b g(x)dx$ exists we have that for any $\varepsilon > 0$ there exists $\delta > 0$ such that for any division $P = \{[v, w]; \xi\}$ of $[a, b]$ with norms $\Delta(P) < \delta$ it holds

$$\tau := \left| \sum_{P} (w - v)g(\xi) - \int_a^b g(x)\,dx \right| < \varepsilon.$$

Notice that $u \odot \int_a^b g(x)\,dx \in \mathbb{R}_{\mathcal{F}}$. Hence by Lemmas 2.1, 2.2(iii) of [17] we get

$$D\left(\sum_{P}(w-v)g(\xi) \odot u, u \odot \int_a^b g(x)\,dx \right) = D\left(u \odot \left(\sum_{P}(w-v)g(\xi) \right), \right.$$
$$\left. u \odot \int_a^b g(x)\,dx \right)$$
$$\leq \tau D(u, \tilde{o}) \leq \varepsilon D(u, \tilde{o}).$$

That is proving the claim. $\square$

Definition 14.5. Let $f : \mathbb{R} \to \mathbb{R}_{\mathcal{F}}$ be such that

(i) $(FR) \int_a^t f(x)\,dx$ exists for every real number $t \geq a$, $a \in \mathbb{R}$, then

$$(FR) \int_a^\infty f(x)\,dx = \lim_{t\to+\infty} (FR) \int_a^t f(x)\,dx$$

provided the limit exists in $\mathbb{R}_{\mathcal{F}}$ in the D-metric.

(ii) If $(FR) \int_t^b f(x)\,dx$ exists for every real number $t \leq b$, $b \in \mathbb{R}$ then

$$(FR) \int_{-\infty}^b f(x)\,dx = \lim_{t\to-\infty} (FR) \int_t^b f(x)\,dx$$

provided the limit exists in $\mathbb{R}_{\mathcal{F}}$ in the D-metric. The *improper* integrals $(FR) \int_a^\infty f(x)\,dx$ and $(FR) \int_{-\infty}^b f(x)\,dx$ are called *convergent* if the corresponding limit exists in $\mathbb{R}_{\mathcal{F}}$ and *divergent* if the limit does not exist in $\mathbb{R}_{\mathcal{F}}$.

(iii) If both $(FR) \int_a^\infty f(x)\,dx$ and $(FR) \int_{-\infty}^a f(x)\,dx$ are convergent for some $a \in \mathbb{R}$, then we define the *improper fuzzy-Riemann integral* over $\mathbb{R}$ as follows:

$$(FR) \int_{-\infty}^\infty f(x)\,dx = (FR) \int_{-\infty}^a f(x)\,dx \oplus (FR) \int_a^{+\infty} f(x)\,dx.$$

In this case we say that $(FR) \int_{-\infty}^\infty f(x)\,dx$ *converges*. Otherwise we say it *diverges*.

Remark 14.6. (i) Let $f \colon \mathbb{R} \to \mathbb{R}_{\mathcal{F}}$ be fuzzy continuous and of compact support $[a, b] \subset \mathbb{R}$ (See Theorem 14.10 next), then

$$(FR) \int_{-\infty}^{\infty} f(x)\, dx = (FR) \int_{a}^{b} f(x)\, dx. \qquad (14.16)$$

(ii) Given that $I := (FR) \int_{a}^{b} f(x)\, dx$, $a \leq b$, exists in $\mathbb{R}_{\mathcal{F}}$ from Definition 2.4 ([17]) we have

$$
\begin{aligned}
\varepsilon \;>\; & D\left(\sum_{P}^{*} (v - u) \odot f(\xi), I \right) \\
=\; & D\left((-1) \odot \left(\sum_{P}^{*} (v - u) \odot f(\xi) \right), (-1) \odot I \right) \\
=\; & D\left(\sum_{P}^{*} (-1) \odot ((v - u) \odot f(\xi)), (-1) \odot I \right) \\
=\; & D\left(\sum_{P}^{*} (u - v) \odot f(\xi), (-1) \odot I \right).
\end{aligned}
$$

That last motivates us to define

$$(FR) \int_{b}^{a} f(x)\, dx := (-1) \odot (FR) \int_{a}^{b} f(x)\, dx. \qquad (14.17)$$

Also

$$(FR) \int_{a}^{a} f(x)\, dx := \tilde{o}. \qquad (14.18)$$

Similarly given that $(FR) \int_{-\infty}^{\infty} f(x)\, dx$ exists in $\mathbb{R}_{\mathcal{F}}$, one can define

$$(FR) \int_{\infty}^{-\infty} f(x)\, dx := (-1) \odot (FR) \int_{-\infty}^{\infty} f(x)\, dx. \qquad (14.19)$$

Linear change of variable is possible and valid in (FR)-integrals.

Theorem 14.7. Let $x := \varphi(t) := mt + \gamma$, $m > 0$, $\gamma \in \mathbb{R}$, $t \in [\alpha, \beta]$. Call $a := m\alpha + \gamma$, $b := m\beta + \gamma$. Then $(FR) \int_{\alpha}^{\beta} f(\varphi(t))\, dt$ exists in $\mathbb{R}_{\mathcal{F}}$ iff $(FR) \int_{a}^{b} f(x)\, dx$ exists in $\mathbb{R}_{\mathcal{F}}$, and

$$(FR) \int_{a}^{b} f(x)\, dx = m \odot (FR) \int_{\alpha}^{\beta} f(\varphi(t))\, dt. \qquad (14.20)$$

Proof. Here $x \in [a, b]$ and φ is an $(1-1)$ and onto map from $[\alpha, \beta]$ to $[a, b]$. Also $f \colon [a, b] \to \mathbb{R}_{\mathcal{F}}$ and $f \circ \varphi \colon [\alpha, \beta] \to \mathbb{R}_{\mathcal{F}}$. If $(FR) \int_{\alpha}^{\beta} f(\varphi(t))\, dt$

exists in $\mathbb{R}_{\mathcal{F}}$ and equals $\frac{1}{m} \odot I$, by Definition 2.4 ([17]) for every $\varepsilon > 0$ there exists $\delta > 0$ such that for any division $P = \{[u, v]; \xi\}$ of $[\alpha, \beta]$ with norm $\Delta(P) < \delta$, we have

$$D\left(\sum_{P}^{*}(v - u) \odot f(\varphi(\xi)), \frac{1}{m} \odot I\right) < \varepsilon.$$

Equivalently we find

$$D\left(\sum_{P}^{*}(mv - mu) \odot f(\varphi(\xi)), I\right) < m\varepsilon =: \varepsilon'$$

and equivalently we have

$$D\left(\sum_{P}^{*}((mv + \gamma) - (mu + \gamma)) \odot f(\varphi(\xi)), I\right) < \varepsilon'.$$

Denoting $v' := mv + \gamma$, $u' := mu + \gamma$, $\xi' := m\xi + \gamma$ we have equivalently,

$$D\left(\sum_{P'}^{*}(v' - u') \odot f(\xi'), I\right) < \varepsilon',$$

for $P' = \{[u', v']; \xi'\}$ any division of $[a, b]$ with norm $\Delta(P') < m\delta =: \delta'$. By Definition 2.4 ([17]) we get equivalently that $(FR) \int_{a}^{b} f(x)\, dx$ exists in $\mathbb{R}_{\mathcal{F}}$ and equals I. $\qquad \square$

The counterpart of the last result comes next.

Theorem 14.8. *Let* $x := \varphi(t) := mt + \gamma$, $m < 0$, $\gamma \in \mathbb{R}$, $t \in [\alpha, \beta]$. *Call* $a := m\beta + \gamma$, $b := m\alpha + \gamma$. *Then* $(FR) \int_{\alpha}^{\beta} f(\varphi(t))\, dt$ *exists in* $\mathbb{R}_{\mathcal{F}}$ *iff* $(FR) \int_{a}^{b} f(x)\, dx$ *exists in* $\mathbb{R}_{\mathcal{F}}$, *and*

$$(FR) \int_{a}^{b} f(x)\, dx = |m| \odot (FR) \int_{\alpha}^{\beta} f(\varphi(t))\, dt. \qquad (14.21)$$

Proof. Similar to Theorem 14.7. $\qquad \square$

Next we present a result regarding linear change of variables for improper (FR)-integrals.

Theorem 14.9. *Let* $x := \varphi(t) := mt + \gamma$, m, γ *are fixed reals and* $t \in \mathbb{R}$. *Then* $(FR) \int_{-\infty}^{\infty} f(\varphi(t))\, dt$ *exists in* $\mathbb{R}_{\mathcal{F}}$ *iff* $(FR) \int_{-\infty}^{\infty} f(x)\, dx$ *exists in* $\mathbb{R}_{\mathcal{F}}$, *and*

$$(FR) \int_{-\infty}^{\infty} f(x)\, dx = |m| \odot (FR) \int_{-\infty}^{\infty} f(\varphi(t))\, dt. \qquad (14.22)$$

Proof. Clearly φ is $(1 - 1)$ and onto from $\mathbb{R} \to \mathbb{R}$.

(i) Here $m > 0$.

If $(FR) \int_{-\infty}^{\infty} f(\varphi(t)) \, dt$ exists in $\mathbb{R}_{\mathcal{F}}$ then $\exists \alpha \in \mathbb{R}$ such that (cf. Definition 14.5(iii))

$$(FR) \int_{-\infty}^{\infty} f(\varphi(t)) \, dt = (FR) \int_{-\infty}^{\alpha} f(\varphi(t)) \, dt \oplus (FR) \int_{\alpha}^{\infty} f(\varphi(t)) \, dt,$$

where the improper integrals

$$(FR) \int_{-\infty}^{\alpha} f(\varphi(t)) \, dt, \quad (FR) \int_{\alpha}^{\infty} f(\varphi(t)) \, dt \quad \text{exist in } \mathbb{R}_{\mathcal{F}}.$$

That is there exist $(FR) \int_{\theta}^{\alpha} f(\varphi(t)) \, dt$ for any $\theta \leq \alpha$ and $(FR) \int_{\alpha}^{\rho} f(\varphi(t)) \, dt$ for any $\rho \geq \alpha$, and in the D-metric we have

$$(FR) \int_{-\infty}^{\alpha} f(\varphi(t)) \, dt = \lim_{\theta \to -\infty} (FR) \int_{\theta}^{\alpha} f(\varphi(t)) \, dt,$$

and

$$(FR) \int_{\alpha}^{\infty} f(\varphi(t)) \, dt = \lim_{\rho \to \infty} (FR) \int_{\alpha}^{\rho} f(\varphi(t)) \, dt.$$

Hence by Theorem 14.7 we obtain

$$\begin{aligned}
(FR) \int_{-\infty}^{\alpha} f(\varphi(t)) \, dt &= \frac{1}{m} \odot \lim_{\theta \to -\infty} (FR) \int_{m\theta+\gamma}^{m\alpha+\gamma} f(x) \, dx \\
&= \frac{1}{m} \odot (FR) \int_{-\infty}^{m\alpha+\gamma} f(x) \, dx,
\end{aligned}$$

and

$$\begin{aligned}
(FR) \int_{\alpha}^{\infty} f(\varphi(t)) \, dt &= \frac{1}{m} \odot \lim_{\rho \to +\infty} (FR) \int_{m\alpha+\gamma}^{m\rho+\gamma} f(x) \, dx \\
&= \frac{1}{m} \odot (FR) \int_{m\alpha+\gamma}^{\infty} f(x) \, dx.
\end{aligned}$$

Consequently we obtain

$$\begin{aligned}
m \odot (FR) \int_{-\infty}^{\infty} f(\varphi(t)) \, dt &= (FR) \int_{-\infty}^{m\alpha+\gamma} f(x) \, dx \oplus (FR) \int_{m\alpha+\gamma}^{\infty} f(x) \, dx \\
&= (FR) \int_{-\infty}^{\infty} f(x) \, dx.
\end{aligned}$$

(ii) The case of $m < 0$ is similar. $\qquad \square$

We need

Theorem 14.10 ([67]). *Let* $f\colon [a,b] \to \mathbb{R}_{\mathcal{F}}$ *be fuzzy continuous with respect to metric* D. *Then* $(FR)\int_a^b f(x)\,dx$ *exists and belongs to* $\mathbb{R}_{\mathcal{F}}$, *furthermore it holds*

$$\left[(FR)\int_a^b f(x)\,dx\right]^r = \left[\int_a^b (f)_-^{(r)}(x)\,dx,\ \int_a^b (f)_+^{(r)}(x)\,dx\right], \quad \forall r \in [0,1].$$
(14.23)

Clearly $f_\pm^{(r)}\colon [a,b] \to \mathbb{R}$ *are continuous functions.*

Denote by $C(\mathbb{R}, \mathbb{R}_{\mathcal{F}})$ the space of fuzzy continuous functions and by $C_b(\mathbb{R}, \mathbb{R}_{\mathcal{F}})$ the space of bounded fuzzy continuous functions on $\mathbb{R}$ with respect to metric D.

We use also the following

Definition 14.11 ([11]). Let $f\colon \mathbb{R} \to \mathbb{R}_{\mathcal{F}}$ be a fuzzy real number valued function.

We define the (*first*) *fuzzy modulus of continuity of* f by

$$\omega_1^{(\mathcal{F})}(f, \delta) := \sup_{\substack{x,y \in \mathbb{R} \\ |x-y| \le \delta}} D(f(x), f(y)), \quad \delta > 0. \tag{14.24}$$

Definition 14.12 ([11]). Let $f\colon \mathbb{R} \to \mathbb{R}_{\mathcal{F}}$. We call f a *uniformly continuous fuzzy real number valued function*, iff for any $\varepsilon > 0$ there exists $\delta > 0$: whenever $|x - y| \le \delta$; $x, y \in \mathbb{R}$, implies that $D(f(x), f(y)) \le \varepsilon$. We denote it as $f \in C_{\mathcal{F}}^U(\mathbb{R})$.

We denote by $C_b^U(\mathbb{R}, \mathbb{R}_{\mathcal{F}})$ the space of uniformly continuous functions from $\mathbb{R} \to \mathbb{R}_{\mathcal{F}}$ that are bounded.

Proposition 14.13 ([11]). *Let* $f \in C_{\mathcal{F}}^U(\mathbb{R})$. *Then* $\omega_1^{(\mathcal{F})}(f, \delta) < +\infty$, *any* $\delta > 0$.

Proposition 14.14 ([11]). *It holds*

(i) $\omega_1^{(\mathcal{F})}(f, \delta)$ *is nonnegative and nondecreasing in* $\delta > 0$, *any* $f\colon \mathbb{R} \to \mathbb{R}_{\mathcal{F}}$.

(ii) $\lim_{\delta \downarrow 0} \omega_1^{(\mathcal{F})}(f, \delta) = \omega_1^{(\mathcal{F})}(f, 0) = 0$, *iff* $f \in C_{\mathcal{F}}^U(\mathbb{R})$.

(iii) $\omega_1^{(\mathcal{F})}(f, \delta_1 + \delta_2) \le \omega_1^{(\mathcal{F})}(f, \delta_1) + \omega_1^{(\mathcal{F})}(f, \delta_2)$, $\delta_1, \delta_2 > 0$, *any* $f\colon \mathbb{R} \to \mathbb{R}_{\mathcal{F}}$.

(iv) $\omega_1^{(\mathcal{F})}(f, n\delta) \le n\omega_1^{(\mathcal{F})}(f, \delta)$, $\delta > 0$, $n \in \mathbb{N}$, *any* $f\colon \mathbb{R} \to \mathbb{R}_{\mathcal{F}}$.

(v) $\omega_1^{(\mathcal{F})}(f, \lambda\delta) \le \lceil \lambda \rceil \omega_1^{(\mathcal{F})}(f, \delta) \le (\lambda + 1)\omega_1^{(\mathcal{F})}(f, \delta)$, $\lambda > 0$, $\delta > 0$, *where* $\lceil \cdot \rceil$ *is the ceiling of the number, any* $f\colon \mathbb{R} \to \mathbb{R}_{\mathcal{F}}$.

(vi) $\omega_1^{(\mathcal{F})}(f \oplus g, \delta) \le \omega_1^{(\mathcal{F})}(f, \delta) + \omega_1^{(\mathcal{F})}(g, \delta)$, $\delta > 0$, *any* $f, g\colon \mathbb{R} \to \mathbb{R}_{\mathcal{F}}$.

(vii) $\omega_1^{(\mathcal{F})}(f, \cdot)$ is continuous on $\mathbb{R}_+$, for $f \in C_{\mathcal{F}}^U(\mathbb{R})$.

It follows a very important and useful representation result for moduli of continuity.

Proposition 14.15. Let $f: \mathbb{R} \to \mathbb{R}_{\mathcal{F}}$ be a fuzzy real number valued function. Assume that $\omega_1^{(\mathcal{F})}(f, \delta)$, $\omega_1(f_-^{(r)}, \delta)$, $\omega_1(f_+^{(r)}, \delta)$ are finite for $\delta > 0$. Then it holds

$$\omega_1^{(\mathcal{F})}(f, \delta) = \sup_{r \in [0,1]} \max\{\omega_1(f_-^{(r)}, \delta), \omega_1(f_+^{(r)}, \delta)\}. \tag{14.25}$$

Formula (14.25) is valid also when f and the moduli of continuity are defined over $[a, b] \subset \mathbb{R}$.

Proof. Let $x, y \in \mathbb{R}: |x - y| \le \delta$. We have from [17], Section 2 that

$$
\begin{aligned}
D(f(x), f(y)) &= \sup_{r \in [0,1]} \max\{|(f(x))_-^{(r)} - (f(y))_-^{(r)}|, |(f(x))_+^{(r)} - (f(y))_+^{(r)}|\} \\
&\le \sup_{r \in [0,1]} \max\{\omega_1(f_-^{(r)}, \delta), \omega_1(f_+^{(r)}, \delta)\}.
\end{aligned}
$$

Thus

$$\omega_1^{(\mathcal{F})}(f, \delta) \le \sup_{r \in [0,1]} \max\{\omega_1(f_-^{(r)}, \delta), \omega_1(f_+^{(r)}, \delta)\}.$$

Next we observe for any $r \in [0, 1]$ and any $x, y \in \mathbb{R}: |x - y| \le \delta$ that

$$\omega_1^{(\mathcal{F})}(f, \delta) \ge D(f(x), f(y)) \ge |(f(x))_-^{(r)} - (f(y))_-^{(r)}|, |(f(x))_+^{(r)} - (f(y))_+^{(r)}|.$$

Therefore

$$\omega_1(f_-^{(r)}, \delta), \omega_1(f_+^{(r)}, \delta) \le \omega_1^{(\mathcal{F})}(f, \delta), \quad \forall r \in [0, 1].$$

Clearly it holds

$$\sup_{r \in [0,1]} \max\{\omega_1(f_-^{(r)}, \delta), \omega_1(f_+^{(r)}, \delta)\} \le \omega_1^{(\mathcal{F})}(f, \delta).$$

We have established formula (14.25). $\qquad\square$

Remark 14.16. (i) Let $f: [a, b] \to \mathbb{R}_{\mathcal{F}}$ be fuzzy continuous. From [17], Section 2 we have

$$D(f(t), \tilde{o}) = \sup_{r \in [0,1]} \max\{|f_-^{(r)}(t)|, |f_+^{(r)}(t)|\}, \quad \forall t \in [a, b]. \tag{14.26}$$

Since f is fuzzy bounded we have that $\exists M > 0$ such that $D(f(t), \tilde{o}) \le M$, $\forall t \in [a, b]$. Since $D(f(t), \tilde{o})$ is continuous in $t \in [a, b]$ for some $t^* \in [a, b]$ we have that

$$D(f(t^*), \tilde{o}) = \sup_{t \in [a,b]} D(f(t), \tilde{o}).$$

Notice that by the principle of iterated suprema we get

$$\sup_{t\in[a,b]} D(f(t),\tilde{o}) \;=\; \sup_{t\in[a,b]}\sup_{r\in[0,1]}\max\{|f_-^{(r)}(t)|,|f_+^{(r)}(t)|\}$$

$$=\; \sup_{r\in[0,1]}\max\sup_{t\in[a,b]}\{|f_-^{(r)}(t)|,|f_+^{(r)}(t)|\}.$$

Clearly we find that

$$\sup_{t\in[a,b]} D(f(t),\tilde{o}) = \sup_{r\in[0,1]}\max\{\|f_-^{(r)}\|_\infty,\|f_+^{(r)}\|_\infty\}. \tag{14.27}$$

We observe here easily that $\forall r\in[0,1]$, $\forall t\in[a,b]$ we derive

$$|f_-^{(r)}(t)|,|f_+^{(r)}(t)| \le D(f(t^*),\tilde{o}) \le M,$$

so we can apply above the principle of iterated suprema.

(ii) Let $f\colon \mathbb{R}\to\mathbb{R}_{\mathcal{F}}$ be fuzzy continuous, then clearly $f_\pm^{(r)}$ are continuous functions from $\mathbb{R}$ into $\mathbb{R}$, $\forall r\in[0,1]$. If $f\in C_{\mathcal{F}}^U(\mathbb{R})$, equivalently (by Proposition 14.14(ii)) we have $\lim_{\delta\downarrow0}\omega_1^{(\mathcal{F})}(f,\delta)=\omega_1^{(\mathcal{F})}(f,0)=0$. Thus by formula (14.25) we get $\lim_{\delta\downarrow0}\omega_1(f_\pm^{(r)},\delta)=\omega_1(f_\pm^{(r)},0)=0$, $\forall r\in[0,1]$, equivalently $f_\pm^{(r)}\in C_U(\mathbb{R})$. I.e. If $f\in C_{\mathcal{F}}^U(\mathbb{R})$ then $f_\pm^{(r)}\in C_U(\mathbb{R})$, $\forall r\in[0,1]$.

We need the following

Lemma 14.17 ([14]). *Let $f\colon\mathbb{R}\to\mathbb{R}_{\mathcal{F}}$ fuzzy continuous and bounded. Let $g\colon J\subseteq\mathbb{R}\to\mathbb{R}_+$ continuous and bounded, where J is an interval. Then $f(x)\odot g(x)$ is fuzzy continuous function $\forall x\in J$.*

Remark 14.18. Here $r\in[0,1]$, $x_i^{(r)},y_i^{(r)}\in\mathbb{R}$, $i=1,\dots,m\in\mathbb{N}$. Suppose that

$$\sup_{r\in[0,1]}\max(x_i^{(r)},y_i^{(r)})\in\mathbb{R},\quad \text{for } i=1,\dots,m. \tag{14.28}$$

We see that

$$x_i^{(r)} \;\le\; \max(x_i^{(r)},y_i^{(r)}) \le \sup_r\max(x_i^{(r)},y_i^{(r)}),$$

$$y_i^{(r)} \;\le\; \max(x_i^{(r)},y_i^{(r)}) \le \sup_r\max(x_i^{(r)},y_i^{(r)}).$$

Hence

$$\sum_{i=1}^m x_i^{(r)} \;\le\; \sum_{i=1}^m \sup_r\max(x_i^{(r)},y_i^{(r)}),$$

$$\sum_{i=1}^m y_i^{(r)} \;\le\; \sum_{i=1}^m \sup_r\max(x_i^{(r)},y_i^{(r)}).$$

Consequently

$$\max\left(\sum_{i=1}^{m} x_i^{(r)}, \sum_{i=1}^{m} y_i^{(r)}\right) \le \sum_{i=1}^{m} \sup_r \max(x_i^{(r)}, y_i^{(r)}).$$

We conclude that

$$\sup_{r \in [0,1]} \max\left(\sum_{i=1}^{m} x_i^{(r)}, \sum_{i=1}^{m} y_i^{(r)}\right) \le \sum_{i=1}^{m} \sup_{r \in [0,1]} \max(x_i^{(r)}, y_i^{(r)}). \qquad (14.29)$$

Inequality (14.29) is used in the proofs of the main results here.

We denote by $C^N(\mathbb{R}, \mathbb{R}_{\mathcal{F}})$, $N \ge 1$, the space of all N-times continuously fuzzy differentiable functions from $\mathbb{R}$ into $\mathbb{R}_{\mathcal{F}}$.

We need the following

Theorem 14.19 ([71]). *Let $f: [a,b] \subseteq \mathbb{R} \to \mathbb{R}_{\mathcal{F}}$ be H-fuzzy differentiable. Let $t \in [a,b]$, $0 \le r \le 1$. Clearly*

$$[f(t)]^r = \left[(f(t))_-^{(r)}, (f(t))_+^{(r)}\right] \subseteq \mathbb{R}. \qquad (14.30)$$

Then $(f(t))_\pm^{(r)}$ are differentiable and

$$[f'^r = \left[((f(t))_-^{(r)})', ((f(t))_+^{(r)})'\right]. \qquad (14.31)$$

I.e.

$$(f'_\pm)^{(r)} = (f_\pm^{(r)})', \quad \forall r \in [0,1]. \qquad (14.32)$$

Remark 14.20. Let $f \in C^N(\mathbb{R}, \mathbb{R}_{\mathcal{F}})$, $N \ge 1$. Then by Theorem 14.19 we obtain

$$[f^{(i)}(t)]^r = \left[((f(t))_-^{(r)})^{(i)}, ((f(t))_+^{(r)})^{(i)}\right],$$

for $i = 0, 1, 2, \ldots, N$ and in particular we have that

$$(f_\pm^{(i)})^{(r)} = (f_\pm^{(r)})^{(i)}, \qquad (14.33)$$

for any $r \in [0,1]$.

14.3 Main results

Let here $f \in C_b^U(\mathbb{R}, \mathbb{B}_{\mathcal{F}})$, and let $\varphi \in C_B(\mathbb{R})$ of compact support $\subseteq [-a,a]$, $a > 0$, $\varphi \ge 0$ and be such that

$$\int_{-\infty}^{\infty} \varphi(x - u)\, du = 1, \quad \forall x \in \mathbb{R}, \qquad (14.34)$$

equivalently

$$\int_{-\infty}^{\infty} \varphi(u)\, du = 1. \tag{14.35}$$

(See Example (ii).)

By Proposition 14.13 we have that $\omega_1^{(\mathcal{F})}(f, \delta) < +\infty$, for any $\delta > 0$. Let $k \in \mathbb{Z}$, $x \in \mathbb{R}$. We introduce the following *fuzzy convolution type operators*,

(i)

$$(B_k f)(x) := (FR) \int_{-\infty}^{\infty} f\left(\frac{u}{2^k}\right) \odot \varphi(2^k x - u)\, du, \tag{14.36}$$

(ii)

$$(L_k f)(x) := (FR) \int_{-\infty}^{\infty} c_k^f(u) \odot \varphi(2^k x - u)\, du, \tag{14.37}$$

where

$$c_k^f(u) := 2^k \odot (FR) \int_{2^{-k} u}^{2^{-k}(u+1)} f(t)\, dt = 2^k \odot (FR) \int_{0}^{2^{-k}} f\left(t + \frac{u}{2^k}\right) dt \tag{14.38}$$

(equality true by Theorems 14.7, 14.10),

(iii)

$$(\Gamma_k f)(x) := (FR) \int_{-\infty}^{\infty} \gamma_k^f(u) \odot \varphi(2^k x - u)\, du, \tag{14.39}$$

where

$$\gamma_k^f(u) := \sum_{j=0}^{n}{}^{*} w_j \odot f\left(\frac{u}{2^k} + \frac{j}{2^k n}\right), n \in \mathbb{N} \text{ fixed, } w_j \geq 0, \sum_{j=0}^{m} w_j = 1, \tag{14.40}$$

(iv) here additionally we assume that φ is even and define

$$(A_k f)(x) := (FR) \int_{-\infty}^{\infty} r_k^f(u) \odot \varphi(2^k x - u)\, du, \tag{14.41}$$

where

$$r_k^f(u) := 2^k \odot (FR) \int_{-\infty}^{\infty} f(t) \odot \varphi(2^k t - u)\, dt. \tag{14.42}$$

We emphasize again that the above defined fuzzy operators A_k, B_k, L_k, Γ_k are the fuzzy analogs of the real convolution operators A_k, B_k, L_k, Γ_k discussed in *Section 14.1. Introduction*, defined in an analogous way.

We present our results

Theorem 14.21. *It holds*

$$D^*(B_k f, f) \leq \omega_1^{(\mathcal{F})}\left(f, \frac{a}{2^k}\right), \tag{14.43}$$

and

$$\omega_1^{(\mathcal{F})}(B_k f, \delta) \leq \omega_1^{(\mathcal{F})}(f, \delta), \quad \forall \delta > 0, \tag{14.44}$$

that is B_k fulfilling the property of Fuzzy Global Smoothness preservation.

Proof of inequality (14.43). We notice

$$D((B_k f)(x), f(x)) \overset{\text{(by (14.15) \& (14.34))}}{=} D\left((FR)\int_{2^k x - a}^{2^k x + a} f\left(\frac{u}{2^k}\right)\right.$$

$$\left.\odot \varphi(2^k x - u)\, du, (FR)\int_{2^k x - a}^{2^k x + a} f(x) \odot \varphi(2^k x - u)\, du\right)$$

(by (2.9) of [17])

$$\leq \int_{2^k x - a}^{2^k x + a} D\left(f\left(\frac{u}{2^k}\right) \odot \varphi(2^k x - u), f(x) \odot \varphi(2^k x - u)\right) du$$

$$= \int_{2^k x - a}^{2^k x + a} \varphi(2^k x - u) D\left(f\left(\frac{u}{2^k}\right), f(x)\right) du$$

$$\leq \int_{2^k x - a}^{2^k x + a} \varphi(2^k x - u) \omega_1^{(\mathcal{F})}\left(f, \left|\frac{u}{2^k} - x\right|\right) du$$

$$\left(\text{notice } -\frac{a}{2^k} \leq x - \frac{u}{2^k} \leq \frac{a}{2^k}\right)$$

$$\leq \int_{2^k x - a}^{2^k x + a} \varphi(2^k x - u) \omega_1^{(\mathcal{F})}\left(f, \frac{a}{2^k}\right) du$$

$$= \left(\int_{2^k x - a}^{2^k x + a} \varphi(2^k x - u)\, du\right) \omega_1^{(\mathcal{F})}\left(f, \frac{a}{2^k}\right) = \omega_1^{(\mathcal{F})}\left(f, \frac{a}{2^k}\right),$$

by $\int_{2^k x - a}^{2^k x + a} \varphi(2^k x - u)\, du = 1$. That is proving

$$D((B_k f)(x), f(x)) \leq \omega_1^{(\mathcal{F})}\left(f, \frac{a}{2^k}\right),$$

hence (14.43) is true.

Proof of inequality (14.44). Let $x, y \in \mathbb{R}$: $|x - y| \leq \delta$. Then we observe that

$$D((B_k f)(x), (B_k f)(y))$$

$$= D\left((FR)\int_{-\infty}^{\infty} f\left(\frac{u}{2^k}\right) \odot \varphi(2^k x - u)\, du, (FR)\int_{-\infty}^{\infty} f\left(\frac{u}{2^k}\right) \odot\right.$$

$$\varphi(2^k y - u)\, du\Big)$$

$$= D\left((FR)\int_{-\infty}^{\infty} f\left(x - \frac{(-u + 2^k x)}{2^k}\right) \odot \varphi(-u + 2^k x)\, du,\right.$$

$$(FR)\int_{-\infty}^{\infty} f\left(y - \frac{(-u + 2^k y)}{2^k}\right) \odot \varphi(-u + 2^k y)\, du\Big)$$

$$(\text{call } \sigma(u) := -u + 2^k x, \rho(u) := -u + 2^k y)$$

$$= D\left((FR)\int_{-\infty}^{\infty} f\left(x - \frac{\sigma(u)}{2^k}\right) \odot \varphi(\sigma(u))\,du,\right.$$

$$(FR)\int_{-\infty}^{\infty} f\left(y - \frac{\rho(u)}{2^k}\right) \odot \varphi(\rho(u))\,du\Bigg)$$

$$\stackrel{(14.22)}{=} D\left((FR)\int_{-\infty}^{\infty} f\left(x - \frac{w}{2^k}\right) \odot \varphi(w)\,dw,\right.$$

$$(FR)\int_{-\infty}^{\infty} f\left(y - \frac{w}{2^k}\right) \odot \varphi(w)\,dw\Bigg)$$

$$= D\left((FR)\int_{-a}^{a} f\left(x - \frac{w}{2^k}\right) \odot \varphi(w)\,dw,\right.$$

$$(FR)\int_{-a}^{a} f\left(y - \frac{w}{2^k}\right) \odot \varphi(w)\,dw\Bigg)$$

(by (2.9) of [17])

$$\leq \int_{-a}^{a} D\left(f\left(x - \frac{w}{2^k}\right) \odot \varphi(w), f\left(y - \frac{w}{2^k}\right) \odot \varphi(w)\right)dw$$

$$= \int_{-a}^{a} \varphi(w) D\left(f\left(x - \frac{w}{2^k}\right), f\left(y - \frac{w}{2^k}\right)\right)dw$$

$$\leq \int_{-a}^{a} \varphi(w)\omega_1^{(\mathcal{F})}(f, |x - y|)\,dw \leq \int_{-a}^{a} \varphi(w)\omega_1^{(\mathcal{F})}(f, \delta)\,dw$$

$$= \left(\int_{-a}^{a} \varphi(w)\,dw\right)\omega_1^{(\mathcal{F})}(f, \delta) \stackrel{(14.35)}{=} 1 \cdot \omega_1^{(\mathcal{F})}(f, \delta) = \omega_1^{(\mathcal{F})}(f, \delta).$$

We have proved that

$$D((B_k f)(x), (B_k f)(y)) \leq \omega_1^{(\mathcal{F})}(f, \delta)$$

which clearly implies (14.44). □

We continue with

Theorem 14.22. *It holds*

$$D^*((L_k f), f) \leq \omega_1^{(\mathcal{F})}\left(f, \frac{a+1}{2^k}\right), \tag{14.45}$$

and

$$\omega_1^{(\mathcal{F})}(L_k f, \delta) \leq \omega_1^{(\mathcal{F})}(f, \delta), \quad \forall \delta > 0, \tag{14.46}$$

that is L_k fulfilling the property of Fuzzy Global Smoothness preservation.

Proof. First we prove that $c_k^f(u)$ is fuzzy continuous in u. We notice that

$$D\left((FR)\int_0^{2^{-k}} f\left(t+\frac{u_m}{2^k}\right)dt,\ (FR)\int_0^{2^{-k}} f\left(t+\frac{u}{2^k}\right)dt\right)$$

$$(\text{by (2.9) of [17]}) \leq \int_0^{2^{-k}} D\left(f\left(t+\frac{u_m}{2^k}\right),f\left(t+\frac{u}{2^k}\right)\right)dt$$

$$\leq \int_0^{2^{-k}} \omega_1^{(\mathcal{F})}\left(f,\frac{|u_m-u|}{2^k}\right)dt$$

$$= 2^{-k}\omega_1^{(\mathcal{F})}\left(f,\frac{|u_m-u|}{2^k}\right) \to 0,\quad \text{as } u_m \to u,$$

where $u_m, u \in \mathbb{R}$. That is proving fuzzy continuity of $c_k^f(u)$. Also we see by (2.9) of [17] that

$$D\left((FR)\int_0^{2^{-k}} f\left(t+\frac{u}{2^k}\right)dt, \tilde{o}\right) \leq \int_0^{2^{-k}} D\left(f\left(t+\frac{u}{2^k}\right),\tilde{o}\right)dt$$

$$\leq M2^{-k} < +\infty,$$

where $M > 0$ such that $D(f(x),\tilde{o}) \leq M$, $\forall x \in \mathbb{R}$. That is proving that $c_k^f(u)$ is also fuzzy bounded.

By Lemma 14.17 now we have that $c_k^f(u) \odot \varphi(2^k x - u) \in C(\mathbb{R}, \mathbb{R}_{\mathcal{F}})$ as a function of u. Again we get that

$$(L_k f)(x) = (FR)\int_{-a+2^k x}^{a+2^k x} c_k^f(u) \odot \varphi(2^k x - u)du,\tag{14.47}$$

and by [17], Section 2

$$D((L_k f)(x), f(x)) = \sup_{r \in [0,1]}\max\{|(L_k f)_-^{(r)}(x)-f_-^{(r)}(x)|,|(L_k f)_+^{(r)}(x)-f_+^{(r)}(x)|\}.\tag{14.48}$$

We see that

$$[c_k^f(u)]^r = 2^k\left[(FR)\int_0^{2^{-k}} f\left(t+\frac{u}{2^k}\right)dt\right]^r$$

$$\overset{(14.23)}{=} \left[2^k\int_0^{2^{-k}} f_-^{(r)}\left(t+\frac{u}{2^k}\right)dt, 2^k\int_0^{2^{-k}} f_+^{(r)}\left(t+\frac{u}{2^k}\right)dt\right]$$

$$= [c_k^{(f_-^{(r)})}(u), c_k^{(f_+^{(r)})}(u)].$$

That is proving

$$(c_k^f(u))_\pm^{(r)} = c_k^{(f_\pm^{(r)})}(u),\quad \forall u \in \mathbb{R}.\tag{14.49}$$

Thus

$$[(L_k f)(x)]^r = \left[(FR) \int_{-a+2^k x}^{a+2^k x} c_k^f(u) \odot \varphi(2^k x - u) \, du \right]^r$$

$$\overset{(14.23)}{=} \left[\int_{-a+2^k x}^{a+2^k x} (c_k^f(u))_-^{(r)} \varphi(2^k x - u) \, du, \right.$$

$$\left. \int_{-a+2^k x}^{a+2^k x} (c_k^f(u))_+^{(r)} \varphi(2^k x - u) \, du \right]$$

$$= \left[\int_{-a+2^k x}^{a+2^k x} c_k^{(f_-^{(r)})}(u) \varphi(2^k x - u) \, du, \right.$$

$$\left. \int_{-a+2^k x}^{a+2^k x} c_k^{(f_+^{(r)})}(u) \varphi(2^k x - u) \, du \right]$$

$$= \left[(L_k(f_-^{(r)}))(x), (L_k(f_+^{(r)}))(x) \right].$$

That is

$$[(L_k f)(x)]^r = [(L_k(f_-^{(r)}))(x), (L_k(f_+^{(r)}))(x)],$$

which gives

$$((L_k f)(x))_\pm^{(r)} = (L_k(f_\pm^{(r)}))(x), \quad \forall r \in [0, 1]. \qquad (14.50)$$

Therefore

$$D((L_k f)(x), f(x)) = \sup_{r \in [0,1]} \max \{ |(L_k(f_-^{(r)}))(x) - f_-^{(r)}(x)|,$$

$$|(L_k(f_+^{(r)}))(x) - f_+^{(r)}(x)| \}$$

$$\overset{(14.9)}{\leq} \sup_{r \in [0,1]} \max \left\{ \omega_1 \left(f_-^{(r)}, \frac{a+1}{2^k} \right), \right.$$

$$\left. \omega_1 \left(f_+^{(r)}, \frac{a+1}{2^k} \right) \right\} \overset{(14.25)}{=} \omega_1^{(\mathcal{F})} \left(f, \frac{a+1}{2^k} \right).$$

We have established that

$$D((L_k f)(x), f(x)) \leq \omega_1^{(\mathcal{F})} \left(f, \frac{a+1}{2^k} \right). \qquad (14.51)$$

From [17], Section 2 we get (14.45).

Next we see that

$$\omega_1^{(\mathcal{F})}(L_k f, \delta) \overset{(14.25)}{=} \sup_{r \in [0,1]} \max \{ \omega_1((L_k f)_-^{(r)}, \delta), \omega_1((L_k f)_+^{(r)}, \delta) \}$$

$$= \sup_{r \in [0,1]} \max \{ \omega_1(L_k(f_-^{(r)}), \delta), \omega_1(L_k(f_+^{(r)}), \delta) \}$$

$$\overset{(14.11)}{\leq} \sup_{r \in [0,1]} \max \{ \omega_1(f_-^{(r)}, \delta), \omega_1(f_+^{(r)}, \delta) \} \overset{(14.25)}{=} \omega_1^{(\mathcal{F})}(f, \delta),$$

proving (14.46).   □

Next we give

Theorem 14.23. *It holds*

$$D^*((\Gamma_k f), f) \leq \omega_1^{(\mathcal{F})}\left(f, \frac{a+1}{2^k}\right), \tag{14.52}$$

and

$$\omega_1^{(\mathcal{F})}(\Gamma_k f, \delta) \leq \omega_1^{(\mathcal{F})}(f, \delta), \quad \forall \delta > 0, \tag{14.53}$$

that is Γ_k fulfilling the property of Fuzzy Global Smoothness preservation.

Proof. Here $\gamma_k^f(u)$ is fuzzy continuous in u. We observe that

$$D\left(\sum_{j=0}^{n}{}^* w_j \odot f\left(\frac{u_m}{2^k} + \frac{j}{2^k n}\right), \sum_{j=0}^{n}{}^* w_j \odot f\left(\frac{u}{2^k} + \frac{j}{2^k n}\right)\right),$$

$$\text{(by (2.1) of [17])} \leq \sum_{j=0}^{n} w_j D\left(f\left(\frac{u_m}{2^k} + \frac{j}{2^k n}\right), f\left(\frac{u}{2^k} + \frac{j}{2^k n}\right)\right)$$

$$\leq \sum_{j=0}^{n} w_j \omega_1^{(\mathcal{F})}\left(f, \frac{|u_m - u|}{2^k}\right) = \omega_1^{(\mathcal{F})}\left(f, \frac{|u_m - u|}{2^k}\right) \to 0,$$

as $u_m \to u$, where $\{u_m\}$, $u \in \mathbb{R}$. That is proving fuzzy continuity of $\gamma_k^f(u)$.
Next we observe

$$D(\gamma_k^f(u), \tilde{o}) = D\left(\sum_{j=0}^{n}{}^* w_j \odot f\left(\frac{u}{2^k} + \frac{j}{2^k n}\right), \tilde{o}\right)$$

$$\leq \sum_{j=0}^{n} w_j D\left(f\left(\frac{u}{2^k} + \frac{j}{2^k n}\right), \tilde{o}\right) \leq \sum_{j=0}^{n} w_j M = M < +\infty,$$

where $M > 0$ such that $D(f(x), \tilde{o}) \leq M$, $\forall x \in \mathbb{R}$. That is $\gamma_k^f(u)$ is fuzzy bounded.

By Lemma 14.17 we get again that $\gamma_k^f(u) \odot \varphi(2^k x - u)$ is fuzzy continuous in u. We have from [17], Section 2, that

$$D((\Gamma_k f)(x), f(x)) = \sup_{r \in [0,1]} \max\{(\Gamma_k f)_-^{(r)}(x) - f_-^{(r)}(x)|, |(\Gamma_k f)_+^{(r)}(x) - f_+^{(r)}(x)|\},$$

$$\tag{14.54}$$

where

$$(\Gamma_k f)(x) = (FR)\int_{-a+2^k x}^{a+2^k x} \gamma_k^f(u) \odot \varphi(2^k x - u)\, du. \tag{14.55}$$

We easily see $\forall r \in [0,1]$ that

$$[\gamma_k^f(u)]^r = \sum_{j=0}^n w_j \left[f_-^{(r)}\left(\frac{u}{2^k} + \frac{j}{2^k n}\right), f_+^{(r)}\left(\frac{u}{2^k} + \frac{j}{2^k n}\right)\right]$$

$$= \left[\gamma_k^{(f_-^{(r)})}(u), \gamma_k^{(f_+^{(r)})}(u)\right].$$

Therefore we obtain

$$[(\Gamma_k f)(x)]^r = \left[(FR)\int_{-a+2^k x}^{a+2^k x} \gamma_k^f(u) \odot \varphi(2^k x - u)\, du\right]^r$$

$$\overset{(14.23)}{=} \left[\int_{-a+2^k x}^{a+2^k x} (\gamma_k^f(u))_-^{(r)}\varphi(2^k x - u)\, du,\right.$$

$$\left.\int_{-a+2^k x}^{a+2^k x} (\gamma_k^f(u))_+^{(r)}\varphi(2^k x - u)\, du\right]$$

$$= \left[\int_{-a+2^k x}^{a+2^k x} \gamma_k^{(f_-^{(r)})}(u)\varphi(2^k x - u)\, du,\right.$$

$$\left.\int_{-a+2^k x}^{a+2^k x} \gamma_k^{(f_+^{(r)})}(u)\varphi(2^k x - u)\, du\right]$$

$$= \left(\Gamma_k(f_-^{(r)})\right)(x), \left(\Gamma_k(f_+^{(r)})\right)(x)].$$

That is, we have proved that

$$((\Gamma_k(f))(x))_\pm^{(r)} = (\Gamma_k(f_\pm^{(r)}))(x), \quad \forall r \in [0,1]. \tag{14.56}$$

Consequently we derive

$$D((\Gamma_k f)(x), f(x)) = \sup_{r\in[0,1]} \max\{|(\Gamma_k(f_-^{(r)}))(x) - f_-^{(r)}(x)|,$$

$$|(\Gamma_k(f)_+^{(r)})(x) - f_+^{(r)}(x)|\}$$

$$\overset{(14.10)}{\leq} \sup_{r\in[0,1]} \max\left\{\omega_1\left(f_-^{(r)}, \frac{a+1}{2^k}\right),\right.$$

$$\left.\omega_1\left(f_+^{(r)}, \frac{a+1}{2^k}\right)\right\} \overset{(14.25)}{=} \omega_1^{(\mathcal{F})}\left(f, \frac{a+1}{2^k}\right).$$

I.e. we get

$$D((\Gamma_k f)(x), f(x)) \leq \omega_1^{(\mathcal{F})}\left(f, \frac{a+1}{2^k}\right), \tag{14.57}$$

hence establishing (14.52).

Next we see that

$$
\begin{aligned}
\omega_1^{(\mathcal{F})}(\Gamma_k f, \delta) \quad &= \quad \sup_{r \in [0,1]} \max\{\omega_1((\Gamma_k f)_-^{(r)}, \delta), \omega_1((\Gamma_k f)_+^{(r)}, \delta)\} \\
&= \quad \sup_{r \in [0,1]} \max\{\omega_1((\Gamma_k(f_-^{(r)})), \delta), \omega_1((\Gamma_k(f_+^{(r)})), \delta)\} \\
&\overset{(14.11)}{\leq} \quad \sup_{r \in [0,1]} \max\{\omega_1(f_-^{(r)}, \delta), \omega_1(f_+^{(r)}, \delta)\} = \omega_1^{(\mathcal{F})}(f, \delta).
\end{aligned}
$$

That is proving (14.53). $\square$

It follows

Theorem 14.24. *It holds*

$$
D^*((A_k f), f) \leq \omega_1^{(\mathcal{F})}\left(f, \frac{a}{2^{k-1}}\right), \tag{14.58}
$$

and

$$
\omega_1^{(\mathcal{F})}(A_k f, \delta) \leq \omega_1^{(\mathcal{F})}(f, \delta), \quad \forall \delta > 0, \tag{14.59}
$$

that is A_k fulfilling the property of Fuzzy Global Smoothness preservation.

Proof. Here we observe the following, $\varphi(2^k x - u) \neq 0$ iff $-a \leq 2^k x - u \leq a$ iff

$$
2^k x - a \leq u \leq 2^k x + a.
$$

Let us fix $u \in [2^k x - a, 2^k x + a]$. Also we have $\varphi(2^k t - u) \neq 0$ iff $-a \leq 2^k t - u \leq a$ iff $\frac{u-a}{2^k} \leq t \leq \frac{u+a}{2^k}$.

Let $t \in \left[\frac{u-a}{2^k}, \frac{u+a}{2^k}\right]$, then $t \in \left[x - \frac{a}{2^{k-1}}, x + \frac{a}{2^{k-1}}\right]$. If $t > x + \frac{a}{2^{k-1}}$ then $\varphi(2^k t - u) = 0$ and if $t < x - \frac{a}{2^{k-1}}$ then $\varphi(2^k t - u) = 0$. Therefore we have

$$
r_k^f(u) = 2^k \odot (FR) \int_{x - \frac{a}{2^{k-1}}}^{x + \frac{a}{2^{k-1}}} f(t) \odot \varphi(2^k t - u) \, dt. \tag{14.60}
$$

Also we have

$$
(A_k f)(x) = (FR) \int_{-a + 2^k x}^{a + 2^k x} r_k^f(u) \odot \varphi(2^k x - u) \, du. \tag{14.61}
$$

We prove that $r_k^f(u)$ is a fuzzy continuous and bounded function in $u \in [2^k x - a, 2^k x + a]$. Let $u_m \in [2^k x - a, 2^k x + a]$ be such that $u_m \to u$, as $m \to +\infty$. Clearly $\varphi(2^k t - u_m) \to \varphi(2^k t - u)$, so $\varphi(2^k t - u)$ is continuous in u.

We notice that

$$D(r_k^f(u_m), r_k^f(u)) = 2^k D\left((FR)\int_{x-\frac{a}{2^{k-1}}}^{x+\frac{a}{2^{k-1}}} f(t) \odot \varphi(2^k t - u_m)\, dt,\right.$$

$$\left.(FR)\int_{x-\frac{a}{2^{k-1}}}^{x+\frac{a}{2^{k-1}}} f(t) \odot \varphi(2^k t - u)\, dt\right)$$

(by Lemma 14.17

$$f(t) \odot \varphi(2^k t - u_m), f(t) \odot \varphi(2^k t - u)$$

are fuzzy continuous functions in t, using also (2.9) of [17] to get)

$$\leq 2^k \int_{x-\frac{a}{2^{k-1}}}^{x+\frac{a}{2^{k-1}}} D(f(t) \odot \varphi(2^k t - u_m), f(t) \odot \varphi(2^k t - u))\, dt$$

(by Lemma 2.1 of [17])

$$\leq 2^k \int_{x-\frac{a}{2^{k-1}}}^{x+\frac{a}{2^{k-1}}} |\varphi(2^k t - u_m) - \varphi(2^k t - u)| D(f(t), \tilde{o})\, dt$$

$$\leq 4aM\omega_1(\varphi\big|_{[-3a,3a]}, |u_m - u|) \to 0, \quad \text{as } m \to +\infty,$$

where $M > 0$ such that $D(f(x), \tilde{o}) \leq M$, $\forall x \in \mathbb{R}$. (Clearly here if $t \in \left[x - \frac{a}{2^{k-1}}, x + \frac{a}{2^{k-1}}\right]$ then $2^k t - u$, $2^k t - u_m \in [-3a, 3a]$.) That is proving $r_k^f(u)$ is fuzzy continuous in $u \in [2^k x - a, 2^k x + a]$.

Next we see that

$$2^k D\left((FR)\int_{x-\frac{a}{2^{k-1}}}^{x+\frac{a}{2^{k-1}}} f(t) \odot \varphi(2^k t - u)\, dt, \tilde{o}\right)$$

$$\text{(by (2.9) of [17])} \leq 2^k \int_{x-\frac{a}{2^{k-1}}}^{x+\frac{a}{2^{k-1}}} D(f(t) \odot \varphi(2^k t - u), \tilde{o})\, dt$$

$$= 2^k \int_{x-\frac{a}{2^{k-1}}}^{x+\frac{a}{2^{k-1}}} D(f(t) \odot \varphi(2^k t - u), f(t) \odot \tilde{o})\, dt$$

$$\text{(by Lemma 2.1 of [17])} \leq 2^k \int_{x-\frac{a}{2^{k-1}}}^{x+\frac{a}{2^{k-1}}} D(f(t), \tilde{o}) \varphi(2^k t - u)\, dt$$

$$\leq 4aML < +\infty,$$

where $L > 0$ be such that $\varphi(x) \leq L$, $\forall x \in \mathbb{R}$. That is proving $r_k^f(u)$ is fuzzy bounded in $u \in [2^k x - a, 2^k x + a]$.

Again by Lemma 14.17 we get that $r_k^f(u) \odot \varphi(2^k x - u)$ is fuzzy continuous in $u \in [2^k x - a, 2^k x + a]$, and to repeat $f(t) \odot \varphi(2^k t - u)$ is fuzzy continuous in $t \in \left[x - \frac{a}{2^{k-1}}, x + \frac{a}{2^{k-1}}\right]$. Furthermore we observe $\forall r \in [0,1]$ that

$$
\begin{aligned}
[r_k^f(u)]^r &= 2^k \left[(FR) \int_{x-\frac{a}{2^{k-1}}}^{x+\frac{a}{2^{k-1}}} f(t) \odot \varphi(2^k t - u)\, dt \right]^r \\
&\overset{(14.23)}{=} \left[\int_{x-\frac{a}{2^{k-1}}}^{x+\frac{a}{2^{k-1}}} f_-^{(r)}(t) \varphi(2^k t - u)\, dt, \right. \\
&\qquad \left. \int_{x-\frac{a}{2^{k-1}}}^{x+\frac{a}{2^{k-1}}} f_+^{(r)}(t) \varphi(2^k t - u)\, dt \right] \\
&= \left[r_k^{(f_-^{(r)})}(u), r_k^{(f_+^{(r)})}(u) \right].
\end{aligned}
$$

That is, we proved

$$
(r_k^f(u))_{\pm}^{(r)} = r_k^{(f_{\pm}^{(r)})}(u). \tag{14.62}
$$

Then

$$
\begin{aligned}
[(A_k f)(x)]^r &= \left[(FR) \int_{2^k x - a}^{2^k x + a} r_k^f(u) \odot \varphi(2^k x - u)\, du \right]^r \\
&\overset{(14.23)}{=} \left[\int_{2^k x - a}^{2^k x + a} (r_k^f(u))_-^{(r)} \varphi(2^k x - u)\, du, \right. \\
&\qquad \left. \int_{2^k x - a}^{2^k x + a} (r_k^f(u))_+^{(r)} \varphi(2^k x - u)\, du \right] \\
&\overset{(14.62)}{=} \left[\int_{2^k x - a}^{2^k x + a} r_k^{(f_-^{(r)})}(u) \varphi(2^k x - u)\, du, \right. \\
&\qquad \left. \int_{2^k x - a}^{2^k x + a} r_k^{(f_+^{(r)})}(u) \varphi(2^k x - u)\, du \right] \\
&= \left[(A_k(f_-^{(r)}))(x), (A_k(f_+^{(r)}))(x) \right].
\end{aligned}
$$

We have established that

$$
((A_k f)(x))_{\pm}^{(r)} = (A_k(f_{\pm}^{(r)}))(x). \tag{14.63}
$$

Next we see that

$$
\begin{aligned}
D((A_k f)(x), f(x)) &= \sup_r \max\{|(A_k f)^{(r)}_-(x) - f^{(r)}_-(x)|, \\
&\qquad |(A_k f)^{(r)}_+(x) - f_+(x)|\} \\
&\overset{(14.63)}{=} \sup_r \max\{|(A_k(f^{(r)}_-))(x) - f^{(r)}_-(x)|, \\
&\qquad |(A_k(f^{(r)}_+))(x) - f^{(r)}_+(x)|\} \\
&\leq \sup_r \max\left\{\omega_1\left(f^{(r)}_-, \frac{a}{2^{k+1}}\right), \omega_1\left(f^{(r)}_+, \frac{a}{2^{k-1}}\right)\right\} \\
&\overset{(14.25)}{=} \omega_1^{(\mathcal{F})}\left(f, \frac{a}{2^{k-1}}\right).
\end{aligned}
$$

Hence

$$
D((A_k f)(x), f(x)) \leq \omega_1^{(\mathcal{F})}\left(f, \frac{a}{2^{k-1}}\right), \tag{14.64}
$$

and from [17], Section 2 we get (14.58).

Finally we treat

$$
\begin{aligned}
\omega_1^{(\mathcal{F})}(A_k f, \delta) &\overset{(14.25)}{=} \sup_{r \in [0,1]} \max\{\omega_1((A_k f)^{(r)}_-, \delta), \omega_1((A_k f)^{(r)}_+, \delta)\} \\
&\overset{(14.63)}{=} \sup_{r \in [0,1]} \max\{\omega_1(A_k(f^{(r)}_-), \delta), \omega_1(A_k(f^{(r)}_+), \delta)\} \\
&\overset{(14.11)}{\leq} \sup_{r \in [0,1]} \max\{\omega_1(f^{(r)}_-, \delta), \omega_1(f^{(r)}_+, \delta)\} \overset{(14.25)}{=} \omega_1^{(\mathcal{F})}(f, \delta),
\end{aligned}
$$

so proving (14.59). $\qquad\square$

In the following three theorems of high order fuzzy approximation to the fuzzy unit operator by the fuzzy operators A_k, B_k, L_k, Γ_k the scaling function φ will be as before in this Section 14.3. However now we take for consideration $f \in C^N(\mathbb{R}, \mathbb{R}_{\mathcal{F}})$, $N \geq 1$, with $f^{(N)} \in C^U_{\mathcal{F}}(\mathbb{R})$ and $f \in C_b(\mathbb{R}, \mathbb{R}_{\mathcal{F}})$. Clearly here by Theorem 14.19, $f^{(r)}_\pm \in C^N(\mathbb{R})$ and of course

$$
(f^{(r)}_\pm)^{(N)} \overset{(14.33)}{=} (f^{(N)}_\pm)^{(r)} \in C_U(\mathbb{R}), \quad \forall r \in [0,1].
$$

We present

Theorem 14.25. *It holds*

$$
D((A_k f)(x), f(x)) \leq \sum_{i=1}^N \frac{D(f^{(i)}(x), \tilde{o})}{i!} \frac{a^i}{2^{i(k-1)}} + \frac{a^N}{N! 2^{N(k-1)}} \omega_1^{(\mathcal{F})}\left(f^{(N)}, \frac{a}{2^{k-1}}\right), \tag{14.65}
$$

$\forall k \in \mathbb{Z}, \forall x \in \mathbb{R}$.

Proof. We have that

$$
D((A_k f)(x), f(x)) \quad = \quad \sup_{r \in [0,1]} \max\{|((A_k f)(x)_{-}^{(r)} - f_{-}^{(r)}(x)|,
$$

$$
|((A_k f)(x))_{+}^{(r)} - f_{+}^{(r)}(x)|\}
$$

$$
\stackrel{(14.63)}{=} \sup_{r \in [0,1]} \max\{|(A_k(f_{-}^{(r)}))(x) - f_{-}^{(r)}(x)|,
$$

$$
|(A_k(f_{+}^{(r)}))(x) - f_{+}^{(r)}(x)|\}
$$

$$
\stackrel{(14.12)}{\leq} \sup_{r \in [0,1]} \max\left\{ \sum_{i=1}^{N} \frac{|(f_{-}^{(r)})^{(i)}(x)|}{i!} \frac{a^i}{2^{i(k-1)}} \right.
$$

$$
+ \frac{a^N}{N! 2^{N(k-1)}} \omega_1\left((f_{-}^{(r)})^{(N)}, \frac{a}{2^{k-1}}\right),
$$

$$
\sum_{i=1}^{N} \frac{|(f_{+}^{(r)})^{(i)}(x)|}{i!} \frac{a^i}{2^{i(k-1)}}
$$

$$
\left. + \frac{a^N}{N! 2^{N(k-1)}} \omega_1\left((f_{+}^{(r)})^{(N)}, \frac{a}{2^{k-1}}\right) \right\}
$$

$$
\stackrel{(14.33)}{=} \sup_{r \in [0,1]} \max\left\{ \sum_{i=1}^{N} \frac{|(f_{-}^{(i)})^{(r)}(x)|}{i!} \frac{a^i}{2^{i(k-1)}} \right.
$$

$$
+ \frac{a^N}{N! 2^{N(k-1)}} \omega_1\left((f_{-}^{(N)})^{(r)}, \frac{a}{2^{k-1}}\right),
$$

$$
\sum_{i=1}^{N} \frac{|(f_{+}^{(i)})^{(r)}(x)|}{i!} \frac{a^i}{2^{i(k-1)}}
$$

$$
\left. + \frac{a^N}{N! 2^{N(k-1)}} \omega_1\left((f_{+}^{(N)})^{(r)}, \frac{a}{2^{k-1}}\right) \right\}
$$

$$
\stackrel{(14.29)}{\leq} \left\{ \sum_{i=1}^{N} \left[\sup_{r \in [0,1]} \max\left\{|(f_{-}^{(i)})^{(r)}(x)|, |(f_{+}^{(i)})^{(r)}(x)|\right\} \right] \right.
$$

$$
\times \frac{a^i}{i! 2^{i(k-1)}}
$$

$$
+ \frac{a^N}{N! 2^{N(k-1)}} \sup_{r \in [0,1]} \max\left\{ \omega_1\left((f_{-}^{(N)})^{(r)}, \frac{a}{2^{k-1}}\right), \right.
$$

$$
\left. \omega_1\left((f_{+}^{(N)})^{(r)}, \frac{a}{2^{k-1}}\right) \right\}
$$

$$
\text{(by (14.25) and (14.26))}
$$

$$
= \sum_{i=1}^{N} \frac{D(f^{(i)}(x), \tilde{o})}{i!} \frac{a^i}{2^{i(k-1)}}
$$

$$
+ \frac{a^N}{N! 2^{N(k-1)}} \omega_1^{(\mathcal{F})}\left(f^{(N)}, \frac{a}{2^{k-1}}\right). \qquad \square
$$

We continue with

Theorem 14.26. *It holds*

$$D((B_k f)(x), f(x)) \le \sum_{i=1}^{N} \frac{D(f^{(i)}(x), \tilde{o})}{i!} \frac{a^i}{2^{ki}} + \frac{a^N}{n! 2^{kN}} \omega_1^{(\mathcal{F})} \left(f^{(N)}, \frac{a}{2^k} \right),$$

(14.66)

$\forall k \in \mathbb{Z}, \forall x \in \mathbb{R}.$

Proof. Using (14.13) and very similar to the proof of Theorem 14.25. □

Finally we give

Theorem 14.27. *It holds*

$$\begin{cases} D((L_k f)(x), f(x)), \\ D((\Gamma_k f)(x), f(x)) \end{cases} \le \sum_{i=1}^{N} \frac{D(f^{(i)}(x), \tilde{o})}{i!} \frac{(a+1)^i}{2^{ki}}$$
$$+ \frac{(a+1)^N}{N! 2^{kN}} \omega_1^{(\mathcal{F})} \left(f^{(N)}, \frac{a+1}{2^k} \right),$$

(14.67)

$\forall k \in \mathbb{Z}, \forall x \in \mathbb{R}.$

Proof. Using (14.14) and very similar to the proof of Theorem 14.25. □

Note 14.28. Since here $f^{(N)} \in C_{\mathcal{F}}^U(\mathbb{R})$, as $k \to +\infty$, we derive

$$\omega_1^{(\mathcal{F})} \left(f^{(N)}, \frac{a}{2^{k-1}} \right), \omega_1^{(\mathcal{F})} \left(f^{(N)}, \frac{a}{2^k} \right), \omega_1^{(\mathcal{F})} \left(f^{(N)}, \frac{a+1}{2^k} \right) \to 0,$$

Thus from (14.65), (14.66) and (14.67) we obtain that $D((A_k f)(x), f(x)) \to 0$, $D((B_k f)(x), f(x)) \to 0$, $D((L_k f)(x), f(x)) \to 0$, and $D((\Gamma_k f)(x), f(x)) \to 0$, pointwise with rates.

15

DEGREE OF APPROXIMATION OF FUZZY NEURAL NETWORK OPERATORS, UNIVARIATE CASE

In this chapter we study the rate of convergence to the unit operator of very specific well described univariate Fuzzy neural network operators of Cardaliaguet–Euvrard and "Squashing" types. These Fuzzy operators arise in a very natural and common way among Fuzzy neural networks. The rates are given through *Jackson type inequalities* involving the Fuzzy modulus of continuity of the engaged Fuzzy valued function or its derivative in the Fuzzy sense. Also several interesting results in Fuzzy real analysis are presented to be used in the proofs of the main results. This chapter is based on [11].

15.1 Background

Let $f: \mathbb{R} \to \mathbb{R}_{\mathcal{F}}$ be a uniformly continuous Fuzzy real valued function. For each $n \in \mathbb{N}$, the neural network we deal with here has the following structure: it is a three-layer feedforward network with one hidden layer. It has one input and one output unit. The hidden layer has $(2n^2+1)$ processing units. To each pair of connecting units (input to each processing unit) we assign the same weight $n^{1-\alpha}$, $0 < \alpha < 1$. The threshold values $\frac{k}{n^\alpha}$ are one for each processing unit k.

G.A. Anastassiou: Fuzzy Mathematics: Approximation The., STUDFUZZ 251, pp. 263–288.
springerlink.com   © Springer-Verlag Berlin Heidelberg 2010

The activation function b (or S) is the same for each processing unit. The Fuzzy weights associated with the output unit are $f\left(\frac{k}{n}\right) \odot \frac{1}{I^{(*)}n^{\alpha}}$, one for each processing unit k, where $I = \int_{-\infty}^{\infty} b(x)dx$ (or $I^* = \int_{-\infty}^{\infty} S(x)dx$), "$\odot$" denotes the scalar Fuzzy multiplication.

The above fully described neural networks give rise to some associated completely described Fuzzy neural network operators of Cardaliaguet–Euvrard and "Squashing" types. We study here thoroughly the Fuzzy pointwise convergence of these operators to the unit operator, we give also some L_p, $p \geq 1$ analogs; see Theorems 15.17, 15.19, 15.24, 15.26 and Corollaries 15.21 and 15.22. This is done with rates through Jackson type inequalities involving Fuzzy moduli of continuity of the engaged Fuzzy functions. On the way to establish these results we produce some new results on Fuzzy Real Analysis, especially see Theorem 15.14, where we show a Fuzzy Taylor's formula with Fuzzy integral remainder.

The real ordinary theory of the above mentioned operators was presented earlier in [6] and [47]. And, of course, this chapter is motivated from there. The monumental revolutionizing work of L. Zadeh [103] is the foundation of this chapter, as well as another strong motivation. Fuzziness in Computer Science and Engineering is one of the main trends today. This way of quantitative approach over Fuzzy neural networks appeared recently in the literature. It determines the rates of convergence precisely in a natural quantitative manner through very tight inequalities using the measurement of smoothness of the engaged Fuzzy functions.

As in Remark 4.4 ([31]) one can show easily that a sequence of operators of the form

$$L_n(f)(x) := \sum_{k=0}^{n} {}^* f(x_{k_n}) \odot w_{n,k}(x), \quad n \in \mathbb{N},$$

($\sum^*$ denotes the fuzzy summation) where $f: \mathbb{R} \to \mathbb{R}_{\mathcal{F}}$, $x_{k_n} \in \mathbb{R}$, $w_{n,k}(x)$ real valued weights, are *linear* over $\mathbb{R}$, i.e.,

$$L_n(\lambda \odot f \oplus \mu \odot g)(x) = \lambda \odot L_n(f)(x) \oplus \mu \odot L_n(g)(x),$$

$\forall \lambda, \mu \in \mathbb{R}$, any $x \in \mathbb{R}$; $f, g: \mathbb{R} \to \mathbb{R}_{\mathcal{F}}$. (Proof based on Lemma 4.1(iv) of [31].)

15.2   Basic Properties

We need the following

Definition 15.1. Let $f : \mathbb{R} \to \mathbb{R}_{\mathcal{F}}$ be a fuzzy real number valued function. We define the (*first*) *fuzzy modulus of continuity of* f by

$$\omega_1^{(\mathcal{F})}(f, \delta) := \sup_{\substack{x,y \in \mathbb{R} \\ |x-y| \le \delta}} D(f(x), f(y)), \quad \delta > 0. \tag{15.1}$$

Definition 15.2. Let $f : \mathbb{R} \to \mathbb{R}_{\mathcal{F}}$. If $D(f(x), \tilde{o}) \le M$, $\forall x \in \mathbb{R}$, $M > 0$, we call f a *bounded fuzzy real number valued function.*

Definition 15.3. Let $f : \mathbb{R} \to \mathbb{R}_{\mathcal{F}}$. We say that f is *continuous at* $a \in \mathbb{R}$ if whenever $x_n \to a$, then $D(f(x_n), f(a)) \to 0$. If f is continuous for every $a \in \mathbb{R}$, then we call f a *continuous fuzzy real number valued function.* We denote it as $f \in C_{\mathcal{F}}(\mathbb{R})$.

Remark 15.4. Let f be bounded from $\mathbb{R}$ into $\mathbb{R}_{\mathcal{F}}$. Then we observe that

$$\omega_1^{(\mathcal{F})}(f, \delta) := \sup_{\substack{x,y \in \mathbb{R} \\ |x-y| \le \delta}} D(f(x), f(y))$$

$$= \sup_{\substack{x,y \in \mathbb{R} \\ |x-y| \le \delta}} D(f(x) \oplus \tilde{o}, f(y) \oplus \tilde{o})$$

$$\le \sup_{\substack{x,y \in \mathbb{R} \\ |x-y| \le \delta}} (D(f(x), \tilde{o}) + D(f(y), \tilde{o})) \le 2M.$$

That is, $\omega_1^{(\mathcal{F})}(f, \delta) < +\infty$.

Definition 15.5. Let $f : \mathbb{R} \to \mathbb{R}_{\mathcal{F}}$. We call f a *uniformly continuous fuzzy real number valued function*, iff for any $\varepsilon > 0$ there exists $\delta > 0$: whenever $|x - y| \le \delta$; $x, y \in \mathbb{R}$, implies that $D(f(x), f(y)) \le \varepsilon$. We denote it as $f \in C_{\mathcal{F}}^U(\mathbb{R})$.

Proposition 15.6. *Let* $f \in C_{\mathcal{F}}^U(\mathbb{R})$. *Then* $\omega_1^{(\mathcal{F})}(f, \delta) < +\infty$, *any* $\delta > 0$.

Proof. Let $\varepsilon_0 > 0$ be arbitrary but fixed. Then there exists $\delta_0 > 0$: $|x-y| \le \delta_0$ implies $D(f(x), f(y)) \le \varepsilon_0 < +\infty$. That is $\omega_1^{(\mathcal{F})}(f, \delta_0) \le \varepsilon_0 < +\infty$.

Let now $\delta > 0$ arbitrary, $x, y \in \mathbb{R}$ such that $|x - y| \le \delta$. Choose $n \in \mathbb{N}$: $n\delta_0 \ge \delta$ and set $x_i := x + \frac{i}{n}(y - x)$, $0 \le i \le n$. Then

$$D(f(x), f(y)) = D \left(f(x) \oplus \sum_{k=1}^{n-1}{}^{*} f(x_i), \sum_{k=1}^{n-1}{}^{*} f(x_i) \oplus f(y) \right)$$

$$\le D(f(x), f(x_1)) + D(f(x_1), f(x_2)) + \cdots$$

$$+ D(f(x_{n-1}), f(y)) \le n\omega_1^{(\mathcal{F})}(f, \delta_0) \le n\varepsilon_0 < +\infty,$$

since $|x_i - x_{i+1}| = \frac{1}{n}|x - y| \le \frac{1}{n}\delta \le \delta_0$, $0 \le i \le n$. Therefore $\omega_1^{(\mathcal{F})}(f, \delta) \le n\varepsilon_0 < +\infty$. $\quad\square$

Denote $f \colon \mathbb{R} \to \mathbb{R}_{\mathcal{F}}$ which is bounded and continuous, as $f \in C_{\mathcal{F}}^B(\mathbb{R})$.

Proposition 15.7. *It holds*

(i) $\omega_1^{(\mathcal{F})}(f, \delta)$ *is nonnegative and nondecreasing in* $\delta > 0$, *any* $f \colon \mathbb{R} \to \mathbb{R}_{\mathcal{F}}$.

(ii) $\lim_{\delta \downarrow 0} \omega_1^{(\mathcal{F})}(f, \delta) = \omega_1^{(\mathcal{F})}(f, 0) = 0$, *iff* $f \in C_{\mathcal{F}}^U(\mathbb{R})$.

(iii) $\omega_1^{(\mathcal{F})}(f, \delta_1 + \delta_2) \le \omega_1^{(\mathcal{F})}(f, \delta_1) + \omega_1^{(\mathcal{F})}(f, \delta_2)$, $\delta_1, \delta_2 > 0$, *any* $f \colon \mathbb{R} \to \mathbb{R}_{\mathcal{F}}$.

(iv) $\omega_1^{(\mathcal{F})}(f, n\delta) \le n\omega_1^{(\mathcal{F})}(f, \delta)$, $\delta > 0$, $n \in \mathbb{N}$, *any* $f \colon \mathbb{R} \to \mathbb{R}_{\mathcal{F}}$.

(v) $\omega_1^{(\mathcal{F})}(f, \lambda\delta) \le \lceil \lambda \rceil \omega_1^{(\mathcal{F})}(f, \delta) \le (\lambda + 1)\omega_1^{(\mathcal{F})}(f, \delta)$, $\lambda > 0$, $\delta > 0$, *where* $\lceil \cdot \rceil$ *is the ceiling of the number, any* $f \colon \mathbb{R} \to \mathbb{R}_{\mathcal{F}}$.

(vi) $\omega_1^{(\mathcal{F})}(f \oplus g, \delta) \le \omega_1^{(\mathcal{F})}(f, \delta) + \omega_1^{(\mathcal{F})}(g, \delta)$, $\delta > 0$, *any* $f, g \colon \mathbb{R} \to \mathbb{R}_{\mathcal{F}}$.

(vii) $\omega_1^{(\mathcal{F})}(f, \cdot)$ *is continuous on* $\mathbb{R}_+$, *for* $f \in C_{\mathcal{F}}^U(\mathbb{R})$.

Proof. (i) is obvious.

(ii) Clearly $\omega_1^{(\mathcal{F})}(f, 0) = 0$.

($\Rightarrow$) Let $\omega_1^{(\mathcal{F})}(f, \delta) \to 0$ as $\delta \downarrow 0$. Then $\forall \varepsilon > 0 \; \exists \delta > 0$, $\omega_1^{(\mathcal{F})}(f, \delta) \le \varepsilon$. I.e., for any $x, y \in \mathbb{R}$: $|x - y| \le \delta$ we get $D(f(x), f(y)) \le \varepsilon$. That is, $f \in C_{\mathcal{F}}^U(\mathbb{R})$.

($\Leftarrow$) Let $f \in C_{\mathcal{F}}^U(\mathbb{R})$. Then $\forall \varepsilon > 0 \; \exists \delta > 0$: whenever $|x - y| \le \delta$; $x, y \in \mathbb{R}$, it implies $D(f(x), f(y)) \le \varepsilon$. I.e., $\forall \varepsilon > 0 \; \exists \delta > 0$: $\omega_1^{(\mathcal{F})}(f, \delta) \le \varepsilon$. That is, $\omega_1^{(\mathcal{F})}(f, \delta) \to 0$ as $\delta \downarrow 0$.

(iii) Let $x_1, x_2 \in \mathbb{R}$ be such that $|x_1 - x_2| \le \delta_1 + \delta_2$. Then there exists $x \in \mathbb{R}$: $|x - x_1| \le \delta_1$ and $|x - x_2| \le \delta_2$.

We have

$$
\begin{aligned}
D(f(x_1), f(x_2)) &= D(f(x_1) \oplus f(x), f(x_2) \oplus f(x)) \\
&\le D(f(x_1), f(x)) + D(f(x), f(x_2)) \\
&\le \omega_1^{(\mathcal{F})}(f, |x_1 - x|) + \omega_1^{(\mathcal{F})}(f, |x - x_2|) \\
&\le \omega_1^{(\mathcal{F})}(f, \delta_1) + \omega_1^{(\mathcal{F})}(f, \delta_2).
\end{aligned}
$$

Therefore (iii) is true.

(iv) and (v) are obvious.

(vi) Notice that

$$
D(f(x) \oplus g(x), f(y) \oplus g(y)) \le D(f(x), f(y)) + D(g(x), g(y)).
$$

That is (vi) is now clear.

(vii) For any $f: \mathbb{R} \to \mathbb{R}_{\mathcal{F}}$ it holds by (iii) that

$$|\omega_1^{(\mathcal{F})}(f, \delta_1 + \delta_2) - \omega_1^{(\mathcal{F})}(f, \delta_1)| \le \omega_1^{(\mathcal{F})}(f, \delta_2).$$

Let now $f \in C_{\mathcal{F}}^U(\mathbb{R})$, then by (ii) $\lim\limits_{\delta_2 \downarrow 0} \omega_1^{(\mathcal{F})}(f, \delta_2) = 0$. That is proving the continuity of $\omega_1^{(\mathcal{F})}(f, \cdot)$ on $\mathbb{R}_+$. □

We mention the following fundamental theorem of Fuzzy calculus

Theorem 15.8 ([52]). *If* $f: [a, b] \to \mathbb{R}_{\mathcal{F}}$ *is differentiable on* $[a, b]$, *then* $f'(x)$ *is (FH)-integrable over* $[a, b]$ *and*

$$f(s) = f(t) \oplus (FH) \int_t^s f'(x)dx, \quad \text{for any } s \ge t, \ s, t \in [a, b].$$

The Fuzzy–Henstock integral $(FH) \int$ is defined in [52], Definition 2.1.

Note. In Theorem 15.8 when $s < t$ the formula is invalid! Since fuzzy real numbers correspond to closed intervals etc.

We need

Corollary 15.9. *Let* $f: [a, b] \to \mathbb{R}_{\mathcal{F}}$ *be fuzzy differentiable on* $[a, b]$, *and the fuzzy derivative* $f': [a, b] \to \mathbb{R}_{\mathcal{F}}$ *is assumed to be fuzzy continuous. Then it holds*

$$f(s) = f(a) \oplus (FR) \int_a^s f'(t)dt, \quad \text{for any } s \in [a, b].$$

Proof. By Corollary 13.2 of [66], p. 644 we get that $(FR) \int_a^s f'(t)dt$, $a \le s \le b$ exists in $\mathbb{R}_{\mathcal{F}}$. Clearly we have

$$(FR) \int_a^s f'(t)dt = (FH) \int_a^s f'(t)dt.$$

By Theorem 3.6 of [52], $f'(t)$ is (FH)-integrable over $[a, b]$, and

$$f(a) \oplus (FR) \int_a^s f'(t)dt = f(a) \oplus (FH) \int_a^s f'(t)dt = f(s). □$$

We need also

Lemma 15.10. *If* $f, g: [a, b] \subseteq \mathbb{R} \to \mathbb{R}_{\mathcal{F}}$ *are continuous, then the function* $F: [a, b] \to \mathbb{R}_+$ *defined by* $F(x) := D(f(x), g(x))$ *is continuous on* $[a, b]$ *and*

$$D\left((FR) \int_a^b f(u)du, (FR) \int_a^b g(u)du\right) \le \int_a^b F(x)dx.$$

Proof. Exactly the same as in Lemma 13.2 (ii) for 2π-periodic functions, see [66], p. 644. □

Lemma 15.11. *Let* $f: [a,b] \to \mathbb{R}_{\mathcal{F}}$ *continuous, then* $D(f(x), \tilde{o}) \le M$, $\forall x \in [a,b]$, $M > 0$, *that is* f *is fuzzy bounded.*

Proof. Let $x_n, x \in [a,b]$ such that $x_n \to x$, as $n \to +\infty$, then $D(f(x_n), f(x)) \to 0$, by continuity of f. But

$$D(f(x_n), f(x)) = \sup_{r \in [0,1]} \max\{|(f(x_n))_{-}^{(r)} - (f(x))_{-}^{(r)}|, |(f(x_n))_{+}^{(r)} - (f(x))_{+}^{(r)}|\}.$$

Hence $|(f(x_n))_{\pm}^{(r)} - (f(x))_{\pm}^{(r)}| \to 0$, all $0 \le r \le 1$, as $n \to +\infty$. That is, $(f(x_n))_{\pm}^{(r)} \to (f(x))_{\pm}^{(r)}$, all $0 \le r \le 1$, as $n \to +\infty$. Hence $(f)_{\pm}^{(r)} \in C([a,b])$, all $0 \le r \le 1$. Consequently, $(f)_{\pm}^{(r)}$ are bounded on $[a,b]$, all $0 \le r \le 1$. Here

$$D(f(x), \tilde{o}) = \sup_{r \in [0,1]} \max\{|(f(x))_{-}^{(r)}|, |(f(x))_{+}^{(r)}|\}.$$

From basic Fuzzy theory we get that

$$(f(x))_{-}^{(0)} \le (f(x))_{-}^{(r)} \le (f(x))_{-}^{(1)},$$

and

$$(f(x))_{+}^{(1)} \le (f(x))_{+}^{(r)} \le (f(x))_{+}^{(0)}.$$

Thus

$$|(f(x))_{-}^{(r)}| \le \max(|(f(x))_{-}^{(0)}|, |f(x))_{-}^{(1)}|),$$

and

$$|(f(x))_{+}^{(r)}| \le \max(|(f(x))_{+}^{(0)}|, |(f(x))_{+}^{(1)}|),$$

all $0 \le r \le 1$. Therefore

$$\begin{aligned} D(f(x), \tilde{o}) &\le \max\{|(f(x))_{\pm}^{(0)}|, |(f(x))_{\pm}^{(1)}|\} \le \max\{\|f_{\pm}^{(0)}\|_{\infty}, \|f_{\pm}^{(1)}\|_{\infty}\} \\ &\le M, \quad \forall x \in [a,b], \end{aligned}$$

for some $M > 0$. I. e. for all $0 \le r \le 1, -M \le (f(x))_{\pm}^{(r)} \le M, \forall x \in [a,b] \iff \chi_{-M} \le f(x) \le \chi_M, f(x) \in \mathbb{R}_{\mathcal{F}}.$ $\square$

Lemma 15.12. *Let* $f: [a,b] \subset \mathbb{R} \to \mathbb{R}_{\mathcal{F}}$ *be continuous. Then*

$$(FR) \int_a^x f(t)dt \quad \text{is a continuous function in } x \in [a,b].$$

Proof. By Corollary 13.2, p. 644 of [66], see also [52], f is (FR) integrable on $[a,b]$ and on its closed subintervals. Using Theorem 2.5 of [52] and a property of D, and without loss of generality by assuming $s_n \ge s$, as

$s_n \to s$, $n \to +\infty$, we have

$$D\left((FR)\int_a^{s_n} f(t)dt, (FR)\int_a^s f(t)dt\right)$$

$$= D\left((FR)\int_a^s f(t)dt \oplus (FR)\int_s^{s_n} f(t)dt, (FR)\int_a^s f(t)dt\right)$$

$$= D\left((FR)\int_s^{s_n} f(t)dt, \tilde{o}\right) = D\left((FR)\int_s^{s_n} f(t)dt, (FR)\int_s^{s_n} \tilde{o}dt\right)$$

$$\overset{(\text{Lemma}15.10)}{\leq} \int_s^{s_n} D(f(t), \tilde{o})dt$$

$$\overset{(\text{Lemma}15.11)}{\leq} \int_s^{s_n} Mdt \leq M(s_n - s) \to 0.$$

That is

$$D\left((FR)\int_a^{s_n} f(t)dt, (FR)\int_a^s f(t)dt\right) \to 0,$$

as $s_n \to s$, $n \to +\infty$. $\qquad\square$

Lemma 15.13. *Let $f \in C_\mathcal{F}(\mathbb{R})$, $r \in \mathbb{N}$. Then the following integrals*

$$(FR)\int_a^{s_{r-1}} f(s_r)ds_r, (FR)\int_a^{s_{r-2}} \left(\int_a^{s_{r-1}} f(s_r)ds_r\right)ds_{r-1},$$

$$\cdots, (FR)\int_a^s \left(\int_a^{s_1} \cdots \left(\int_a^{s_{r-2}} \left(\int_a^{s_{r-1}} f(s_r)ds_r\right)ds_{r-1}\right)\cdots\right)ds_1,$$

are continuous functions in $s_{r-1}, s_{r-2}, \ldots, s$, respectively.
  Here $s_{r-1}, s_{r-2}, \ldots, s \geq a$ and all are real numbers.

Proof. By Lemma 15.12. $\qquad\square$

  We present the following new interesting result, which is the Fuzzy Taylor's formula.

Theorem 15.14. *Let $T := [x_0, x_0 + \beta] \subset \mathbb{R}$, with $\beta > 0$. We assume that $f^{(i)} : T \to \mathbb{R}_\mathcal{F}$ are differentiable for all $i = 0, 1, \ldots, n - 1$, for any $x \in T$. (I.e., there exist in $\mathbb{R}_\mathcal{F}$ the H-differences $f^{(i)}(x + h) - f^{(i)}(x)$, $f^{(i)}(x) - f^{(i)}(x-h)$, $i = 0, 1, \ldots, n-1$ for all small $0 < h < \beta$. Furthermore there exist $f^{(i+1)}(x) \in \mathbb{R}_\mathcal{F}$ such that the limits in D-distance exist and*

$$f^{(i+1)}(x) = \lim_{h\to 0^+} \frac{f^{(i)}(x+h) - f^{(i)}(x)}{h} = \lim_{h\to 0^+} \frac{f^{(i)}(x) - f^{(i)}(x-h)}{h},$$

for all $i = 0, 1, \ldots, n - 1$.) Also we assume that $f^{(i)}$, $i = 0, 1, \ldots, n$ are continuous on T in the fuzzy sense. Then for $s \geq a$; $s, a \in T$ we obtain

$$f(s) = f(a) \oplus f'(a) \odot (s - a) \oplus f''(a) \odot \frac{(s - a)^2}{2!}$$

$$\oplus \cdots \oplus f^{(n-1)}(a) \odot \frac{(s - a)^{n-1}}{(n - 1)!} \oplus R_n(a, s),$$

where

$$R_n(a,s) := (FR) \int_a^s \left(\int_a^{s_1} \cdots \left(\int_a^{s_{n-1}} f^{(n)}(s_n)ds_n \right) ds_{n-1} \right) \cdots \right) ds_1.$$

Here $R_n(a,s) \in C_{\mathcal{F}}(T)$ as a function of s.

Note. (1) This formula is invalid when $s < a$, as it is based on Theorem 3.6 of [52].

(2) This Fuzzy Taylor formula is also valid with Fuzzy–Henstock integral remainder where now we can drop the assumptions that $f^{(i)}$, $i = 0, 1, \ldots, n$ are fuzzy continuous on T. The proof is totally the same, and it is now based on Theorem 3.6 of [52]. Clearly again the remainder $R_n(a,s)$ exists in $\mathbb{R}_{\mathcal{F}}$.

Proof (of Theorem 15.14). By Corollary 15.9 we have

$$f(s) = f(a) \oplus (FR) \int_a^s f'(t)dt,$$

with f' being (FR)-integrable on T,

$$f'(t) = f'(a) \oplus (FR) \int_a^t f''(\gamma)d\gamma,$$

with f'' being (FR)-integrable on T. Therefore

$$(FR) \int_a^s f'(t)dt = (FR) \int_a^s f'(a)dt \oplus (FR) \int_a^s (FR) \left(\int_a^t f''(\gamma)d\gamma \right) dt$$

$$= f'(a) \odot (s-a) \oplus (FR) \int_a^s \left(\int_a^t f''(\gamma)d\gamma \right) dt.$$

Clearly the last double integral belongs to $\mathbb{R}_{\mathcal{F}}$ by Lemma 15.12. That is

$$f(s) = f(a) \oplus f'(a) \odot (s-a) \oplus (FR) \int_a^s \left(\int_a^t f''(\gamma)d\gamma \right) dt.$$

But also next similarly we have

$$f''(\gamma) = f''(a) \oplus (FR) \int_a^\gamma f'''(\ell)d\ell,$$

with f''' being (FR)-integrable on T, and

$$\int_a^t f''(\gamma)d\gamma = f''(a) \odot (t-a) \oplus (FR) \int_a^t \left(\int_a^\gamma f'''(\ell)d\ell \right) d\gamma.$$

Furthermore,

$$\int_a^s \left(\int_a^t f''(\gamma)d\gamma \right) dt = f''(a) \odot \int_a^s (t-a)dt$$

$$\oplus (FR) \int_a^s \left(\int_a^t \left(\int_a^\gamma f'''(\ell)d\ell \right) d\gamma \right) dt.$$

Clearly the last triple integral belongs to $\mathbb{R}_{\mathcal{F}}$ by Lemma 15.12. Hence it holds

$$f(s) = f(a) \oplus f'(a) \odot (s-a) \oplus f''(a) \odot \frac{(s-a)^2}{2}$$
$$\oplus (FR) \int_a^s \left(\int_a^t \left(\int_a^\gamma f'''(\ell)d\ell \right) d\gamma \right) dt.$$

The remainder of the last formula by Lemma 15.12 is a continuous function in s. Etc. □

15.3 Main Results

We need (see also [6], [47])

Definition 15.15. A function $b\colon \mathbb{R} \to \mathbb{R}$ is said to be *bell-shaped* if b belongs to L^1 and its integral is nonzero, if it is nondecreasing on $(-\infty, a)$ and nonincreasing on $[a, +\infty)$, where a belongs to $\mathbb{R}$. In particular $b(x)$ is a nonnegative number and at a b takes a global maximum; it is the center of the bell-shaped function. A bell-shaped function is said to be *centered* if its center is zero. The function $b(x)$ may have jump discontinuities. In this article we consider only centered bell-shaped functions of compact support $[-T, T]$, $T > 0$. Denote $I := \int_{-T}^T b(t)dt$. Notice that $I > 0$.

Examples 15.16.

(1) $b(x)$ can be the characteristic function on $[-1, 1]$.

(2) $b(x)$ can be the hat function on $[-1, 1]$, i.e.,

$$b(x) = \begin{cases} 1+x, & -1 \le x \le 0, \\ 1-x, & 0 < x \le 1, \\ 0, & \text{elsewhere.} \end{cases}$$

Here we consider functions $f\colon \mathbb{R} \to \mathbb{R}_{\mathcal{F}}$ that are either continuous and bounded, i.e., $f \in C_{\mathcal{F}}^B(\mathbb{R})$, or uniformly continuous, i.e., $f \in C_{\mathcal{F}}^U(\mathbb{R})$, both in the fuzzy sense.

In this chapter we study first the pointwise convergence with rates over the real line, to the *fuzzy unit operator* of the *fuzzy univariate Cardaliaguet–Euvrard neural network operators*,

$$(F_n(f))(x) := f_n(x) := \sum_{k=-n^2}^{n^2} {}^* f\left(\frac{k}{n}\right) \odot \frac{b\left(n^{1-\alpha}\left(x-\frac{k}{n}\right)\right)}{In^\alpha}, \tag{15.2}$$

where $0 < \alpha < 1$ and $x \in \mathbb{R}$, $n \in \mathbb{N}$.

The above are linear operators over $\mathbb{R}$. These operators in the ordinary real case were thoroughly studied in [6], [47]. So as in [6], we consider without loss of generality and for simplicity that $n \geq \max(T + |x|, T^{-1/\alpha})$. In this case we have that $-n^2 \leq nx - Tn^\alpha \leq nx + Tn^\alpha \leq n^2$, and $\text{card}(k) \geq 1$. Furthermore, it holds $\text{card}(k) \to +\infty$, as $n \to +\infty$. Set $b^* := b(0)$, the maximum of $b(x)$. Denote by $[\cdot]$ the integral part of a number.

Next we present the first main result.

Theorem 15.17. *Let $x \in \mathbb{R}$, $T > 0$, and $n \in \mathbb{N}$ such that $n \geq \max(T + |x|, T^{-1/\alpha})$. Then*

$$
D(f_n(x), f(x)) \leq \left| \sum_{k=\lceil nx - Tn^\alpha \rceil}^{[nx + Tn^\alpha]} \frac{b(n^{1-\alpha}(x - \frac{k}{n}))}{In^\alpha} - 1 \right| \tag{15.3}
$$
$$
\cdot D(f(x), \tilde{o}) + \frac{b^*}{I}\left(2T + \frac{1}{n^\alpha}\right) \omega_1^{(\mathcal{F})}\left(f, \frac{T}{n^{1-\alpha}}\right).
$$

Proof. As in [6] we have that

$$
\sum_{k=\lceil nx - Tn^\alpha \rceil}^{[nx + Tn^\alpha]} \frac{b(n^{1-\alpha}(x - \frac{k}{n}))}{In^\alpha} \leq \frac{b^*}{I}\left(2T + \frac{1}{n^\alpha}\right). \tag{15.4}
$$

Next we estimate

$$
D(f_n(x), f(x)) = D\left(\sum_{k=-n^2}^{n^2}{}^* f\left(\frac{k}{n}\right) \odot \frac{b(n^{1-\alpha}(x - \frac{k}{n}))}{In^\alpha}, f(x)\right)
$$
$$
= D\left(\sum_{k=-n^2}^{\lceil nx - Tn^\alpha \rceil - 1}{}^* f\left(\frac{k}{n}\right) \odot \frac{b(n^{1-\alpha}(x - \frac{k}{n}))}{In^\alpha}\right.
$$
$$
\oplus \sum_{k=[nx + Tn^\alpha]+1}^{n^2}{}^* f\left(\frac{k}{n}\right) \odot \frac{b(n^{1-\alpha}(x - \frac{k}{n}))}{In^\alpha}
$$
$$
\left.\oplus \sum_{k=\lceil nx - Tn^\alpha \rceil}^{[nx + Tn^\alpha]}{}^* f\left(\frac{k}{n}\right) \odot \frac{b(n^{1-\alpha}(x - \frac{k}{n}))}{In^\alpha}, f(x)\right)
$$
$$
= D\left(\sum_{k=\lceil nx - Tn^\alpha \rceil}^{[nx + Tn^\alpha]}{}^* f\left(\frac{k}{n}\right) \odot \frac{b(n^{1-\alpha}(x - \frac{k}{n}))}{In^\alpha}, f(x)\right)
$$

(by b having compact support $[-T, T]$)

$$= D\left(\sum_{k=\lceil nx-Tn^\alpha\rceil}^{[nx+Tn^\alpha]}{}^* f\left(\frac{k}{n}\right) \odot \frac{b\left(n^{1-\alpha}\left(x-\frac{k}{n}\right)\right)}{In^\alpha}\right.$$

$$\oplus f(x) \odot \sum_{k=\lceil nx-Tn^\alpha\rceil}^{[nx+Tn^\alpha]} \frac{b\left(n^{1-\alpha}\left(x-\frac{k}{n}\right)\right)}{In^\alpha}, f(x)$$

$$\left.\oplus f(x) \odot \sum_{k=\lceil nx-Tn^\alpha\rceil}^{[nx+Tn^\alpha]} \frac{b\left(n^{1-\alpha}\left(x-\frac{k}{n}\right)\right)}{In^\alpha}\right)$$

$$\leq \gamma + D\left(f(x) \odot \sum_{k=\lceil nx-Tn^\alpha\rceil}^{[nx+Tn^\alpha]} \frac{b\left(n^{1-\alpha}\left(x-\frac{k}{n}\right)\right)}{In^\alpha}, f(x)\right)$$

$$=: \quad \otimes,$$

where

$$\gamma := \quad D\left(\sum_{k=\lceil nx-Tn^\alpha\rceil}^{[nx+Tn^\alpha]}{}^* f\left(\frac{k}{n}\right) \odot \frac{b\left(n^{1-\alpha}\left(x-\frac{k}{n}\right)\right)}{In^\alpha}, \right.$$

$$\left. f(x) \odot \sum_{k=\lceil nx-Tn^\alpha\rceil}^{[nx+Tn^\alpha]} \frac{b\left(n^{1-\alpha}\left(x-\frac{k}{n}\right)\right)}{In^\alpha}\right). \tag{15.5}$$

Here we observe that

$$D\left(f(x) \odot \sum_{k=\lceil nx-Tn^\alpha\rceil}^{[nx+Tn^\alpha]} \frac{b\left(n^{1-\alpha}\left(x-\frac{k}{n}\right)\right)}{In^\alpha}, f(x) \odot 1\right)$$

$$\leq \left|\sum_{k=\lceil nx-Tn^\alpha\rceil}^{[nx+Tn^\alpha]} \frac{b\left(n^{1-\alpha}\left(x-\frac{k}{n}\right)\right)}{In^\alpha} - 1\right| D(f(x), \tilde{o}). \tag{15.6}$$

The last is true by Lemma 2.2 of [31].

Next we see that

$$\gamma = D\left(\sum_{k=\lceil nx-Tn^\alpha\rceil}^{[nx+Tn^\alpha]}{}^* f\left(\frac{k}{n}\right) \odot \frac{b\left(n^{1-\alpha}\left(x-\frac{k}{n}\right)\right)}{In^\alpha}, \right.$$

$$\sum_{k=\lceil nx-Tn^\alpha\rceil}^{[nx+Tn^\alpha]}{}^* f(x) \odot \frac{b\left(n^{1-\alpha}\left(x-\frac{k}{n}\right)\right)}{In^\alpha}\right)$$

$$\leq \sum_{k=\lceil nx-Tn^\alpha \rceil}^{\lceil nx+Tn^\alpha \rceil} D\left(f\left(\frac{k}{n}\right) \odot \frac{b\left(n^{1-\alpha}\left(x-\frac{k}{n}\right)\right)}{In^\alpha}, \right.$$

$$\left. f(x) \odot \frac{b\left(n^{1-\alpha}\left(x-\frac{k}{n}\right)\right)}{In^\alpha} \right)$$

$$= \sum_{k=\lceil nx-Tn^\alpha \rceil}^{\lceil nx+Tn^\alpha \rceil} \frac{b\left(n^{1-\alpha}\left(x-\frac{k}{n}\right)\right)}{In^\alpha} D\left(f\left(\frac{k}{n}\right), f(x) \right)$$

$$\leq \sum_{k=\lceil nx-Tn^\alpha \rceil}^{\lceil nx+Tn^\alpha \rceil} \frac{b\left(n^{1-\alpha}\left(x-\frac{k}{n}\right)\right)}{In^\alpha} \omega_1^{(\mathcal{F})}\left(f, \left|\frac{k}{n}-x\right|\right)$$

$$\leq \left(\sum_{k=\lceil nx-Tn^\alpha \rceil}^{\lceil nx+Tn^\alpha \rceil} \frac{b\left(n^{1-\alpha}\left(x-\frac{k}{n}\right)\right)}{In^\alpha} \right) \omega_1^{(\mathcal{F})}\left(f, \frac{T}{n^{1-\alpha}}\right).$$

That is, we have that

$$\gamma \leq \left(\sum_{k=\lceil nx-Tn^\alpha \rceil}^{\lceil nx+Tn^\alpha \rceil} \frac{b\left(n^{1-\alpha}\left(x-\frac{k}{n}\right)\right)}{In^\alpha} \right) \omega_1^{(\mathcal{F})}\left(f, \frac{T}{n^{1-\alpha}}\right). \tag{15.7}$$

Finally, applying (15.5), (15.6) and (15.7) into $\otimes$, we obtain

$$D(f_n(x), f(x)) \leq \left| \sum_{k=\lceil nx-Tn^\alpha \rceil}^{\lceil nx+Tn^\alpha \rceil} \frac{b\left(n^{1-\alpha}\left(x-\frac{k}{n}\right)\right)}{In^\alpha} - 1 \right| \cdot D(f(x), \tilde{o}) \tag{15.8}$$

$$+ \left(\sum_{k=\lceil nx-Tn^\alpha \rceil}^{\lceil nx+Tn^\alpha \rceil} \frac{b\left(n^{1-\alpha}\left(x-\frac{k}{n}\right)\right)}{In^\alpha} \right) \cdot \omega_1^{(\mathcal{F})}\left(f, \frac{T}{n^{1-\alpha}}\right).$$

Using now (15.4) into (15.8) we get (15.3). $\qquad\qquad \square$

Remark 15.18. By Lemma 2.1 of [9], p. 64 we obtain that

$$\lim_{n\to+\infty} \left| \sum_{k=\lceil nx-Tn^\alpha \rceil}^{\lceil nx+Tn^\alpha \rceil} \frac{b\left(n^{1-\alpha}\left(x-\frac{k}{n}\right)\right)}{In^\alpha} - 1 \right| = 0,$$

any $x \in \mathbb{R}$. Let $f \in C_{\mathcal{F}}^U(\mathbb{R})$, then $\lim_{n\to+\infty} \omega_1^{(\mathcal{F})}\left(f, \frac{T}{n^{1-\alpha}}\right) = 0$. Therefore from (15.3) we get

$$\lim_{n\to+\infty} D(f_n(x), f(x)) = 0.$$

That is, $f_n(x) \to f(x)$, pointwise with rates, as $n \to +\infty$, where $x \in \mathbb{R}$. We denote by $C_{\mathcal{F}}^{NU}(\mathbb{R}) := \{f: \mathbb{R} \to \mathbb{R}_{\mathcal{F}}$, such that all the derivatives $f^{(i)}: \mathbb{R} \to \mathbb{R}_{\mathcal{F}}$, $i = 0, 1, \dots, N$ exist and all are uniformly continuous in the fuzzy sense}, $N \in \mathbb{N}$. Also denote by $C_{\mathcal{F}}^{NB}(\mathbb{R}) := \{f \colon \mathbb{R} \to \mathbb{R}_{\mathcal{F}}, \text{ such that all } f^{(i)} \colon \mathbb{R} \to \mathbb{R}_{\mathcal{F}}, i = 0, 1, \ldots, N \text{ exist and all are bounded and continuous in the fuzzy sense}\}, N \in \mathbb{N}.$

Now we present the second main result.

Theorem 15.19. *Let* $x \in \mathbb{R}$, $T > 0$, $n \in \mathbb{N}$ *such that* $n \geq \max(T + |x|, T^{-1/\alpha})$. *Let* $f \in C_{\mathcal{F}}^{NU}(\mathbb{R})$, *or* $f \in C_{\mathcal{F}}^{NB}(\mathbb{R})$, $N \in \mathbb{N}$. *Then*

$$
\begin{aligned}
D(f_n(x), f(x)) \leq & \left| \sum_{k=\lceil nx - Tn^{\alpha} \rceil}^{\lceil nx + Tn^{\alpha} \rceil} \frac{b\left(n^{1-\alpha}\left(x - \frac{k}{n}\right)\right)}{In^{\alpha}} - 1 \right| \left[D(f(x), \tilde{o}) \right. \\
& \left. + \omega_1^{(\mathcal{F})}\left(f, \frac{T}{n^{1-\alpha}}\right) \right] + \omega_1^{(\mathcal{F})}\left(f, \frac{T}{n^{1-\alpha}}\right) \\
& + \frac{b^*}{I}\left(2T + \frac{1}{n^{\alpha}}\right) \left\{ \sum_{j=1}^{N-1} \frac{2^j T^j}{j! n^{(1-\alpha)j}} \right. \\
& \left. \cdot \left[D(f^{(j)}(x), \tilde{o}) + \omega_1^{(\mathcal{F})}\left(f^{(j)}, \frac{T}{n^{1-\alpha}}\right) \right] \right\} \quad (15.9) \\
& + \frac{2^N T^N b^*}{N! In^{(1-\alpha)N}}\left(2T + \frac{1}{n^{\alpha}}\right) \left\{ D(f^{(N)}(x), \tilde{o}) \right. \\
& \left. + 3\omega_1^{(\mathcal{F})}\left(f^{(N)}, \frac{T}{n^{1-\alpha}}\right) \right\}.
\end{aligned}
$$

Remark 15.20. By Lemma 2.1 of [9], p. 64 we obtain that

$$
\lim_{n \to +\infty} \left| \sum_{k=\lceil nx - Tn^{\alpha} \rceil}^{\lceil nx + Tn^{\alpha} \rceil} \frac{b\left(n^{1-\alpha}\left(x - \frac{k}{n}\right)\right)}{In^{\alpha}} - 1 \right| = 0,
$$

any $x \in \mathbb{R}$. Let $f \in C_{\mathcal{F}}^{NU}(\mathbb{R})$, then

$$
\lim_{n \to +\infty} \omega_1^{(\mathcal{F})}\left(f^{(j)}, \frac{T}{n^{1-\alpha}}\right) = 0, \quad j = 0, 1, \ldots, N.
$$

Therefore from (15.9) we get

$$
\lim_{n \to +\infty} D(f_n(x), f(x)) = 0.
$$

That is, $f_n(x) \to f(x)$, pointwise with rates, as $n \to +\infty$, $x \in \mathbb{R}$.

Proof (of Theorem 15.19). Because b is of compact support $[-T, T]$ we have that

$$
f_n(x) = \sum_{k=\lceil nx - Tn^{\alpha} \rceil}^{\lceil nx + Tn^{\alpha} \rceil *} f\left(\frac{k}{n}\right) \odot \frac{b\left(n^{1-\alpha}\left(x - \frac{k}{n}\right)\right)}{In^{\alpha}}.
$$

Clearly the terms in the sum (15.2) are nonzero iff $n^{1-\alpha}\left|x-\frac{k}{n}\right|\le T$, iff

$$-\frac{T}{n^{1-\alpha}}\le x-\frac{k}{n}\le\frac{T}{n^{1-\alpha}}.$$

Hence in here $x-\frac{T}{n^{1-\alpha}}\le\frac{k}{n}$, $x\in\mathbb{R}$.

Using the fuzzy Taylor formula (Theorem 15.14) we have

$$f\left(\frac{k}{n}\right)=\sum_{j=0}^{N-1}{}^{*}f^{(j)}\left(x-\frac{T}{n^{1-\alpha}}\right)\odot\frac{\left(\frac{k}{n}-x+\frac{T}{n^{1-\alpha}}\right)^{j}}{j!}\oplus R_{N}\left(x-\frac{T}{n^{1-\alpha}},\frac{k}{n}\right),$$

where

$$R_{N}\left(x-\frac{T}{n^{1-\alpha}},\frac{k}{n}\right)\quad:\ =(FR)\int_{x-\frac{T}{n^{1-\alpha}}}^{\frac{k}{n}}\left(\int_{x-\frac{T}{n^{1-\alpha}}}^{s_{1}}\cdots\right.$$

$$\left.\left(\int_{x-\frac{T}{n^{1-\alpha}}}^{s_{N-1}}f^{(N)}(s_{N})ds_{N}\right)ds_{N-1}\cdots\right)ds_{1}.$$

Then

$$f\left(\frac{k}{n}\right)\odot\frac{b\left(n^{1-\alpha}\left(x-\frac{k}{n}\right)\right)}{In^{\alpha}}\ =\ \sum_{j=0}^{N-1}{}^{*}f^{(j)}\left(x-\frac{T}{n^{1-\alpha}}\right)\odot\frac{\left(\frac{k}{n}-x+\frac{T}{n^{1-\alpha}}\right)^{j}}{j!}$$

$$\odot\frac{b\left(n^{1-\alpha}\left(x-\frac{k}{n}\right)\right)}{In^{\alpha}}$$

$$\oplus R_{N}\left(x-\frac{T}{n^{1-\alpha}},\frac{k}{n}\right)\odot\frac{b\left(n^{1-\alpha}\left(x-\frac{k}{n}\right)\right)}{In^{\alpha}}.$$

Therefore we have

$$\sum_{k=\lceil nx-Tn^{\alpha}\rceil}^{\lceil nx+Tn^{\alpha}\rceil}{}^{*}f\left(\frac{k}{n}\right)\odot\frac{b\left(n^{1-\alpha}\left(x-\frac{k}{n}\right)\right)}{In^{\alpha}}$$

$$=\sum_{k=\lceil nx-Tn^{\alpha}\rceil}^{\lceil nx+Tn^{\alpha}\rceil}{}^{*}\sum_{j=0}^{N-1}{}^{*}f^{(j)}\left(x-\frac{T}{n^{1-\alpha}}\right)\odot\frac{\left(\frac{k}{n}-x+\frac{T}{n^{1-\alpha}}\right)^{j}}{j!}$$

$$\odot\frac{b\left(n^{1-\alpha}\left(x-\frac{k}{n}\right)\right)}{In^{\alpha}}\oplus\sum_{k=\lceil nx-Tn^{\alpha}\rceil}^{\lceil nx+Tn^{\alpha}\rceil}{}^{*}R_{N}\left(x-\frac{T}{n^{1-\alpha}},\frac{k}{n}\right)$$

$$\odot\frac{b\left(n^{1-\alpha}\left(x-\frac{k}{n}\right)\right)}{In^{\alpha}}.$$

That is,

$$
\begin{aligned}
f_n(x) \;=\; & \sum_{j=0}^{N-1}{}^{*}\; \sum_{k=\lceil nx-Tn^\alpha \rceil}^{\lfloor nx+Tn^\alpha \rfloor}{}^{*} f^{(j)}\left(x - \frac{T}{n^{1-\alpha}}\right) \odot \frac{\left(\frac{k}{n} - x + \frac{T}{n^{1-\alpha}}\right)^j}{j!} \\
& \odot \frac{b\left(n^{1-\alpha}\left(x - \frac{k}{n}\right)\right)}{In^\alpha} \\
& \oplus \sum_{k=\lceil nx-Tn^\alpha \rceil}^{\lfloor nx+Tn^\alpha \rfloor}{}^{*} R_N\left(x - \frac{T}{n^{1-\alpha}}, \frac{k}{n}\right) \odot \frac{b\left(n^{1-\alpha}\left(x - \frac{k}{n}\right)\right)}{In^\alpha}.
\end{aligned}
$$

Next we estimate

$$
D(f_n(x), f(x)) \le D\left(f_n(x),\; \sum_{k=\lceil nx-Tn^\alpha \rceil}^{\lfloor nx+Tn^\alpha \rfloor}{}^{*} f\left(x - \frac{T}{n^{1-\alpha}}\right) \odot \right.
$$
$$
\left. \frac{b\left(n^{1-\alpha}\left(x - \frac{k}{n}\right)\right)}{In^\alpha} \right)
$$
$$
+ D\left(f\left(x - \frac{T}{n^{1-\alpha}}\right) \odot \sum_{k=\lceil nx-Tn^\alpha \rceil}^{\lfloor nx+Tn^\alpha \rfloor} \frac{b\left(n^{1-\alpha}\left(x - \frac{k}{n}\right)\right)}{In^\alpha}, \right.
$$
$$
\left. f\left(x - \frac{T}{n^{1-\alpha}}\right) \odot 1 \right) + D\left(f\left(x - \frac{T}{n^{1-\alpha}}\right), f(x)\right)
$$

$$
\begin{aligned}
&\text{(by Lemma 1.2)} \\
&\quad\;\; \le \\
&\text{\& (properties of } D\text{)}
\end{aligned}
\quad
D\left(\sum_{j=1}^{N-1}{}^{*}\; \sum_{k=\lceil nx-Tn^\alpha \rceil}^{\lfloor nx+Tn^\alpha \rfloor}{}^{*} f^{(j)}\left(x - \frac{T}{n^{1-\alpha}}\right) \right.
$$
$$
\odot \frac{\left(\frac{k}{n} - x + \frac{T}{n^{1-\alpha}}\right)^j}{j!} \odot \frac{b\left(n^{1-\alpha}\left(x - \frac{k}{n}\right)\right)}{In^\alpha} \oplus \sum_{k=\lceil nx-Tn^\alpha \rceil}^{\lfloor nx+Tn^\alpha \rfloor}{}^{*} R_N\left(x - \frac{T}{n^{1-\alpha}}, \frac{k}{n}\right)
$$
$$
\left. \odot \frac{b\left(n^{1-\alpha}\left(x - \frac{k}{n}\right)\right)}{In^\alpha}, \tilde{o}\right) + \left| \sum_{k=\lceil nx-Tn^\alpha \rceil}^{\lfloor nx+Tn^\alpha \rfloor} \frac{b\left(n^{1-\alpha}\left(x - \frac{k}{n}\right)\right)}{In^\alpha} - 1 \right|
$$
$$
\cdot D\left(f\left(x - \frac{T}{n^{1-\alpha}}\right), \tilde{o}\right) + \omega_1^{(\mathcal{F})}\left(f, \frac{T}{n^{1-\alpha}}\right)
$$
$$
\le \sum_{j=1}^{N-1}\; \sum_{k=\lceil nx-Tn^\alpha \rceil}^{\lfloor nx+Tn^\alpha \rfloor} \frac{\left(\frac{k}{n} - x + \frac{T}{n^{1-\alpha}}\right)^j}{j!} \cdot \frac{b\left(n^{1-\alpha}\left(x - \frac{k}{n}\right)\right)}{In^\alpha}.
$$

$$D\left(f^{(j)}\left(x-\frac{T}{n^{1-\alpha}}\right),\tilde{o}\right)$$

$$+\sum_{k=\lceil nx-Tn^{\alpha}\rceil}^{[nx+Tn^{\alpha}]}\frac{b\left(n^{1-\alpha}\left(x-\frac{k}{n}\right)\right)}{In^{\alpha}}D\left(R_N\left(x-\frac{T}{n^{1-\alpha}},\frac{k}{n}\right),\tilde{o}\right)$$

$$+\left|\sum_{k=\lceil nx-Tn^{\alpha}\rceil}^{[nx+Tn^{\alpha}]}\frac{b\left(n^{1-\alpha}\left(x-\frac{k}{n}\right)\right)}{In^{\alpha}}-1\right|\left(D(f(x),\tilde{o})+\omega_1^{(\mathcal{F})}\left(f,\frac{T}{n^{1-\alpha}}\right)\right)$$

$$+\omega_1^{(\mathcal{F})}\left(f,\frac{T}{n^{1-\alpha}}\right)$$

$$\leq\sum_{j=1}^{N-1}\frac{2^jT^j}{j!n^{(1-\alpha)j}}\frac{b^*}{I}\left(2T+\frac{1}{n^{\alpha}}\right)\cdot\left(D(f^{(j)}(x),\tilde{o})+\omega_1^{(\mathcal{F})}\left(f^{(j)},\frac{T}{n^{1-\alpha}}\right)\right)$$

$$+\sum_{k=\lceil nx-Tn^{\alpha}\rceil}^{[nx+Tn^{\alpha}]}\frac{b\left(n^{1-\alpha}\left(x-\frac{k}{n}\right)\right)}{In^{\alpha}}D\left(R_N\left(x-\frac{T}{n^{1-\alpha}},\frac{k}{n}\right),\tilde{o}\right)$$

$$+\left|\sum_{k=\lceil nx-Tn^{\alpha}\rceil}^{[nx+Tn^{\alpha}]}\frac{b\left(n^{1-\alpha}\left(x-\frac{k}{n}\right)\right)}{In^{\alpha}}-1\right|\cdot\left(D(f(x),\tilde{o})+\omega_1^{(\mathcal{F})}\left(f,\frac{T}{n^{1-\alpha}}\right)\right)$$

$$+\omega_1^{(\mathcal{F})}\left(f,\frac{T}{n^{1-\alpha}}\right)=:(*).$$

In the following we work on

$$D\left(R_N\left(x-\frac{T}{n^{1-\alpha}},\frac{k}{n}\right),\tilde{o}\right)$$

$$=D\left(R_N\left(x-\frac{T}{n^{1-\alpha}},\frac{k}{n}\right)\oplus f^{(N)}\left(x-\frac{T}{n^{1-\alpha}}\right)\odot\frac{\left(\frac{k}{n}-x+\frac{T}{n^{1-\alpha}}\right)^N}{N!},\right.$$

$$\left.f^{(N)}\left(x-\frac{T}{n^{1-\alpha}}\right)\odot\frac{\left(\frac{k}{n}-x+\frac{T}{n^{1-\alpha}}\right)^N}{N!}\right)$$

$$\leq D\left(R_N\left(x-\frac{T}{n^{1-\alpha}},\frac{k}{n}\right),f^{(N)}\left(x-\frac{T}{n^{1-\alpha}}\right)\odot\frac{\left(\frac{k}{n}-x+\frac{T}{n^{1-\alpha}}\right)^N}{N!}\right)$$

$$+D\left(f^{(N)}\left(x-\frac{T}{n^{1-\alpha}}\right)\odot\frac{\left(\frac{k}{n}-x+\frac{T}{n^{1-\alpha}}\right)^N}{N!},\tilde{o}\right).$$

Hence

$$D\left(R_N\left(x - \frac{T}{n^{1-\alpha}}, \frac{k}{n}\right), \tilde{o}\right)$$
$$\leq D\left(R_N\left(x - \frac{T}{n^{1-\alpha}}, \frac{k}{n}\right), f^{(N)}\left(x - \frac{T}{n^{1-\alpha}}\right) \odot \frac{\left(\frac{k}{n} - x + \frac{T}{n^{1-\alpha}}\right)^N}{N!}\right)$$
$$+ \frac{2^N T^N}{N! n^{N(1-\alpha)}}\left(D(f^{(N)}(x), \tilde{o}) + \omega_1^{(\mathcal{F})}\left(f^{(N)}, \frac{T}{n^{1-\alpha}}\right)\right). \qquad (15.10)$$

Finally we estimate

$$D\left(R_N\left(x - \frac{T}{n^{1-\alpha}}, \frac{k}{n}\right), f^{(N)}\left(x - \frac{T}{n^{1-\alpha}}\right) \odot \frac{\left(\frac{k}{n} - x + \frac{T}{n^{1-\alpha}}\right)^N}{N!}\right)$$
$$= D\left((FR)\int_{x-\frac{T}{n^{1-\alpha}}}^{\frac{k}{n}}\left(\int_{x-\frac{T}{n^{1-\alpha}}}^{s_1}\cdots\left(\int_{x-\frac{T}{n^{1-\alpha}}}^{s_{N-1}} f^{(N)}(s_N)ds_N\right)\right.\right.$$
$$ds_{N-1}\cdots)\,ds_1),$$
$$f^{(N)}\left(x - \frac{T}{n^{1-\alpha}}\right) \odot \int_{x-\frac{T}{n^{1-\alpha}}}^{\frac{k}{n}}\left(\int_{x-\frac{T}{n^{1-\alpha}}}^{s_1}\cdots\left(\int_{x-\frac{T}{n^{1-\alpha}}}^{s_{N-1}} 1ds_N\right)\right.$$
$$\left. ds_{N-1}\cdots)\,ds_1\right)$$

$$= D\left((FR)\int_{x-\frac{T}{n^{1-\alpha}}}^{\frac{k}{n}}\left(\int_{x-\frac{T}{n^{1-\alpha}}}^{s_1}\cdots\left(\int_{x-\frac{T}{n^{1-\alpha}}}^{s_{N-1}} f^{(N)}(s_N)ds_N\right)\right.\right.$$
$$ds_{N-1}\cdots)\,ds_1, (FR)\int_{x-\frac{T}{n^{1-\alpha}}}^{\frac{k}{n}}\left(\int_{x-\frac{T}{n^{1-\alpha}}}^{s_1}\cdots\right.$$
$$\left.\left(\int_{x-\frac{T}{n^{1-\alpha}}}^{s_{N-1}} f^{(N)}\left(x - \frac{T}{n^{1-\alpha}}\right)ds_N\right)ds_{N-1}\cdots\right)ds_1\right)$$

$$(\text{by Lemmas 15.10,15.13}) \quad \overset{\leq}{} \int_{x-\frac{T}{n^{1-\alpha}}}^{\frac{k}{n}} \left(\int_{x-\frac{T}{n^{1-\alpha}}}^{s_1} \cdots \left(\int_{x-\frac{T}{n^{1-\alpha}}}^{s_{N-1}} \right. \right.$$

$$D\left(f^{(N)}(s_N), f^{(N)}\left(x - \frac{T}{n^{1-\alpha}} \right) \right) ds_N \Bigg) ds_{N-1} \cdots \Bigg) ds_1$$

$$\leq \int_{x-\frac{T}{n^{1-\alpha}}}^{\frac{k}{n}} \left(\int_{x-\frac{T}{n^{1-\alpha}}}^{s_1} \cdots \left(\int_{x-\frac{T}{n^{1-\alpha}}}^{s_{N-1}} \omega_1^{(\mathcal{F})} \left(f^{(N)}, \left| s_N - x + \frac{T}{n^{1-\alpha}} \right| \right) ds_N \right) \right.$$

$$\cdots ds_{N-1} \cdots \Bigg) ds_1$$

$$\leq \int_{x-\frac{T}{n^{1-\alpha}}}^{\frac{k}{n}} \int_{x-\frac{T}{n^{1-\alpha}}}^{s_1} \cdots \left(\int_{x-\frac{T}{n^{1-\alpha}}}^{s_{N-1}} 1 ds_N \right) ds_{N-1} \cdots \Bigg) ds_1 \Bigg)$$

$$\cdot \omega_1^{(\mathcal{F})} \left(f^{(N)}, \frac{k}{n} - x + \frac{T}{n^{1-\alpha}} \right) \leq \frac{\left(\frac{k}{n} - x + \frac{T}{n^{1-\alpha}} \right)^N}{N!} \cdot \omega_1^{(\mathcal{F})} \left(f^{(N)}, \frac{2T}{n^{1-\alpha}} \right)$$

$$\leq \frac{2^N T^N}{N! n^{N(1-\alpha)}} \omega_1^{(\mathcal{F})} \left(f^{(N)}, \frac{2T}{n^{1-\alpha}} \right).$$

That is, we get that

$$D\left(R_N \left(x - \frac{T}{n^{1-\alpha}}, \frac{k}{n} \right), f^{(N)} \left(x - \frac{T}{n^{1-\alpha}} \right) \odot \frac{\left(\frac{k}{n} - x + \frac{T}{n^{1-\alpha}} \right)^N}{N!} \right)$$

$$\leq \frac{2^N T^N}{N! n^{N(1-\alpha)}} \omega_1^{(\mathcal{F})} \left(f^{(N)}, \frac{2T}{n^{1-\alpha}} \right). \tag{15.11}$$

Consequently we obtain

$$D\left(R_N \left(x - \frac{T}{n^{1-\alpha}}, \frac{k}{n} \right), \tilde{o} \right) \tag{15.12}$$

$$\leq \frac{2^N T^N}{N! n^{N(1-\alpha)}} \left\{ D(f^{(N)}(x), \tilde{o}) + 3\omega_1^{(\mathcal{F})} \left(f^{(N)}, \frac{T}{n^{1-\alpha}} \right) \right\}.$$

Therefore, by (15.12) and (15.4), we have

$$\sum_{k=\lceil nx-Tn^\alpha \rceil}^{\lceil nx+Tn^\alpha \rceil} \frac{b\left(n^{1-\alpha}\left(x - \frac{k}{n} \right) \right)}{In^\alpha} D\left(R_N \left(x - \frac{T}{n^{1-\alpha}}, \frac{k}{n} \right), \tilde{o} \right) \tag{15.13}$$

$$\leq \frac{2^N T^N b^*}{N! In^{N(1-\alpha)}} \left(2T + \frac{1}{n^\alpha} \right) \left\{ D(f^{(N)}(x), \tilde{o}) + 3\omega_1^{(\mathcal{F})} \left(f^{(N)}, \frac{T}{n^{1-\alpha}} \right) \right\}.$$

At last using (15.13) into (∗) we have completed the proof of the theorem.
□

Corollary 15.21 (to Theorem 15.17). *Let $b(x)$ be a centered bell-shaped continuous function on $\mathbb{R}$ of compact support $[-T, T]$, $T > 0$. Let $x \in$*

$[-T^*, T^*]$, $T^* > 0$, and $n \in \mathbb{N}$ be such that $n \geq \max(T + T^*, T^{-1/\alpha})$, $0 < \alpha < 1$. Consider $p \geq 1$. Then

$$\|D(f_n(x), f(x))\|_{p, [-T^*, T^*]} \leq (\|D(f(x), \tilde{o})\|_{\infty, [-T^*, T^*]})$$

$$\cdot \left\| \sum_{k=\lceil nx-Tn^\alpha \rceil}^{[nx+Tn^\alpha]} \frac{b\left(n^{1-\alpha}\left(x - \frac{k}{n}\right)\right)}{In^\alpha} - 1 \right\|_{p, [-T^*, T^*]}$$

$$+ \frac{b^*}{I}\left(2T + \frac{1}{n^\alpha}\right) 2^{1/p} T^{*1/p} \omega_1^{(\mathcal{F})}\left(f, \frac{T}{n^{1-\alpha}}\right), \qquad (15.14)$$

where $I := \int_{-T}^{T} b(t)dt$. From (15.14), when $f \in C_{\mathcal{F}}^U(\mathbb{R})$, we get the L_p convergence of f_n to f with rates.

Proof. Since f is fuzzy continuous on $[-T^*, T^*]$ it is fuzzy bounded there (Lemma 15.11), therefore

$$\|D(f(x), \tilde{o})\|_{\infty, [-T^*, T^*]} < +\infty.$$

So from Theorem 15.17, we have that

$$D(f_n(x), f(x)) \leq (\|D(f(x), \tilde{o})\|_{\infty, [-T^*, T^*]})$$

$$\cdot \left| \sum_{k=\lceil nx-Tn^\alpha \rceil}^{[nx+Tn^\alpha]} \frac{b\left(n^{1-\alpha}\left(x - \frac{k}{n}\right)\right)}{In^\alpha} - 1 \right|$$

$$+ \frac{b^*}{I}\left(2T + \frac{1}{n^\alpha}\right) \omega_1^{(\mathcal{F})}\left(f, \frac{T}{n^{1-\alpha}}\right). \qquad (15.15)$$

Inequality (15.14) now comes by integration of (15.15) and the properties of L_p norm. As in [9], p. 75, using the bounded convergence theorem we get that

$$\lim_{n \to \infty} \left\| \sum_{k=\lceil nx-Tn^\alpha \rceil}^{[nx+Tn^\alpha]} \frac{b\left(n^{1-\alpha}\left(x - \frac{k}{n}\right)\right)}{In^\alpha} - 1 \right\|_{p, [-T^*, T^*]} = 0. \qquad \square$$

Corollary 15.22 (to Theorem 15.19). Let $b(x)$ be a centered bell-shaped continuous function on $\mathbb{R}$ of compact support $[-T, T]$, $T > 0$. Let $x \in [-T^*, T^*]$, $T^* > 0$, and $n \in \mathbb{N}$ be such that

$$n \geq \max(T + T^*, T^{-1/\alpha}), \quad 0 < \alpha < 1.$$

Consider $p \geq 1$. Then

$$\|D(f_n(x), f(x))\|_{p,[-T^*,T^*]}$$

$$\leq \left(\|D(f(x), \tilde{o})\|_{\infty,[-T^*,T^*]} + \omega_1^{(\mathcal{F})}\left(f, \frac{T}{n^{1-\alpha}}\right)\right)$$

$$\cdot \left\|\sum_{k=\lceil nx - Tn^\alpha \rceil}^{[nx + Tn^\alpha]} \frac{b(n^{1-\alpha}(x - \frac{k}{n}))}{I n^\alpha} - 1\right\|_{p,[-T^*,T^*]}$$

$$+ \omega_1^{(\mathcal{F})}\left(f, \frac{T}{n^{1-\alpha}}\right) 2^{1/p} T^{*1/p} + \frac{b^*}{I}\left(2T + \frac{1}{n^\alpha}\right)\left\{\sum_{j=1}^{N-1} \frac{2^j T^j}{j! n^{(1-\alpha)j}}\right.$$

$$\cdot \left(\|D(f^{(j)}(x), \tilde{o})\|_{p,[-T^*,T^*]} + \omega_1^{(\mathcal{F})}\left(f^{(j)}, \frac{T}{n^{1-\alpha}}\right) 2^{1/p} T^{*1/p}\right)\right\}$$

$$+ \frac{2^N T^N b^*}{N! I n^{(1-\alpha)N}}\left(2T + \frac{1}{n^\alpha}\right)\left\{\|D(f^{(N)}(x), \tilde{o})\|_{p,[-T^*,T^*]}\right.$$

$$\left. + 3\omega_1^{(\mathcal{F})}\left(f^{(N)}, \frac{T}{n^{1-\alpha}}\right) 2^{1/p} T^{*1/p}\right\}. \tag{15.16}$$

When $f \in C_{\mathcal{F}}^{NU}(\mathbb{R})$, from inequality (15.16) we obtain the L_p convergence of f_n to f with rates.

Proof. Similar to Corollary 15.21.   □

We need (see also [6], [47])

Definition 15.23. Let the nonnegative function $S: \mathbb{R} \to \mathbb{R}$, S has compact support $[-T, T]$, $T > 0$, and is nondecreasing there and it can be continuous only on either $(-\infty, T]$ or $[-T, T]$. S can have jump discontinuities. We call S the "squashing function".

Let $f: \mathbb{R} \to \mathbb{R}_{\mathcal{F}}$ be either fuzzy uniformly continuous ($f \in C_{\mathcal{F}}^U(\mathbb{R})$), or fuzzy continuous and bounded ($f \in C_{\mathcal{F}}^B(\mathbb{R})$). Assume that

$$I^* := \int_{-T}^{T} S(t)dt > 0.$$

Clearly

$$\max_{x \in [-T,T]} S(x) = S(T).$$

For $x \in \mathbb{R}$ we define the "univariate fuzzy squashing operator"

$$(G_n(f))(x) := \sum_{k=-n^2}^{n^2} {}^* f\left(\frac{k}{n}\right) \odot \frac{S(n^{1-\alpha}(x - \frac{k}{n}))}{I^* n^\alpha}, \tag{15.17}$$

$0 < \alpha < 1$ and $n \in \mathbb{N}$: $n \geq \max(T + |x|, T^{-1/\alpha})$. The above neural network operators are also linear operators over $\mathbb{R}$. These operators in the ordinary real case were thoroughly studied in [6], [47].

It is clear to see again that

$$(G_n(f))(x) = \sum_{k=\lceil nx-Tn^\alpha \rceil}^{[nx+Tn^\alpha]*} f\left(\frac{k}{n}\right) \odot \frac{S\left(n^{1-\alpha}\left(x - \frac{k}{n}\right)\right)}{I^* n^\alpha}. \tag{15.18}$$

Here we study the pointwise convergence with rates of $(G_n(f))(x) \to f(x)$, as $n \to +\infty$, $x \in \mathbb{R}$.

Theorem 15.24. *Under the above terms and assumptions we obtain*

$$D(G_n(f)(x), f(x)) \leq \left| \sum_{k=\lceil nx-Tn^\alpha \rceil}^{[nx+Tn^\alpha]} \frac{S\left(n^{1-\alpha}\left(x - \frac{k}{n}\right)\right)}{I^* n^\alpha} - 1 \right| D(f(x), \tilde{o})$$

$$+ \frac{S(T)}{I^*}\left(2T + \frac{1}{n^\alpha}\right) \omega_1^{(\mathcal{F})}\left(f, \frac{T}{n^{1-\alpha}}\right). \tag{15.19}$$

Proof. Notice that

$$\sum_{k=\lceil nx-Tn^\alpha \rceil}^{[nx+Tn^\alpha]} 1 \leq (2Tn^\alpha + 1). \tag{15.20}$$

We have that

$$D((G_n(f))(x), f(x)) =$$

$$D\left(\sum_{k=\lceil nx-Tn^\alpha \rceil}^{[nx+Tn^\alpha]*} f\left(\frac{k}{n}\right) \odot \frac{S\left(n^{1-\alpha}\left(x - \frac{k}{n}\right)\right)}{I^* n^\alpha}, f(x) \right)$$

$$\leq D\left(\sum_{k=\lceil nx-Tn^\alpha \rceil}^{[nx+Tn^\alpha]*} f\left(\frac{k}{n}\right) \odot \frac{S\left(n^{1-\alpha}\left(x - \frac{k}{n}\right)\right)}{I^* n^\alpha}, \right.$$

$$\left. f(x) \odot \sum_{k=\lceil nx-Tn^\alpha \rceil}^{[nx+Tn^\alpha]} \frac{S\left(n^{1-\alpha}\left(x - \frac{k}{n}\right)\right)}{I^* n^\alpha} \right)$$

$$+ D\left(f(x) \odot \sum_{k=\lceil nx-Tn^\alpha \rceil}^{[nx+Tn^\alpha]} \frac{S\left(n^{1-\alpha}\left(x - \frac{k}{n}\right)\right)}{I^* n^\alpha}, f(x) \right)$$

$$\leq \left| \sum_{k=\lceil nx-Tn^\alpha \rceil}^{[nx+Tn^\alpha]} \frac{S\left(n^{1-\alpha}\left(x - \frac{k}{n}\right)\right)}{I^* n^\alpha} - 1 \right| D(f(x), \tilde{o}) + \gamma =: (*),$$

where

$$\gamma := \quad D\left(\sum_{k=\lceil nx-Tn^{\alpha}\rceil}^{[nx+Tn^{\alpha}]*} f\left(\frac{k}{n}\right) \odot \frac{S\left(n^{1-\alpha}\left(x-\frac{k}{n}\right)\right)}{I^{*}n^{\alpha}}, \right.$$

$$\left. f(x) \odot \sum_{k=\lceil nx-Tn^{\alpha}\rceil}^{[nx+Tn^{\alpha}]} \frac{S\left(n^{1-\alpha}\left(x-\frac{k}{n}\right)\right)}{I^{*}n^{\alpha}} \right).$$

Thus

$$\gamma \leq \sum_{k=\lceil nx-Tn^{\alpha}\rceil}^{[nx+Tn^{\alpha}]} D\left(f\left(\frac{k}{n}\right) \odot \frac{S\left(n^{1-\alpha}\left(x-\frac{k}{n}\right)\right)}{I^{*}n^{\alpha}}, \right.$$

$$\left. f(x) \odot \frac{S\left(n^{1-\alpha}\left(x-\frac{k}{n}\right)\right)}{I^{*}n^{\alpha}} \right)$$

$$= \sum_{k=\lceil nx-Tn^{\alpha}\rceil}^{[nx+Tn^{\alpha}]} \frac{S\left(n^{1-\alpha}\left(x-\frac{k}{n}\right)\right)}{I^{*}n^{\alpha}} D\left(f\left(\frac{k}{n}\right), f(x) \right)$$

$$\leq \sum_{k=\lceil nx-Tn^{\alpha}\rceil}^{[nx+Tn^{\alpha}]} \frac{S\left(n^{1-\alpha}\left(x-\frac{k}{n}\right)\right)}{I^{*}n^{\alpha}} \omega_{1}^{(\mathcal{F})}\left(f, \left|\frac{k}{n}-x\right| \right)$$

$$\leq \left(\sum_{k=\lceil nx-Tn^{\alpha}\rceil}^{[nx+Tn^{\alpha}]} \frac{S\left(n^{1-\alpha}\left(x-\frac{k}{n}\right)\right)}{I^{*}n^{\alpha}} \right) \omega_{1}^{(\mathcal{F})}\left(f, \frac{T}{n^{1-\alpha}} \right)$$

$$\leq \left(\frac{S(T)}{I^{*}n^{\alpha}} \right) \omega_{1}^{(\mathcal{F})}\left(f, \frac{T}{n^{1-\alpha}} \right) \left(\sum_{k=\lceil nx-Tn^{\alpha}\rceil}^{[nx+Tn^{\alpha}]} 1 \right)$$

$$\overset{(15.20)}{\leq} \frac{S(T)}{I^{*}n^{\alpha}} \omega_{1}^{(\mathcal{F})}\left(f, \frac{T}{n^{1-\alpha}} \right) (2Tn^{\alpha}+1).$$

I.e.,

$$\gamma \leq \frac{S(T)}{I^{*}} \left(2T + \frac{1}{n^{\alpha}} \right) \omega_{1}^{(\mathcal{F})}\left(f, \frac{T}{n^{1-\alpha}} \right). \qquad (15.21)$$

Using (15.21) into $(*)$ we have established (15.19). $\qquad \square$

Remark 15.25. From Lemma 2.2, p. 79, [9], we have that

$$\lim_{n\to+\infty} \left| \sum_{k=\lceil nx-Tn^{\alpha}\rceil}^{[nx+Tn^{\alpha}]} \frac{S\left(n^{1-\alpha}\left(x-\frac{k}{n}\right)\right)}{I^{*}n^{\alpha}} - 1 \right| = 0. \qquad (15.22)$$

Let $f \in C_{\mathcal{F}}^{U}(\mathbb{R})$, then from (15.19) as $n \to +\infty$, we obtain the pointwise convergence with rates of $(G_n(f)(x))$ to $f(x)$, where $x \in \mathbb{R}$.

As a related final main result we give

Theorem 15.26. *Let* $x \in \mathbb{R}$, $T > 0$, $n \in \mathbb{N}$ *such that* $n \geq \max(T + |x|,$ $T^{-1/\alpha})$. *Let* $f \in C_{\mathcal{F}}^{NU}(\mathbb{R})$ *or* $f \in C_{\mathcal{F}}^{NB}(\mathbb{R})$, $N \in \mathbb{N}$. *Then*

$$
D(G_n(f)(x), f(x)) \leq \left| \sum_{k=\lceil nx-Tn^\alpha \rceil}^{\lceil nx+Tn^\alpha \rceil} \frac{S\left(n^{1-\alpha}\left(x - \frac{k}{n}\right)\right)}{I^* n^\alpha} - 1 \right|
$$

$$
\cdot \left(D(f(x), \tilde{o}) + \omega_1^{(\mathcal{F})}\left(f, \frac{T}{n^{1-\alpha}}\right) \right) + \omega_1^{(\mathcal{F})}\left(f, \frac{T}{n^{1-\alpha}}\right)
$$

$$
+ \frac{S(T)}{I^*}\left(2T + \frac{1}{n^\alpha}\right) \left\{ \sum_{j=1}^{N-1} \frac{2^j T^j}{j! n^{(1-\alpha)j}} \left[D(f^{(j)}(x), \tilde{o}) \right. \right.
$$

$$
\left. + \omega_1^{(\mathcal{F})}\left(f^{(j)}, \frac{T}{n^{1-\alpha}}\right) \right] \right\} + \frac{2^N T^N S(T)}{N! I^* n^{(1-\alpha)N}}\left(2T + \frac{1}{n^\alpha}\right)
$$

$$
\cdot \left\{ D(f^{(N)}(x), \tilde{o}) + 3\omega_1^{(\mathcal{F})}\left(f^{(N)}, \frac{T}{n^{1-\alpha}}\right) \right\}. \tag{15.23}
$$

Note. When $f \in C_{\mathcal{F}}^{NU}(\mathbb{R})$ and $n \to +\infty$ from (15.23) we get the pointwise convergence with rates of $(G_n f)(x) \to f(x)$, $x \in \mathbb{R}$.

Proof (of Theorem 15.26). Again we see, that the terms in the sum (15.18) are nonzero iff $n^{1-\alpha}\left|x - \frac{k}{n}\right| \leq T$, iff

$$
-\frac{T}{n^{1-\alpha}} \leq x - \frac{k}{n} \leq \frac{T}{n^{1-\alpha}}.
$$

Therefore here $x - \frac{T}{n^{1-\alpha}} \leq \frac{k}{n}$, $x \in \mathbb{R}$. Using again the fuzzy Taylor formula (Theorem 15.14) we have

$$
f\left(\frac{k}{n}\right) = \sum_{j=0}^{N-1}{}^* f^{(j)}\left(x - \frac{T}{n^{1-\alpha}}\right) \odot \frac{\left(\frac{k}{n} - x + \frac{T}{n^{1-\alpha}}\right)^j}{j!} \oplus R_N\left(x - \frac{T}{n^{1-\alpha}}, \frac{k}{n}\right),
$$

where

$$
R_N\left(x - \frac{T}{n^{1-\alpha}}, \frac{k}{n}\right) :=
$$

$$
(FR)\int_{x-\frac{T}{n^{1-\alpha}}}^{\frac{k}{n}} \left(\int_{x-\frac{T}{n^{1-\alpha}}}^{s_1} \left(\int_{x-\frac{T}{n^{1-\alpha}}}^{s_{N-1}} f^{(N)}(s_N) ds_N \right) \cdots \right) ds_1.
$$

Thus

$$
\sum_{k=\lceil nx-Tn^\alpha \rceil}^{* \, [nx+Tn^\alpha]} f\left(\frac{k}{n}\right) \odot \frac{S\left(n^{1-\alpha}\left(x-\frac{k}{n}\right)\right)}{I^*n^\alpha}
$$

$$
= \sum_{k=\lceil nx-Tn^\alpha \rceil}^{* \, [nx+Tn^\alpha]} \sum_{j=0}^{* \, N-1} f^{(j)}\left(x-\frac{T}{n^{1-\alpha}}\right) \odot \frac{\left(\frac{k}{n}-x+\frac{T}{n^{1-\alpha}}\right)^j}{j!}
$$

$$
\odot \frac{S\left(n^{1-\alpha}\left(x-\frac{k}{n}\right)\right)}{I^*n^\alpha} \oplus \sum_{k=\lceil nx-Tn^\alpha \rceil}^{* \, [nx+Tn^\alpha]} R_N\left(x-\frac{T}{n^{1-\alpha}},\frac{k}{n}\right) \odot \frac{S\left(n^{1-\alpha}\left(x-\frac{k}{n}\right)\right)}{I^*n^\alpha}.
$$

Hence

$$
(G_n f)(x) = \sum_{j=0}^{* \, N-1} \sum_{k=\lceil nx-Tn^\alpha \rceil}^{* \, [nx+Tn^\alpha]} f^{(j)}\left(x-\frac{T}{n^{1-\alpha}}\right) \odot \frac{\left(\frac{k}{n}-x+\frac{T}{n^{1-\alpha}}\right)^j}{j!}
$$

$$
\odot \frac{S\left(n^{1-\alpha}\left(x-\frac{k}{n}\right)\right)}{I^*n^\alpha}
$$

$$
\oplus \sum_{k=\lceil nx-Tn^\alpha \rceil}^{* \, [nx+Tn^\alpha]} R_N\left(x-\frac{T}{n^{1-\alpha}},\frac{k}{n}\right) \odot \frac{S\left(n^{1-\alpha}\left(x-\frac{k}{n}\right)\right)}{I^*n^\alpha}.
$$

Next we estimate

$$
D((G_n(f))(x), f(x)) \le
$$

$$
D\left((G_n(f))(x), \sum_{k=\lceil nx-Tn^\alpha \rceil}^{* \, [nx+Tn^\alpha]} f\left(x-\frac{T}{n^{1-\alpha}}\right) \odot \frac{S\left(n^{1-\alpha}\left(x-\frac{k}{n}\right)\right)}{I^*n^\alpha}\right)
$$

$$
+ D\left(f\left(x-\frac{T}{n^{1-\alpha}}\right) \odot \sum_{k=\lceil nx-Tn^\alpha \rceil}^{[nx+Tn^\alpha]} \frac{S\left(n^{1-\alpha}\left(x-\frac{k}{n}\right)\right)}{I^*n^\alpha},\right.
$$

$$
\left. f\left(x-\frac{T}{n^{1-\alpha}}\right) \odot 1\right) + D\left(f\left(x-\frac{T}{n^{1-\alpha}}\right), f(x)\right)
$$

$$
\le D\left(\sum_{j=1}^{* \, N-1} \sum_{k=\lceil nx-Tn^\alpha \rceil}^{* \, [nx+Tn^\alpha]} f^{(j)}\left(x-\frac{T}{n^{1-\alpha}}\right) \odot \frac{\left(\frac{k}{n}-x+\frac{T}{n^{1-\alpha}}\right)^j}{j!}\right.
$$

$$
\odot \frac{S\left(n^{1-\alpha}\left(x-\frac{k}{n}\right)\right)}{I^*n^\alpha}
$$

$$
\left. \oplus \sum_{k=\lceil nx-Tn^\alpha \rceil}^{* \, [nx+Tn^\alpha]} R_N\left(x-\frac{T}{n^{1-\alpha}},\frac{k}{n}\right) \odot \frac{S\left(n^{1-\alpha}\left(x-\frac{k}{n}\right)\right)}{I^*n^\alpha}, \tilde{o}\right)
$$

$$+ \left| \sum_{k=\lceil nx-Tn^\alpha \rceil}^{[nx+Tn^\alpha]} \frac{S\left(n^{1-\alpha}\left(x - \frac{k}{n}\right)\right)}{I^* n^\alpha} - 1 \right| \cdot D\left(f\left(x - \frac{T}{n^{1-\alpha}}\right), \tilde{o}\right)$$

$$+ \omega_1^{(\mathcal{F})}\left(f, \frac{T}{n^{1-\alpha}}\right)$$

$$\leq \sum_{j=1}^{N-1} \sum_{k=\lceil nx-Tn^\alpha \rceil}^{[nx+Tn^\alpha]} \frac{\left(\frac{k}{n} - x + \frac{T}{n^{1-\alpha}}\right)^j}{j!} \frac{S\left(n^{1-\alpha}\left(x - \frac{k}{n}\right)\right)}{I^* n^\alpha}$$

$$D\left(f^{(j)}\left(x - \frac{T}{n^{1-\alpha}}\right), \tilde{o}\right)$$

$$+ \sum_{k=\lceil nx-Tn^\alpha \rceil}^{[nx+Tn^\alpha]} \frac{S\left(n^{1-\alpha}\left(x - \frac{k}{n}\right)\right)}{I^* n^\alpha} D\left(R_N\left(x - \frac{T}{n^{1-\alpha}}, \frac{k}{n}\right), \tilde{o}\right)$$

$$+ \left| \sum_{k=\lceil nx-Tn^\alpha \rceil}^{[nx+Tn^\alpha]} \frac{S\left(n^{1-\alpha}\left(x - \frac{k}{n}\right)\right)}{I^* n^\alpha} - 1 \right| \left[D(f(x), \tilde{o}) + \omega_1^{(\mathcal{F})}\left(f, \frac{T}{n^{1-\alpha}}\right) \right]$$

$$+ \omega_1^{(\mathcal{F})}\left(f, \frac{T}{n^{1-\alpha}}\right)$$

$$\leq \left\{ \sum_{j=1}^{N-1} \frac{2^j T^j}{j! n^{(1-\alpha)j}} \left[D(f^{(j)}(x), \tilde{o}) + \omega_1^{(\mathcal{F})}\left(f^{(j)}, \frac{T}{n^{1-\alpha}}\right) \right] \right\}$$

$$\frac{S(T)}{I^*}\left(2T + \frac{1}{n^\alpha}\right)$$

$$+ \sum_{k=\lceil nx-Tn^\alpha \rceil}^{[nx+Tn^\alpha]} \frac{S\left(n^{1-\alpha}\left(x - \frac{k}{n}\right)\right)}{I^* n^\alpha} D\left(R_N\left(x - \frac{T}{n^{1-\alpha}}, \frac{k}{n}\right), \tilde{o}\right)$$

$$+ \left| \sum_{k=\lceil nx-Tn^\alpha \rceil}^{[nx+Tn^\alpha]} \frac{S\left(n^{1-\alpha}\left(x - \frac{k}{n}\right)\right)}{I^* n^\alpha} - 1 \right|$$

$$\cdot \left[D(f(x), \tilde{o}) + \omega_1^{(\mathcal{F})}\left(f, \frac{T}{n^{1-\alpha}}\right) \right] + \omega_1^{(\mathcal{F})}\left(f, \frac{T}{n^{1-\alpha}}\right) =: (*).$$

Using (15.12) and (15.20) we get

$$\sum_{k=\lceil nx-Tn^\alpha \rceil}^{[nx+Tn^\alpha]} \frac{S\left(n^{1-\alpha}\left(x - \frac{k}{n}\right)\right)}{I^* n^\alpha} D\left(R_N\left(x - \frac{T}{n^{1-\alpha}}, \frac{k}{n}\right), \tilde{o}\right) \qquad (15.24)$$

$$\leq \frac{2^N T^N S(T)}{N! I^* n^{N(1-\alpha)}}\left(2T + \frac{1}{n^\alpha}\right) \left\{ D(f^{(N)}(x), \tilde{o}) + 3\omega_1^{(\mathcal{F})}\left(f^{(N)}, \frac{T}{n^{1-\alpha}}\right) \right\}.$$

Finally using (15.24) into (*) we obtain (15.23). □

Note. In Chapter 16, see Theorems 16.13, 16.14 and Corollary 16.15, we present an improvement over Theorems 15.19, 15.26 and Corollary 15.22

here, respectively. The reason for this improvement and simplification is that there, in Chapter 16, we use real analysis results, however here we use only the fuzzy method and setting. We feel both have their own merits and so we present them.

16

HIGHER DEGREE OF FUZZY APPROXIMATION BY FUZZY WAVELET TYPE AND NEURAL NETWORK OPERATORS

In this chapter are studied in terms of fuzzy high approximation to the unit several basic sequences of fuzzy wavelet type operators and fuzzy neural network operators. These operators are fuzzy analogs of earlier studied real ones. The produced results generalize earlier real ones into the fuzzy setting. Here the high order fuzzy pointwise convergence with rates to the fuzzy unit operator is established through fuzzy inequalities involving the fuzzy modulus of continuity of the Nth order ($N \geq 1$) H-fuzzy derivative of the engaged fuzzy number valued function. At the end we present a related L_p result for fuzzy neural network operators. This chapter is based on [16].

16.1 Introduction

We need the following results. They motivate this chapter and we generalize them into the Fuzzy setting. Theorems 16.1–16.4 deal with wavelet type operators.

Theorem 16.1 ([9], [8]). *Let $f \in C^N(\mathbb{R})$, $N \geq 1$, $x \in \mathbb{R}$ and $k \in \mathbb{Z}$. Let φ be a bounded function of compact support $\subseteq [-a, a]$, $a > 0$ such that*

$\sum_{j=-\infty}^{\infty} \varphi(x-j) = 1$ *all* $x \in \mathbb{R}$. *Suppose* $\varphi \geq 0$. *Call*

$$(B_k(f))(x) := \sum_{j=-\infty}^{\infty} f\left(\frac{j}{2^k}\right) \varphi(2^k x - j). \qquad (16.1)$$

Then

$$|(B_k(f))(x) - f(x)| \leq \sum_{i=1}^{N} \frac{|f^{(i)}(x)|}{i!} \frac{a^i}{2^{ki}} + \frac{a^N}{2^{kN} N!} \omega_1\left(f^{(N)}, \frac{a}{2^k}\right), \quad (16.2)$$

which is attained by constant function.

Theorem 16.2 ([9], [8]). *Same assumptions as in Theorem 16.1. Additionally, suppose that* φ *is Lebesgue measurable (then* $\int_{-\infty}^{+\infty} \varphi(x)dx = 1$*). Define*

$$\varphi_{kj}(x) := \quad 2^{k/2} \varphi(2^k x - j) \quad \text{all } k, j \in \mathbb{Z},$$

$$\langle f, \varphi_{kj} \rangle := \quad \int_{-\infty}^{\infty} f(t) \varphi_{kj}(t) dt, \qquad (16.3)$$

and

$$(A_k(f))(x) := \sum_{j=-\infty}^{\infty} \langle f, \varphi_{kj} \rangle \varphi_{kj}(x)$$

$$= \sum_{j=-\infty}^{\infty} \left(\int_{-\infty}^{\infty} f\left(\frac{u}{2^k}\right) \varphi(u-j)du \right) \varphi(2^k x - j). \qquad (16.4)$$

Then

$$|(A_k(f))(x) - f(x)| \leq \sum_{i=1}^{N} \frac{|f^{(i)}(x)|}{i!} \frac{a^i}{2^{i(k-1)}} + \frac{2^N}{N! 2^{N(k-1)}} \omega_1\left(f^{(N)}, \frac{a}{2^{k-1}}\right), \qquad (16.5)$$

which is attained by constants.

Theorem 16.3 ([9], [8]). *Same assumptions as in Theorem 16.1. Define*

$$C_k(f)(x) := \sum_{j=-\infty}^{\infty} \gamma_{kj}(f) \varphi(2^k x - j)$$

$$:= \sum_{j=-\infty}^{\infty} \left(2^k \int_{2^{-k}j}^{2^{-k}(j+1)} f(t)dt \right) \varphi(2^k x - j). \qquad (16.6)$$

That is,

$$\gamma_{kj}(f) := 2^k \int_{2^{-k}j}^{2^{-k}(j+1)} f(t)dt = 2^k \int_{0}^{2^{-k}} f\left(t + \frac{j}{2^k}\right) dt. \qquad (16.7)$$

Then $(x \in \mathbb{R}, \ k \in \mathbb{Z})$

$$|(C_k(f))(x) - f(x)| \le \sum_{i=1}^{N} \frac{|f^{(i)}(x)|(a+1)^i}{i!2^{ki}} + \frac{(a+1)^N}{2^{kN}N!}\omega_1\left(f^{(N)}, \frac{a+1}{2^k}\right),$$

(16.8)

which is attained by constants.

Theorem 16.4 ([9], [8]). *Same assumptions as in Theorem 16.1. Define* $(k, j \in \mathbb{Z}, \ x \in \mathbb{R})$

$$(D_k f)(x) := \sum_{j=-\infty}^{\infty} \delta_{kj}(f)\varphi(2^k x - j),$$

(16.9)

where

$$\delta_{kj}(f) := \sum_{\tilde{r}=0}^{n} w_{\tilde{r}} f\left(\frac{j}{2^k} + \frac{\tilde{r}}{2^k n}\right),$$

(16.10)

$n \in \mathbb{N}, \ w_{\tilde{r}} \ge 0, \ \sum_{r=0}^{n} w_{\tilde{r}} = 1.$ *Then*

$$|(D_k f)(x) - f(x)| \le \sum_{i=1}^{N} \frac{|f^{(i)}(x)|}{i!} \frac{(a+1)^i}{2^{ki}} + \frac{(a+1)^N}{2^{kN}N!}\omega_1\left(f^{(N)}, \frac{(a+1)}{2^k}\right),$$

(16.11)

which is attained by constants.

Example of φ's:
  (i)

$$\varphi(x) := \chi_{\left[-\frac{1}{2},\frac{1}{2}\right)}(x) = \begin{cases} 1, & x \in \left[-\frac{1}{2}, \frac{1}{2}\right) \\ \\ 0, & \text{elsewhere}, \end{cases}$$

(16.12)

the characteristic function;
  (ii)

$$\varphi(x) := \begin{cases} 1-x, & 0 \le x \le 1 \\ 1+x, & -1 \le x \le 0 \\ 0, & \text{elsewhere}, \end{cases}$$

(16.13)

the hat function.

The next in this section come from [9], [6], [47].

Here we consider functions $f \colon \mathbb{R} \to \mathbb{R}$ that are continuous.

We mention the pointwise convergence with rates over the real line, to the unit operator, of the *univariate Cardaliaguet–Euvrard neural network operators* (see [9], [6], [47])

$$(F_n(f))(x) := f_n(x) := \sum_{k=-n^2}^{n^2} \frac{f(k/n)}{In^\alpha} \cdot b\left(n^{1-\alpha} \cdot \left(x - \frac{k}{n}\right)\right),$$

(16.14)

where $0 < \alpha < 1$ and $x \in \mathbb{R}$, $n \in \mathbb{N}$, $I := \int_{-T}^{T} b(t)dt$.

Denote by $[\cdot]$ the integral part of a number and by $\lceil \cdot \rceil$ its ceiling. Set $b^* := b(0)$ the maximum of $b(x)$ which is a centered bell-shaped function, of compact support $[-T, T]$, $T > 0$, see Definition 4.1 of [17].

We have

Theorem 16.5. Let $x \in \mathbb{R}$, $T > 0$, and $n \in \mathbb{N}$ such that $n \geq \max(T + |x|, T^{-1/\alpha})$. Let $f \in C^N(\mathbb{R})$, $N \in \mathbb{N}$, such that $f^{(N)}$ is a uniformly continuous function. Then

$$|f_n(x) - f(x)| = \left| \sum_{k=\lceil nx-Tn^\alpha \rceil}^{[nx+Tn^\alpha]} \frac{f(k/n)}{I \cdot n^\alpha} \cdot b\left(n^{1-\alpha} \cdot \left(x - \frac{k}{n}\right)\right) - f(x) \right|$$

$$\leq |f(x)| \cdot \left| \sum_{k=\lceil nx-Tn^\alpha \rceil}^{[nx+Tn^\alpha]} \frac{1}{I \cdot n^\alpha} \cdot b\left(n^{1-\alpha} \cdot \left(x - \frac{k}{n}\right)\right) - 1 \right|$$

$$+ \frac{b^*}{I} \cdot \left(2T + \frac{1}{n^\alpha}\right) \cdot \left(\sum_{j=1}^{N} \frac{|f^{(j)}(x)| \cdot T^j}{n^{j(1-\alpha)} \cdot j!} \right) \qquad (16.15)$$

$$+ \omega_1\left(f^{(N)}, \frac{T}{n^{1-\alpha}}\right) \cdot \frac{T^N}{N! \cdot n^{N(1-\alpha)}} \cdot \frac{b^*}{I} \cdot \left(2T + \frac{1}{n^\alpha}\right).$$

Let S be a squashing function of compact support $[-T, T]$, $T > 0$, see Definition 4.2 of [17]. Let $f: \mathbb{R} \to \mathbb{R}$ be continuous. Suppose that

$$I^* := \int_{-T}^{T} S(t)dt > 0.$$

For $x \in \mathbb{R}$ we define the "univariate squashing neural network operator" (see [9], [6], [47])

$$(G_n(f))(x) := \sum_{k=-n^2}^{n^2} \frac{f(k/n)}{I^* \cdot n^\alpha} \cdot S\left(n^{1-\alpha} \cdot \left(x - \frac{k}{n}\right)\right), \qquad (16.16)$$

$0 < \alpha < 1$ and $n \in \mathbb{N}$: $n \geq \max(T + |x|, T^{-1/\alpha})$. It is easy to see that

$$(G_n(f))(x) = \sum_{k=\lceil nx-Tn^\alpha \rceil}^{[nx+Tn^\alpha]} \frac{f(k/n)}{I^* \cdot n^\alpha} \cdot S\left(n^{1-\alpha} \cdot \left(x - \frac{k}{n}\right)\right). \qquad (16.17)$$

Here we mention the pointwise convergence with rates of $(G_n(f))(x) \to f(x)$, as $n \to +\infty$, $x \in \mathbb{R}$.

Theorem 16.6. Let $x \in \mathbb{R}$, $T > 0$, and $n \in \mathbb{N}$, such that $n \geq \max(T + |x|, T^{-1/\alpha})$. Let $f \in C^N(\mathbb{R})$, $N \in \mathbb{N}$, such that $f^{(N)}$ is a uniformly contin-

uous function. Then

$$|(G_n(f))(x) - f(x)| \leq |f(x)|$$

$$\left| \sum_{k=\lceil nx - Tn^\alpha \rceil}^{\lfloor nx + Tn^\alpha \rfloor} \frac{1}{I^* \cdot n^\alpha} \cdot S\left(n^{1-\alpha} \cdot \left(x - \frac{k}{n}\right)\right) - 1 \right|$$

$$+ \frac{S(T)}{I^*} \cdot \left(2T + \frac{1}{n^\alpha}\right) \cdot \left(\sum_{j=1}^{N} \frac{|f^{(j)}(x)| \cdot T^j}{j! \cdot n^{j \cdot (1-\alpha)}}\right)$$

$$+ \omega_1\left(f^{(N)}, \frac{T}{n^{1-\alpha}}\right) \cdot \frac{T^N}{N! \cdot n^{N \cdot (1-\alpha)}} \qquad (16.18)$$

$$\cdot \frac{S(T)}{I^*} \cdot \left(2T + \frac{1}{n^\alpha}\right).$$

In this chapter we study the same operators B_k, A_k, C_k, D_k F_n, G_n, $k \in \mathbb{Z}$, $n \in \mathbb{N}$, respectively, when they act on *fuzzy real number valued functions*, proving their high order approximation to the unit operator. So according to the context one can understand clearly, when these operators are applied to real valued functions, and when they are applied to fuzzy real valued functions. Here we apply real analysis results into the fuzzy setting.

Remark 16.7. (i) Here $C_U(\mathbb{R})$ denotes the uniformly continuous functions from $\mathbb{R}$ into $\mathbb{R}$. If $f \in C_{\mathcal{F}}^U(\mathbb{R})$, then $f_{\pm}^{(r)} \in C_U(\mathbb{R})$, $\forall r \in [0,1]$. Also one has $\omega_1(f_{\pm}^{(r)}, \delta) < +\infty$ for any $\delta > 0$.

(ii) If $f : \mathbb{R} \to \mathbb{R}_{\mathcal{F}}$ is fuzzy continuous then $f_{\pm}^{(r)} : \mathbb{R} \to \mathbb{R}$ are continuous, $\forall r \in [0,1]$.

Note. Let $f \in C^N(\mathbb{R}, \mathbb{R}_{\mathcal{F}})$, $N \geq 1$. Then by Theorem 8 of [15], [71], we have $f_{\pm}^{(r)} \in C^N(\mathbb{R})$, for any $r \in [0,1]$.

Definition 16.8. Denote by $C_{\mathcal{F}}^{NU}(\mathbb{R}) := \{f : \mathbb{R} \to \mathbb{R}_{\mathcal{F}} \mid$ such that all fuzzy derivatives $f^{(i)} : \mathbb{R} \to \mathbb{R}_{\mathcal{F}}$, $i = 0, 1, \ldots, N$ exist and are fuzzy continuous, furthermore $f^{(N)}$ is fuzzy uniformly continuous from $\mathbb{R}$ into $\mathbb{R}_{\mathcal{F}}\}$, $N \geq 1$.

16.2   Main results

We present the first main result.

Theorem 16.9. *Let* $f \in C_{\mathcal{F}}^{NU}(\mathbb{R})$, $N \geq 1$, $x \in \mathbb{R}$ *and* $k \in \mathbb{Z}$. *Let the scaling function* $\varphi(x)$ *a real valued bounded function with* supp $\varphi(x) \subseteq [-a, a]$, $0 < a < +\infty$, $\varphi(x) \geq 0$, *such that* $\sum_{j=-\infty}^{\infty} \varphi(x - j) \equiv 1$ *on* $\mathbb{R}$. *Consider the* *fuzzy wavelet type operator*

$$(B_k f)(x) := \sum_{j=-\infty}^{\infty *} f\left(\frac{j}{2^k}\right) \odot \varphi(2^k x - j). \qquad (16.19)$$

Then

$$D((B_k f)(x), f(x)) \le \sum_{i=1}^{N} \frac{D(f^{(i)}(x), \tilde{o})}{i!} \frac{a^i}{2^{ki}} + \frac{a^N}{2^{kN} N!} \omega_1^{(\mathcal{F})}\left(f^{(N)}, \frac{a}{2^k}\right) =: \lambda_1(x).$$
$$(16.20)$$

As $k \to +\infty$ *we get* $\omega_1^{(\mathcal{F})}\left(f^{(N)}, \frac{a}{2^k}\right) \to 0$ *and* $D((B_k f)(x), f(x)) \to 0$, *i.e.*
$\lim_{k \to +\infty} B_k f = f$, *pointwise with rates.*

Proof. Since φ is of compact support (16.19) is a finite sum. Thus for $r \in [0, 1]$ we have

$$
\begin{aligned}
[(B_k f)(x)]^r &= \sum_{j=-\infty}^{\infty} \left[f\left(\frac{j}{2^k}\right)\right]^r \varphi(2^k x - j) \\
&= \sum_{j=-\infty}^{\infty} \left[\left(f\left(\frac{j}{2^k}\right)\right)_-^{(r)}, \left(f\left(\frac{j}{2^k}\right)\right)_+^{(r)}\right] \varphi(2^k x - j) \\
&= \left[\sum_{j=-\infty}^{\infty} \left(f\left(\frac{j}{2^k}\right)\right)_-^{(r)} \varphi(2^k x - j),\right. \\
&\qquad \left. \sum_{j=-\infty}^{\infty} \left(f\left(\frac{j}{2^k}\right)\right)_+^{(r)} \varphi(2^k x - j)\right] \\
&= [(B_k(f)_-^{(r)})(x), (B_k(f)_+^{(r)})(x)].
\end{aligned}
$$

I.e.

$$(B_k f)_\pm^{(r)} = B_k(f_\pm^{(r)}), \quad \forall r \in [0, 1]. \qquad (16.21)$$

We observe by Section 2 of [17] that

$$
\begin{aligned}
D((B_k f)(x), f(x)) &= \sup_{r \in [0,1]} \max\{|(B_k f)_-^{(r)}(x) - f_-^{(r)}(x)|, \\
&\qquad |(B_k f)_+^{(r)}(x) - f_+^{(r)}(x)|\} \\
&\overset{(16.21)}{=} \sup_{r \in [0,1]} \max\{|(B_k(f_-^{(r)}))(x) - f_-^{(r)}(x)|, \\
&\qquad |(B_k(f_+^{(r)}))(x) - f_+^{(r)}(x)|\}.
\end{aligned}
$$

At this point we notice by Theorem 8 of [15] that for any $r \in [0, 1]$ the real function $f_\pm^{(r)} \in C^N(\mathbb{R})$ and $(f_\pm^{(r)})^{(N)} \in C_U(\mathbb{R})$. Therefore we can apply

Theorem 16.1 (16.2) to obtain

$$D((B_k f)(x), f(x))$$

$$\leq \sup_{r \in [0,1]} \max \left\{ \left(\sum_{i=1}^{N} \frac{|(f_-^{(r)}(x))^{(i)}|}{i!} \frac{a^i}{2^{ki}} + \frac{a^N}{2^{kN} N!} \omega \left((f_-^{(r)})^{(N)}, \frac{a}{2^k} \right) \right), \right.$$

$$\left. \left(\sum_{i=1}^{N} \frac{|(f_+^{(r)}(x))^{(i)}|}{i!} \frac{a^i}{2^{ki}} + \frac{a^N}{2^{kN} N!} \omega_1 \left((f_+^{(r)})^{(N)}, \frac{a}{2^k} \right) \right) \right\}$$

$$(\text{by (33) of [15]}) = \sup_{r \in [0,1]} \max \left\{ \left(\sum_{i=1}^{N} \frac{|(f_-^{(i)}(x))^{(r)}|}{i!} \frac{a^i}{2^{ki}} \right. \right.$$

$$\left. + \frac{a^N}{2^{kN} N!} \omega_1 \left((f_-^{(N)})^{(r)}, \frac{a}{2^k} \right) \right),$$

$$\left. \left(\sum_{i=1}^{N} \frac{|(f_+^{(i)}(x))^{(r)}|}{i!} \frac{a^i}{2^{ki}} + \frac{a^N}{2^{kN} N!} \omega_1 \left((f_+^{(N)})^{(r)}, \frac{a}{2^k} \right) \right) \right\}$$

$$(\text{by (29) of [15]}) \leq \sum_{i=1}^{N} \frac{a^i}{i! 2^{ki}} \sup_{r \in [0,1]} \max \left(|(f_-^{(i)}(x))^{(r)}|, |(f_+^{(i)}(x))^{(r)}| \right)$$

$$+ \frac{a^N}{2^{kN} N!} \sup_{r \in [0,1]} \max \left\{ \omega_1 \left((f_-^{(N)})^{(r)}, \frac{a}{2^k} \right), \omega_1 \left(f_+^{(N)})^{(r)}, \frac{a}{2^k} \right) \right\}$$

$$(\text{by (25) \& (26) of [15]}) = \sum_{i=1}^{N} \frac{a^i}{i! 2^{ki}} D(f^{(i)}(x), \tilde{o})$$

$$+ \frac{a^N}{2^{kN} N!} \omega_1^{(\mathcal{F})} \left(f^{(N)}, \frac{a}{2^k} \right).$$

We have established (16.20). □

We continue with

Theorem 16.10. *Let* $f \in C_{\mathcal{F}}^{NU}(\mathbb{R}) \cap C_b(\mathbb{R}, \mathbb{R}_{\mathcal{F}})$, $N \geq 1$, $x \in \mathbb{R}$ *and* $k \in \mathbb{Z}$. *Let the scaling function* $\varphi(x)$ *a real valued function with* supp $\varphi(x) \subseteq [-a, a]$, $0 < a < +\infty$, φ *is continuous on* $[-a, a]$, $\varphi(x) \geq 0$, *such that* $\sum_{j=-\infty}^{\infty} \varphi(x - j) = 1$ *on* $\mathbb{R}$ *(then* $\int_{-\infty}^{\infty} \varphi(x) dx = 1$*). Define*

$$\varphi_{kj}(t) := 2^{k/2} \varphi(2^k t - j), \quad \text{for } k, j \in \mathbb{Z}, \ t \in \mathbb{R}, \quad (16.22)$$

$$\langle f, \varphi_{kj} \rangle := (FR) \int_{\frac{j-a}{2^k}}^{\frac{j+a}{2^k}} f(t) \odot \varphi_{kj}(t) dt, \quad (16.23)$$

and the fuzzy wavelet type operator

$$(A_k f)(x) := \sum_{j=-\infty}^{\infty} {}^* \langle f, \varphi_{kj} \rangle \odot \varphi_{kj}(x), \quad x \in \mathbb{R}. \quad (16.24)$$

Then

$$D((A_k f)(x), f(x)) \leq \sum_{i=1}^{N} \frac{D(f^{(i)}(x), \tilde{o})}{i!} \frac{a^i}{2^{i(k-1)}} \tag{16.25}$$

$$+ \frac{a^N}{N! 2^{N(k-1)}} \omega_1^{(\mathcal{F})} \left(f^{(N)}, \frac{a}{2^{k-1}} \right)$$

$$= : \lambda_2(x).$$

As $k \to +\infty$ we get $\lim\limits_{k \to +\infty} A_k f = f$, pointwise with rates in the D-metric.

Proof. Since φ is of compact support (16.24) is a finite sum. For any $r \in [0,1]$ by Theorem 16.9 and Lemma 2 of [15] we notice that

$$[\langle f, \varphi_{kj} \rangle]^r = \left[\int_{\frac{j-a}{2^k}}^{\frac{j+a}{2^k}} (f(t) \odot \varphi_{kj}(t))_-^{(r)} dt, \int_{\frac{j-a}{2^k}}^{\frac{j+a}{2^k}} (f(t) \odot \varphi_{kj}(t))_+^{(r)} dt \right]$$

$$= \left[\int_{\frac{j-a}{2^k}}^{\frac{j+a}{2^k}} (f(t))_-^{(r)} \varphi_{kj}(t) dt, \int_{\frac{j-a}{2^k}}^{\frac{j+a}{2^k}} (f(t))_+^{(r)} \varphi_{kj}(t) dt \right].$$

We see that

$$[(A_k f)(x)]^r = \sum_{j=-\infty}^{\infty} [\langle f, \varphi_{kj} \rangle]^r \varphi_{kj}(x)$$

$$= \sum_{j=-\infty}^{\infty} \left[\int_{\frac{j-a}{2^k}}^{\frac{j+a}{2^k}} (f(t))_-^{(r)} \varphi_{kj}(t) dt, \int_{\frac{j-a}{2^k}}^{\frac{j+a}{2^k}} (f(t))_+^{(r)} \varphi_{kj}(t) dt \right] \varphi_{kj}(x)$$

$$= \left[\sum_{j=-\infty}^{\infty} \left(\int_{\frac{j-a}{2^k}}^{\frac{j+a}{2^k}} (f(t))_-^{(r)} \varphi_{kj}(t) dt \right) \varphi_{kj}(x), \right.$$

$$\left. \sum_{j=-\infty}^{\infty} \left(\int_{\frac{j-a}{2^k}}^{\frac{j+a}{2^k}} (f(t))_+^{(r)} \varphi_{kj}(t) dt \right) \varphi_{kj}(x) \right]$$

$$= [(A_k(f_-^{(r)}))(x), (A_k(f_+^{(r)}))(x)].$$

That is we prove that

$$(A_k f)_{\pm}^{(r)} = A_k(f_{\pm}^{(r)}), \quad \forall r \in [0,1]. \tag{16.26}$$

We observe from Section 2 of [17] that

$$D((A_k f)(x), f(x)) = \sup_{r \in [0,1]} \max \{ |(A_k f)_-^{(r)}(x) - f_-^{(r)}(x)|,$$

$$|(A_k f)_+^{(r)}(x) - f_+^{(r)}(x)| \}$$

$$\overset{(16.26)}{=} \sup_{r \in [0,1]} \max \{ |(A_k(f_-^{(r)}))(x) - f_-^{(r)}(x)|,$$

$$|(A_k(f_+^{(r)}))(x) - f_+^{(r)}(x)| \}.$$

Here by Theorem 8 of [15] for any $r \in [0,1]$ $f_\pm^{(r)} \in C^N(\mathbb{R})$ and $(f_\pm^{(r)})^{(N)} \in C_U(\mathbb{R})$. Hence by applying Theorem 16.2 (16.5) we get

$$D((A_k f)(x), f(x))$$

$$\leq \sup_{r \in [0,1]} \max\left\{ \left(\sum_{i=1}^{N} \frac{|(f_-^{(r)}(x))^{(i)}|}{i!} \frac{a^i}{2^{i(k-1)}} \right. \right.$$

$$\left. + \frac{a^N}{N!2^{N(k-1)}} \omega_1 \left((f_-^{(r)})^{(N)}, \frac{a}{2^{k-1}} \right) \right),$$

$$\left(\sum_{i=1}^{N} \frac{|(f_+^{(r)}(x))^{(i)}|}{i!} \frac{a^i}{2^{i(k-1)}} \right.$$

$$\left. \left. + \frac{a^N}{N!2^{N(k-1)}} \omega_1 \left((f_+^{(r)})^{(N)}, \frac{a}{2^{k-1}} \right) \right) \right\}$$

$$\text{(by (33) of [15])} = \sup_{r \in [0,1]} \max\left\{ \left(\sum_{i=1}^{N} \frac{|(f_-^{(i)}(x))^{(r)}|}{i!} \frac{a^i}{2^{i(k-1)}} \right. \right.$$

$$\left. + \frac{a^N}{N!2^{N(k-1)}} \omega_1 \left((f_-^{(N)})^{(r)}, \frac{a}{2^{k-1}} \right) \right),$$

$$\left. \left(\sum_{i=1}^{N} \frac{|(f_+^{(i)}(x))^{(r)}|}{i!} \frac{a^i}{2^{i(k-1)}} + \frac{a^N}{N!2^{N(k-1)}} \omega_1 \left((f_+^{(N)})^{(r)}, \frac{a}{2^{k-1}} \right) \right) \right\}$$

$$\text{(by (29) of [15])} \leq \sum_{i=1}^{N} \frac{a^i}{i!2^{i(k-1)}} \sup_{r \in [0,1]} \max\{ |(f_-^{(i)}(x))^{(r)}|, |(f_+^{(i)}(x))^{(r)}| \}$$

$$+ \frac{a^N}{N!2^{N(k-1)}} \sup_{r \in [0,1]} \max\left\{ \omega_1 \left((f_-^{(N)})^{(r)}, \frac{a}{2^{k-1}} \right), \right.$$

$$\left. \omega_1 \left((f_+^{(N)})^{(r)}, \frac{a}{2^{k-1}} \right) \right\}$$

$$\text{(by (24) \& (25) of [15])} = \sum_{i=1}^{N} \frac{a^i}{i!2^{i(k-1)}} D(f^{(i)}(x), \tilde{o})$$

$$+ \frac{a^N}{N!2^{N(k-1)}} \omega_1^{(\mathcal{F})} \left(f^{(N)}, \frac{a}{2^{k-1}} \right).$$

We have proven (16.25).   □

Next we give

Theorem 16.11. *All assumptions here are as in Theorem 16.9. For $k \in \mathbb{Z}$, $x \in \mathbb{R}$ we define the fuzzy wavelet type operator*

$$(C_k f)(x) := \sum_{j=-\infty}^{\infty}{}^* \left(2^k \odot (FR) \int_0^{2^{-k}} f\left(t + \frac{j}{2^k} \right) dt \right) \odot \varphi(2^k x - j).$$

$$(16.27)$$

Then

$$D((C_k f)(x), f(x)) \le \sum_{i=1}^{N} \frac{D(f^{(i)}(x), \tilde{o})}{i!} \frac{(a+1)^i}{2^{ki}} +$$

$$\frac{(a+1)^N}{2^{kN} N!} \omega_1^{(\mathcal{F})}\left(f^{(N)}, \frac{a+1}{2^k}\right) =: \lambda_3(x). \tag{16.28}$$

As $k \to +\infty$ *we get* $\lim_{k \to +\infty} C_k f = f$, *pointwise with rates in the D-metric.*

Proof. Since φ is of compact support (16.27) is a finite sum. So for $r \in [0,1]$ we observe that

$$[(C_k f)(x)]^r = \sum_{j=-\infty}^{\infty} \left[2^k \odot (FR) \int_0^{2^{-k}} f\left(t + \frac{j}{2^k}\right) dt \right]^r \varphi(2^k x - j)$$

$$= \sum_{j=-\infty}^{\infty} \left[2^k \odot (FR) \int_{2^{-k}j}^{2^{-k}(j+1)} f(t) dt \right]^r \varphi(2^k x - j)$$

$$\text{(by (23) of [15])} \quad = \sum_{j=-\infty}^{\infty} \left[2^k \int_{2^{-k}j}^{2^{-k}(j+1)} (f)_-^{(r)}(t) dt, 2^k \int_{2^{-k}j}^{2^{-k}(j+1)} (f)_+^{(r)}(t) dt \right]$$

$$\varphi(2^k x - j)$$

$$= \left[\sum_{j=-\infty}^{\infty} \left(2^k \int_{2^{-k}j}^{2^{-k}(j+1)} (f)_-^{(r)}(t) dt \right) \varphi(2^k x - j), \right.$$

$$\left. \sum_{j=-\infty}^{\infty} \left(2^k \int_{2^{-k}j}^{2^{-k}(j+1)} (f)_+^{(r)}(t) dt \right) \varphi(2^k x - j) \right]$$

$$= [(C_k(f_-^{(r)}))(x), (C_k(f_+^{(r)}))(x)].$$

That is, for any $r \in [0,1]$ we proved that

$$(C_k f)_\pm^{(r)} = C_k(f_\pm^{(r)}). \tag{16.29}$$

Then by Section 2 of [17]

$$D((C_k f)(x), f(x)) \quad = \quad \sup_{r \in [0,1]} \max\{|((C_k f)(x))_-^{(r)} - f_-^{(r)}(x)|,$$

$$|((C_k f)(x))_+^{(r)} - f_+^{(r)}(x)|\}$$

$$\overset{(16.29)}{=} \quad \sup_{r \in [0,1]} \max\{|(C_k(f_-^{(r)}))(x) - f_-^{(r)}(x)|,$$

$$|(C_k(f_+^{(r)}))(x) - f_+^{(r)}(x)|\}.$$

Again by Theorem 8 of [15] for any $r \in [0,1]$ the real functions $f_\pm^{(r)} \in C^N(\mathbb{R})$ and $(f_\pm^{(r)})^{(N)} \in C_U(\mathbb{R})$. Hence by Theorem 16.3 (16.8) we get

$D((C_k f)(x), f(x))$

$$\leq \sup_{r\in[0,1]} \max\left\{\left(\sum_{i=1}^{N} \frac{|(f_-^{(r)}(x))^{(i)}|}{i!} \frac{(a+1)^i}{2^{ki}}\right. \right.$$
$$+ \frac{(a+1)^N}{2^{kN} N!} \omega_1\left((f_-^{(r)})^{(N)}, \frac{a+1}{2^k}\right)\bigg),$$
$$\left.\left(\sum_{i=1}^{N} \frac{|(f_+^{(r)}(x))^{(i)}|}{i!} \frac{(a+1)^i}{2^{ki}} + \frac{(a+1)^N}{2^{kN} N!} \omega_1\left((f_+^{(r)})^{(N)}, \frac{a+1}{2^k}\right)\right)\right\}$$

$$(\text{by (33) of [15]}) = \sup_{r\in[0,1]} \max\left\{\left(\sum_{i=1}^{N} \frac{|(f_-^{(i)}(x))^{(r)}|}{i!} \frac{(a+1)^i}{2^{ki}}\right.\right.$$
$$+ \frac{(a+1)^N}{2^{kN} N!} \omega_1\left((f_-^{(N)})^{(r)}, \frac{a+1}{2^k}\right)\bigg),$$
$$\left.\left(\sum_{i=1}^{N} \frac{|(f_+^{(i)}(x))^{(r)}|}{i!} \frac{(a+1)^i}{2^{ki}} + \frac{(a+1)^N}{2^{kN} N!} \omega_1\left((f_+^{(N)})^{(r)}, \frac{a+1}{2^k}\right)\right)\right\}$$

$$(\text{by (29) of [15]}) \leq \sum_{i=1}^{N} \frac{(a+1)^i}{i! 2^{ki}} \sup_{r\in[0,1]} \max\{|(f_-^{(i)}(x))^{(r)}|, |(f_+^{(i)}(x))^{(r)}|\}$$
$$+ \frac{(a+1)^N}{2^{kN} N!} \sup_{r\in[0,1]} \max\left\{\omega_1\left((f_-^{(N)})^{(r)}, \frac{a+1}{2^k}\right),\right.$$
$$\left.\omega_1\left((f_+^{(N)})^{(r)}, \frac{a+1}{2^k}\right)\right\}$$

$$(\text{by (25) \& (26) of [15]}) = \sum_{i=1}^{N} \frac{(a+1)^i}{i! 2^{ki}} D(f^{(i)}(x), \tilde{o})$$
$$+ \frac{(a+1)^N}{2^{kN} N!} \omega_1^{(\mathcal{F})}\left(f^{(N)}, \frac{a+1}{2^k}\right).$$

That is proving (16.28). □

The last fuzzy wavelet type result follows.

Theorem 16.12. *All assumptions here are as in Theorem 16.9. For $k \in \mathbb{Z}$, $x \in \mathbb{R}$ we define the fuzzy wavelet type operator*

$$(D_k f)(x) := \sum_{j=-\infty}^{\infty}{}^* \delta_{kj}(f) \odot \varphi(2^k x - j), \qquad (16.30)$$

where

$$\delta_{kj}(f) := \sum_{\tilde{r}=0}^{n}{}^* w_{\tilde{r}} \odot f\left(\frac{j}{2^k} + \frac{\tilde{r}}{2^k n}\right), \quad n \in \mathbb{N}, \ w_{\tilde{r}} \geq 0, \ \sum_{\tilde{r}=0}^{n} w_{\tilde{r}} = 1. \ (16.31)$$

Then

$$D((D_k f)(x), f(x)) \leq \sum_{i=1}^{N} \frac{D(f^{(i)}(x), \tilde{o})}{i!} \frac{(a+1)^i}{2^{ki}} \tag{16.32}$$

$$+ \frac{(a+1)^N}{2^{kN} N!} \omega_1^{(\mathcal{F})} \left(f^{(N)}, \frac{a+1}{2^k} \right)$$

$$= : \lambda_4(x).$$

As $k \to +\infty$ we get $\lim_{k \to +\infty} D_k f = f$, pointwise with rates in the D-metric.

Proof. Since φ is of compact support (16.30) is a finite sum. So for $r \in [0, 1]$ we observe that

$$[(D_k f)(x)]^r = \sum_{j=-\infty}^{\infty} [\delta_{kj}(f)]^r \varphi(2^k x - j)$$

$$= \sum_{j=-\infty}^{\infty} \left(\sum_{\tilde{r}=0}^{n} w_{\tilde{r}} \left[\left(f \left(\frac{j}{2^k} + \frac{\tilde{r}}{2^k n} \right) \right)_{-}^{(r)}, \left(f \left(\frac{j}{2^k} + \frac{\tilde{r}}{2^k n} \right) \right)_{+}^{(r)} \right] \right)$$

$$\varphi(2^k x - j)$$

$$= \left[\sum_{j=-\infty}^{\infty} \left(\sum_{\tilde{r}=0}^{n} w_{\tilde{r}} \left(f \left(\frac{j}{2^k} + \frac{\tilde{r}}{2^k n} \right) \right)_{-}^{(r)} \right) \varphi(2^k x - j), \right.$$

$$\left. \sum_{j=-\infty}^{\infty} \left(\sum_{\tilde{r}=0}^{n} w_{\tilde{r}} \left(f \left(\frac{j}{2^k} + \frac{\tilde{r}}{2^k n} \right) \right)_{+}^{(r)} \right) \varphi(2^k x - j) \right]$$

$$= [(D_k(f_{-}^{(r)}))(x), (D_k(f_{+}^{(r)}))(x)].$$

That is, we prove that

$$(D_k f)_{\pm}^{(r)} = D_k(f_{\pm}^{(r)}), \quad \forall r \in [0, 1]. \tag{16.33}$$

Next we see by Section 2 of [17] that

$$
\begin{aligned}
D((D_k f)(x), f(x)) &= \sup_{r \in [0,1]} \max\{|((D_k f)(x))_{-}^{(r)} - f_{-}^{(r)}(x)|, \\
&\qquad |((D_k f)(x))_{+}^{(r)} - f_{+}^{(r)}(x)|\} \\
&\overset{(16.33)}{=} \sup_{r \in [0,1]} \max\{|(D_k(f_{-}^{(r)}))(x) - f_{-}^{(r)}(x)|, \\
&\qquad |(D_k(f_{+}^{(r)}))(x) - f_{+}^{(r)}(x)|\}.
\end{aligned}
$$

Using Theorem 16.4 (16.11) and acting as in the proof of Theorem 16.11 we obtain (16.32). □

In the next let b be a bell-shaped function as in Definition 4.1 of [17]. Let $f \in C(\mathbb{R}, \mathbb{R}_{\mathcal{F}})$ we define the *fuzzy univariate Cardaliaguet–Euvrard neural network operators* by

$$(F_n(f))(x) := \sum_{k=-n^2}^{n^2}{}^{*} f\left(\frac{k}{n}\right) \odot \frac{b(n^{1-\alpha}(x - \frac{k}{n}))}{In^{\alpha}}, \qquad (16.34)$$

where $0 < \alpha < 1$ and $x \in \mathbb{R}$, $n \in \mathbb{N}$. Since b is of compact support $[-T, T]$, $T > 0$ we obtain

$$(F_n(f))(x) = \sum_{k=\lceil nx-Tn^{\alpha}\rceil}^{\lceil nx+Tn^{\alpha}\rceil}{}^{*} f\left(\frac{k}{n}\right) \odot \frac{b(n^{1-\alpha}(x - \frac{k}{n}))}{In^{\alpha}}, \qquad (16.35)$$

for $n \geq T + |x|$.

We present the fuzzy pointwise convergence with rates to the unit of operators F_n.

Theorem 16.13. *Let $x \in \mathbb{R}$, $T > 0$, $n \in \mathbb{N}$ such that $n \geq \max(T + |x|, T^{-1/\alpha})$. Let $f \in C_{\mathcal{F}}^{NU}(\mathbb{R})$, $N \geq 1$. Then*

$$D((F_n(f))(x), f(x)) \leq D(f(x), \tilde{o}) \cdot \left| \sum_{k=\lceil nx-Tn^{\alpha}\rceil}^{\lceil nx+Tn^{\alpha}\rceil} \frac{b(n^{1-\alpha}(x - \frac{k}{n}))}{In^{\alpha}} - 1 \right|$$

$$+ \frac{b^{*}}{I}\left(2T + \frac{1}{n^{\alpha}}\right)\left(\sum_{j=1}^{N} D(f^{(j)}(x), \tilde{o})\frac{T^j}{n^{j(1-\alpha)}j!}\right)$$

$$+ \omega_1^{(\mathcal{F})}\left(f^{(N)}, \frac{T}{n^{1-\alpha}}\right)\frac{T^N}{N!n^{N(1-\alpha)}}\frac{b^{*}}{I}\left(2T + \frac{1}{n^{\alpha}}\right), \qquad (16.36)$$

where $b^{*} := b(0)$.

By Lemma 2.1 ([9], p. 64) we have

$$\lim_{n\to+\infty} \sum_{k=\lceil nx-Tn^{\alpha}\rceil}^{\lceil nx+Tn^{\alpha}\rceil} \frac{b(n^{1-\alpha}(x - \frac{k}{n}))}{In^{\alpha}} = 1, \qquad x \in \mathbb{R}.$$

Since $f^{(N)} \in C_{\mathcal{F}}^{U}(\mathbb{R})$ we get that

$$\lim_{n\to+\infty} \omega_1^{(\mathcal{F})}\left(f^{(N)}, \frac{T}{n^{1-\alpha}}\right) = 0.$$

Consequently from (16.36) as $n \to +\infty$ we derive

$$D((F_n(f))(x), f(x)) \to 0, \qquad x \in \mathbb{R}.$$

Proof of theorem 16.13. Notice that (16.35) is a finite sum. Thus

$$
\begin{aligned}
[(F_n(f))(x)]^r &= \sum_{k=\lceil nx-Tn^\alpha\rceil}^{[nx+Tn^\alpha]} \left[f\left(\frac{k}{n}\right)\right]^r \frac{b\left(n^{1-\alpha}\left(x-\frac{k}{n}\right)\right)}{In^\alpha} \\
&= \left[\sum_{k=\lceil nx-Tn^\alpha\rceil}^{[nx+Tn^\alpha]} f_-^{(r)}\left(\frac{k}{n}\right) \frac{b\left(n^{1-\alpha}\left(x-\frac{k}{n}\right)\right)}{In^\alpha}, \right. \\
&\quad \left. \sum_{k=\lceil nx-Tn^\alpha\rceil}^{[nx+Tn^\alpha]} f_+^{(r)}\left(\frac{k}{n}\right) \frac{b\left(n^{1-\alpha}\left(x-\frac{k}{n}\right)\right)}{In^\alpha} \right] \\
&= \left[(F_n(f_-^{(r)}))(x), (F_n(f_+^{(r)}))(x)\right].
\end{aligned}
$$

I.e. for any $r \in [0,1]$ we have

$$
(F_n(f))_\pm^{(r)} = F_n(f_\pm^{(r)}). \tag{16.37}
$$

Next we observe from Section 2 of [17] that

$$
\begin{aligned}
D((F_n f)(x), f(x)) &= \sup_{r\in[0,1]} \max\{|((F_n f)(x)_-^{(r)} - f_-^{(r)}(x)|, \\
&\qquad |((F_n f)(x))_+^{(r)} - f_+^{(r)}(x)|\} \\
&\overset{(16.37)}{=} \sup_{r\in[0,1]} \max\{|F_n(f_-^{(r)})(x) - f_-^{(r)}(x)|, \\
&\qquad |F_n(f_+^{(r)}) - f_+^{(r)}(x)|\}.
\end{aligned}
$$

Since $f \in C_{\mathcal{F}}^{NU}(\mathbb{R})$ we have for any $r \in [0,1]$ that $f_\pm^{(r)} \in C^N(\mathbb{R})$ and $(f_\pm^{(r)})^{(N)} \in C_U(\mathbb{R})$. Therefore we can apply Theorem 16.5 (16.15) to obtain

$$
\begin{aligned}
&D((F_n f)(x), f(x)) \\
&\leq \sup_{r\in[0,1]} \max\left\{ \left(|f_-^{(r)}(x)| \cdot \left| \sum_{k=\lceil nx-Tn^\alpha\rceil}^{[nx+Tn^\alpha]} \frac{b\left(n^{1-\alpha}\left(x-\frac{k}{n}\right)\right)}{In^\alpha} - 1 \right| \right. \right. \\
&\quad + \frac{b^*}{I}\left(2T + \frac{1}{n^\alpha}\right)\left(\sum_{j=1}^{N} \frac{|(f_-^{(r)})^{(j)}(x)|T^j}{n^{j(1-\alpha)}j!}\right) \\
&\quad \left. + \omega_1\left((f_-^{(r)})^{(N)}, \frac{T}{n^{1-\alpha}}\right) \cdot \frac{T^N}{N!n^{N(1-\alpha)}} \frac{b^*}{I}\left(2T + \frac{1}{n^\alpha}\right)\right), \\
&\quad \left(|f_+^{(r)}(x)| \cdot \left| \sum_{k=\lceil nx-Tn^\alpha\rceil}^{[nx+Tn^\alpha]} \frac{b\left(n^{1-\alpha}\left(x-\frac{k}{n}\right)\right)}{In^\alpha} - 1 \right| \right.
\end{aligned}
$$

$$+ \frac{b^*}{I}\left(2T + \frac{1}{n^\alpha}\right)\left(\sum_{j=1}^{N} \frac{|(f_+^{(r)})^{(j)}(x)|T^j}{n^{j(1-\alpha)}j!}\right)$$

$$+ \omega_1\left((f_+^{(r)})^{(N)}, \frac{T}{n^{1-\alpha}}\right)\frac{T^N}{N!n^{N(1-\alpha)}}\frac{b^*}{I}\left(2T + \frac{1}{n^\alpha}\right)\right)\Bigg\}$$

(by (33) of [15]) $= \sup_{r\in[0,1]} \max\left\{\left(|f_-^{(r)}(x)| \cdot \right.\right.$

$$\left|\sum_{k=\lceil nx - Tn^\alpha\rceil}^{\lfloor nx + Tn^\alpha\rfloor} \frac{b\left(n^{1-\alpha}\left(x - \frac{k}{n}\right)\right)}{In^\alpha} - 1\right|$$

$$+ \frac{b^*}{I}\left(2T + \frac{1}{n^\alpha}\right)\left(\sum_{j=1}^{N} \frac{|(f_-^{(j)})^{(r)}(x)|T^j}{n^{j(1-\alpha)}j!}\right) + \omega_1\left((f_-^{(N)})^{(r)}, \frac{T}{n^{1-\alpha}}\right)$$

$$\cdot\frac{T^N}{N!n^{N(1-\alpha)}}\frac{b^*}{I}\left(2T + \frac{1}{n^\alpha}\right)\right), \left(|f_+^{(r)}(x)| \cdot\right.$$

$$\left|\sum_{k=\lceil nx - Tn^\alpha\rceil}^{\lfloor nx + Tn^\alpha\rfloor} \frac{b\left(n^{1-\alpha}\left(x - \frac{k}{n}\right)\right)}{In^\alpha} - 1\right|$$

$$+ \frac{b^*}{I}\left(2T + \frac{1}{n^\alpha}\right)\left(\sum_{j=1}^{N} \frac{|(f_+^{(j)})^{(r)}(x)|T^j}{n^{j(1-\alpha)}j!}\right)$$

$$+ \omega_1\left((f_+^{(N)})^{(r)}, \frac{T}{n^{1-\alpha}}\right)\frac{T^N}{N!n^{N(1-\alpha)}}\frac{b^*}{I}\left(2T + \frac{1}{n^\alpha}\right)\right)\Bigg\}$$

(by (29) of [15]) $\leq \left(\sup_{r\in[0,1]} \max\{|f_-^{(r)}(x)|, |f_+^{(r)}(x)|\}\right) \cdot$

$$\left|\sum_{k=\lceil nx - Tn^\alpha\rceil}^{\lfloor nx + Tn^\alpha\rfloor} \frac{b\left(n^{1-\alpha}\left(x - \frac{k}{n}\right)\right)}{In^\alpha} - 1\right|$$

$$+ \frac{b^*}{I}\left(2T + \frac{1}{n^\alpha}\right)\left\{\sum_{j=1}^{N}\left(\sup_{r\in[0,1]} \max\{|(f_-^{(j)})^{(r)}(x)|, |(f_+^{(j)})^{(r)}(x)|\}\right)\right.$$

$$\left.\frac{T^j}{n^{j(1-\alpha)}j!}\right\}$$

$$+ \left(\sup_{r\in[0,1]} \max\left\{\omega_1\left((f_-^{(N)})^{(r)}, \frac{T}{n^{1-\alpha}}\right), \omega_1\left((f_+^{(N)})^{(r)}, \frac{T}{n^{1-\alpha}}\right)\right\}\right)$$

$$\cdot\frac{T^N}{N!n^{N(1-\alpha)}}\frac{b^*}{I}\cdot\left(2T + \frac{1}{n^\alpha}\right)$$

$$\text{(by (25) \& (26) of [15])} = D(f(x), \tilde{o}) \left| \sum_{k=\lceil nx-Tn^\alpha \rceil}^{\lceil nx+Tn^\alpha \rceil} \frac{b\big(n^{1-\alpha}\big(x - \frac{k}{n}\big)\big)}{In^\alpha} - 1 \right|$$

$$+ \frac{b^*}{I}\left(2T + \frac{1}{n^\alpha}\right)\left(\sum_{j=1}^{N} D(f^{(j)}(x), \tilde{o}) \cdot \frac{T^j}{n^{j(1-\alpha)}j!}\right)$$

$$+ \omega_1^{(\mathcal{F})}\left(f^{(N)}, \frac{T}{n^{1-\alpha}}\right)\frac{T^N}{N!n^{N(1-\alpha)}}\frac{b^*}{I}\cdot\left(2T + \frac{1}{n^\alpha}\right).$$

That is proving (16.36). □

In the following let S be a *squashing function* as in Definition 4.2 of [17]. Clearly $\max S(x) = S(T)$, $x \in [-T, T]$. Assume that $I^* := \int_{-T}^{T} S(t)dt > 0$. Let $f \in C(\mathbb{R}, \mathbb{R}_\mathcal{F})$ we define the *fuzzy univariate squashing neural network operators* by

$$(G_n(f))(x) := \sum_{k=-n^2}^{n^2} {}^* f\left(\frac{k}{n}\right) \odot \frac{S\big(n^{1-\alpha}\big(x - \frac{k}{n}\big)\big)}{I^*n^\alpha}, \qquad (16.38)$$

$x \in \mathbb{R}$, $0 < \alpha < 1$ and $n \in \mathbb{N}: n \geq \max(T + |x|, T^{-1/\alpha})$. It is easy to see that

$$(G_n(f))(x) = \sum_{k=\lceil nx-Tn^\alpha \rceil}^{\lceil nx+Tn^\alpha \rceil} {}^* f\left(\frac{k}{n}\right) \odot \frac{S\big(n^{1-\alpha}\big(x - \frac{k}{n}\big)\big)}{I^*n^\alpha}. \qquad (16.39)$$

We present the fuzzy pointwise convergence with rates to the unit of operators G_n.

Theorem 16.14. *Let* $x \in \mathbb{R}$, $T > 0$, $n \in \mathbb{N}$ *such that* $n \geq \max(T + |x|, T^{-1/\alpha})$. *Let* $f \in C_\mathcal{F}^{NU}(\mathbb{R})$, $N \geq 1$. *Then*

$$D((G_n(f))(x), f(x)) \leq D(f(x), \tilde{o}) \cdot \left| \sum_{k=\lceil nx-Tn^\alpha \rceil}^{\lceil nx+Tn^\alpha \rceil} \frac{S\big(n^{1-\alpha}\big(x - \frac{k}{n}\big)\big)}{I^*n^\alpha} - 1 \right|$$

$$+ \frac{S(T)}{I^*}\left(2T + \frac{1}{n^\alpha}\right)\left(\sum_{j=1}^{N} D(f^{(j)}(x), \tilde{o})\frac{T^j}{n^{j(1-\alpha)}j!}\right)$$

$$+ \omega_1^{(\mathcal{F})}\left(f^{(N)}, \frac{T}{n^{1-\alpha}}\right)\frac{T^N}{N!n^{N(1-\alpha)}}\frac{S(T)}{I^*}\left(2T + \frac{1}{n^\alpha}\right). \quad (16.40)$$

By Lemma 2.2 ([9], p. 79) we have

$$\lim_{n \to +\infty} \sum_{k=\lceil nx-Tn^\alpha \rceil}^{\lceil nx+Tn^\alpha \rceil} \frac{S\big(n^{1-\alpha}\big(x - \frac{k}{n}\big)\big)}{I^*n^\alpha} = 1, \quad x \in \mathbb{R}.$$

Since $f^{(N)} \in C_{\mathcal{F}}^U(\mathbb{R})$ we get that

$$\lim_{n \to +\infty} \omega_1^{(\mathcal{F})}\left(f^{(N)}, \frac{T}{n^{1-\alpha}}\right) = 0.$$

Consequently from (16.40) as $n \to +\infty$ we obtain $D((G_n(f))(x), f(x)) \to 0$, $x \in \mathbb{R}$.

Proof of theorem 16.14. Notice that (16.39) is a finite sum. Then one finds easily that

$$(G_n(f))_{\pm}^{(r)} = G_n(f_{\pm}^{(r)}), \tag{16.41}$$

for all $r \in [0, 1]$.

Next we see again by Section 2 of [17] that

$$
\begin{aligned}
D((G_n f)(x), f(x)) &= \sup_{r \in [0,1]} \max\{|((G_n f)(x))_-^{(r)} - f_-^{(r)}(x)|, \\
&\qquad |((G_n f)(x))_+^{(r)} - f_+^{(r)}(x)|\} \\
&\overset{(16.41)}{=} \sup_{r \in [0,1]} \max\{|(G_n(f_-^{(r)}))(x) - f_-^{(r)}(x)|, \\
&\qquad |((G_n(f_+^{(r)}))(x) - f_+^{(r)}(x)|\}.
\end{aligned}
$$

Then the proof follows like in the proof of Theorem 16.13. We make use of Theorem 16.6 (16.18) for $f_{\pm}^{(r)}$. We again use in order (16.33), (16.29), (16.26) and (16.25) of [15] and finally produce (16.40). $\qquad\square$

Also we give

Corollary 16.15 (to Theorem 16.13). *Let $b(x)$ be a centered bell-shaped continuous function on $\mathbb{R}$ of compact support $[-T, T]$, $T > 0$. Let $x \in [-T^*, T^*]$, $T^* > 0$, and $n \in \mathbb{N}$ such that*

$$n \geq \max(T + T^*, T^{-1/\alpha}), \quad 0 < \alpha < 1.$$

Consider $p \geq 1$. Then we obtain

$$
\begin{aligned}
&\|D((F_n(f))(x), f(x))\|_{p,[-T^*,T^*]} \\
&\leq \left(\|D(f(x), \tilde{o})\|_{\infty,[-T^*,T^*]}\right) \cdot \left\| \sum_{k=\lceil nx-Tn^\alpha \rceil}^{\lceil nx+Tn^\alpha \rceil} \frac{b(n^{1-\alpha}(x - \frac{k}{n}))}{In^\alpha} - 1 \right\|_{p,[-T^*,T^*]} \\
&\quad + \frac{b^*}{I}\left(2T + \frac{1}{n^\alpha}\right)\left(\sum_{j=1}^N \|D(f^{(j)}(x), \tilde{o})\|_{p,[-T^*,T^*]} \cdot \frac{T^j}{n^{j(1-\alpha)}j!}\right) \\
&\quad + \omega_1^{(\mathcal{F})}\left(f^{(N)}, \frac{T}{n^{1-\alpha}}\right) \cdot \frac{T^N 2^{1/p} T^{*1/p}}{N! n^{N(1-\alpha)}} \frac{b^*}{I}\left(2T + \frac{1}{n^\alpha}\right). \tag{16.42}
\end{aligned}
$$

From inequality (16.42) as $n \to +\infty$ we obtain the L_p-fuzzy convergence of $F_n(f)$ to f with rates.

Proof. Since f if fuzzy continuous on $[-T^*, T^*]$ it is fuzzy bounded there (see Lemma 2 of [11]), therefore

$$\|D(f(x), \tilde{o})\|_{\infty, [-T^*, T^*]} < +\infty.$$

So from Theorem 16.13, we have that

$$
\begin{aligned}
D((F_n(f))(x), f(x)) \leq &\ \|D(f(x), \tilde{o})\|_{\infty, [-T^*, T^*]} \cdot \\
&\left| \sum_{k=\lceil nx - Tn^\alpha \rceil}^{\lfloor nx + Tn^\alpha \rfloor} \frac{b(n^{1-\alpha}(x - \frac{k}{n}))}{In^\alpha} - 1 \right| \\
&+ \frac{b^*}{I} \left(2T + \frac{1}{n^\alpha} \right) \left(\sum_{j=1}^{N} D(f^{(j)}(x), \tilde{o}) \frac{T^j}{n^{j(1-\alpha)} j!} \right) \\
&+ \omega_1^{(\mathcal{F})} \left(f^{(N)}, \frac{T}{n^{1-\alpha}} \right) \frac{T^N}{N! n^{N(1-\alpha)}} \frac{b^*}{I} \left(2T + \frac{1}{n^\alpha} \right).
\end{aligned}
\tag{16.43}
$$

Inequality (16.42) now comes by integration of (16.43) and the properties of L_p norm. As in [9], p. 75, using the bounded convergence theorem we get that

$$
\lim_{n \to +\infty} \left\| \sum_{k=\lceil nx - Tn^\alpha \rceil}^{\lfloor nx + Tn^\alpha \rfloor} \frac{b(n^{1-\alpha}(x - \frac{k}{n}))}{In^\alpha} - 1 \right\|_{p, [-T^*, T^*]} = 0. \qquad \square
$$

Note. Theorems 16.13, 16.14 and Corollary 16.15 simplify and improve the author's earlier corresponding results in [11], namely from there see Theorems 3, 5, and Corollary 2, respectively.

A different direct fuzzy method, not using real analysis results, is applied there [11]. See also here Chapter 15.

We also give the following related results

Theorem 16.16. All assumptions as in Theorem 16.9. Let $(B_k f)(x)$ as in (16.19), $(C_k f)(x)$ as in (16.27), $(D_k f)(x)$ as in (16.30). Then

(i)

$$E_{1k}(x) = D((B_k f)(x), (D_k f)(x)) \leq \lambda_1(x) + \lambda_4(x), \tag{16.44}$$

(ii)

$$E_{2k}(x) = D((B_k f)(x), (C_k f)(x)) \leq \lambda_1(x) + \lambda_3(x), \tag{16.45}$$

(iii)

$$E_{3k}(x) = D\left(\left(C_k f\right)(x), \left(D_k f\right)(x)\right) \le \lambda_3(x) + \lambda_4(x). \qquad (16.46)$$

Notice $E_{1k}(x)$, $E_{2k}(x)$, $E_{3k}(x) \to 0$, pointwise with rates, as $k \to \infty$.

Proof. We use the triangle inequality property of metric distance D and (16.20), (16.28) and (16.32) of Theorems 16.9, 16.11 and 16.12, respectively. Indeed we have

(i)

$$
\begin{aligned}
D\left(\left(B_k f\right)(x), \left(D_k f\right)(x)\right) &\le D\left(\left(B_k f\right)(x), f(x)\right) \\
&\quad + D\left(\left(D_k f\right)(x), f(x)\right) \\
&\le \lambda_1(x) + \lambda_4(x),
\end{aligned}
$$

(ii)

$$
\begin{aligned}
D\left(\left(B_k f\right)(x), \left(C_k f\right)(x)\right) &\le D\left(\left(B_k f\right)(x), f(x)\right) \\
&\quad + D\left(\left(C_k f\right)(x), f(x)\right) \\
&\le \lambda_1(x) + \lambda_3(x),
\end{aligned}
$$

(iii)

$$
\begin{aligned}
D\left(\left(C_k f\right)(x), \left(D_k f\right)(x)\right) &\le D\left(\left(C_k f\right)(x), f(x)\right) \\
&\quad + D\left(\left(D_k f\right)(x), f(x)\right) \\
&\le \lambda_3(x) + \lambda_4(x).
\end{aligned}
$$

$\square$

Finally we give

Theorem 16.17. All assumptions as in Theorem 16.10. Let $(A_k f)(x)$ as in (16.24), $(B_k f)(x)$ as in (16.19), $(C_k f)(x)$ as in (16.27) and $(D_k f)(x)$ as in (16.30). Then

(i)

$$E_{4k}(x) = D\left(\left(A_k f\right)(x), \left(B_k f\right)(x)\right) \le \lambda_1(x) + \lambda_2(x), \qquad (16.47)$$

(ii)

$$E_{5k}(x) = D\left(\left(A_k f\right)(x), \left(C_k f\right)(x)\right) \le \lambda_2(x) + \lambda_3(x), \qquad (16.48)$$

(iii)

$$E_{6k}(x) = D\left(\left(A_k f\right)(x), \left(D_k f\right)(x)\right) \le \lambda_2(x) + \lambda_4(x). \qquad (16.49)$$

Notice $E_{4k}(x)$, $E_{5k}(x)$, $E_{6k}(x) \to 0$, pointwise with rates, as $k \to \infty$.

Proof. Similar to Theorem 16.16. It is based on (16.20) of Theorem 16.9, (16.25) of Theorem 16.10, (16.28) of Theorems 16.11, and (16.32) of Theorem 16.12. Of course we apply again the triangle inequality property of D.   □

Note. We notice that (16.44) of Theorem 16.16 improves (13.33) of Theorem 13.21. Also (16.46) of Theorem 16.16 improves (13.34) of Theorem 13.22. Furthermore (16.45) of Theorem 16.16 improves (13.35) of Theorem 13.23. Similarly can be improved Corollaries 13.27, 13.28, 13.29, accordingly. The reason for this improvement is that in the proofs here we use real and fuzzy methods, while in Chapter 13 we use only fuzzy methods.

17

FUZZY RANDOM KOROVKIN
THEOREMS AND INEQUALITIES

Here we study the fuzzy random positive linear operators acting on fuzzy random continuous functions. We establish a series of fuzzy random Shisha–Mond type inequalities of L^q-type
$1 \leq q < \infty$ and related fuzzy random Korovkin type theorems,
regarding the fuzzy random q-mean convergence of fuzzy random positive linear operators to the fuzzy random unit operator
for various cases. All convergences are with rates and are given
using the above fuzzy random inequalities involving the fuzzy
random modulus of continuity of the engaged fuzzy random
function. The assumptions for the Korovkin theorems are minimal and of natural realization, fulfilled by almost all example
fuzzy random positive linear operators. The astonishing fact is
that the real Korovkin test functions assumptions are enough
for the conclusions of the fuzzy random Korovkin theory. We
give at the end applications. This chapter follows [22].

17.1   Introduction

Motivation for this chapter are [4], [18], [24], [32], [17], [66], [96]. We introduce the concept of fuzzy random positive linear operator and we prove
the results for a very large general class of such operators. Most of the

G.A. Anastassiou: Fuzzy Mathematics: Approximation The., STUDFUZZ 251, pp. 309–345.
springerlink.com   © Springer-Verlag Berlin Heidelberg 2010

summation and integration operators fall into this class. To do that we are greatly helped by the fuzzy Riesz representation theorem developed in [24]. The surprising fact is that the basic assumptions of real Korovkin theory for the test functions 1, id, id^2 carry over here and they are the only ones needed. Of course a natural realization condition is required in the fuzzy random setting to prove the fuzzy random q-mean convergence. But first we establish a series of fuzzy random Shisha–Mond type inequalities for various important cases. These contain the fuzzy random modulus of continuity of the involved function.

So this chapter is basically the study with rates and quantitavely for the fuzzy random q-mean convergence of a sequence of very general and abstract fuzzy random positive linear operators to the fuzzy random unit operator. Linearity and positivity here are the analogs of the real case. Finally we give applications to fuzzy random Bernstein operators.

We need the following definitions.

Definition 17.1. (See also [66, Definition 13.16, p. 654]). Let $(X, \mathcal{B}, P)$ be a probability space. A *fuzzy-random variable is a $\mathcal{B}$-measurable mapping* $g \colon X \to \mathbb{R}_{\mathcal{F}}$, i.e., for any open set $U \subseteq \mathbb{R}_{\mathcal{F}}$, in the topology of $\mathbb{R}_{\mathcal{F}}$ generated by the metric D, we have

$$g^{-1}(U) = \{s \in X; \; g(s) \in U\} \in \mathcal{B}. \tag{17.1}$$

The set of all fuzzy-random variables is denoted by $\mathcal{L}_{\mathcal{F}}(X, \mathcal{B}, P)$. Let $g_n, g \in \mathcal{L}_{\mathcal{F}}(X, \mathcal{B}, P)$, $n \in \mathbb{N}$, and $0 < q < +\infty$. We say,

$$g_n(s) \underset{n \to +\infty}{\overset{\text{``}q\text{-mean''}}{\longrightarrow}} g(s), \tag{17.2}$$

if

$$\lim_{n \to +\infty} \int_X \left(D(g_n(s), g(s)) \right)^q P(ds) = 0. \tag{17.3}$$

Definition 17.2. (See [66, p. 654, Definition 13.17].) Let $(T, \mathcal{T})$ be a topological space. A mapping $f \colon T \to \mathcal{L}_{\mathcal{F}}(X, \mathcal{B}, P)$ will be called *fuzzy-random function* (or *fuzzy-stochastic process*) on T. We denote $f(t)(s) = f(t, s)$, $t \in T$, $s \in X$.

Remark 17.3. (See [66, p. 655].) Any usual fuzzy real function $f \colon T \to \mathbb{R}_{\mathcal{F}}$ can be identified with the degenerate fuzzy-random function $f(t, s) = f(t)$, $\forall t \in T$, $s \in X$.

Remark 17.4. (See [66, p. 655].) Fuzzy-random functions that coincide with probability one, for each $t \in T$, will be considered equivalent.

Remark 17.5. (see [66, p. 655].) Let $f, g \colon T \to \mathcal{L}_{\mathcal{F}}(X, \mathcal{B}, P)$. Then, $f \oplus g$ and $k \odot f$ are defined pointwise, i.e.,

$$\begin{aligned}
(f \oplus g)(t, s) &= f(t, s) \oplus g(t, s), \\
(k \odot f)(t, s) &= k \odot f(t, s), \quad t \in T, \quad s \in X, \quad k \in \mathbb{R}.
\end{aligned}$$

Definition 17.6. (See also [66, Definition 13.18, pp. 655–656].) For a fuzzy-random function $f\colon [a,b] \to \mathcal{L}_{\mathcal{F}}(X, \mathcal{B}, P)$, we define the (first) fuzzy-random modulus of continuity

$$\Omega_1^{(\mathcal{F})}(f, \delta)_{L^q} = \sup\left\{\left(\int_X D^q\big(f(x,s), f(y,s)\big) P(ds)\right)^{1/q} ; \ x, y \in [a,b],\right.$$
$$\left. |x-y| \le \delta\right\}, \quad 0 < \delta, \quad 1 \le q < \infty. \tag{17.4}$$

Definition 17.7. Here, $1 \le q < \infty$. Let $f\colon [a,b] \to \mathcal{L}_{\mathcal{F}}(X, \mathcal{B}, P)$ be a fuzzy random function. We call f a (q-mean) uniformly continuous fuzzy random function over $[a,b]$ iff $\forall \varepsilon > 0 \ \exists \delta > 0$: whenever $|x - y| \le \delta$, $x, y \in [a, b]$, implies that

$$\int_X \big(D(f(x,s), f(y,s))\big)^q P(ds) \le \varepsilon. \tag{17.5}$$

We denote it as $f \in C_{\mathcal{F}R}^{Uq}([a,b])$.

We need

Proposition 17.8. Let $f \in C_{\mathcal{F}R}^{Uq}([a,b])$. Then, $\Omega_1^{(\mathcal{F})}(f, \delta)_{L^q} < \infty$, any $\delta > 0$.

Proof. Let $\varepsilon_0 > 0$ be arbitrary, but fixed. Then, there exists $\delta_0 > 0$: $|x - y| \le \delta_0$, $x, y \in [a,b]$ which implies

$$\int_X \big(D(f(x,s), f(y,s))\big)^q P(ds) \le \varepsilon_0 < \infty.$$

That is, $\Omega_1^{(\mathcal{F})}(f, \delta_0)_{L^q} \le \varepsilon_0^{1/q} < \infty$. Let now $\delta > 0$ arbitrary, $x, y \in [a, b]$, such that $|x-y| \le \delta$. Choose $n \in \mathbb{N}$: $n\delta_0 \ge \delta$ and set $x_i := x + (i/n)(y - x)$, $0 \le i \le n$. Then,

$$D\big(f(x,s), f(y,s)\big) \le D\big(f(x,s), f(x_1, s)\big) \quad + \quad D\big(f(x_1, s), f(x_2, s)\big)$$
$$+ \quad \cdots + D\big(f(x_{n-1}, s), f(y, s)\big).$$

Consequently,

$$\left(\int_X \big(D(f(x,s), f(y,s))\big)^q P(ds)\right)^{1/q}$$
$$\le \left(\int_X \big(D(f(x,s), f(x_1, s))\big)^q P(ds)\right)^{1/q}$$
$$+ \cdots + \left(\int_X \big(D(f(x_{n-1}, s), f(y, s))\big)^q P(ds)\right)^{1/q}$$
$$\le n\Omega_1^{(\mathcal{F})}(f, \delta_0)_{L^q} \le n\varepsilon_0^{1/q} < \infty,$$

since $|x_i - x_{i+1}| \leq (1/n)|x - y| \leq (1/n)\delta \leq \delta_0$, $0 \leq i \leq n$. Therefore, $\Omega_1^{(\mathcal{F})}(f, \delta)_{L^q} \leq n\varepsilon_0^{1/q} < \infty$. $\qquad\square$

Proposition 17.9. *Let* $f, g \colon [a, b] \to \mathcal{L}_{\mathcal{F}}(X, \mathcal{B}, P)$ *be fuzzy random functions,* $[a, b] \subset \mathbb{R}$. *The following hold.*

(i) $\Omega_1^{(\mathcal{F})}(f, \delta)_{L^q}$ *be nonnegative and nondecreasing in* $\delta > 0$.

(ii) $\lim_{\delta \downarrow 0} \omega_1^{(\mathcal{F})}(f, \delta)_{L^q} = \Omega_1^{(\mathcal{F})}(f, 0)_{L^q} = 0$, *iff* $f \in C_{\mathcal{F}R}^{Uq}([a, b])$.

(iii) $\Omega_1^{(\mathcal{F})}(f, \delta_1 + \delta_2)_{L^q} \leq \Omega_1^{(\mathcal{F})}(f, \delta_1)_{L^q} + \Omega_1^{(\mathcal{F})}(f, \delta_2)_{L^q}$, $\delta_1, \delta_2 > 0$.

(iv) $\Omega_1^{(\mathcal{F})}(f, n\delta)_{L^q} \leq n\Omega_1^{(\mathcal{F})}(f, \delta)_{L^q}$, $\delta > 0$, $n \in \mathbb{N}$.

(v) $\Omega_1^{(\mathcal{F})}(f, \lambda\delta)_{L^q} \leq \lceil\lambda\rceil\Omega_1^{(\mathcal{F})}(f, \delta)_{L^q} \leq (\lambda + 1)\Omega_1^{(\mathcal{F})}(f, \delta)_{L^q}$, $\lambda > 0$, $\delta > 0$, *where* $\lceil\cdot\rceil$ *is the ceiling of the number.*

(vi) $\Omega_1^{(\mathcal{F})}(f \oplus g, \delta)_{L^q} \leq \Omega_1^{(\mathcal{F})}(f, \delta)_{L^q} + \Omega_1^{(\mathcal{F})}(g, \delta)_{L^q}$, $\delta > 0$. *Here,* $f \oplus g$ *is a fuzzy random function.*

(vii) $\Omega_1^{(\mathcal{F})}(f, \cdot)_{L^q}$ *is continuous on* $\mathbb{R}_+$, *for* $f \in C_{\mathcal{F}R}^{Uq}([a, b])$.

Proof. The proof is obvious.

Proposition 17.10 (see [17]). *Let* f, g *be fuzzy random variables from* $X \to \mathbb{R}_{\mathcal{F}}$. *Then, we have the following.*

(i) *Let* $c \in \mathbb{R}$, *then* $c \odot f$ *is a fuzzy random variable.*

(ii) $f \oplus g$ *is a fuzzy random variable.*

For the definition of general fuzzy integral we follow [75] next.

Definition 17.11. Let (Ω, Σ, μ) be a complete σ-finite measure space. We call $F \colon \Omega \to \mathbb{R}_{\mathcal{F}}$ *measurable* iff $\forall$ closed $B \subseteq \mathbb{R}$ the function $F^{-1}(B) \colon \Omega \to [0, 1]$ defined by

$$F^{-1}(B)(\omega) := \sup_{x \in B} F(\omega)(x), \quad \text{all } \omega \in \Omega$$

is measurable, see [46], [75].

Notice here that the concept of measurability is different than the $\mathcal{B}$-measurability of Definition 17.1.

Theorem 17.12 ([75]). *For* $F \colon \Omega \to \mathbb{R}_{\mathcal{F}}$, $F(\omega) = \{(F_-^{(r)}(\omega), F_+^{(r)}(\omega)) \mid 0 \leq r \leq 1\}$, *the following are equivalent.*

(1) F *is measurable,*

(2) $\forall r \in [0, 1]$, $F_-^{(r)}$, $F_+^{(r)}$ *are measurable.*

Following [75], given that for each $r \in [0, 1]$, $F_-^{(r)}$, $F_+^{(r)}$ are integrable we have that the parametrized representation

$$\left\{ \left(\int_A F_-^{(r)} \, d\mu, \int_A F_+^{(r)} \, d\mu \right) \mid 0 \leq r \leq 1 \right\}$$

is a fuzzy real number for each $A \in \Sigma$.

The last fact leads to

Definition 17.13 ([75]). A measurable function $F \colon \Omega \to \mathbb{R}_{\mathcal{F}}$,

$$F(\omega) = \left\{ (F_-^{(r)}(\omega), F_+^{(r)}(\omega)) \mid 0 \leq r \leq 1 \right\}$$

is called *integrable* if for each $r \in [0, 1]$, $F_{\pm}^{(r)}$ are integrable, or equivalently, if $F_{\pm}^{(0)}$ are integrable. In this case, the fuzzy integral of F over $A \in \Sigma$ is defined by

$$\int_A F \, d\mu := \left\{ \left(\int_A F_-^{(r)} \, d\mu, \int_A F_+^{(r)} \, d\mu \right) \mid 0 \leq r \leq 1 \right\}.$$

By [14], F is integrable iff $\omega \to \|F(\omega)\|_{\mathcal{F}}$ is real-valued integrable.

We need also

Theorem 17.14 ([75]). *Let $F, G \colon \Omega \to \mathbb{R}_{\mathcal{F}}$ be integrable. Then*

(1) *Let $a, b \in \mathbb{R}$, then $a \odot F + b \odot G$ is integrable and for each $A \in \Sigma$,*

$$\int_A (a \odot F \oplus b \odot G) \, d\mu = a \odot \int_A F \, d\mu \oplus b \odot \int_A G \, d\mu;$$

(2) *$D(F, G)$ is a real-valued integrable function and for each $A \in \Sigma$,*

$$D\left(\int_A F \, d\mu, \int_A G \, d\mu \right) \leq \int_A D(F, G) \, d\mu.$$

In particular,

$$\left\| \int_A F \, d\mu \right\|_{\mathcal{F}} \leq \int_A \|F\|_{\mathcal{F}} \, d\mu.$$

We need

Definition 17.15. Let U open or compact $\subseteq (M, d)$ metric space and $f \colon U \to \mathbb{R}_{\mathcal{F}}$. We say that f is *fuzzy continuous at $x_0 \in U$* iff whenever $x_n \to x_0$, then $D(f(x_n), f(x_0)) \to 0$. If f is continuous for every $x_0 \in U$, we then call f a *fuzzy continuous real number valued function*. We denote the related space by $C_{\mathcal{F}}(U)$. Similarly one defines $C_{\mathcal{F}}([a, b])$, $[a, b] \subseteq \mathbb{R}$.

Definition 17.16. Let $L\colon C_{\mathcal{F}}(U) \hookrightarrow C_{\mathcal{F}}(U)$, where U is open or compact $\subseteq (M, d)$ metric space, such that

$$L(c_1 \odot f \oplus c_2 \odot g) = c_1 \odot L(f) \oplus c_2 \odot L(g), \quad \forall c_1, c_2 \in \mathbb{R}.$$

We call L a *fuzzy linear operator.*

We give the following example of a fuzzy linear operator, etc.

Definition 17.17. Let $f\colon [0,1] \to \mathbb{R}_{\mathcal{F}}$ be a fuzzy real function. The fuzzy algebraic polynomial defined by

$$B_n^{(\mathcal{F})}(f)(x) = \sum_{k=0}^{n}{}^{*} \binom{n}{k} x^k (1-x)^{n-k} \odot f\left(\frac{k}{n}\right), \quad \forall x \in [0, 1],$$

will be called the *fuzzy Bernstein operator.* Here $\sum^{*}$ stands for the fuzzy summation.

We also need

Definition 17.18. Let $f, g\colon U \to \mathbb{R}_{\mathcal{F}}, U \subseteq (M, d)$ metric space. We denote $f \succeq g$, iff $f(x) \succeq g(x)$, $\forall x \in U$, iff $f_+^{(r)}(x) \geq g_+^{(r)}(x)$ and $f_-^{(r)}(x) \geq g_-^{(r)}(x)$, $\forall x \in U$, $\forall r \in [0,1]$, iff $f_+^{(r)} \geq g_+^{(r)}$ and $f_-^{(r)} \geq g_-^{(r)}$, $\forall r \in [0,1]$, where $[f(x)]^r = [f_-^{(r)}(x), f_+^{(r)}(x)]$.

We give

Definition 17.19. Let $L\colon C_{\mathcal{F}}(U) \hookrightarrow C_{\mathcal{F}}(U)$ be a fuzzy linear operator, U open or compact $\subseteq (M, d)$ metric space. We say that L is *positive,* iff whenever $f, g \in C_{\mathcal{F}}(U)$ are such that $f \succeq g$ then $L(f) \succeq L(g)$, iff

$$(L(f))_+^{(r)} \geq (L(g))_+^{(r)}$$

and

$$(L(f))_-^{(r)} \geq (L(g))_-^{(r)}, \quad \forall r \in [0, 1].$$

Here we denote

$$[L(f)]^r = [(L(f))_-^{(r)}, (L(f))_+^{(r)}], \quad \forall r \in [0, 1].$$

An example of a fuzzy positive linear operator is the fuzzy Bernstein operator on the domain $[0, 1]$, etc. For more see [18], [24], [32].

We mention

Assumption 17.20 (see [24]). Let L be a fuzzy positive linear operator from $C_{\mathcal{F}}(K)$, K compact $\subseteq (M, d)$ metric space, into itself. Here we *assume* that there exists a positive linear operator $\tilde{L}$ from $C(K)$ into itself with the property

$$(Lf)_{\pm}^{(r)} = \tilde{L}(f_{\pm}^{(r)}), \tag{17.6}$$

respectively, for all $r \in [0, 1]$, $\forall f \in C_{\mathcal{F}}(K)$.

As an example again we mention the fuzzy Bernstein operator and the real Bernstein operator fulfilling the above assumption on $[0, 1]$, etc.

We apply the following *Fuzzy Riesz Representation Theorem.*

Theorem 17.21 (see [24]). *Let L be a fuzzy positive linear operator from $C_{\mathcal{F}}(K)$ into itself as in Assumption 17.20, K compact $\subseteq (M, d)$ metric space. Then for each $x \in K$ there exists a unique positive finite completed Borel measure μ_x on K such that*

$$(Lf)(x) = \int_K f(t)\mu_x(dt), \quad \forall f \in C_{\mathcal{F}}(K).$$

17.2 Auxilliary Material

We apply in proofs

Remark 17.22. Let $f \colon [a, b] \to \mathcal{L}_{\mathcal{F}}(X, \mathcal{B}, P)$, $[a, b] \subset \mathbb{R}$ be a fuzzy random function. Then by Proposition 17.9(v) we get

$$\Omega_1^{(\mathcal{F})}(f, |x - y|)_{L^q} \le \left\lceil \frac{|x - y|}{\delta} \right\rceil \Omega_1^{(\mathcal{F})}(f, \delta)_{L^q}, \quad \forall x, y \in [a, b] \text{ any } \delta > 0.$$

$$(17.7)$$

The main function space we are going to work on in the chapter is defined as follows.

Definition 17.23. Let $(X, \mathcal{B}, P)$ be a probability space, $[a, b] \subset \mathbb{R}$, and the fuzzy random function $f \colon [a, b] \times X \to \mathbb{R}_{\mathcal{F}}$ such that $f(t, \omega)$ is *fuzzy continuous in $t \in [a, b]$ uniformly with respect to ω in X*. I.e. $\forall \varepsilon > 0 \ \exists \delta > 0$ such that whenever $|x - y| \le \delta$; $x, y \in [a, b]$, then

$$D\big(f(x, \omega), f(y, \omega)\big) \le \varepsilon, \quad \forall \omega \in X.$$

We denote the space of all these functions by $C^U_{\mathcal{F}R}([a, b])$.

One can easily see that if $f \in C^U_{\mathcal{F}R}([a, b])$ then for each $\omega \in X$ we have that $f(\cdot, \omega) \in C_{\mathcal{F}}([a, b])$ and f is q-mean uniformly continuous in $t \in [a, b]$, i.e. $f \in C^{Uq}_{\mathcal{F}R}([a, b])$, any $1 \le q < +\infty$, see Definition 17.7.

We mention

Definition 17.24. Let $L^* \colon C^U_{\mathcal{F}R}([a, b]) \hookrightarrow C^U_{\mathcal{F}R}([a, b])$ such that

$$L^*(c_1 \odot f_1 \oplus c_2 \odot f_2) = c_1 \odot L^*(f_1) \oplus c_2 \odot L^*(f_2), \quad \forall c_1, c_2 \in \mathbb{R}.$$

We call L^* a *fuzzy random linear operator* on $C^U_{\mathcal{F}R}([a, b])$.

The following motivate this chapter.

Example 17.25 (see [66], p. 656). For $f: [0,1] \to \mathcal{L}_{\mathcal{F}}(X, B, P)$, the fuzzy random polynomials defined by

$$B_n^{(\mathcal{F})}(f)(x,\omega) := \sum_{k=0}^n {}^* \binom{n}{k} x^k (1-x)^{n-k} \odot f\left(\frac{k}{n}, \omega\right), \quad x \in [0,1], \ \omega \in X$$

will be called a Bernstein-type. Clearly $B_n^{(\mathcal{F})}(\cdot)(x,\omega)$ is a fuzzy random linear operator, $n \in \mathbb{N}$.

We have

Theorem 17.26 (see [66], p. 656). *For $f: [0,1] \to L_{\mathcal{F}}(X, \mathcal{B}, P)$ we have the estimate*

$$\int_X D\big(B_n^{(\mathcal{F})}(f)(x,\omega), f(x,\omega)\big) P(d\omega) \leq \frac{3}{2} \Omega_1^{(\mathcal{F})}\left(f; \frac{1}{\sqrt{n}}\right)_{L^1}, \qquad (17.8)$$

$\forall x \in [0,1]$, $n \in \mathbb{N}$. *If moreover f satisfies the condition*

$$\lim_{\delta \downarrow 0} \Omega_1^{(\mathcal{F})}(f, \delta)_{L_1} = 0,$$

then

$$\underset{n \to +\infty}{\overset{\text{``1-mean''}}{B_n^{(\mathcal{F})}(f)(x,\omega) \quad \longrightarrow \quad f(x,\omega),}}$$

uniformly with respect to $x \in [0,1]$.

We mention

Definition 17.27. Let $L^*: C_{\mathcal{F}R}^U([a,b]) \hookrightarrow C_{\mathcal{F}R}^U([a,b])$ be a fuzzy random linear operator. We call L^* a *positive* fuzzy random linear operator iff whenever we have $f, g \in C_{\mathcal{F}R}^U([a,b])$ such that $f \gtrsim g$, i.e. $f(x,\omega) \gtrsim g(x,\omega)$ for all $(x,\omega) \in [a,b] \times X$ then $L^* f \gtrsim L^* g$, i.e. $(L^* f)(x,\omega) \gtrsim (L^* g)(x,\omega)$ for all $(x,\omega) \in [a,b] \times X$, iff $(L^* f)_+^{(r)}(x,\omega) \geq (L^* g)_+^{(r)}(x,\omega)$ and

$$(L^* f)_-^{(r)}(x,\omega) \geq (L^* g)_-^{(r)}(x,\omega), \quad \forall r \in [0,1], \ \forall (x,\omega) \in [a,b] \times X.$$

Here we denote

$$[L^*(f)(x,\omega)]^r = [(L^* f)_-^{(r)}(x,\omega), (L^* f)_+^{(r)}(x,\omega)], \quad \forall r \in [0,1],$$
$$\forall (x,\omega) \in [a,b] \times X.$$

An example of a positive fuzzy random linear operator is $B_n^{(\mathcal{F})}(\cdot)(x,\omega)$, etc.

We give the useful

Remark 17.28. Let L be a fuzzy positive linear operator from $C_{\mathcal{F}}([a,b])$ into itself. We *suppose* that there exists a positive linear operator $\tilde{L}$ from $C([a,b])$ into itself with the property

$$(Lf)_\pm^{(r)} = \tilde{L}(f_\pm^{(r)}), \qquad (17.9)$$

respectively, $\forall r \in [0,1]$, $\forall f \in C_{\mathcal{F}}([a,b])$. Then by Theorem 17.21, $\forall t \in [a,b]$ there exists a unique positive finite completed Borel measure μ_t on $[a,b]$ such that

$$(Lf)(t) = \int_{[a,b]} f(s)\mu_t(ds), \quad \forall f \in C_{\mathcal{F}}([a,b]). \tag{17.10}$$

Consequently for $f \in C_{\mathcal{F}R}^U([a,b])$ and since $f(\cdot,\omega) \in C_{\mathcal{F}}([a,b])$, $\forall \omega \in X$, we get that

$$L\big(f(\cdot,\omega)\big)(t) = \int_{[a,b]} f(s,\omega)\mu_t(ds), \quad \forall t \in [a,b], \ \ \forall \omega \in X. \tag{17.11}$$

Of course here by (17.9) we have

$$\big(L(f(\cdot,\omega))\big)_{\pm}^{(r)}(t) = \tilde{L}\big(f_{\pm}^{(r)}(\cdot,\omega)\big)(t), \quad \forall t \in [a,b], \ \forall \omega \in X, \ \forall r \in [0,1]. \tag{17.12}$$

We call

$$m_t := \mu_t([a,b]) \geq 0. \tag{17.13}$$

By setting

$$M(f)(t,\omega) := L\big(f(\cdot,\omega)\big)(t), \tag{17.14}$$

that is

$$M(f)(t,\omega) = \int_{[a,b]} f(s,\omega)\mu_t(ds), \tag{17.15}$$

from Theorem 17.14 (1) we have that

$$M(c_1 \odot f \oplus c_2 \odot g)(t,\omega) = c_1 \odot M(f)(t,\omega) \oplus c_2 \odot M(g)(t,\omega), \tag{17.16}$$

$\forall (t,\omega) \in [a,b] \times X$, $\forall f,g \in C_{\mathcal{F}R}^U([a,b])$, $\forall c_1, c_2 \in \mathbb{R}$.

Let $C_{\mathcal{F}R}([a,b]) := \{f \colon [a,b] \times X \to \mathbb{R}_{\mathcal{F}}$: such that $f(t,\omega)$ is fuzzy continuous in $t \in [a,b]$ and $\mathcal{B}$-measurable in $\omega \in X\}$. Additionally we assume here that $M(f)(t,\omega)$ is $\mathcal{B}$-measurable in $\omega \in X$. Then

$$M(f) \in C_{\mathcal{F}R}([a,b]), \quad \forall f \in C_{\mathcal{F}R}^U([a,b]).$$

That is M is a *fuzzy random linear operator* from $C_{\mathcal{F}R}^U([a,b])$ into $C_{\mathcal{F}R}([a,b])$. Thus by (17.12) we have

$$(M(f))_{\pm}^{(r)}(t,\omega) = \big(L(f(\cdot,\omega))\big)_{\pm}^{(r)}(t) = \tilde{L}\big(f_{\pm}^{(r)}(\cdot,\omega)\big)(t). \tag{17.17}$$

Let $f,g \in C_{\mathcal{F}}^U([a,b])$ such that $f \succsim g$ iff $f_-^{(r)} \geq g_-^{(r)}$ and $f_+^{(r)} \geq g_+^{(r)}$, $\forall r \in [0,1]$. Then

$$\tilde{L}\big(f_-^{(r)}(\cdot,\omega)\big)(t) \geq \tilde{L}\big(g_-^{(r)}(\cdot,\omega)\big)(t)$$

and

$$\tilde{L}\big(f_+^{(r)}(\cdot,\omega)\big)(t) \geq \tilde{L}\big(g_+^{(r)}(\cdot,\omega)\big)(t).$$

That is $(M(f))_{-}^{(r)} \geq (M(g))_{-}^{(r)}$ and $(M(f))_{+}^{(r)} \geq (M(g))_{+}^{(r)}$, $\forall r \in [0,1]$. Hence M is a *positive* fuzzy random linear operator.

For example we notice that

$$
B_n^{(\mathcal{F})}\big(f(\cdot,\omega)\big)(x) = \sum_{k=0}^{n}{}^{*} \binom{n}{k} x^k (1-x)^{n-k} \odot f\left(\frac{k}{n},\omega\right) = B_n^{(\mathcal{F})}(f)(x,\omega),
$$

(17.18)

$\forall x \in [0,1]$, $\forall \omega \in X$, $\forall f \in C_{\mathcal{F}R}^{U}([0,1])$, the last fulfills all the above theory. So fuzzy operators like L, M are quite common, e.g. summation, integral operators in the fuzzy sense, therefore we study their approximation properties next.

Clearly, by Theorem 5 of [24], any positive linear operator $\tilde{L}$ from $C([a,b])$ into itself induces a unique positive fuzzy linear operator L acting on $C_{\mathcal{F}}([a,b])$, which in turn generates by (17.14) a unique positive fuzzy random linear operator M acting on $C_{\mathcal{F}R}^{U}([a,b])$, so the class of M's is very rich.

17.3   Main Results

We will use the following

Proposition 17.29. *Let $(X, \mathcal{B}, P)$ be a probability space, $[a,b] \subset \mathbb{R}$, $f \in C_{\mathcal{F}R}^{U}([a,b])$. Let L a fuzzy positive linear operator from $C_{\mathcal{F}}([a,b])$ into itself for which there exists a positive linear operator $\tilde{L}$ from $C([a,b])$ into itself such that*

$$
(Lg)_{\pm}^{(r)} = \tilde{L}(g_{\pm}^{(r)}),
$$

(17.19)

respectively, $\forall r \in [0,1]$, $\forall g \in C_{\mathcal{F}}([a,b])$. We consider the positive fuzzy random linear operator M acting on $C_{\mathcal{F}R}^{U}([a,b])$ defined by

$$
M(f)(t,\omega) := L\big(f(\cdot,\omega)\big)(t), \quad \forall (t,\omega) \in [a,b] \times X, \ \forall f \in C_{\mathcal{F}R}^{U}([a,b]).
$$

(17.20)

We assume that $M(f)(t,\omega)$ is $\mathcal{B}$-measurable in $\omega \in X$. That is $M(f) \in C_{\mathcal{F}R}([a,b])$. Then

$$
D\big(M(f)(t,\omega), f(t,\omega)\big) \leq \int_{[a,b]} D\big(f(s,\omega), f(t,\omega)\big)\mu_t(ds)
$$

(17.21)

$$
+ |m_t - 1| D\big(f(t,\omega), \tilde{o}\big), \quad \forall (t,\omega) \in [a,b] \times X,
$$

where μ_t is as in (17.10) and m_t as in (17.13).

Proof. We observe that the $\mathcal{B}$-measurable function (see Remark 13.39, p. 654, [66])

$$D\big(M(f)(t,\omega), f(t,\omega)\big) \stackrel{(17.15)}{=} D\left(\int_{[a,b]} f(s,\omega)\mu_t(ds), f(t,\omega)\right)$$

$$\leq D\left(\int_{[a,b]} f(s,\omega)\mu_t(ds), f(t,\omega) \odot m_t\right) + D\big(f(t,\omega) \odot m_t, f(t,\omega)\big)$$

$$= D\left(\int_{[a,b]} f(s,\omega)\mu_t(ds), \int_{[a,b]} f(t,\omega)\mu_t(ds)\right) + D\big(f(t,\omega) \odot m_t, f(t,\omega)\big)$$

$$\leq \text{(by Theorem 17.14(2) and Lemma 1.2)}$$

$$\int_{[a,b]} D\big(f(s,\omega), f(t,\omega)\big)\mu_t(ds) + |m_t - 1|D\big(f(t,\omega), \tilde{o}\big).$$

Here notice that

$$f(t,\omega) \odot m_t \;=\; \left\{ \big(m_t(f(t,\omega))^{(r)}_-, m_t(f(t,\omega))^{(r)}_+\big) \mid 0 \leq r \leq 1\right\}$$

$$=\; \left\{ \left(\int_{[a,b]} (f(t,\omega))^{(r)}_- d\mu_t, \int_{[a,b]} (f(t,\omega))^{(r)}_+ d\mu_t\right) \mid 0 \leq r \leq 1\right\}$$

$$=\; \int_{[a,b]} f(t,\omega)\mu_t(ds).$$

$\square$

By Remark 2 of [24] we trivially see that

$$m_t = \tilde{L}(1)(t) \geq 0, \quad \forall t \in [a,b]. \tag{17.22}$$

We give the first main result.

Theorem 17.30. *Assume all terms and assumptions of Proposition 17.29 and*

$$\int_X D\big(f(t,\omega), \tilde{o}\big)\, dP(\omega) < \infty, \quad \forall t \in [a,b].$$

Then

$$\int_X D\big(M(f)(t,\omega), f(t,\omega)\big)dP(\omega)$$

$$\leq |\tilde{L}(1)(t) - 1|\left(\int_X D\big(f(t,\omega), \tilde{o}\big)dP(\omega)\right) \tag{17.23}$$

$$+ \left(\tilde{L}(1)(t) + \sqrt{\tilde{L}(1)(t)}\right)\Omega_1^{(\mathcal{F})}\big(f, (\tilde{L}((\cdot - t)^2)(t))^{1/2}\big)_{L^1}, \quad \forall t \in [a,b],$$

and

$$\sup_{t\in[a,b]}\left(\int_X D\big(M(f)(t,\omega), f(t,\omega)\big)dP(\omega)\right) \tag{17.24}$$

$$\leq \|\tilde{L}(1)-1\|_\infty \sup_{t\in[a,b]}\left(\int_X D\big(f(t,\omega), \tilde{o}\big)dP(\omega)\right)$$

$$+ \|\tilde{L}(1)+\sqrt{\tilde{L}(1)}\|_\infty \Omega_1^{(\mathcal{F})}\big(f, \|\tilde{L}((\cdot-t)^2)(t)\|_\infty^{1/2}\big)_{L^1}.$$

Proof. Integrating (17.21) we obtain

$$\int_X D\big(M(f)(t,\omega), f(t,\omega)\big)dP(\omega)$$

$$\leq \int_X\left(\int_{[a,b]} D\big(f(s,\omega), f(t,\omega)\big)\mu_t(ds)\right)dP(\omega)$$

$$+ |m_t-1|\left(\int_X D\big(f(t,\omega), \tilde{o}\big)dP(\omega)\right)=$$

(by $D\geq 0$ and the facts that $D\big(f(s,\omega), f(t,\omega)\big)$ is continuous in $s\in[a,b]$, by Lemma 1 of [11], also it is a real random variable in ω, by Remark 13.39 of [66], p. 654 and thus by Proposition 3.3(i) of [17] it is jointly measurable in (s,ω), and then being able to use Tonelli–Fubini's theorem, p. 104 of [62] and thus see both double integrals make sense)

$$\int_{[a,b]}\left(\int_X D\big(f(s,\omega), f(t,\omega)\big)dP(\omega)\right)\mu_t(ds)$$

$$+ |m_t-1|\left(\int_X D\big(f(t,\omega), \tilde{o}\big)dP(\omega)\right)$$

$$(h>0)$$

$$\leq \int_{[a,b]} \Omega_1^{(\mathcal{F})}\left(f, \frac{|s-t|}{h}h\right)_{L^1}\mu_t(ds)$$

$$+ |m_t-1|\left(\int_X D\big(f(t,\omega), \tilde{o}\big)dP(\omega)\right) \quad \text{(by (17.7))}$$

$$\leq \Omega_1^{(\mathcal{F})}(f,h)_{L^1}\int_{[a,b]}\left\lceil\frac{|s-t|}{h}\right\rceil\mu_t(ds)$$

$$+ |m_t-1|\left(\int_X D\big(f(t,\omega), \tilde{o}\big)dP(\omega)\right)$$

$$\leq |m_t - 1| \left(\int_X D\big(f(t,\omega), \tilde{o}\big) dP(\omega) \right)$$

$$+ \left(\int_{[a,b]} \left(1 + \frac{|s-t|}{h} \right) \mu_t(ds) \right) \Omega_1^{(\mathcal{F})}(f,h)_{L^1}$$

$$= |m_t - 1| \left(\int_X D\big(f(t,\omega), \tilde{o}\big) dP(\omega) \right)$$

$$+ \left(m_t + \frac{1}{h} \int_{[a,b]} |s-t| \mu_t(ds) \right) \Omega_1^{(\mathcal{F})}(f,h)_{L^1},$$

(by Cauchy–Schwarz inequality)

$$\leq |m_t - 1| \left(\int_X D\big(f(t,\omega), \tilde{o}\big) dP(\omega) \right)$$

$$+ \left(m_t + \frac{1}{h}\sqrt{m_t} \left(\int_{[a,b]} (s-t)^2 \mu_t(ds) \right)^{1/2} \right) \Omega_1^{(\mathcal{F})}(f,h)_{L^1}$$

(by choosing

$$h := \left(\int_{[a,b]} (s-t)^2 \mu_t(ds) \right)^{1/2} = \big(\tilde{L}((\cdot - t)^2)(t) \big)^{1/2} > 0,$$

for > 0 it is enough to assume $\mu_t([a,b] - \{t\}) > 0)$

$$\leq |m_t - 1| \left(\int_X D\big(f(t,\omega), \tilde{o}\big) dP(\omega) \right)$$

$$+ (m_t + \sqrt{m_t}) \Omega_1^{(\mathcal{F})} \left(f, \big(\tilde{L}((\cdot - t)^2)(t) \big)^{1/2} \right)_{L^1},$$

by using (17.22) we have established (17.23). One can easily see that if

$$\tilde{L}((\cdot - t)^2)(t) = 0$$

then again (17.23) is valid. Clearly by Remark 13.39, p. 654, [66] $D\big(f(t,\omega), \tilde{o}\big)$ is a real random variable in $\omega \in X$, for each $t \in [a,b]$.

Next we notice that

$$|D(f(x,\omega), \tilde{o}) - D(f(y,\omega), \tilde{o})| \leq D(f(x,\omega), f(y,\omega)) \; \forall x, y \in [a,b], \; \forall \omega \in X.$$

Hence $\forall \varepsilon > 0 \; \exists \delta > 0$ such that whenever $x, y \in [a,b]$ with $|x-y| \leq \delta$ then

$$\left| \int_X D\big(f(x,\omega), \tilde{o}\big) P(d\omega) - \int_X D\big(f(y,\omega), \tilde{o}\big) P(d\omega) \right|$$

$$\leq \int_X D\big(f(x,\omega), f(y,\omega)\big) P(d\omega) \leq \varepsilon,$$

because $f \in C_{\mathcal{F}R}^{U1}([a,b])$ by $f \in C_{\mathcal{F}R}^{U}([a,b])$. Therefore the function

$$F(x) := \int_X D\big(f(x,\omega),\tilde{o}\big)P(d\omega),$$

is continuous in $x \in [a,b]$ and hence is bounded, i.e. $\|F(x)\|_\infty < \infty$, making (17.24) valid. $\qquad\square$

We need the following

Proposition 17.31. *All here as in Proposition 17.29 and*

$$\int_X \big(D(f(t,\omega),\tilde{o})\big)^q P(d\omega) < \infty, \quad q > 1, \quad \forall t \in [a,b].$$

Then

$$\left(\int_X \big(D(M(f)(t,\omega),f(t,\omega))\big)^q dP(\omega)\right)^{1/q}$$

$$\leq |m_t - 1| \left(\int_X \big(D(f(t,\omega),\tilde{o})\big)^q P(d\omega)\right)^{1/q} \qquad (17.25)$$

$$+ m_t^{1-\frac{1}{q}} \left(\int_{[a,b]} \left(1 + \frac{|s-t|}{h}\right)^q d\mu_t(s)\right)^{1/q} \Omega_1^{(\mathcal{F})}(f,h)_{L^q},$$

$h > 0$, $\forall t \in [a,b]$.

Proof. Let $q > 1$ then by (17.21) we have

$$\left(\int_X D\big(M(f)(t,\omega),f(t,\omega)\big)^q dP(\omega)\right)^{1/q}$$

$$\leq \left(\int_X \left(\int_{[a,b]} D\big(f(s,\omega),f(t,\omega)\big)\mu_t(ds)\right)^q P(d\omega)\right)^{1/q} + \theta =: (*),$$

where

$$\theta := |m_t - 1| \left(\int_X \big(D(f(t,\omega),\tilde{o})\big)^q P(d\omega)\right)^{1/q}. \qquad (17.26)$$

Let $p > 1$ such that $\frac{1}{p} + \frac{1}{q} = 1$. Hence by Hölder's inequality we have

$$(*) \quad \leq \quad m_t^{1/p} \left(\int_X \left(\int_{[a,b]} D^q\big(f(s,\omega), f(t,\omega)\big) \mu_t(ds) \right) P(d\omega) \right)^{1/q} + \theta$$

(by Tonelli-Fubini's theorem as in the proof of Theorem 17.30)

$$= \quad m_t^{1/p} \left(\int_{[a,b]} \left(\int_X D^q\big(f(s,\omega), f(t,\omega)\big) P(d\omega) \right) \mu_t(ds) \right)^{1/q} + \theta$$

(let $h > 0$)

$$\leq \quad m_t^{1/p} \left(\int_{[a,b]} \left(\Omega_1^{(\mathcal{F})}\left(f, \frac{|s-t|}{h} h\right)_{L^q} \right)^q \mu_t(ds) \right)^{1/q} + \theta$$

$$\overset{(17.7)}{\leq} \quad m_t^{1/p} \left(\int_{[a,b]} \left(1 + \frac{|s-t|}{h}\right)^q \mu_t(ds) \right)^{1/q} \Omega_1^{(\mathcal{F})}(f,h)_{L^q} + \theta.$$

$\square$

We examine two cases and we give

Theorem 17.32. *Here we assume all as in Proposition 17.31.*
1) *Let* $q \in \mathbb{N} - \{1\}$. *Then*

$$\left(\int_X D^q\big(M(f)(t,\omega), f(t,\omega)\big) P(d\omega) \right)^{1/q}$$

$$\leq |\tilde{L}(1)(t) - 1| \left(\int_X D^q\big(f(t,\omega), \tilde{o}\big) P(d\omega) \right)^{1/q}$$

$$+ (\tilde{L}(1)(t))^{1 - \frac{1}{q}} \left(\sum_{k=0}^{q} \binom{q}{k} (\tilde{L}(1)(t))^{1-\frac{k}{q}} \right)^{1/q}$$

$$\times \Omega_1^{(\mathcal{F})}\big(f, ((\tilde{L}(1 \cdot -t|^q))(t))^{1/q}\big)_{L^q}, \qquad (17.27)$$

$\forall t \in [a,b]$.
2) *Let* $q > 1$ *real. Then*

$$\left(\int_X D^q\big(M(f)(t,\omega), f(t,\omega)\big) P(d\omega) \right)^{1/q}$$

$$\leq |\tilde{L}(1)(t) - 1| \left(\int_X D^q\big(f(t,\omega), \tilde{o}\big) P(d\omega) \right)^{1/q}$$

$$+ 2^{1-\frac{1}{q}} (\tilde{L}(1)(t))^{1 - \frac{1}{q}} (\tilde{L}(1)(t) + 1)^{1/q}$$

$$\times \Omega_1^{(\mathcal{F})}\big(f, ((\tilde{L}(| \cdot -t|^q))(t))^{1/q}\big)_{L^q}, \qquad (17.28)$$

$\forall t \in [a,b]$.

When $q \in \mathbb{N} - \{1\}$ then (17.27) is sharper than (17.28). Furthermore we have

3) Let $q \in \mathbb{N} - \{1\}$. Then

$$
\sup_{t \in [a,b]} \left(\int_X D^q \big(M(f)(t,\omega), f(t,\omega) \big) P(d\omega) \right)^{1/q}
$$

$$
\leq \|\tilde{L}(1) - 1\|_\infty \sup_{t \in [a,b]} \left(\int_X D^q \big(f(t,\omega), \tilde{o} \big) P(d\omega) \right)^{1/q}
$$

$$
+ \|\tilde{L}(1)\|_\infty^{1 - \frac{1}{q}} \left(\left\| \sum_{k=0}^{q} \binom{q}{k} (\tilde{L}(1))^{1 - \frac{k}{q}} \right\|_\infty \right)^{1/q}
$$

$$
\times \Omega_1^{(\mathcal{F})} \big(f, \|(\tilde{L}(| \cdot -t|^q))(t)\|_\infty^{1/q} \big)_{L^q}. \tag{17.29}
$$

4) Let $q > 1$ real. Then

$$
\sup_{t \in [a,b]} \left(\int_X D^q \big(M(f)(t,\omega), f(t,\omega) \big) P(d\omega) \right)^{1/q}
$$

$$
\leq \|\tilde{L}(1) - 1\|_\infty \sup_{t \in [a,b]} \left(\int_X D^q \big(f(t,\omega), \tilde{o} \big) P(d\omega) \right)^{1/q}
$$

$$
+ 2^{1 - \frac{1}{q}} \|\tilde{L}(1)\|_\infty^{1 - \frac{1}{q}} \|\tilde{L}(1) + 1\|_\infty^{1/q}
$$

$$
\times \Omega_1^{(\mathcal{F})} \big(f, \|(\tilde{L}(| \cdot -t|^q))(t)\|_\infty^{1/q} \big)_{L^q}. \tag{17.30}
$$

When $q \in \mathbb{N} - \{1\}$ inequality (17.29) is sharper than (17.30).

Proof. 1) We notice that

$$
\int_{[a,b]} \left(1 + \frac{|s-t|}{h} \right)^q d\mu_t(s) = \int_{[a,b]} \left(\sum_{k=0}^{q} \binom{q}{k} \frac{|s-t|^k}{h^k} \right) d\mu_t(s)
$$

$$
= m_t + \sum_{k=1}^{q-1} \binom{q}{k} \frac{1}{h^k} \int_{[a,b]} |s-t|^k d\mu_t(s) + \frac{1}{h^q} \int_{[a,b]} |s-t|^q d\mu_t(s).
$$

Also we see for $k = 1, \ldots, q-1$ that

$$
\int_{[a,b]} |s-t|^k d\mu_t(s) \leq m_t^{1 - \frac{k}{q}} \left(\int_{[a,b]} |s-t|^q d\mu_t(s) \right)^{k/q}.
$$

Hence

$$\int_{[a,b]} \left(1 + \frac{|s-t|}{h}\right)^q d\mu_t(s)$$

$$\leq m_t + \sum_{k=1}^{q-1} \binom{q}{k} \frac{1}{h^k} m_t^{1-\frac{k}{q}} \left(\int_{[a,b]} |s-t|^q d\mu_t(s)\right)^{k/q}$$

$$+ \frac{1}{h^q} \int_{[a,b]} |s-t|^q d\mu_t(s)$$

$$= \sum_{k=0}^{q} \binom{q}{k} \frac{m_t^{1-(k/q)}}{h^k} \left(\int_{[a,b]} |s-t|^q d\mu_t(s)\right)^{k/q}$$

$\big($by choosing

$$h := \left(\int_{[a,b]} |s-t|^q d\mu_t(s)\right)^{1/q}$$

$$= \left((\tilde{L}(|\cdot - t|^q))(t)\right)^{1/q} > 0)$$

$$= \sum_{k=0}^{q} \binom{q}{k} m_t^{1-\frac{k}{q}}. \tag{17.31}$$

That is, we got that

$$\int_{[a,b]} \left(1 + \frac{|s-t|}{h}\right)^q d\mu_t(s) \leq \sum_{k=0}^{q} \binom{q}{k} m_t^{1-\frac{k}{q}}.$$

Hence proving (17.27) with the use of (17.25). The inequality (17.27) is true easily if our choice is easily

$$h^q := \int_{[a,b]} |s-t|^q d\mu_t(s) = 0.$$

2) The function x^q is convex for $x \geq 0$, $q > 1$. Therefore

$$\left(\frac{1 + \frac{|s-t|}{h}}{2}\right)^q \leq \frac{1 + \frac{|s-t|^q}{h^q}}{2}, \quad h > 0.$$

I.e.

$$\left(1 + \frac{|s-t|}{h}\right)^q \leq 2^{q-1} \left(1 + \frac{|s-t|^q}{h^q}\right), \quad \forall s, t \in [a,b].$$

Hence

$$\left(\int_{[a,b]} \left(1 + \frac{|s-t|}{h} \right)^q d\mu_t(s) \right)^{1/q}$$

$$\leq 2^{1-\frac{1}{q}} \left(\int_{[a,b]} \left(1 + \frac{|s-t|^q}{h^q} \right) d\mu_t(s) \right)^{1/q}$$

$$= 2^{1-\frac{1}{q}} \left(m_t + \frac{1}{h^q} \left(\int_{[a,b]} |s-t|^q d\mu_t(s) \right) \right)^{1/q}$$

$\Big($by choosing again

$$h := \left(\int_{[a,b]} |s-t|^q d\mu_t(s) \right)^{1/q} > 0\Big)$$

$$= 2^{1-\frac{1}{q}} (m_t + 1)^{1/q}.$$

That is, we got that

$$\left(\int_{[a,b]} \left(1 + \frac{|s-t|}{h} \right)^q d\mu_t(s) \right)^{1/q} \leq 2^{1-\frac{1}{q}} (m_t + 1)^{1/q}.$$

Using (17.25) and the last estimate we obtain (17.28). Again if our above choice is $h = 0$ then (17.28) is still valid.

When $q \in \mathbb{N} - \{1\}$ and $m_t > 0$ we would like to prove that

$$\left(\sum_{k=0}^{q} \binom{q}{k} m_t^{1-\frac{k}{q}} \right)^{1/q} \leq 2^{1-\frac{1}{q}} (m_t + 1)^{1/q}, \qquad (17.32)$$

hence (17.27) is better than (17.28). Notice that (17.32) is trivially true when $m_t = 0$.

Equivalently we need valid

$$\sum_{k=0}^{q} \binom{q}{k} m_t^{1-\frac{k}{q}} \leq 2^{q-1} (m_t + 1)$$

$$\Leftrightarrow$$

$$\sum_{k=0}^{q} \binom{q}{k} m_t^{-k/q} \leq 2^{q-1} (1 + m_t^{-1}).$$

By calling $z := m_t^{-1} > 0$, equivalently we need true

$$\sum_{k=0}^{q} \binom{q}{k} z^{k/q} \leq 2^{q-1} (1 + z)$$

$$\Leftrightarrow$$

$$(1 + z^{1/q})^q \leq 2^{q-1} (1 + z).$$

The last is true by the convexity of z^q, $z \geq 0$, $q \in \mathbb{N} - \{1\}$. If $m_t = 0$, then both (17.27) and (17.28) are trivially the same.

Deriving (17.29), (17.30) from (17.27), (17.28), respectively, it is easy. Clearly $D^q(f(t,\omega), \tilde{o})$ is a real random variable in $\omega \in X$, $\forall t \in [a,b]$. Additionally we notice that $\forall \varepsilon > 0$ $\exists \delta > 0$ such that whenever $x, y \in [a,b]$ with $|x - y| \leq \delta$ then

$$\left| \left(\int_X D^q(f(x,\omega), \tilde{o}) dP(\omega) \right)^{1/q} - \left(\int_X D^q(f(y,\omega), \tilde{o}) dP(\omega) \right)^{1/q} \right|$$

$$\leq \left(\int_X D^q(f(x,\omega), f(y,\omega)) dP(\omega) \right)^{1/q} \leq \varepsilon, \quad \text{by } f \in C^U_{\mathcal{F}R}([a,b]).$$

Hence proving that the function

$$G(x) := \left(\int_X D^q(f(x,\omega), \tilde{o}) dP(\omega) \right)^{1/q},$$

is continuous in $x \in [a,b]$. Therefore $\|G\|_\infty < \infty$, making the inequalities (17.29), (17.30) valid. $\qquad \square$

Similar general results using a different initial estimate follow.

Lemma 17.33. *Let* $f : [a,b] \rightarrow \mathcal{L}_F(X, \mathcal{B}, P)$ *be fuzzy random function,* $1 \leq q < \infty$, $\delta > 0$. *Then*

$$\Omega_1^{(\mathcal{F})}(f, |x-y|)_{L^q} \leq \left(1 + \frac{(x-y)^2}{\delta^2} \right) \Omega_1^{(\mathcal{F})}(f, \delta)_{L^q}, \quad \forall x, y \in [a,b]. \quad (17.33)$$

Proof. We have by (17.7) that

$$\Omega_1^{(\mathcal{F})}(f, |x-y|)_{L^q} \leq \left(1 + \frac{|x-y|}{\delta} \right) \Omega_1^{(\mathcal{F})}(f, \delta)_{L^q}. \quad (17.34)$$

Let $|x - y| > \delta$, thus $\frac{|x-y|}{\delta} > 1$. Then

$$\text{R.H.S.}(17.34) \leq \left(1 + \frac{(x-y)^2}{\delta^2} \right) \Omega_1^{(\mathcal{F})}(f, \delta)_{L^q}.$$

Let $|x - y| \leq \delta$ then

$$\Omega_1^{(\mathcal{F})}(f, |x-y|) \leq \Omega_1^{(\mathcal{F})}(f, \delta)_{L^q} \leq \left(1 + \frac{(x-y)^2}{\delta^2} \right) \Omega_1^{(\mathcal{F})}(f, \delta)_{L^q}.$$

$\qquad \square$

We present

Theorem 17.34. *Assume all terms and assumptions of Proposition 17.29 and*

$$\int_X D(f(t,\omega), \tilde{o}) dP(\omega) < \infty, \quad \forall t \in [a,b].$$

Then
1)

$$\int_X D\big(M(f), (t, \omega), f(t, \omega)\big) dP(\omega)$$

$$\leq |\tilde{L}(1)(t) - 1| \left(\int_X D\big(f(t, \omega), \tilde{o}\big) dP(\omega) \right)$$

$$+ (\tilde{L}(1)(t) + 1)\Omega_1^{(\mathcal{F})} \left(f, \big(\tilde{L}((\cdot - t)^2)(t)\big)^{1/2} \right)_{L^1}, \quad (17.35)$$

$\forall t \in [a, b]$, *and*
2)

$$\sup_{t \in [a, b]} \int_X D\big(M(f)(t, \omega), f(t, \omega)\big) dP(\omega)$$

$$\leq \|\tilde{L}(1) - 1\|_\infty \sup_{t \in [a, b]} \left(\int_X D\big(f(t, \omega), \tilde{o}\big) dP(\omega) \right)$$

$$+ \|\tilde{L}(1) + 1\|_\infty \Omega_1^{(\mathcal{F})} \big(f, \|\tilde{L}((\cdot - t)^2)(t)\|_\infty^{1/2}\big)_{L^1}. \quad (17.36)$$

Proof. Initially from the proof of Theorem 17.30 we have

$$\int_X D\big(M(f)(t, \omega), f(t, \omega)\big) dP(\omega) \leq |m_t - 1| \left(\int_X D\big(f(t, \omega), \tilde{o}\big) dP(\omega) \right)$$

$$+ \int_{[a, b]} \Omega_1^{(\mathcal{F})}(f, |s - t|)_{L^1} \mu_t(s)$$

$$(\text{let } h > 0)$$

$$\overset{(17.33)}{\leq} |m_t - 1| \left(\int_X D\big(f(t, \omega), \tilde{o}\big) dP(\omega) \right)$$

$$+ \left(\int_{[a, b]} \left(1 + \frac{(s - t)^2}{h^2} \right) \mu_t(ds) \right) \Omega_1^{(\mathcal{F})}(f, h)_{L^1}$$

$$= |m_t - 1| \left(\int_X D\big(f(t, \omega), \tilde{o}\big) dP(\omega) \right)$$

$$+ \left(m_t + \frac{1}{h^2} \int_{[a, b]} (s - t)^2 \mu_t(ds) \right) \Omega_1^{(\mathcal{F})}(f, h)_{L^1}$$

taking

$$h := \left(\int_{[a, b]} (s - t)^2 \mu_t(dt) \right)^{1/2} = \big(\tilde{L}((\cdot - t)^2)(t)\big)^{1/2} > 0$$

$$(17.37))$$

$$= |m_t - 1| \left(\int_X D\big(f(t, \omega), \tilde{o}\big) dP(\omega) \right) + (m_t + 1)\Omega_1^{(\mathcal{F})} \left(f, \big(\tilde{L}((\cdot - t)^2)(t)\big)^{1/2} \right)_{L^1},$$

$\forall t \in [a,b]$. That is proving (17.35).

The above choice (17.37) of h if $h = 0$ makes again (17.35) valid. Inequality (17.36) is now clear.   □

Finally we get the very useful

Theorem 17.35. *Assume all terms and assumptions of Proposition 17.29 and*

$$\int_X D\big(f(t,\omega),\tilde{o}\big)dP(\omega) < \infty, \quad \forall t \in [a,b].$$

Then
1)

$$\int_X D\big(M(f)(t,\omega), f(t,\omega)\big)dP(\omega)$$

$$\leq |\tilde{L}(1)(t) - 1| \left(\int_X D\big(f(t,\omega),\tilde{o}\big)dP(\omega) \right)$$

$$+ \min \left\{ (\tilde{L}(1)(t) + \sqrt{\tilde{L}(1)(t)}), (\tilde{L}(1)(t) + 1) \right\}$$

$$\Omega_1^{(\mathcal{F})}\big(f, (\tilde{L}((\cdot - t)^2)(t))^{1/2}\big)_{L^1}, \quad \forall t \in [a,b], \quad (17.38)$$

and
2)

$$\sup_{t\in[a,b]} \left(\int_X D\big(M(f)(t,\omega), f(t,\omega)\big)dP(\omega) \right)$$

$$\leq \|\tilde{L}(1) - 1\|_\infty \sup_{t\in[a,b]} \left(\int_X D\big(f(t,\omega),\tilde{o}\big)dP(\omega) \right)$$

$$+ \min \left\{ \|\tilde{L}(1) + \sqrt{\tilde{L}(1)}\|_\infty, \|\tilde{L}(1) + 1\|_\infty \right\}$$

$$\Omega_1^{(\mathcal{F})}\big(f, \|\tilde{L}((\cdot - t)^2)(t)\|_\infty^{1/2}\big)_{L^1}. \quad (17.39)$$

Proof. Obvious from Theorems 17.30 and 17.34.   □

The corresponding results for $q > 1$ follow.

Theorem 17.36. *Here we assume all as in Proposition 17.31. Let $q \in \mathbb{N} - \{1\}$. Then*

1)

$$\left(\int_X D^q\big(M(f)(t,\omega), f(t,\omega)\big) dP(\omega) \right)^{1/q}$$

$$\leq |\tilde{L}(1)(t) - 1| \left(\int_X \big(D(f(t,\omega), \tilde{o})\big)^q dP(\omega) \right)^{1/q}$$

$$+ (\tilde{L}(1)(t))^{1-\frac{1}{q}} \left(\sum_{k=0}^q \binom{q}{k} \big((\tilde{L}(1))(t)\big)^{1-\frac{k}{q}} \right)^{1/q}$$

$$\Omega_1^{(\mathcal{F})}\big(f, ((\tilde{L}(\cdot - t)^{2q})(t))^{1/2q}\big)_{L^q}, \qquad (17.40)$$

$\forall t \in [a,b]$.

And also holds

2)

$$\sup_{t \in [a,b]} \left(\int_X D\big(M(f)(t,\omega), f(t,\omega)\big)^q dP(\omega) \right)^{1/q}$$

$$\leq \|\tilde{L}(1) - 1\|_\infty \sup_{t \in [a,b]} \left(\int_X D^q\big(f(t,\omega), \tilde{o}\big) dP(\omega) \right)^{1/q} \qquad (17.41)$$

$$+ \|\tilde{L}(1)\|_\infty^{1-\frac{1}{q}} \left(\left\| \sum_{k=0}^q \binom{q}{k} (\tilde{L}(1))^{1-\frac{k}{q}} \right\|_\infty \right)^{1/q}$$

$$\Omega_1^{(\mathcal{F})}\big(f, \|(\tilde{L}(\cdot - t)^{2q})(t)\|_\infty^{1/2q}\big)_{L^q}.$$

Let $q > 1$ real. Then

3)

$$\left(\int_X D\big(M(f)(t,\omega), f(t,\omega)\big)^q dP(\omega) \right)^{1/q}$$

$$\leq |\tilde{L}(1)(t) - 1| \left(\int_X \big(D(f(t,\omega), \tilde{o})\big)^q dP(\omega) \right)^{1/q} \qquad (17.42)$$

$$+ 2^{1-\frac{1}{q}} (\tilde{L}(1)(t))^{1-\frac{1}{q}} (\tilde{L}(1)(t) + 1)^{1/q} \Omega_1^{(\mathcal{F})}\big(f, ((\tilde{L}(\cdot - t)^{2q})(t))^{1/2q}\big)_{L^q},$$

$\forall t \in [a,b]$. *And also holds*

4)

$$\sup_{t \in [a,b]} \left(\int_X D\big(M(f)(t,\omega), f(t,\omega)\big)^q dP(\omega) \right)^{1/q}$$

$$\leq \|\tilde{L}(1) - 1\|_\infty \sup_{t \in [a,b]} \left(\int_X D^q(f(t,\omega), \tilde{o}) dP(\omega) \right)^{1/q} \qquad (17.43)$$

$$+ 2^{1-\frac{1}{q}} \|\tilde{L}(1)\|_\infty^{1-\frac{1}{q}} \|\tilde{L}(1) + 1\|_\infty^{1/q} \Omega_1^{(\mathcal{F})}\big(f, \|(\tilde{L}(\cdot - t)^{2q})(t)\|_\infty^{1/2q}\big)_{L^q}.$$

When $q \in \mathbb{N} - \{1\}$ then inequality (17.40) is sharper than (17.42) and inequality (17.41) is sharper than (17.43).

Note 17.37. We see later that inequalities (17.40)–(17.43) and/or inequalities (17.27)–(17.30) can be used to prove "q-mean" convergence with rates of a sequence of M's to unit operator I.

Proof. Initially from the proof of Proposition 17.31 we obtain

$$\left(\int_X \left(D(M(f)(t,\omega), f(t,\omega)) \right)^q dP(\omega) \right)^{1/q}$$

$$\leq \theta + \left(\int_X \left(\int_{[a,b]} D\big(f(s,\omega), f(t,\omega)\big) \mu_t(dt) \right)^q dP(\omega) \right)^{1/q}$$

$$(\theta \text{ as in } (17.26))$$

$$\leq \theta + m_t^{1-\frac{1}{q}} \left(\int_{[a,b]} \big(\Omega_1^{(\mathcal{F})}(f, |s-t|)_{L^q} \big)^q d\mu_t(s) \right)^{1/q}$$

$$(\text{let } h > 0)$$

$$\overset{(\text{by } (17.33))}{\leq} \theta + m_t^{1-\frac{1}{q}} \left(\int_{[a,b]} \left(1 + \frac{(s-t)^2}{h^2} \right)^q d\mu_t(s) \right)^{1/q} \Omega_1^{(\mathcal{F})}(f,h)_{L^q}$$

$$= \; : (\xi).$$

1) Let $q \in \mathbb{N} - \{1\}$. We observe that

$$\int_{[a,b]} \left(1 + \frac{(s-t)^2}{h^2} \right)^q d\mu_t(s) = \int_{[a,b]} \left(\sum_{k=0}^{q} \binom{q}{k} \frac{(s-t)^{2k}}{h^{2k}} \right) d\mu_t(s)$$

$$= m_t + \sum_{k=1}^{q-1} \binom{q}{k} \frac{1}{h^{2k}} \left(\int_{[a,b]} (s-t)^{2k} d\mu_t(s) \right) + \frac{1}{h^{2q}} \left(\int_{[a,b]} (s-t)^{2q} d\mu_t(s) \right).$$

For $k = 1, \ldots, q-1$, $\frac{q}{k} > 1$ and by Hölder's inequality we have

$$\int_{[a,b]} (s-t)^{2k} d\mu_t(s) \leq m_t^{1-\frac{k}{q}} \left(\int_{[a,b]} (s-t)^{2q} d\mu_t(s) \right)^{k/q}.$$

Hence

$$\int_{[a,b]} \left(1 + \frac{(s-t)^2}{h^2} \right)^q d\mu_t(s) \leq \sum_{k=0}^{q} \binom{q}{k} \frac{m_t^{1-(k/q)}}{h^{2k}} \left(\int_{[a,b]} (s-t)^{2q} d\mu_t(s) \right)^{k/q}$$

(by choosing

$$h := \left(\int_{[a,b]} (s-t)^{2q} d\mu_t(s) \right)^{1/2q}$$

$$= ((\tilde{L}((\cdot - t)^{2q}))(t))^{1/2q} > 0)$$

$$\le \sum_{k=0}^{q} \binom{q}{k} m_t^{1-\frac{k}{q}}. \tag{17.44}$$

That is

$$\left(\int_{[a,b]} \left(1 + \frac{(s-t)^2}{h^2} \right)^q d\mu_t(s) \right)^{1/q} \le \left(\sum_{k=0}^{q} \binom{q}{k} m_t^{1-\frac{k}{q}} \right)^{1/q}. \tag{17.45}$$

Thus by (17.44) and (17.45) we have

$$(\xi) \le \theta + (\tilde{L}(1)(t))^{1-\frac{1}{q}} \left(\sum_{k=0}^{q} \binom{q}{k} (\tilde{L}(1)(t))^{1-\frac{k}{q}} \right)^{1/q}$$

$$\times \, \Omega_1^{(\mathcal{F})} \left(f, ((\tilde{L}((\cdot - t)^{2q}))(t))^{1/2q} \right)_{L^q}, \quad \forall t \in [a,b].$$

That is establishing (17.40). When the choice (17.44) of $h = 0$ then again (17.40) is trivially valid.

2) Let now $q > 1$ real, then again by convexity of x^q, $x \ge 0$ we have

$$\left(1 + \frac{(s-t)^2}{h^2} \right)^q \le 2^{q-1} \left(1 + \frac{(s-t)^{2q}}{h^{2q}} \right), \quad h > 0, \; \forall s,t \in [a,b].$$

Hence

$$(\xi) \le \theta + m_t^{1-\frac{1}{q}} 2^{1-\frac{1}{q}} \left(\int_{[a,b]} \left(1 + \frac{(s-t)^{2q}}{h^{2q}} \right) d\mu_t(s) \right)^{1/q} \Omega_1^{(\mathcal{F})}(f,h)_{L^q}$$

$$= \theta + 2^{1-\frac{1}{q}} m_t^{1-\frac{1}{q}} \left[m_t + \frac{1}{h^{2q}} \int_{[a,b]} (s-t)^{2q} d\mu_t(s) \right]^{1/q} \Omega_1^{(\mathcal{F})}(f,h)_{L^q}$$

$$\text{(let } h > 0 \text{ as in (17.44))}$$

$$= \theta + 2^{1-\frac{1}{q}} m_t^{1-\frac{1}{q}} (m_t + 1)^{1/q} \Omega_1^{(\mathcal{F})}(f,h)_{L^q}$$

$$= \theta + 2^{1-\frac{1}{q}} (\tilde{L}(1)(t))^{1-\frac{1}{q}} (\tilde{L}(1)(t) + 1)^{1/q} \Omega_1^{(\mathcal{F})} \left(f, ((\tilde{L}((\cdot - t)^{2q}))(t))^{1/2q} \right)_{L^q},$$

$\forall t \in [a,b]$. That is proving (17.42).

When the choice (17.44) for $h = 0$ then inequality (17.42) is trivially valid. Inequalities (17.41) and (17.43) derive easily from (17.40) and (17.42), respectively, and they are valid, similarly, as inequalities (17.29) and (17.30). The comparison of inequalities is the same as in Theorem 17.32. $\quad \square$

Finally we derive

Theorem 17.38. *Here we suppose all as in Proposition 17.31. Let $q \in \mathbb{N} - \{1\}$. Then*
1)

$$\left(\int_X D^q \big(M(f)(t,\omega), f(t,\omega) \big) dP(\omega) \right)^{1/q}$$

$$\leq |\tilde{L}(1)(t) - 1| \left(\int_X D^q \big(f(t,\omega), \tilde{o} \big) dP(\omega) \right)^{1/q}$$

$$+ (\tilde{L}(1)(t))^{1-\frac{1}{q}} \left(\sum_{k=0}^q \binom{q}{k} ((\tilde{L}(1))(t))^{1-\frac{k}{q}} \right)^{1/q}$$

$$\times \min\{\Omega_1^{(\mathcal{F})} \big(f, ((\tilde{L}(|\cdot - t|^q))(t))^{1/q} \big)_{L^q},$$

$$\Omega_1^{(\mathcal{F})} \big(f, ((\tilde{L}(\cdot - t)^{2q})(t))^{1/2q} \big)_{L^q}\}, \qquad (17.46)$$

$\forall t \in [a,b]$.
And also holds
2)

$$\sup_{t \in [a,b]} \left(\int_X D^q \big(M(f)(t,\omega), f(t,\omega) \big) dP(\omega) \right)^{1/q}$$

$$\leq \|\tilde{L}(1) - 1\|_\infty \sup_{t \in [a,b]} \left(\int_X D^q \big(f(t,\omega), \tilde{o} \big) dP(\omega) \right)^{1/q}$$

$$+ \|\tilde{L}(1)\|_\infty^{1-\frac{1}{q}} \left(\left\| \sum_{k=0}^q \binom{q}{k} (\tilde{L}(1))^{1-\frac{k}{q}} \right\|_\infty \right)^{1/q}$$

$$\times \min\{\Omega_1^{(\mathcal{F})} \big(f, \|(\tilde{L}(|\cdot - t|^q))(t)\|_\infty^{1/q} \big)_{L^q},$$

$$\Omega_1^{(\mathcal{F})} \big(f, \|(\tilde{L}(\cdot - t)^{2q})(t)\|_\infty^{1/2q} \big)_{L^q}\}. \qquad (17.47)$$

Let $q > 1$ real. Then
3)

$$\left(\int_X D \big(M(f)(t,\omega), f(t,\omega) \big)^q dP(\omega) \right)^{1/q}$$

$$\leq |\tilde{L}(1)(t) - 1| \left(\int_X \big(D(f(t,\omega), \tilde{o}) \big)^q dP(\omega) \right)^{1/q}$$

$$+ 2^{1-\frac{1}{q}} (\tilde{L}(1)(t))^{1-\frac{1}{q}} (\tilde{L}(1)(t) + 1)^{1/q}$$

$$\times \min\{\Omega_1^{(\mathcal{F})} \big(f, ((\tilde{L}(|\cdot - t|^q))(t))^{1/q} \big)_{L^q},$$

$$\Omega_1^{(\mathcal{F})} \big(f, ((\tilde{L}(\cdot - t)^{2q})(t))^{1/2q} \big)_{L^q}, \}, \qquad (17.48)$$

$\forall t \in [a, b]$.

And also holds

4)

$$\sup_{t \in [a,b]} \left(\int_X D^q \big(M(f)(t, \omega), f(t, \omega) \big) P(d\omega) \right)^{1/q}$$

$$\leq \|\tilde{L}(1) - 1\|_\infty \sup_{t \in [a,b]} \left(\int_X D^q \big(f(t, \omega), \tilde{o} \big) P(d\omega) \right)^{1/q}$$

$$+ 2^{1-\frac{1}{q}} \|\tilde{L}(1)\|_\infty^{1-\frac{1}{q}} \|\tilde{L}(1) + 1\|_\infty^{1/q}$$

$$\times \min \{ \Omega_1^{(\mathcal{F})} \big(f, \|(\tilde{L}(|\cdot - t|^q))(t)\|_\infty^{1/q} \big)_{L^q},$$

$$\Omega_1^{(\mathcal{F})} \big(f, \|(\tilde{L}(\cdot - t)^{2q})(t)\|_\infty^{1/2q} \big)_{L^q} \}. \tag{17.49}$$

When $q \in \mathbb{N} - \{1\}$ then inequality (17.46) is sharper than (17.48) and (17.47) sharper than (17.49).

Proof. By Theorems 17.32 and 17.36.   □

We give

Corollary 17.39. All here as in Proposition 17.29 and

$$\int_X D^2 \big(f(t, \omega), \tilde{o} \big) dP(\omega) < \infty, \quad \forall t \in [a, b].$$

Then

1)

$$\left(\int_X D^2 \big(M(f)(t, \omega), f(t, \omega) \big) P(d\omega) \right)^{1/2} \tag{17.50}$$

$$\leq |\tilde{L}(1)(t) - 1| \left(\int_X D^2 \big(f(t, \omega), \tilde{o} \big) P(d\omega) \right)^{1/2}$$

$$+ (\tilde{L}(1)(t))^{1/2} \big(\tilde{L}(1)(t) + 2(\tilde{L}(1)(t))^{1/2} + 1 \big)^{1/2}$$

$$\Omega_1^{(\mathcal{F})} \big(f, ((\tilde{L}((\cdot - t)^2))(t))^{1/2} \big)_{L^2}, \quad \forall t \in [a, b].$$

and

2)

$$\sup_{t \in [a,b]} \left(\int_X D^2 \big(M(f)(t, \omega), f(t, \omega) \big) P(d\omega) \right)^{1/2}$$

$$\leq \|\tilde{L}(1) - 1\|_\infty \sup_{t \in [a,b]} \left(\int_X D^2 \big(f(t, \omega), \tilde{o} \big) P(d\omega) \right)^{1/2} \tag{17.51}$$

$$+ \|\tilde{L}(1)\|_\infty^{1/2} \|\tilde{L}(1) + 2(\tilde{L}(1))^{1/2} + 1\|_\infty^{1/2}$$

$$\Omega_1^{(\mathcal{F})} \big(f, \|(\tilde{L}((\cdot - t)^2))(t)\|_\infty^{1/2} \big)_{L^q}.$$

Proof. By Theorem 17.32, inequalities (17.27) and (17.29).   □

All inequalities presented in this chapter are of Shisha–Mond type (see [96]) in the fuzzy-random sense. We will derive next some Fuzzy-Random Korovkin Theorems regarding the spaces of functions

$$K_q([a,b]) := \left\{ f \in C^U_{\mathcal{F}R}([a,b]) \colon \int_X D^q\big(f(t,\omega),\tilde{o}\big)dP(\omega) < \infty, \ \forall t \in [a,b] \right\},$$

where $1 \le q < \infty$. We observe that if $1 \le k < \infty$ such that $k \le q$ then

$$K_q([a,b]) \subseteq K_k([a,b]).$$

For the above purpose we need to put together the following assumptions and settings.

Assumption 17.40. *Let $(X, \mathcal{B}, P)$ be a probability space, $[a,b] \subset \mathbb{R}$, $f \in C^U_{\mathcal{F}R}([a,b])$. Let $\{L_n\}_{n\in\mathbb{N}}$ be a sequence of fuzzy positive linear operators from $C_\mathcal{F}([a,b])$ into itself for which there exists a corresponding sequence of positive linear operators $\{\tilde{L}\}_{n\in\mathbb{N}}$ from $C([a,b])$ into itself such that*

$$(L_n g)^{(r)}_\pm = \tilde{L}_n(g^{(r)}_\pm), \tag{17.52}$$

respectively, $\forall r \in [0,1]$, $\forall n \in \mathbb{N}$, $\forall g \in C_\mathcal{F}([a,b])$. We then consider the sequence of positive fuzzy random linear operators $\{M_n\}_{n\in\mathbb{N}}$ from $C^U_{\mathcal{F}R}([a,b])$ into $C_{\mathcal{F}R}([a,b])$ defined by

$$M_n(f)(t,\omega) := L_n(f(\cdot,\omega))(t), \tag{17.53}$$

$\forall (t,\omega) \in [a,b] \times X$, $\forall n \in \mathbb{N}$, $\forall f \in C^U_{\mathcal{F}R}([a,b])$. Here I is the fuzzy random unit operator, i.e. $I(f)(t,\omega) = f(t,\omega)$, $\forall (t,\omega) \in [a,b] \times X$. We assume also that $\{\tilde{L}_n(1)\}_{n\in\mathbb{N}}$ is bounded.

From Theorem 17.35 we have

Corollary 17.41. *Here all are as in Assumption 17.40, and*

$$\int_X D\big(f(t,\omega),\tilde{o}\big)dP(\omega) < \infty, \quad \forall t \in [a,b].$$

Then
1)

$$\int_X D\big(M_n(f)(t,\omega), f(t,\omega)\big)dP(\omega)$$

$$\le |\tilde{L}_n(1)(t) - 1| \left(\int_X D\big(f(t,\omega),\tilde{o}\big)dP(\omega) \right)$$

$$+ \min \left\{ \big(\tilde{L}_n(1)(t) + \sqrt{\tilde{L}_n(1)(t)}\big), \big(\tilde{L}_n(1)(t) + 1\big) \right\}$$

$$\times \Omega_1^{(\mathcal{F})}\big(f, (\tilde{L}_n((\cdot - t)^2)(t))^{1/2}\big)_{L^1}, \tag{17.54}$$

$\forall t \in [a,b], \quad \forall n \in \mathbb{N},$
  and
  2)

$$\sup_{t\in[a,b]} \left(\int_X D\big(M_n(f)(t,\omega), f(t,\omega)\big) dP(\omega) \right)$$

$$\leq \|\tilde{L}_n(1) - 1\|_\infty \sup_{t\in[a,b]} \left(\int_X D\big(f(t,\omega), \tilde{o}\big) dP(\omega) \right)$$

$$+ \min\left\{ \|\tilde{L}_n(1) + \sqrt{\tilde{L}_n(1)}\|_\infty, \|\tilde{L}_n(1) + 1\|_\infty \right\}$$

$$\times \Omega_1^{(\mathcal{F})}\big(f, \|\tilde{L}_n((\cdot - t)^2)(t)\|_\infty^{1/2}\big)_{L^1}, \qquad (17.55)$$

$\forall n \in \mathbb{N}.$
  From Corollary 17.39 we obtain

Corollary 17.42. *Here all are as in Assumption 17.40, and*

$$\int_X D^2\big(f(t,\omega), \tilde{o}\big) dP(\omega) < \infty, \quad \forall t \in [a,b].$$

Then
  1)

$$\left(\int_X D^2\big(M_n(f)(t,\omega), f(t,\omega)\big) P(d\omega) \right)^{1/2}$$

$$\leq |\tilde{L}_n(1)(t) - 1| \left(\int_X D^2\big(f(t,\omega), \tilde{o}\big) P(d\omega) \right)^{1/2}$$

$$+ (\tilde{L}_n(1)(t))^{1/2} \big(\tilde{L}_n(1)(t) + 2(\tilde{L}_n(1)(t))^{1/2} + 1\big)^{1/2}$$

$$\times \Omega_1^{(\mathcal{F})}\big(f, ((\tilde{L}_n((\cdot - t)^2))(t))^{1/2}\big)_{L^2}, \qquad (17.56)$$

$\forall t \in [a,b], \quad \forall n \in \mathbb{N}.$
  And also holds
  2)

$$\sup_{t\in[a,b]} \left(\int_X D^2\big(M_n(f)(t,\omega), f(t,\omega)\big) P(d\omega) \right)^{1/2}$$

$$\leq \|\tilde{L}_n(1) - 1\|_\infty \sup_{t\in[a,b]} \left(\int_X D^2\big(f(t,\omega), \tilde{o}\big) P(d\omega) \right)^{1/2}$$

$$+ \|\tilde{L}_n(1)\|_\infty^{1/2} \|\tilde{L}_n(1) + 2(\tilde{L}_n(1))^{1/2} + 1\|_\infty^{1/2}$$

$$\times \Omega_1^{(\mathcal{F})}\big(f, \|(\tilde{L}_n((\cdot - t)^2))(t)\|_\infty^{1/2}\big)_{L^2}, \qquad (17.57)$$

$\forall n \in \mathbb{N}.$

Note 17.43. One sees from [96] that

$$\left\|\left(\tilde{L}_n((\cdot - t)^2)\right)(t)\right\|_\infty \;\leq\; \left\|\tilde{L}_n(x^2)(t) - t^2\right\|_\infty \tag{17.58}$$
$$+ 2c\|L_n(x)(t) - t\|_\infty + c^2\|L_n(1)(t) - 1\|_\infty,$$

where $c := \max(|a|,|b|)$, $\forall n \in \mathbb{N}$. Then one from the above fuzzy random Shisha–Mond type inequalities (17.55) and (17.57) derives the following basic fuzzy random Korovkin Theorems, see also [78].

Theorem 17.44. *Here all are as in Assumption 17.40. Furthermore assume that*

$$\tilde{L}_n(1) \xrightarrow{u} 1, \quad \tilde{L}_n(id) \xrightarrow{u} id, \quad \tilde{L}_n(id^2) \xrightarrow{u} id^2,$$

as $n \to \infty$, where u means uniformly and id is the identity map. Then

$$\lim_{n\to\infty} \left\| \int_X D\big(M_n(f)(t,\omega), f(t,\omega)\big)dP(\omega) \right\|_{\infty,t} = 0, \quad \forall f \in K_1([a,b]).$$
$$\tag{17.59}$$

I.e.

$$M_n(f)(t,\omega) \;\overset{\text{``1-mean''}}{\underset{n\,\to\,\infty}{\longrightarrow}}\; f(t,\omega), \tag{17.60}$$

uniformly, $\forall f \in K_1([a,b])$, that is uniformly $M_n \overset{\text{``1-mean''}}{\longrightarrow} I$, as $n \to \infty$, on $K_1([a,b])$.

Proof. Using (17.55), (17.58) and Proposition 17.9(ii). $\qquad\qquad\square$

We continue with the second basic fuzzy random Korovkin theorem.

Theorem 17.45. *Here all are as in Assumption 17.40. Furthermore suppose that*

$$\tilde{L}_n(1) \xrightarrow{u} 1, \quad \tilde{L}_n(id) \xrightarrow{u} id, \quad \tilde{L}_n(id^2) \xrightarrow{u} id^2, \quad as\ n \to \infty.$$

Then

$$\lim_{n\to\infty} \left\| \int_X D^2\big(M_n(f)(t,\omega), f(t,\omega)\big)P(d\omega) \right\|_{\infty,t} = 0, \quad \forall f \in K_2([a,b]).$$
$$\tag{17.61}$$

I.e.

$$M_n(f)(t,\omega) \overset{\text{``2-mean''}}{\longrightarrow} f(t,\omega), \tag{17.62}$$

uniformly, $\forall f \in K_2([a,b])$, that is uniformly $M_n \overset{\text{``2-mean''}}{\longrightarrow} I$, as $n \to \infty$, on $K_2([a,b])$.

Proof. From (17.57), (17.58) and Proposition 17.9(ii). $\qquad\qquad\square$

The related general fuzzy random Korovkin theorem follows.

Theorem 17.46. *Here all are as in Assumption 17.40, $q > 2$. Furthermore we assume that*

(i) $\tilde{L}_n(1) \xrightarrow[n \to \infty]{u} 1,$

and

(ii)
$$\lim_{n \to \infty} \|(\tilde{L}_n(|\cdot - t|^q))(t)\|_\infty = 0, \qquad (17.63)$$

or

(ii)' $\lim_{n \to \infty} \|(\tilde{L}_n(\cdot - t)^{2q})(t)\|_\infty = 0.$

Then

$$\lim_{n \to \infty} \left\| \int_X D^q\big(M_n(f)(t,\omega), f(t,\omega)\big)P(d\omega) \right\|_{\infty,t} = 0, \quad \forall f \in K_q([a,b]). \qquad (17.64)$$

I.e.

$$M_n(f)(t,\omega) \xrightarrow{\text{``}q\text{-mean''}} f(t,\omega), \qquad (17.65)$$

uniformly, $\forall f \in K_q([a,b])$, *that is uniformly* $M_n \xrightarrow{\text{``}q\text{-mean''}} I$, *as* $n \to \infty$, *on* $K_q([a,b])$.

Proof. By (17.30) or (17.43) and Proposition 17.9(ii). In fact (ii)' implies (ii). So one can use (ii) or (ii)' as long as it is easier to be verified. $\quad\square$

The case $m_t = \tilde{L}(1)(t) = 1$, $\forall t \in [a,b]$ is a very important and common one. Then all results of the chapter simplify a lot as follows.

Proposition 17.47. *All here as in Proposition 17.29 and* $m_t = 1$, $\forall t \in [a,b]$. *Then*

$$D\big(M(f)(t,\omega), f(t,\omega)\big) \le \int_{[a,b]} D\big(f(s,\omega), f(t,\omega)\big)\mu_t(ds), \quad \forall(t,\omega) \in [a,b]\times X, \qquad (17.66)$$

where μ_t *is as in (17.10).*

Proof. We notice that the B-measurable function

$$D\big(M(f)(t,\omega), f(t,\omega)\big) \overset{(17.15)}{=} D\left(\int_{[a,b]} f(s,\omega)\mu_t(ds), \int_{[a,b]} f(t,\omega)\mu_t(ds) \right)$$

$$\underset{\text{(by Theorem 17.14(2))}}{\le} \int_{[a,b]} D\big(f(s,\omega), f(t,\omega)\big)\mu_t(ds).$$

$\square$

Thus we obtain

Theorem 17.48. *All here as in Proposition 17.47. Then*

1)

$$\int_X D\big(M(f)(t,\omega), f(t,\omega)\big)dP(\omega) \le 2\Omega_1^{(\mathcal{F})}\big(f, (\tilde{L}((\cdot-t)^2)(t))^{1/2}\big)_{L^1}, \quad \forall t \in [a,b],$$

(17.67)

and

2)

$$\sup_{t\in[a,b]} \int_X D\big(M(f)(t,\omega), f(t,\omega)\big)dP(\omega) \le 2\Omega_1^{(\mathcal{F})}\big(f, \|\tilde{L}((\cdot-t)^2)(t)\|_\infty^{1/2}\big)_{L^1}.$$

(17.68)

Proof. By integrating (17.66), then it follows, in a simpler way, as the proof of Theorem 17.30. ☐

Also we have

Proposition 17.49. *All here as in Proposition 17.47, $q > 1$. Then*

$$\left(\int_X \big(D(M(f)(t,\omega), f(t,\omega))^q dP(\omega)\right)^{1/q}$$

$$\le \left(\int_{[a,b]} \left(1 + \frac{|s-t|}{h}\right)^q d\mu_t(s)\right)^{1/q} \Omega_1^{(\mathcal{F})}(f,h)_{L^q}, \quad (17.69)$$

$h > 0$, $\forall t \in [a,b]$.

Proof. Using (17.66) in exactly the same but simpler manner as in the proof of Proposition 17.31. ☐

We present

Theorem 17.50. *All here as in Proposition 17.47, $q > 1$. Then*

1)

$$\left(\int_X D^q\big(M(f)(t,\omega), f(t,\omega)\big)P(d\omega)\right)^{1/q} \quad (17.70)$$

$$\le 2\Omega_1^{(\mathcal{F})}\big(f, ((\tilde{L}(|\cdot-t|^q))(t))^{1/q}\big)_{L^q}, \quad \forall t \in [a,b],$$

and

2)

$$\sup_{t\in[a,b]} \left(\int_X D^q\big(M(f)(t,\omega), f(t,\omega)\big)P(d\omega)\right)^{1/q}$$

$$\le 2\Omega_1^{(\mathcal{F})}\big(f, \|(\tilde{L}(|\cdot-t|^q))(t)\|_\infty^{1/q}\big)_{L^q}. \quad (17.71)$$

Proof. We use (17.69) and it follows similarly as the proof of Theorem 17.32. ☐

We also give

Theorem 17.51. *Here all as in Proposition 17.29, $m_t = 1$, $\forall t \in [a, b]$ and $q > 1$. Then*
1)

$$\left(\int_X D^q \big(M(f)(t, \omega), f(t, \omega) \big) dP(\omega) \right)^{1/q}$$
$$\leq 2\Omega_1^{(\mathcal{F})} \big(f, ((\tilde{L}(\cdot - t)^{2q})(t))^{1/2q} \big)_{L^q}, \qquad (17.72)$$

$\forall t \in [a, b]$,
and
2)

$$\sup_{t \in [a, b]} \left(\int_X D^q \big(M(f)(t, \omega), f(t, \omega) \big) dP(\omega) \right)^{1/q}$$
$$\leq 2\Omega_1^{(\mathcal{F})} \big(f, \|(\tilde{L}(\cdot - t)^{2q})(t)\|_\infty^{1/2q} \big)_{L^q}. \qquad (17.73)$$

Proof. Similar to the proof of Theorem 17.36.   □

We derive

Theorem 17.52. *Here all as in Proposition 17.29, $m_t = 1$, $\forall t \in [a, b]$ and $q > 1$. Then*
1)

$$\left(\int_X D^q \big(M(f)(t, \omega), f(t, \omega) \big) dP(\omega) \right)^{1/q} \qquad (17.74)$$
$$\leq 2 \min \{ \Omega_1^{(\mathcal{F})} \big(f, ((\tilde{L}(|\cdot - t|^q))(t))^{1/q} \big)_{L^q},$$
$$\Omega_1^{(\mathcal{F})} \big(f, ((\tilde{L}(\cdot - t)^{2q})(t))^{1/2q} \big)_{L^q} \}, \quad \forall t \in [a, b],$$

and
2)

$$\sup_{t \in [a, b]} \left(\int_X D^q \big(M(f)(t, \omega), f(t, \omega) \big) dP(\omega) \right)^{1/q}$$
$$\leq 2 \min \{ \Omega_1^{(\mathcal{F})} \big(f, \|(\tilde{L}(|\cdot - t|^q))(t)\|_\infty^{1/q} \big)_{L^q},$$
$$\Omega_1^{(\mathcal{F})} \big(f, \|(\tilde{L}(\cdot - t)^{2q})(t)\|_\infty^{1/2q} \big)_{L^q} \}. \qquad (17.75)$$

Proof. From Theorems 17.50 and 17.51.   □

We have

Corollary 17.53. *All here as in Proposition 17.29, $m_t = 1$, $\forall t \in [a, b]$. Then*

1)

$$\left(\int_X D^2\big(M(f)(t,\omega),f(t,\omega)\big)P(d\omega)\right)^{1/2}$$
$$\leq 2\Omega_1^{(\mathcal{F})}\big(f,((\tilde{L}((\cdot-t)^2))(t))^{1/2}\big)_{L^2}, \quad \forall t \in [a,b], \quad (17.76)$$

and
2)

$$\sup_{t\in[a,b]}\left(\int_X D^2\big(M(f)(t,\omega),f(t,\omega)\big)P(d\omega)\right)^{1/2}$$
$$\leq 2\Omega_1^{(\mathcal{F})}\big(f,\|(\tilde{L}((\cdot-t)^2))(t)\|_\infty^{1/2}\big)_{L^2}. \qquad (17.77)$$

Proof. By Theorem 17.50, $q = 2$. □

Corollary 17.54. *Here all as in Assumption 17.40, $m_t = 1$, $\forall t \in [a,b]$. Then*
1)

$$\int_X D\big(M_n(f)(t,\omega),f(t,\omega)\big)dP(\omega)$$
$$\leq 2\Omega_1^{(\mathcal{F})}\big(f,(\tilde{L}_n((\cdot-t)^2)(t))^{1/2}\big)_{L^1}, \qquad (17.78)$$

$\forall t \in [a,b]$, $\forall n \in \mathbb{N}$,
    and
2)

$$\sup_{t\in[a,b]}\left(\int_X D\big(M_n(f)(t,\omega),f(t,\omega)\big)dP(\omega)\right)$$
$$\leq 2\Omega_1^{(\mathcal{F})}\big(f,\|(\tilde{L}_n((\cdot-t)^2))(t)\|_\infty^{1/2}\big)_{L^1}, \quad \forall n \in \mathbb{N}. \ (17.79)$$

Proof. By Theorem 17.48. □

Corollary 17.55. *Here all as in Assumption 17.40, $m_t = 1$, $\forall t \in [a,b]$. Then*
1)

$$\left(\int_X D^2\big(M_n(f)(t,\omega),f(t,\omega)\big)P(d\omega)\right)^{1/2}$$
$$\leq 2\Omega_1^{(\mathcal{F})}\big(f,((\tilde{L}_n((\cdot-t)^2))(t))^{1/2}\big)_{L^2}, \qquad (17.80)$$

$\forall t \in [a,b]$, $\forall n \in \mathbb{N}$,
    and

2)

$$\sup_{t\in[a,b]} \left(\int_X D^2 \big(M_n(f)(t,\omega), f(t,\omega)\big) P(d\omega) \right)^{1/2}$$
$$\leq 2\Omega_1^{(\mathcal{F})} \big(f, \|(\tilde{L}_n((\cdot - t)^2))(t)\|_\infty^{1/2}\big)_{L^2}, \qquad (17.81)$$

$\forall n \in \mathbb{N}$.

Proof. By Theorem 17.50, $q = 2$.   □

We give now the following fuzzy random Korovkin Theorems for the case of $\tilde{L}(1)(t) = 0$, $\forall t \in [a, b]$.

Theorem 17.56. *Here all are as in Assumption 17.40. Furthermore suppose that*

$$\tilde{L}_n(1)(t) = 1, \quad \forall t \in [a,b], \quad \tilde{L}_n(id) \overset{u}{\to} id, \quad \tilde{L}_n(id^2) \overset{u}{\to} id^2, \quad as\ n \to \infty.$$

Then

$$\lim_{n\to\infty} \left\| \int_X D\big(M_n(f)(t,\omega), f(t,\omega)\big) dP(\omega) \right\|_{\infty,t} = 0, \quad \forall f \in C^U_{\mathcal{F}R}([a,b]).$$
$$(17.82)$$

I.e.

$$M_n(f)(t,\omega) \overset{\text{"1-mean"}}{\longrightarrow} f(t,\omega), \qquad (17.83)$$

uniformly, $\forall f \in C^U_{\mathcal{F}R}([a,b])$, that is uniformly $M_n \overset{\text{"1-mean"}}{\longrightarrow} I$, as $n \to \infty$, on $C^U_{\mathcal{F}R}([a,b])$.

Proof. Using (17.79) and (17.58) and Proposition 17.9(ii).   □

We continue with

Theorem 17.57. *Here all are as in Assumption 17.40. Furthermore assume that*

$$\tilde{L}_n(1)(t) = 1, \quad \forall t \in [a,b], \quad \tilde{L}_n(id) \overset{u}{\to} id, \quad \tilde{L}_n(id^2) \overset{u}{\to} id^2, \quad as\ n \to \infty.$$

Then

$$\lim_{n\to\infty} \left\| \int_X D^2\big(M_n(f)(t,\omega), f(t,\omega)\big) P(d\omega) \right\|_{\infty,t} = 0, \quad \forall f \in C^U_{\mathcal{F}R}([a,b]).$$
$$(17.84)$$

I.e.

$$M_n(f)(t,\omega) \overset{\text{"2-mean"}}{\longrightarrow} f(t,\omega), \qquad (17.85)$$

uniformly, $\forall f \in C^U_{\mathcal{F}R}([a,b])$, that is uniformly $M_n \overset{\text{"2-mean"}}{\longrightarrow} I$, as $n \to \infty$, on $C^U_{\mathcal{F}R}([a,b])$. Notice here that $M_n \overset{\text{"2-mean"}}{\longrightarrow} I$ implies $M_n \overset{\text{"1-mean"}}{\longrightarrow} I$, uniformly, i.e. Theorem 17.56.

Proof. Using (17.81) and (17.58) and Proposition 17.9(ii).   □

Finally we present

Theorem 17.58. *Here all are as in Assumption 17.40, $\tilde{L}_n(1)(t) = 1$, $\forall t \in [a,b]$ and $q > 2$. We assume further that*

(i) $\lim\limits_{n\to\infty} \|(\tilde{L}_n(|\cdot -t|^q))(t)\|_\infty = 0$,

or

(ii)''

$$\lim\limits_{n\to\infty} \|(\tilde{L}_n(\cdot - t)^{2q})(t)\|_\infty = 0. \tag{17.86}$$

Then

$$\lim\limits_{n\to\infty} \left\| \left\| \int_X D^q\big(M_n(f)(t,\omega), f(t,\omega)\big) P(d\omega) \right\| \right\|_{\infty,t} = 0, \quad \forall f \in C^U_{\mathcal{F}R}([a,b]). \tag{17.87}$$

I.e.

$$M_n(f)(t,\omega) \xrightarrow{\text{``q-mean''}} f(t,\omega), \tag{17.88}$$

uniformly, $\forall f \in C^U_{\mathcal{F}R}([a,b])$ that is uniformly $M_n \xrightarrow{\text{``q-mean''}} I$, as $n \to \infty$, on $C^U_{\mathcal{F}R}([a,b])$.

Proof. By (17.71) or (17.73) and Propositoin 17.9(ii).   □

Remark 17.59. 1) Notice here from (17.87), that $M_n \xrightarrow{\text{``$q$-mean''}} I$ implies $M_n \xrightarrow{\text{``k-mean''}} I$, uniformly for any $1 \le k \le q < \infty$ on $C^U_{\mathcal{F}R}([a,b])$.

2) In the case of $m_t = 1$, $\forall t \in [a,b]$, all presented results here *did not require* the condition

$$\int_X D^q\big(f(t,\omega), \tilde{o}\big) P(d\omega) < \infty, \quad \forall t \in [a,b], \ 1 \le q < \infty,$$

as they did the earlier ones for general $m_t \ge 0$.

3) One can do related research for other domains other than $[a,b]$, e.g. $[0,+\infty)$, multivariate domains in $\mathbb{R}^k$, $k > 1$ and K compact convex subset of a metric space. Of course not all results can pass through there.

17.4 Application

We consider here the fuzzy random Bernstein polynomials

$$B_n^{(\mathcal{F})}(f)(x,\omega) = \sum_{k=0}^n {}^* \binom{n}{k} x^k (1-x)^{n-k} \odot f\left(\frac{k}{n},\omega\right),$$

$\forall x \in [0,1]$, $\forall \omega \in X$, $\forall f \in C_{\mathcal{F}R}^{U}([0,1])$, $\forall n \in \mathbb{N}$, see (17.18). We apply first here (17.79) for

$$M_n(f)(t,\omega) = B_n^{(\mathcal{F})}(f)(t,\omega), \quad \forall (t,\omega) \in [a,b] \times X,$$

and $\tilde{L}_n = B_n$ the real Bernstein operator

$$B_n(g)(x) = \sum_{k=0}^{n} \binom{n}{k} x^k (1-x)^{n-k} g\left(\frac{k}{n}\right), \quad \forall g \in C([0,1]), \ \forall x \in [0,1], \ \forall n \in \mathbb{N}.$$

Clearly

$$B_n((\cdot - t)^2)(t) = \frac{t(1-t)}{n}, \quad t \in [0,1].$$

Hence

$$\left\| B_n((\cdot - t)^2)(t) \right\|_{\infty}^{1/2} \le \frac{1}{2\sqrt{n}}, \quad \forall n \in \mathbb{N}.$$

Notice also that $(B_n(1))(t) = 1$, $\forall t \in [0,1]$.

Clearly here $B_n^{(\mathcal{F})}(f)(t,\omega)$ fulfill Assumption 17.40. Thus by (17.79) we obtain

$$\sup_{t \in [0,1]} \left(\int_X D\big(B_n^{(\mathcal{F})}(f)(t,\omega), f(t,\omega)\big) dP(\omega) \right)$$

$$\le 2\Omega_1^{(\mathcal{F})}\left(f, \frac{1}{2\sqrt{n}} \right)_{L^1}, \tag{17.89}$$

$\forall f \in C_{\mathcal{F}R}^{U}([0,1])$, $\forall n \in \mathbb{N}$.

Similarly, from (17.81) we obtain

$$\sup_{t \in [0,1]} \left(\int_X D^2\big(B_n^{(\mathcal{F})}(f)(t,\omega), f(t,\omega)\big) dP(\omega) \right)^{1/2}$$

$$\le 2\Omega_1^{(\mathcal{F})}\left(f, \frac{1}{2\sqrt{n}} \right)_{L^2}, \quad \forall f \in C_{\mathcal{F}R}^{U}([0,1]), \tag{17.90}$$

$\forall n \in \mathbb{N}$.

Finally, from (17.75) for $q > 2$ we obtain

$$\sup_{t \in [0,1]} \left(\int_X D^q\big(B_n^{(\mathcal{F})}(f)(t,\omega), f(t,\omega)\big) dP(\omega) \right)^{1/q}$$

$$\le 2 \min\{ \Omega_1^{(\mathcal{F})}\big(f, \|(B_n(|\cdot - t|^q))(t)\|_{\infty}^{1/q}\big)_{L^q},$$

$$\Omega_1^{(\mathcal{F})}\big(f, \|(B_n(\cdot - t)^{2q})(t)\|_{\infty}^{1/2q}\big)_{L^q}, \}, \tag{17.91}$$

$\forall f \in C_{\mathcal{F}R}^{U}([0,1])$, $\forall n \in \mathbb{N}$.

In particular, if f is additionally of Lipschitz type, i.e.

$$\int_X D\big(f(x,\omega), f(y,\omega)\big) P(d\omega) \le \theta |x-y|, \quad \theta > 0, \quad \forall x,y \in [0,1], \quad (17.92)$$

then

$$\Omega_1^{(\mathcal{F})}(f,\delta)_{L^1} \le \theta \cdot \delta, \quad \delta > 0, \quad (17.93)$$

and

$$\Omega_1^{(\mathcal{F})}\left(f, \frac{1}{2\sqrt{n}}\right)_{L^1} \le \frac{\theta}{2\sqrt{n}}, \quad \forall n \in \mathbb{N}. \quad (17.94)$$

Hence

$$\sup_{t \in [0,1]} \left(\int_X D\big(B_n^{(\mathcal{F})}(f)(t,\omega), f(t,\omega)\big) dP(\omega)\right) \le \frac{\theta}{\sqrt{n}}, \quad \forall n \in \mathbb{N}, \quad (17.95)$$

$\forall f \in C_{\mathcal{F}R}^U([0,1])$ which is of Lipschitz type (17.92).

Inequality (17.95) improves the corresponding inequality from (17.8), since over there we only get

$$\sup_{x \in [0,1]} \left(\int_X D\big(B_n^{(\mathcal{F})}(f)(x,\omega), f(x,\omega)\big) P(d\omega)\right) \le \frac{3\theta}{2\sqrt{n}}, \quad \forall n \in \mathbb{N}. \quad (17.96)$$

18

FUZZY-RANDOM NEURAL NETWORK APPROXIMATION OPERATORS, UNIVARIATE CASE

In this chapter we study the rate of pointwise convergence in the q-mean to the Fuzzy-Random unit operator of very precise univariate Fuzzy-Random neural network operators of Cardaliaguet–Euvrard and "Squashing" types. These Fuzzy-Random operators arise in a natural and common way among Fuzzy-Random neural networks. These rates are given through *Probabilistic-Jackson type inequalities* involving the Fuzzy-Random modulus of continuity of the engaged Fuzzy-Random function or its Fuzzy derivatives. Also several interesting results in Fuzzy-Random Analysis are given of independent merit, which are used then in the proofs of the main results of the chapter. This chapter follows [17].

18.1 Introduction

Let $(X, \mathcal{B}, P)$ be a probability space. Consider the set of all fuzzy-random variables $\mathcal{L}_{\mathcal{F}}(X, \mathcal{B}, P)$. Let $f \colon \mathbb{R} \to \mathcal{L}_{\mathcal{F}}(X, \mathcal{B}, P)$ be a *fuzzy-random function* or *fuzzy-stochastic process*. Here for $t \in \mathbb{R}$, $s \in X$ we denote $(f(t))(s) = f(t, s)$ and actually we have $f \colon \mathbb{R} \times X \to \mathbb{R}_{\mathcal{F}}$, where $\mathbb{R}_{\mathcal{F}}$ is the set of fuzzy real numbers. Let $1 \leq q < +\infty$. Here we consider only fuzzy-random functions f which are *(q-mean) uniformly continuous over* $\mathbb{R}$. For each $n \in \mathbb{N}$, the fuzzy-random neural network we deal with has the following structure:

G.A. Anastassiou: Fuzzy Mathematics: Approximation The., STUDFUZZ 251, pp. 347–372.
springerlink.com

It is a three-layer feed forward network with one hidden layer. It has one input unit and one output unit. The hidden layer has $(2n^2 + 1)$ processing units. To each pair of connecting units (input to each processing unit) we assign the same weight $n^{1-\alpha}$, $0 < \alpha < 1$. The threshold values $\frac{k}{n^\alpha}$ are one for each processing unit k. The activation function b (or S) is the same for each processing unit. The Fuzzy-Random weights associated with the output unit are $f\left(\frac{k}{n}, s\right) \odot \frac{1}{I^{(*)}n^\alpha}$, one for each processing unit k, where $I = \int_{-\infty}^{\infty} b(x)dx$ (or $I^* = \int_{-\infty}^{\infty} S(x)dx$), "$\odot$" denotes the scalar Fuzzy multiplication.

The above precisely described Fuzzy-Random neural networks induce some completely described Fuzzy-Random neural network operators of Cardaliaguet–Euvrard and "Squashing" types.

We study here thoroughly the Fuzzy-Random pointwise convergence (in q-mean) of these operators to the unit operator. See Theorems 18.21, 18.24, 18.27, 18.29. This is done with rates through Probabilistic-Jackson type inequalities involving Fuzzy-Random moduli of continuity of the engaged Fuzzy- Random function and its Fuzzy derivatives.

On the way to establish these main results we produce some independent and interesting results for Fuzzy-Random Analysis. The real ordinary theory of the above mentioned operators was presented earlier in [6], [9] and [47]. And the fuzzy case was treated in [11], see also here Chapters 15, 16. Of course this chapter is strongly motivated from there and is a continuation.

The monumental revolutionizing work of L. Zadeh [103] is the foundation of this chapter, as well as another strong motivation. Fuzziness in Computer Science and Engineering seems one of the main trends today. Also Fuzziness has penetrated many areas of Mathematics and Statistics.

The approach here is quantitative and new on the topic, started in [6], [9] and continued in [11]. It determines precisely the rates of convergence through natural very tight inequalities using the measurement of smoothness of the engaged Fuzzy-Random functions.

As in Remark 4.4 ([31]) one can show easily that a sequence of operators of the form

$$L_n(f)(x) := \sum_{k=0}^{n} {}^* f(x_{k_n}) \odot w_{n,k}(x), \quad n \in \mathbb{N}, \tag{18.1}$$

($\sum^*$ denotes the fuzzy summation) where $f: \mathbb{R} \to \mathbb{R}_\mathcal{F}$, $x_{k_n} \in \mathbb{R}$, $w_{n,k}(x)$ real valued weights, are *linear* over $\mathbb{R}$, i.e.,

$$L_n(\lambda \odot f \oplus \mu \odot g)(x) = \lambda \odot L_n(f)(x) \oplus \mu \odot L_n(g)(x), \tag{18.2}$$

$\forall \lambda, \mu \in \mathbb{R}$, any $x \in \mathbb{R}$; $f, g: \mathbb{R} \to \mathbb{R}_\mathcal{F}$. (Proof based on Lemma 1.3(iv).)
We need

Lemma 18.1 ([11]). *Let $f \colon \mathbb{R} \to \mathbb{R}_{\mathcal{F}}$ continuous, $r \in \mathbb{N}$. Then the following integrals*

$$(FR) \int_a^{s_{r-1}} f(s_r) ds_r, (FR) \int_a^{s_{r-2}} \left(\int_a^{s_{r-1}} f(s_r) ds_r \right) ds_{r-1},$$

$$\dots, (FR) \int_a^s \left(\int_a^{s_1} \cdots \left(\int_a^{s_{r-2}} \left(\int_a^{s_{r-1}} f(s_r) ds_r \right) ds_{r-1} \right) \cdots \right) ds_1,$$

$$(18.3)$$

are continuous functions in $s_{r-1}, s_{r-2}, \dots, s$, respectively.
   Here $s_{r-1}, s_{r-2}, \dots, s \geq a$ and all are real numbers.
   Additionally we have

Lemma 18.2. *Let $f \colon [a,b] \to \mathbb{R}_{\mathcal{F}}$ have an existing fuzzy derivative f' at $c \in [a,b]$. Then f is fuzzy continuous at c.*

Proof. By the assumption we have that

$$\lim_{h \to 0^+} D\left(\frac{f(c+h) - f(c)}{h}, f'(c) \right) = 0 \tag{18.4}$$

and

$$\lim_{h \to 0^+} D\left(\frac{f(c) - f(c-h)}{h}, f'(c) \right) = 0. \tag{18.5}$$

Furthermore here is assumed that the H-differences $f(c+h) - f(c)$, $f(c) - f(c-h)$ exist for small $h \colon 0 < h < b - a$.
   Let $z := f(c+h) - f(c)$, then $f(c+h) = f(c) \oplus z$. Hence

$$
\begin{aligned}
D(f(c+h), f(c)) &= D(f(c) \oplus z, f(c)) = D(z, \tilde{o}) \tag{18.6} \\
&= D(f(c+h) - f(c), \tilde{o}) \\
&= hD\left(\frac{f(c+h) - f(c)}{h}, \tilde{o} \right).
\end{aligned}
$$

But

$$D\left(\frac{f(c+h) - f(c)}{h}, \tilde{o} \right) - D(f'(c), \tilde{o}) \leq D\left(\frac{f(c+h) - f(c)}{h}, f'(c) \right).$$

$$(18.7)$$

Hence

$$D(f(c+h), f(c)) \leq h \left\{ D(f'(c), \tilde{o}) + D\left(\frac{f(c+h) - f(c)}{h}, f'(c) \right) \right\}.$$

Consequently

$$\lim_{h \to 0^+} D(f(c+h), f(c)) \leq \tag{18.8}$$

$$\left(\lim_{h \to 0^+} h \right) \left\{ D(f'(c), \tilde{o}) + \lim_{h \to 0^+} D\left(\frac{f(c+h) - f(c)}{h}, f'(c) \right) \right\} = 0.$$

Therefore

$$\lim_{h\to 0^+} D(f(c+h), f(c)) = 0.$$

That is, f is right continuous at c. One can prove f is left continuous at c. We have proved that f is fuzzy continuous at $c \in [a, b]$. $\qquad\square$

We need the Fuzzy Taylor formula

Theorem 18.3 ([11]). *Let $T := [x_0, x_0 + \beta] \subset \mathbb{R}$, with $\beta > 0$. We suppose that $f^{(i)}: T \to \mathbb{R}_\mathcal{F}$ are differentiable for all $i = 0, 1, \ldots, n-1$, for any $x \in T$. (I.e., there exist in $\mathbb{R}_\mathcal{F}$ the H-differences $f^{(i)}(x+h) - f^{(i)}(x)$, $f^{(i)}(x) - f^{(i)}(x-h)$, $i = 0, 1, \ldots, n-1$ for small h: $0 < h < \beta$. Furthermore there exist $f^{(i+1)}(x) \in \mathbb{R}_\mathcal{F}$ such that the limits in D-distance exist and*

$$f^{(i+1)}(x) = \lim_{h\to 0^+} \frac{f^{(i)}(x+h) - f^{(i)}(x)}{h} = \lim_{h\to 0^+} \frac{f^{(i)}(x) - f^{(i)}(x-h)}{h},$$

for all $i = 0, 1, \ldots, n-1$.) Also we suppose that $f^{(i)}$, $i = 0, 1, \ldots, n$ are continuous on T in the fuzzy sense. Then for $s \geq a$; $s, a \in T$ we obtain

$$f(s) = f(a) \oplus f'(a) \odot (s-a) \oplus f''(a) \odot \frac{(s-a)^2}{2!}$$

$$\oplus \cdots \oplus f^{(n-1)}(a) \odot \frac{(s-a)^{n-1}}{(n-1)!} \oplus R_n(a, s), \tag{18.9}$$

where

$$R_n(a, s) := (FR) \int_a^s \left(\int_a^{s_1} \cdots \left(\int_a^{s_{n-1}} f^{(n)}(s_n) ds_n \right) ds_{n-1} \cdots \right) ds_1. \tag{18.10}$$

Here $R_n(a, s) \in C_\mathcal{F}(T)$ as a function of s.

Note. This formula is invalid when $s < a$, as it is totally based on Corollary 1.12.

18.2   Basic properties

We need

Definition 18.4 (see also [66], Definition 13.16, p. 654). Let $(X, \mathcal{B}, P)$ be a probability space. A *fuzzy-random variable* is a $\mathcal{B}$-measurable mapping $g: X \to \mathbb{R}_\mathcal{F}$ (i.e., for any open set $U \subseteq \mathbb{R}_\mathcal{F}$, in the topology of $\mathbb{R}_\mathcal{F}$ generated by the metric D, we have

$$g^{-1}(U) = \{s \in X; g(s) \in U\} \in \mathcal{B}). \tag{18.11}$$

The set of all fuzzy-random variables is denoted by $\mathcal{L}_\mathcal{F}(X, \mathcal{B}, P)$. Let $g_n, g \in \mathcal{L}_\mathcal{F}(X, \mathcal{B}, P)$, $n \in \mathbb{N}$ and $0 < q < +\infty$. We say $g_n(s) \xrightarrow[n \to +\infty]{\text{"q-mean"}} g(s)$ if

$$\lim_{n \to +\infty} \int_X (D(g_n(s), g(s)))^q P(ds) = 0. \tag{18.12}$$

Remark 18.5 (see [66], p. 654). If $f, g \in \mathcal{L}_\mathcal{F}(X, \mathcal{B}, P)$, let us denote $F \colon X \to \mathbb{R}_+ \cup \{0\}$ by $F(s) = D(f(x), g(s))$, $s \in X$. Here, F is $\mathcal{B}$-measurable, because $F = G \circ H$, where $G(u, v) = D(u, v)$ is continuous on $\mathbb{R}_\mathcal{F} \times \mathbb{R}_\mathcal{F}$, and $H \colon X \to \mathbb{R}_\mathcal{F} \times \mathbb{R}_\mathcal{F}$, $H(s) = (f(s), g(s))$, $s \in X$, is $\mathcal{B}$-measurable. This shows that the above convergence in q-mean makes sense.

Definition 18.6 (see [66], p. 654, Definition 13.17). Let $(T, \mathcal{T})$ be a topological space. A mapping $f \colon T \to \mathcal{L}_\mathcal{F}(X, \mathcal{B}, P)$ will be called *fuzzy-random function* (or *fuzzy-stochastic process*) on T. We denote $f(t)(s) = f(t, s)$, $t \in T$, $s \in X$.

Remark 18.7 (see [66], p. 655). Any usual fuzzy real function $f \colon T \to \mathbb{R}_\mathcal{F}$ can be identified with the degenerate fuzzy-random function $f(t, s) = f(t)$, $\forall t \in T$, $s \in X$.

Remark 18.8 (see [66], p. 655). Fuzzy-random functions that coincide with probability one for each $t \in T$ will be considered equivalent.

Remark 18.9 (see [66], p. 655). Let $f, g \colon T \to \mathcal{L}_\mathcal{F}(X, \mathcal{B}, P)$. Then $f \oplus g$ and $k \odot f$ are defined pointwise, i.e.,

$$\begin{aligned} (f \oplus g)(t, s) &= f(t, s) \oplus g(t, s), \\ (k \odot f)(t, s) &= k \odot f(t, s), \quad t \in T, \ s \in X. \end{aligned}$$

Definition 18.10 (see also Definition 13.18, pp. 655–656, [66]). For a fuzzy-random function $f \colon \mathbb{R} \to \mathcal{L}_\mathcal{F}(X, \mathcal{B}, P)$ we define the (first) fuzzy-random modulus of continuity

$$\Omega_1^{(\mathcal{F})}(f, \delta)_{L^q} = \sup \left\{ \left(\int_X D^q(f(x, s), f(y, s)) P(ds) \right)^{1/q} ; \right.$$

$$\left. x, y \in \mathbb{R}, |x - y| \le \delta \right\},$$

$0 < \delta$, $1 \le q < \infty$.

Definition 18.11. Here $1 \le q < +\infty$. Let $f \colon \mathbb{R} \to \mathcal{L}_\mathcal{F}(X, \mathcal{B}, P)$ be a fuzzy random function. We call f a (*q-mean*) *uniformly continuous fuzzy random function over* $\mathbb{R}$, iff $\forall \varepsilon > 0 \ \exists \delta > 0$: whenever $|x - y| \le \delta$; $x, y \in \mathbb{R}$, implies that

$$\int_X (D(f(x, s), f(y, s)))^q P(ds) \le \varepsilon.$$

We denote it as $f \in C_{\mathcal{FR}}^{Uq}(\mathbb{R})$.

Proposition 18.12. Let $f \in C_{\mathcal{FR}}^{Uq}(\mathbb{R})$. Then $\Omega_1^{(\mathcal{F})}(f, \delta)_{L^q} < \infty$, any $\delta > 0$.

Proof. Let $\varepsilon_0 > 0$ be arbitrary but fixed. Then there exists $\delta_0 > 0$: $|x-y| \le \delta_0$ implies

$$\int_X (D(f(x,s), f(y,s)))^q \, P(ds) \le \varepsilon_0 < \infty.$$

That is, $\Omega_1^{(\mathcal{F})}(f, \delta_0)_{L^q} \le \varepsilon_0^{1/q} < \infty$. Let now $\delta > 0$ arbitrary, $x, y \in \mathbb{R}$ such that $|x - y| \le \delta$. Choose $n \in \mathbb{N}$: $n\delta_0 \ge \delta$ and set $x_i := x + \frac{i}{n}(y - x)$, $0 \le i \le n$. Then

$$
\begin{aligned}
D(f(x,s), f(y,s)) \le\ & D(f(x,s), f(x_1,s)) + D(f(x_1,s), f(x_2,s)) \\
& + \cdots + D(f(x_{n-1},s), f(y,s)).
\end{aligned}
$$

Consequently

$$
\left(\int_X (D(f(x,s), f(y,s)))^q \, P(ds) \right)^{1/q}
$$

$$
\le \left(\int_X (D(f(x,s), f(x_1,s)))^q \, P(ds) \right)^{1/q} + \cdots
$$

$$
+ \left(\int_X (D(f(x_{n-1},s), f(y,s)))^q \, P(ds) \right)^{1/q}
$$

$$
\le n\Omega_1^{(\mathcal{F})}(f, \delta_0)_{L^q} \le n\varepsilon_0^{1/q} < \infty,
$$

since $|x_i - x_{i+1}| = \frac{1}{n}|x-y| \le \frac{1}{n}\delta \le \delta_0$, $0 \le i \le n$. Therefore $\Omega_1^{(\mathcal{F})}(f, \delta)_{L^q} \le n\varepsilon_0^{1/q} < \infty$. $\qquad\square$

Proposition 18.13. Let $f, g \colon \mathbb{R} \to \mathcal{L}_{\mathcal{F}}(X, \mathcal{B}, P)$ be fuzzy random functions. It holds

(i) $\Omega_1^{(\mathcal{F})}(f, \delta)_{L^q}$ is nonnegative and nondecreasing in $\delta > 0$.

(ii) $\lim_{\delta \downarrow 0} \omega_1^{(\mathcal{F})}(f, \delta)_{L^q} = \Omega_1^{(\mathcal{F})}(f, 0)_{L^q} = 0$, iff $f \in C_{\mathcal{FR}}^{Uq}(\mathbb{R})$.

(iii) $\Omega_1^{(\mathcal{F})}(f, \delta_1 + \delta_2)_{L^q} \le \Omega_1^{(\mathcal{F})}(f, \delta_1)_{L^q} + \Omega_1^{(\mathcal{F})}(f, \delta_2)_{L^q}$, $\delta_1, \delta_2 > 0$.

(iv) $\Omega_1^{(\mathcal{F})}(f, n\delta)_{L^q} \le n\Omega_1^{(\mathcal{F})}(f, \delta)_{L^q}$, $\delta > 0$, $n \in \mathbb{N}$.

(v) $\Omega_1^{(\mathcal{F})}(f, \lambda\delta)_{L^q} \le \lceil \lambda \rceil \Omega_1^{(\mathcal{F})}(f, \delta)_{L^q} \le (\lambda + 1)\Omega_1^{(\mathcal{F})}(f, \delta)_{L^q}$, $\lambda > 0$, $\delta > 0$, where $\lceil \cdot \rceil$ is the ceiling of the number.

(vi) $\Omega_1^{(\mathcal{F})}(f \oplus g, \delta)_{L^q} \le \Omega_1^{(\mathcal{F})}(f, \delta)_{L^q} + \Omega_1^{(\mathcal{F})}(g, \delta)_{L^q}$, $\delta > 0$. Here $f \oplus g$ is a fuzzy random function.

(vii) $\Omega_1^{(\mathcal{F})}(f,\cdot)_{L^q}$ is continuous on $\mathbb{R}_+$, for $f \in C_{\mathcal{F}R}^{Uq}(\mathbb{R})$.

Proof. Obvious.   □

Also we have

Proposition 18.14. (i) *Let $Y(t,\omega)$ be a real valued stochastic process such that Y is continuous in $t \in [a,b]$. Then Y is jointly measurable in (t,ω).*

(ii) *Further assume that the expectation $(E|Y|)(t) \in C([a,b])$, or more generally $\int_a^b (E|Y|)(t)dt$ makes sense and is finite. Then*

$$E\left(\int_a^b Y(t,\omega)dt\right) = \int_a^b (EY)(t)dt.$$

Proof. Clearly $\int_a^b Y(t,\omega)dt$ is a real valued random variable. Also by the assumption on $E|Y|$ we obtain that

$$\int_a^b (E|Y|)(t)dt < \infty.$$

Here we assume that $\omega \in S$, S is a probability space.

Also here Y is continuous in t, and measurable in ω and takes values in $\mathbb{R}$. Clearly Y is a "Caratheodory function", see [2], p. 156, Definition 20.14. Hence by Theorem 20.15, p. 156, [2] we get that Y is (jointly) measurable on $[a,b] \times S$. In the last we took into account the following: let $f : \mathcal{X} \times \mathcal{Y} \to \mathbb{R}$ be measurable on $\mathcal{X} \times \mathcal{Y}$, and $g : \mathcal{Y} \times \mathcal{X} \to \mathbb{R}$ be such that $g(y,x) := f(x,y)$. Then g is measurable on $\mathcal{Y} \times \mathcal{X}$.

From the above we derive that $Y(t,\omega)$ is integrable on $[a,b] \times S$. The claim now is proved by applying Fubini's theorem.   □

According to [62], p. 94 we have the following

Definition 18.15. Let $(Y,\mathcal{T})$ be a topological space, with its σ-algebra of Borel sets $\mathcal{B} := \mathcal{B}(Y,\mathcal{T})$ generated by $\mathcal{T}$. If $(X,\mathcal{S})$ is a measurable space, a function $f : X \to Y$ is called *measurable* iff $f^{-1}(B) \in \mathcal{S}$ for all $B \in \mathcal{B}$.

By Theorem 4.1.6 of [62], p. 89 f as above is measurable iff

$$f^{-1}(C) \in \mathcal{S} \quad \text{for all } C \in \mathcal{T}.$$

We would need

Theorem 18.16 (see [62], p. 95). *Let $(X,\mathcal{S})$ be a measurable space and (Y,d) be a metric space. Let f_n be measurable functions from X into Y such that for all $x \in X$, $f_n(x) \to f(x)$ in Y. Then f is measurable. I.e., $\lim_{n\to\infty} f_n = f$ is measurable.*

We need also

Proposition 18.17. *Let f,g be fuzzy random variables from S into $\mathbb{R}_{\mathcal{F}}$. Then*

(i) *Let $c \in \mathbb{R}$, then $c \odot f$ is a fuzzy random variable.*

(ii) *$f \oplus g$ is a fuzzy random variable.*

Proof. Obvious.   □

Finally we need

Proposition 18.18. *Let $f: [a, b] \to L_{\mathcal{F}}(X, \mathcal{B}, P)$ be a fuzzy-random function. We assume that $f(t, s)$ is fuzzy continuous in $t \in [a, b]$, $s \in X$. Then $(FR) \int_a^b f(t, s)dt$ exists and is a fuzzy-random variable.*

Proof. It comes easily from: Definition 1.9 and Corollary 13.2 of p. 644, [66] about Fuzzy-Riemann integral, Proposition 18.17, and Theorem 18.16. □

18.3   Main results

We need (see also [6], [11], [47])

Definition 18.19. A function $b: \mathbb{R} \to \mathbb{R}$ is said to be *bell-shaped* if b belongs to L^1 and its integral is nonzero, if it is nondecreasing on $(-\infty, a)$ and nonincreasing on $[a, +\infty)$, where a belongs to $\mathbb{R}$. In particular $b(x)$ is a nonnegative number and at a b takes a global maximum; it is the center of the bell-shaped function. A bell-shaped function is said to be *centered* if its center is zero. The function $b(x)$ may have jump discontinuities. In this chapter we consider only centered bell-shaped functions of compact support $[-T, T]$, $T > 0$. Denote $I := \int_{-T}^{T} b(t)dt$. Notice that $I > 0$.

Example 18.20.

(1) $b(x)$ can be the characteristic function on $[-1, 1]$.

(2) $b(x)$ can be the hat function on $[-1, 1]$, i.e.,

$$b(x) = \begin{cases} 1 + x, & -1 \leq x \leq 0, \\ 1 - x, & 0 < x \leq 1, \\ 0, & \text{elsewhere.} \end{cases}$$

Here we consider functions $f \in C_{\mathcal{F}R}^{Uq}(\mathbb{R})$.

In this chapter we study among others in q-mean the pointwise convergence with rates over the real line, to the *fuzzy-random unit operator* of the *fuzzy-random univariate Cardaliaguet–Euvrard neural network operators*,

$$(F_n(f))(x, s) := f_n(x, s) := \sum_{k=-n^2}^{n^2}{}^* f\left(\frac{k}{n}, s\right) \odot \frac{b\left(n^{1-\alpha}\left(x - \frac{k}{n}\right)\right)}{In^\alpha}, \quad (18.13)$$

where $0 < \alpha < 1$, $x \in \mathbb{R}$, $s \in X$ and $n \in \mathbb{N}$.

The above are linear operators over $\mathbb{R}$. These operators in the ordinary real case were thoroughly studied in [9], [47] and in the fuzzy sense were studied in [11]. So as in [6], [11], for convenience we consider that $n \geq \max(T+|x|, T^{-1/\alpha})$. This implies that $-n^2 \leq nx - Tn^\alpha \leq nx + Tn^\alpha \leq n^2$, and $\mathrm{card}(k) \geq 1$. Furthermore, it holds that $\mathrm{card}(k) \to +\infty$, as $n \to +\infty$. Set $b^* := b(0)$, the maximum of $b(x)$. Denote by $[\cdot]$ the integral part of a number.

Next we give the first main result.

Theorem 18.21. *Let $x \in \mathbb{R}$, $s \in X$, $T > 0$, and $1 \leq q < +\infty$. Here $n \in \mathbb{N}$ is such that $n \geq \max(T+|x|, T^{-1/\alpha})$. Assume that $\int_X D^q(f(x,s), \tilde{o})P(ds) < +\infty$. Then*

$$\left(\int_X D^q(f_n(x,s), f(x,s))P(ds) \right)^{1/q}$$

$$\leq \left| \sum_{k=\lceil nx-Tn^\alpha \rceil}^{[nx+Tn^\alpha]} \frac{b\left(n^{1-\alpha}\left(x - \frac{k}{n}\right)\right)}{In^\alpha} - 1 \right| \left(\int_X D^q(f(x,s), \tilde{o})P(ds) \right)^{1/q}$$

$$+ \frac{b^*}{I}\left(2T + \frac{1}{n^\alpha}\right) \Omega_1^{(\mathcal{F})}\left(f, \frac{T}{n^{1-\alpha}}\right)_{L^q}.$$

$$(18.14)$$

Proof. All parts of inequality (18.14) make sense because of the notions and basic results developed in Section 18.2. As in [6] we have that

$$\sum_{k=\lceil nx-Tn^\alpha \rceil}^{[nx+Tn^\alpha]} \frac{b\left(n^{1-\alpha}\left(x - \frac{k}{n}\right)\right)}{In^\alpha} \leq \frac{b^*}{I}\left(2T + \frac{1}{n^\alpha}\right). \qquad (18.15)$$

We see that

$$D(f_n(x,s), f(x,s)) = D\left(\sum_{k=\lceil nx-Tn^\alpha \rceil}^{[nx+Tn^\alpha]}{}^* f\left(\frac{k}{n}, s\right) \odot \right.$$

$$\left. \frac{b\left(n^{1-\alpha}\left(x - \frac{k}{n}\right)\right)}{In^\alpha}, f(x,s) \right)$$

$$\leq D\left(\sum_{k=\lceil nx-Tn^\alpha \rceil}^{[nx+Tn^\alpha]}{}^* f\left(\frac{k}{n}, s\right) \odot \frac{b\left(n^{1-\alpha}\left(x - \frac{k}{n}\right)\right)}{In^\alpha}, \right.$$

$$\left. \sum_{k=\lceil nx-Tn^\alpha \rceil}^{[nx+Tn^\alpha]}{}^* f(x,s) \odot \frac{b\left(n^{1-\alpha}\left(x - \frac{k}{n}\right)\right)}{In^\alpha} \right)$$

$$+ D\left(f(x,s) \odot \sum_{k=\lceil nx-Tn^\alpha \rceil}^{[nx+Tn^\alpha]} \frac{b\left(n^{1-\alpha}\left(x - \frac{k}{n}\right)\right)}{In^\alpha}, f(x,s) \odot 1 \right)$$

(by Lemma 1.2)

$$\leq \left| \sum_{k=\lceil nx-Tn^\alpha \rceil}^{\lfloor nx+Tn^\alpha \rfloor} \frac{b\left(n^{1-\alpha}\left(x-\frac{k}{n}\right)\right)}{In^\alpha} - 1 \right| D(f(x,s),\tilde{o})$$

$$+ \sum_{k=\lceil nx-Tn^\alpha \rceil}^{\lfloor nx+Tn^\alpha \rfloor} \frac{b\left(n^{1-\alpha}\left(x-\frac{k}{n}\right)\right)}{In^\alpha} D\left(f\left(\frac{k}{n},s\right),f(x,s)\right).$$

That is,

$$D(f_n(x,s),f(x,s)) \leq \left| \sum_{k=\lceil nx-Tn^\alpha \rceil}^{\lfloor nx+Tn^\alpha \rfloor} \frac{b\left(n^{1-\alpha}\left(x-\frac{k}{n}\right)\right)}{In^\alpha} - 1 \right| D(f(s,x),\tilde{o})$$

$$+ \sum_{k=\lceil nx-Tn^\alpha \rceil}^{\lfloor nx+Tn^\alpha \rfloor} \frac{b\left(n^{1-\alpha}\left(x-\frac{k}{n}\right)\right)}{In^\alpha} D\left(f\left(\frac{k}{n},s\right),f(x,s)\right).$$

Therefore

$$\left(\int_X D^q(f_n(x,s),f(x,s))P(ds) \right)^{1/q}$$

$$\leq \left| \sum_{k=\lceil nx-Tn^\alpha \rceil}^{\lfloor nx+Tn^\alpha \rfloor} \frac{b\left(n^{1-\alpha}\left(x-\frac{k}{n}\right)\right)}{In^\alpha} - 1 \right| \cdot \left(\int_X D^q(f(x,s),\tilde{o})P(ds) \right)^{1/q}$$

$$+ \sum_{k=\lceil nx-Tn^\alpha \rceil}^{\lfloor nx+Tn^\alpha \rfloor} \frac{b\left(n^{1-\alpha}\left(x-\frac{k}{n}\right)\right)}{In^\alpha} \left(\int_X D^q\left(f\left(\frac{k}{n},s\right),f(x,s)\right)P(ds) \right)^{1/q}$$

$$\leq \left| \sum_{k=\lceil nx-Tn^\alpha \rceil}^{\lfloor nx+Tn^\alpha \rfloor} \frac{b\left(n^{1-\alpha}\left(x-\frac{k}{n}\right)\right)}{In^\alpha} - 1 \right| \left(\int_X D^q(f(x,s),\tilde{o})P(ds) \right)^{1/q}$$

$$+ \sum_{k=\lceil nx-Tn^\alpha \rceil}^{\lfloor nx+Tn^\alpha \rfloor} \frac{b\left(n^{1-\alpha}\left(x-\frac{k}{n}\right)\right)}{In^\alpha} \Omega_1^{(\mathcal{F})}\left(f,\left|\frac{k}{n}-x\right|\right)_{L^q}$$

$$\leq \left| \sum_{k=\lceil nx-Tn^\alpha \rceil}^{\lfloor nx+Tn^\alpha \rfloor} \frac{b\left(n^{1-\alpha}\left(x-\frac{k}{n}\right)\right)}{In^\alpha} - 1 \right| \left(\int_X D^q(f(x,s),\tilde{o})P(ds) \right)^{1/q}$$

$$+ \left(\sum_{k=\lceil nx-Tn^\alpha \rceil}^{\lfloor nx+Tn^\alpha \rfloor} \frac{b\left(n^{1-\alpha}\left(x-\frac{k}{n}\right)\right)}{In^\alpha} \right) \Omega_1^{(\mathcal{F})}\left(f,\frac{T}{n^{1-\alpha}}\right)_{L^q}$$

(by (18.15)) $$\leq \left| \sum_{k=\lceil nx-Tn^\alpha \rceil}^{\lfloor nx+Tn^\alpha \rfloor} \frac{b\left(n^{1-\alpha}\left(x-\frac{k}{n}\right)\right)}{In^\alpha} - 1 \right| \cdot$$

$$\left(\int_X D^q(f(x,s),\tilde{o})P(ds) \right)^{1/q} + \frac{b^*}{I}\left(2T+\frac{1}{n^\alpha}\right)\Omega_1^{(\mathcal{F})}\left(f,\frac{T}{n^{1-\alpha}}\right)_{L^q}. \quad \square$$

Remark 18.22. By Lemma 2.1 of [9], p. 64 we get that

$$\lim_{n\to+\infty}\left|\sum_{k=\lceil nx-Tn^\alpha\rceil}^{[nx+Tn^\alpha]}\frac{b\left(n^{1-\alpha}\left(x-\frac{k}{n}\right)\right)}{In^\alpha}-1\right|=0,$$

any $x \in \mathbb{R}$. Let $f \in C_{\mathcal{FR}}^{Uq}(\mathbb{R})$, then $\lim_{n\to+\infty}\Omega_1^{(\mathcal{F})}\left(f,\frac{T}{n^{1-\alpha}}\right)_{L^q}=0$, by Proposition 18.13(ii). Therefore from (18.4) we obtain

$$\lim_{n\to+\infty}\int_X D^q(f_n(x,s),f(x,s))P(ds)=0.$$

That is, $f_n(x,s)\ \underset{\substack{\text{"}q\text{-mean"}\\ n\to+\infty}}{\longrightarrow}\ f(x,s)$, pointwise in $x \in \mathbb{R}$ with rates.

We need also (see also [6], [47])

Definition 18.23. Let the nonnegative function $S\colon \mathbb{R} \to \mathbb{R}$, S has compact support $[-T,T]$, $T > 0$, and is nondecreasing there and it can be continuous only on either $(-\infty,T]$ or $[-T,T]$. S can have jump discontinuities. We call S the "squashing function".

Suppose that

$$I^* := \int_{-T}^{T} S(t)dt > 0.$$

Clearly

$$\max_{x\in[-T,T]} S(x) = S(T).$$

Here we consider again

$$f \in C_{\mathcal{FR}}^{Uq}(\mathbb{R}).$$

For $x \in \mathbb{R}$, $s \in X$ we define the "*univariate fuzzy-random squashing operator*"

$$(G_n(f))(x,s) := \sum_{k=-n^2}^{n^2} f\left(\frac{k}{n},s\right)\odot\frac{S\left(n^{1-\alpha}\left(x-\frac{k}{n}\right)\right)}{I^*n^\alpha},\tag{18.16}$$

$0 < \alpha < 1$ and $n \in \mathbb{N}$: $n \geq \max(T+|x|,T^{-1/\alpha})$.

The above neural network operators are also linear operators in $\mathbb{R}$. These operators in the ordinary real case were thoroughly studied in [6], [47], and in the fuzzy sense were studied also in [11].

It is clear to see again that

$$(G_n(f))(x,s) = \sum_{k=\lceil nx-Tn^\alpha\rceil}^{[nx+Tn^\alpha]} f\left(\frac{k}{n},s\right)\odot\frac{S\left(n^{1-\alpha}\left(x-\frac{k}{n}\right)\right)}{I^*n^\alpha}.\tag{18.17}$$

Here we study the pointwise convergence in $x \in \mathbb{R}$ with rates of

$(G_n(f))(x,s) \overset{\text{"q-mean"}}{\longrightarrow} f(x,s)$, $1 \le q < +\infty$, as $n \to +\infty$.

We present

Theorem 18.24. *Let* $x \in \mathbb{R}$, $s \in X$, $T > 0$ *and* $1 \le q < +\infty$. *Let* $n \in \mathbb{N}$ *such that* $n \ge \max(T + |x|, T^{-1/\alpha})$. *Suppose that*

$$\int_X D^q(f(x,s), \tilde{o}) P(ds) < +\infty.$$

Then

$$\left(\int_X D^q((G_n f)(x,s), f(x,s)) P(ds) \right)^{1/q}$$

$$\le \left| \sum_{k=\lceil nx-Tn^\alpha \rceil}^{\lfloor nx+Tn^\alpha \rfloor} \frac{S\left(n^{1-\alpha}\left(x - \frac{k}{n}\right)\right)}{I^* n^\alpha} - 1 \right| \left(\int_X D^q(f(x,s), \tilde{o}) P(ds) \right)^{1/q}$$

$$+ \frac{S(T)}{I^*} \left(2T + \frac{1}{n^\alpha} \right) \Omega_1^{(\mathcal{F})}\left(f, \frac{T}{n^{1-\alpha}} \right)_{L^q}.$$

$$(18.18)$$

Proof. All parts of inequality (18.18) make sense because of the notions and basic results presented in Section 18.2.

Notice that

$$\sum_{k=\lceil nx-Tn^\alpha \rceil}^{\lfloor nx+Tn^\alpha \rfloor} 1 \le (2Tn^\alpha + 1). \qquad (18.19)$$

We observe that

$$D((G_n(f))(x,s), f(x,s))$$

$$= D\left(\sum_{k=\lceil nx-Tn^\alpha \rceil}^{\lfloor nx+Tn^\alpha \rfloor *} f\left(\frac{k}{n}, s\right) \odot \frac{S\left(n^{1-\alpha}\left(x - \frac{k}{n}\right)\right)}{I^* n^\alpha}, f(x,s) \right)$$

$$\le D\left(\sum_{k=\lceil nx-Tn^\alpha \rceil}^{\lfloor nx+Tn^\alpha \rfloor *} f\left(\frac{k}{n}, s\right) \odot \frac{S\left(n^{1-\alpha}\left(x - \frac{k}{n}\right)\right)}{I^* n^\alpha}, \right.$$

$$\left. \sum_{k=\lceil nx-Tn^\alpha \rceil}^{\lfloor nx+Tn^\alpha \rfloor *} f(x,s) \odot \frac{S\left(n^{1-\alpha}\left(x - \frac{k}{n}\right)\right)}{I^* n^\alpha} \right)$$

$$+ D\left(f(x,s) \odot \sum_{k=\lceil nx-Tn^\alpha \rceil}^{\lfloor nx+Tn^\alpha \rfloor} \frac{S\left(n^{1-\alpha}\left(x - \frac{k}{n}\right)\right)}{I^* n^\alpha}, f(x,s) \odot 1 \right)$$

$$
\text{(by Lemma 1.2)} \leq \left| \sum_{k=\lceil nx-Tn^\alpha \rceil}^{\lceil nx+Tn^\alpha \rceil} \frac{S\left(n^{1-\alpha}\left(x-\frac{k}{n}\right)\right)}{I^* n^\alpha} - 1 \right| D(f(x,s),\tilde{o})
$$

$$
+ \sum_{k=\lceil nx-Tn^\alpha \rceil}^{\lceil nx+Tn^\alpha \rceil} \frac{S\left(n^{1-\alpha}\left(x-\frac{k}{n}\right)\right)}{I^* n^\alpha} D\left(f\left(\frac{k}{n},s\right),f(x,s)\right).
$$

That is,

$$
D((G_n(f))(x,s),f(x,s)) \leq \left| \sum_{k=\lceil nx-Tn^\alpha \rceil}^{\lceil nx+Tn^\alpha \rceil} \frac{S\left(n^{1-\alpha}\left(x-\frac{k}{n}\right)\right)}{I^* n^\alpha} - 1 \right| \cdot
$$
$$
D(f(x,s),\tilde{o})
$$
$$
+ \sum_{k=\lceil nx-Tn^\alpha \rceil}^{\lceil nx+Tn^\alpha \rceil} \frac{S\left(n^{1-\alpha}\left(x-\frac{k}{n}\right)\right)}{I^* n^\alpha} \cdot
$$
$$
D\left(f\left(\frac{k}{n},s\right),f(x,s)\right).
$$

Hence

$$
\left(\int_X D^q((G_n(f))(x,s),f(x,s))P(ds) \right)^{1/q}
$$

$$
\leq \left| \sum_{k=\lceil nx-Tn^\alpha \rceil}^{\lceil nx+Tn^\alpha \rceil} \frac{S\left(n^{1-\alpha}\left(x-\frac{k}{n}\right)\right)}{I^* n^\alpha} - 1 \right| \left(\int_X D^q(f(x,s),\tilde{o})P(ds) \right)^{1/q}
$$

$$
+ \sum_{k=\lceil nx-Tn^\alpha \rceil}^{\lceil nx+Tn^\alpha \rceil} \frac{S\left(n^{1-\alpha}\left(x-\frac{k}{n}\right)\right)}{I^* n^\alpha} \cdot
$$

$$
\left(\int_X D^q\left(f\left(\frac{k}{n},s\right),f(x,s)\right)P(ds) \right)^{1/q}
$$

$$
\leq \left| \sum_{k=\lceil nx-Tn^\alpha \rceil}^{\lceil nx+Tn^\alpha \rceil} \frac{S\left(n^{1-\alpha}\left(x-\frac{k}{n}\right)\right)}{I^* n^\alpha} - 1 \right| \left(\int_X D^q(f(x,s),\tilde{o})P(ds) \right)^{1/q}
$$

$$
+ \sum_{k=\lceil nx-Tn^\alpha \rceil}^{\lceil nx+Tn^\alpha \rceil} \frac{S\left(n^{1-\alpha}\left(x-\frac{k}{n}\right)\right)}{I^* n^\alpha} \Omega_1^{(\mathcal{F})}\left(f,\left|\frac{k}{n}-x\right|\right)_{L^q}
$$

$$
\leq \left| \sum_{k=\lceil nx-Tn^\alpha \rceil}^{\lceil nx+Tn^\alpha \rceil} \frac{S\left(n^{1-\alpha}\left(x-\frac{k}{n}\right)\right)}{I^* n^\alpha} - 1 \right| \left(\int_X D^q(f(x,s),\tilde{o})P(ds) \right)^{1/q}
$$

$$
+ \left(\sum_{k=\lceil nx-Tn^\alpha \rceil}^{\lceil nx+Tn^\alpha \rceil} \frac{S\left(n^{1-\alpha}\left(x-\frac{k}{n}\right)\right)}{I^* n^\alpha} \right) \Omega_1^{(\mathcal{F})}\left(f,\frac{T}{n^{1-\alpha}}\right)_{L^q}
$$

$$\text{(by (18.19))} \leq \left| \sum_{k=\lceil nx-Tn^\alpha \rceil}^{\lceil nx+Tn^\alpha \rceil} \frac{S\left(n^{1-\alpha}\left(x-\frac{k}{n}\right)\right)}{I^* n^\alpha} - 1 \right| \cdot$$

$$\left(\int_X D^q(f(x,s),\tilde{o}) P(ds) \right)^{1/q}$$

$$+ \frac{S(T)}{I^*} \left(2T + \frac{1}{n^\alpha} \right) \Omega_1^{(\mathcal{F})} \left(f, \frac{T}{n^{1-\alpha}} \right)_{L^q} . \qquad \square$$

Remark 18.25. From Lemma 2.2, p. 79, [9], we have that

$$\lim_{n\to+\infty} \left| \sum_{k=\lceil nx-Tn^\alpha \rceil}^{\lceil nx+Tn^\alpha \rceil} \frac{S\left(n^{1-\alpha}\left(x-\frac{k}{n}\right)\right)}{I^* n^\alpha} - 1 \right| = 0,$$

any $x \in \mathbb{R}$.

Let $f \in C_{\mathcal{F}R}^{Uq}(\mathbb{R})$, then $\lim_{n\to+\infty} \Omega_1^{(\mathcal{F})} \left(f, \frac{T}{n^{1-\alpha}} \right)_{L^q} = 0$, by Proposition 18.13(ii). Thus from (18.18) we obtain

$$\lim_{n\to+\infty} \int_X D^q((G_n f)(x,s), f(x,s)) P(ds) = 0.$$

That is, $(G_n f)(x,s) \overset{\text{``q-mean''}}{\underset{n\to+\infty}{\longrightarrow}} f(x,s)$, pointwise in $x \in \mathbb{R}$ with rates.

Next we study further the fuzzy random operators F_n, G_n in 1-mean, their pointwise convergence with rates to the fuzzy-random unit operator again. However, this time the domain of operators is the fuzzy-random differentiable functions. That fact is reflected in the derived inequalities.

Assumptions 18.26. Let $x \in \mathbb{R}$, $s \in X$; where $(X, \mathcal{B}, P)$ is a probability space, and $n \in \mathbb{N}$ such that

$$n \geq \max(T + |x|, T^{-1/\alpha}), \quad T > 0, \ 0 < \alpha < 1.$$

Let $f \colon \mathbb{R} \to \mathcal{L}_{\mathcal{F}}(X, \mathcal{B}, P)$ be a fuzzy random function, such that the fuzzy derivative in x

$$f^{(j)} \colon \mathbb{R} \to \mathcal{L}_{\mathcal{F}}(X, \mathcal{B}, P), \quad j = 1, \ldots, N,$$

exist and are fuzzy random functions. (I.e., $f^{(j)}(x,s)$, for all $j = 0, 1, \ldots, N$ is $\mathcal{B}$-measurable in s.)

We suppose that $f^{(N)}(x,s)$ is fuzzy continuous in x, and $D(f^{(N)}(x,s),\tilde{o}) \leq M$, $\forall (x,s) \in \mathbb{R} \times X$, where $M > 0$. We also assume that

$$\int_X D(f^{(j)}(x,s),\tilde{o}) P(ds) < +\infty, \quad \text{all } j = 0, 1, \ldots, N.$$

We finally suppose that

$$f^{(j)} \in C^{U1}_{\mathcal{F}R}(\mathbb{R}), \quad \text{all } j = 0, 1, \ldots, N.$$

We give

Theorem 18.27. *All here as in Assumptions 18.26. Then*

$$\int_X D(f_n(x,s), f(x,s))P(ds) \leq \left| \sum_{k=\lceil nx-Tn^\alpha \rceil}^{[nx+Tn^\alpha]} \frac{b\left(n^{1-\alpha}\left(x - \frac{k}{n}\right)\right)}{In^\alpha} - 1 \right|$$

$$\cdot \left[\int_X D(f(x,s), \tilde{o})P(ds) + \Omega_1^{(\mathcal{F})}\left(f, \frac{T}{n^{1-\alpha}}\right)_{L^1} \right]$$

$$+ \Omega_1^{(\mathcal{F})}\left(f, \frac{T}{n^{1-\alpha}}\right)_{L^1} + \frac{b^*}{I}\left(2T + \frac{1}{n^\alpha}\right)$$

$$\cdot \left\{ \sum_{j=1}^{N-1} \frac{2^j T^j}{j! n^{(1-\alpha)j}} \cdot \left[\int_X D(f^{(j)}(x,s), \tilde{o})P(ds) + \right. \right. \tag{18.20}$$

$$\left. \left. \Omega_1^{(\mathcal{F})}\left(f^{(j)}, \frac{T}{n^{1-\alpha}}\right)_{L^1}\right]\right\}$$

$$+ \frac{2^N T^N b^*}{N! In^{N(1-\alpha)}}\left(2T + \frac{1}{n^\alpha}\right) \cdot$$

$$\left[\int_X D(f^{(N)}(x,s), \tilde{o})P(ds) + 3\Omega_1^{(\mathcal{F})}\left(f^{(N)}, \frac{T}{n^{1-\alpha}}\right)_{L^1}\right].$$

Remark 18.28. By Lemma 2.1 of [9], p. 64 we get that

$$\lim_{n \to +\infty} \left| \sum_{k=\lceil nx-Tn^\alpha \rceil}^{[nx+Tn^\alpha]} \frac{b\left(n^{1-\alpha}\left(x - \frac{k}{n}\right)\right)}{In^\alpha} - 1 \right| = 0, \quad \text{any } x \in \mathbb{R}.$$

Since $f^{(j)} \in C^{U1}_{\mathcal{F}R}(\mathbb{R}), j = 0, 1, \ldots, N$, we have that $\lim_{n \to +\infty} \Omega_1^{(\mathcal{F})}\left(f^{(j)}, \frac{T}{n^{1-\alpha}}\right)_{L^1} = 0$, by Proposition 18.13(ii). Thus from (18.20) we find that

$$\lim_{n \to +\infty} \int_X D(f_n(x,s), f(x,s))P(ds) = 0.$$

That is, $f_n(x,s) \xrightarrow[n \to +\infty]{\text{"1-mean"}} f(x,s)$, pointwise in $x \in \mathbb{R}$ with rates.

Proof (of Theorem 18.27). All parts of inequality (18.20) make sense and are finite, by Assumptions 18.26 and results of Section 18.2. Because b has compact support $[-T, T]$ we have that

$$f_n(x,s) = \sum_{k=\lceil nx-Tn^\alpha \rceil}^{[nx+Tn^\alpha]*} f\left(\frac{k}{n}, s\right) \odot \frac{b\left(n^{1-\alpha}\left(x - \frac{k}{n}\right)\right)}{In^\alpha}.$$

Clearly the terms in the sum (18.13) are non-zero iff $n^{1-\alpha}\left|x - \frac{k}{n}\right| \leq T$, iff

$$-\frac{T}{n^{1-\alpha}} \leq x - \frac{k}{n} \leq \frac{T}{n^{1-\alpha}}, \quad x \in \mathbb{R}.$$

Hence here we have $x - \frac{T}{n^{1-\alpha}} \leq \frac{k}{n}$. By Lemma 18.2 all $f^{(j)}$, $j = 0, 1, \ldots, N-1$ are fuzzy continuous. Using the fuzzy Taylor formula (Theorem 18.3) we obtain

$$f\left(\frac{k}{n}, s\right) = \sum_{j=0}^{N-1}{}^{*} f^{(j)}\left(x - \frac{T}{n^{1-\alpha}}, s\right) \odot \frac{\left(\frac{k}{n} - x + \frac{T}{n^{1-\alpha}}\right)^{j}}{j!} \oplus R_N\left(x - \frac{T}{n^{1-\alpha}}, \frac{k}{n}, s\right),$$

where

$$R_N\left(x - \frac{T}{n^{1-\alpha}}, \frac{k}{n}, s\right) :=$$

$$(FR)\int_{x-\frac{T}{n^{1-\alpha}}}^{\frac{k}{n}}\left(\int_{x-\frac{T}{n^{1-\alpha}}}^{s_1}\left(\int_{x-\frac{T}{n^{1-\alpha}}}^{s_{N-1}} f^{(N)}(s_N, s)ds_N\right) ds_{N-1} \ldots\right) ds_1$$

is a fuzzy random variable by Proposition 18.18, Lemma 1.15 and Lemma 18.1.

Easily we derive

$$\sum_{k=\lceil nx-Tn^{\alpha}\rceil}^{[nx+Tn^{\alpha}]}{}^{*} f\left(\frac{k}{n}, s\right) \odot \frac{b\left(n^{1-\alpha}\left(x - \frac{k}{n}\right)\right)}{In^{\alpha}}$$

$$= \sum_{k=\lceil nx-Tn^{\alpha}\rceil}^{[nx+Tn^{\alpha}]}{}^{*} \sum_{j=0}^{N-1}{}^{*} f^{(j)}\left(x - \frac{T}{n^{1-\alpha}}, s\right) \odot \frac{\left(\frac{k}{n} - x + \frac{T}{n^{1-\alpha}}\right)^{j}}{j!}$$

$$\odot \frac{b\left(n^{1-\alpha}\left(x - \frac{k}{n}\right)\right)}{In^{\alpha}}$$

$$\oplus \sum_{k=\lceil nx-Tn^{\alpha}\rceil}^{[nx+Tn^{\alpha}]}{}^{*} R_N\left(x - \frac{T}{n^{1-\alpha}}, \frac{k}{n}, s\right)$$

$$\odot \frac{b\left(n^{1-\alpha}\left(x - \frac{k}{n}\right)\right)}{In^{\alpha}}.$$

That is,

$$
\begin{aligned}
f_n(x,s) \;=\; & \sum_{j=0}^{N-1}{}^{*}\; \sum_{k=\lceil nx-Tn^{\alpha}\rceil}^{[nx+Tn^{\alpha}]}{}^{*}\; f^{(j)}\left(x-\frac{T}{n^{1-\alpha}},s\right) \odot \frac{\left(\frac{k}{n}-x+\frac{T}{n^{1-\alpha}}\right)^{j}}{j!} \\
& \odot \frac{b\left(n^{1-\alpha}\left(x-\frac{k}{n}\right)\right)}{In^{\alpha}} \\
& \oplus \sum_{k=\lceil nx-Tn^{\alpha}\rceil}^{[nx+Tn^{\alpha}]}{}^{*}\; R_N\left(x-\frac{T}{n^{1-\alpha}},\frac{k}{n},s\right) \\
& \odot \frac{b\left(n^{1-\alpha}\left(x-\frac{k}{n}\right)\right)}{In^{\alpha}}.
\end{aligned}
$$

Next we estimate

$$
\begin{aligned}
D(f_n(x,s),f(x,s)) \leq\; & D\left(f_n(x,s),\; \sum_{k=\lceil nx-Tn^{\alpha}\rceil}^{[nx+Tn^{\alpha}]}{}^{*}\; f\left(x-\frac{T}{n^{1-\alpha}},s\right)\odot \right. \\
& \left. \frac{b\left(n^{1-\alpha}\left(x-\frac{k}{n}\right)\right)}{In^{\alpha}}\right) \\
& +D\left(f\left(x-\frac{T}{n^{1-\alpha}},s\right)\odot \sum_{k=\lceil nx-Tn^{\alpha}\rceil}^{[nx+Tn^{\alpha}]} \frac{b\left(n^{1-\alpha}\left(x-\frac{k}{n}\right)\right)}{In^{\alpha}},\right. \\
& \left. f\left(x-\frac{T}{n^{1-\alpha}},s\right)\odot 1\right) + D\left(f\left(x-\frac{T}{n^{1-\alpha}},s\right),f(x,s)\right) \\
\text{(by Lemma 1.2)} \leq\; & D\left(\sum_{j=1}^{N-1}{}^{*}\; \sum_{k=\lceil nx-Tn^{\alpha}\rceil}^{[nx+Tn^{\alpha}]}{}^{*}\; f^{(j)}\left(x-\frac{T}{n^{1-\alpha}},s\right)\right. \\
& \odot \frac{\left(\frac{k}{n}-x+\frac{T}{n^{1-\alpha}}\right)^{j}}{j!}\odot \frac{b\left(n^{1-\alpha}\left(x-\frac{k}{n}\right)\right)}{In^{\alpha}} \\
& \oplus \sum_{k=\lceil nx-Tn^{\alpha}\rceil}^{[nx+Tn^{\alpha}]}{}^{*}\; R_N\left(x-\frac{T}{n^{1-\alpha}},\frac{k}{n},s\right) \\
& \left. \odot \frac{b\left(n^{1-\alpha}\left(x-\frac{k}{n}\right)\right)}{In^{\alpha}},\tilde{o}\right) + \left|\sum_{k=\lceil nx-Tn^{\alpha}\rceil}^{[nx+Tn^{\alpha}]} \frac{b\left(n^{1-\alpha}\left(x-\frac{k}{n}\right)\right)}{In^{\alpha}}-1\right| \\
& \cdot D\left(f\left(x-\frac{T}{n^{1-\alpha}},s\right),\tilde{o}\right) + D\left(f\left(x-\frac{T}{n^{1-\alpha}},s\right),f(x,s)\right)
\end{aligned}
$$

$$\leq \sum_{j=1}^{N-1} \sum_{k=\lceil nx-Tn^\alpha \rceil}^{[nx+Tn^\alpha]} \frac{\left(\frac{k}{n} - x + \frac{T}{n^{1-\alpha}}\right)^j}{j!} \cdot \frac{b\left(n^{1-\alpha}\left(x - \frac{k}{n}\right)\right)}{In^\alpha} \cdot$$

$$D\left(f^{(j)}\left(x - \frac{T}{n^{1-\alpha}}, s\right), \tilde{o}\right)$$

$$+ \sum_{k=\lceil nx-Tn^\alpha \rceil}^{[nx+Tn^\alpha]} \frac{b\left(n^{1-\alpha}\left(x - \frac{k}{n}\right)\right)}{In^\alpha} D\left(R_N\left(x - \frac{T}{n^{1-\alpha}}, \frac{k}{n}, s\right), \tilde{o}\right)$$

$$+ \left| \sum_{k=\lceil nx-Tn^\alpha \rceil}^{[nx+Tn^\alpha]} \frac{b\left(n^{1-\alpha}\left(x - \frac{k}{n}\right)\right)}{In^\alpha} - 1 \right|$$

$$\cdot D\left(f\left(x - \frac{T}{n^{1-\alpha}}, s\right), \tilde{o}\right) + D\left(f\left(x - \frac{T}{n^{1-\alpha}}, s\right), f(x, s)\right).$$

Consequently we have

$$\int_X D(f_n(x, s), f(x, s)) P(ds)$$

$$\leq \sum_{j=1}^{N-1} \sum_{k=\lceil nx-Tn^\alpha \rceil}^{[nx+Tn^\alpha]} \frac{\left(\frac{k}{n} - x + \frac{T}{n^{1-\alpha}}\right)^j}{j!} \frac{b\left(n^{1-\alpha}\left(x - \frac{k}{n}\right)\right)}{In^\alpha}$$

$$\cdot \int_X D\left(f^{(j)}\left(x - \frac{T}{n^{1-\alpha}}, s\right), \tilde{o}\right) P(ds)$$

$$+ \sum_{k=\lceil nx-Tn^\alpha \rceil}^{[nx+Tn^\alpha]} \frac{b\left(n^{1-\alpha}\left(x - \frac{k}{n}\right)\right)}{In^\alpha} \int_X D\left(R_N\left(x - \frac{T}{n^{1-\alpha}}, \frac{k}{n}, s\right), \tilde{o}\right) P(ds)$$

$$+ \left| \sum_{k=\lceil nx-Tn^\alpha \rceil}^{[nx+Tn^\alpha]} \frac{b\left(n^{1-\alpha}\left(x - \frac{k}{n}\right)\right)}{In^\alpha} - 1 \right| \cdot \int_X D\left(f\left(x - \frac{T}{n^{1-\alpha}}, s\right), \tilde{o}\right) P(ds)$$

$$+ \int_X D\left(f\left(x - \frac{T}{n^{1-\alpha}}, s\right), f(x, s)\right) P(ds).$$

I.e., it makes sense by Remark 18.5 and Proposition 18.17 the integral

$$\int_X D(f_n(x,s), f(x,s))P(ds)$$

$$(\text{by } (18.15)) \leq \frac{b^*}{I}\left(2T + \frac{1}{n^\alpha}\right) \cdot \left\{ \sum_{j=1}^{N-1} \frac{2^j T^j}{j! n^{(1-\alpha)j}} \right.$$

$$\left. \cdot \left[\int_X D(f^{(j)}(x,s), \tilde{o})P(ds) + \Omega_1^{(\mathcal{F})}\left(f^{(j)}, \frac{T}{n^{1-\alpha}}\right)_{L^1} \right] \right\}$$

$$+ \sum_{k=\lceil nx - Tn^\alpha \rceil}^{[nx+Tn^\alpha]} \frac{b\left(n^{1-\alpha}\left(x - \frac{k}{n}\right)\right)}{In^\alpha} \int_X D\left(R_N\left(x - \frac{T}{n^{1-\alpha}}, \frac{k}{n}, s\right), \tilde{o}\right)P(ds)$$

$$+ \left| \sum_{k=\lceil nx - Tn^\alpha \rceil}^{[nx+Tn^\alpha]} \frac{b\left(n^{1-\alpha}\left(x - \frac{k}{n}\right)\right)}{In^\alpha} - 1 \right| \qquad (18.21)$$

$$\cdot \left[\int_X D(f(x,s), \tilde{o})P(ds) + \Omega_1^{(\mathcal{F})}\left(f, \frac{T}{n^{1-\alpha}}\right)_{L^1} \right] + \Omega_1^{(\mathcal{F})}\left(f, \frac{T}{n^{1-\alpha}}\right)_{L^1} =: (*).$$

In the following we work on

$$\int_X D\left(R_N\left(x - \frac{T}{n^{1-\alpha}}, \frac{k}{n}, s\right)\tilde{o}\right)P(ds)$$

$$\leq \int_X D\left(R_N\left(x - \frac{T}{n^{1-\alpha}}, \frac{k}{n}, s\right), f^{(N)}\left(x - \frac{T}{n^{1-\alpha}}, s\right) \odot \right.$$

$$\left. \frac{\left(\frac{k}{n} - x + \frac{T}{n^{1-\alpha}}\right)^N}{N!}\right)P(ds)$$

$$+ \int_X D\left(f^{(N)}\left(x - \frac{T}{n^{1-\alpha}}, s\right) \odot \frac{\left(\frac{k}{n} - x + \frac{T}{n^{1-\alpha}}\right)^N}{N!}, \tilde{o}\right)P(ds),$$

the last makes sense by Remark 18.5 and Proposition 18.17(i).
That is,

$$\int_X D\left(R_N\left(x - \frac{T}{n^{1-\alpha}}, \frac{k}{n}, s\right)\tilde{o}\right)P(ds)$$

$$\leq \int_X D\left(R_N\left(x - \frac{T}{n^{1-\alpha}}, \frac{k}{n}, s\right), f^{(N)}\left(x - \frac{T}{n^{1-\alpha}}, s\right) \odot \frac{\left(\frac{k}{n} - x + \frac{T}{n^{1-\alpha}}\right)^N}{N!}\right).$$

$$P(ds) + \frac{2^N T^N}{N! n^{N(1-\alpha)}}\left(\int_X D(f^{(N)}(x,s), \tilde{o})P(ds) + \Omega_1^{(\mathcal{F})}\left(f^{(N)}, \frac{T}{n^{1-\alpha}}\right)_{L^1}\right).$$

Finally we estimate

$$\int_X D\left(R_N\left(x-\frac{T}{n^{1-\alpha}},\frac{k}{n},s\right),f^{(N)}\left(x-\frac{T}{n^{1-\alpha}},s\right)\odot\frac{\left(\frac{k}{n}-x+\frac{T}{n^{1-\alpha}}\right)^N}{N!}\right)\cdot$$

$$P(ds)$$

$$=\int_X D\left((FR)\int_{x-\frac{T}{n^{1-\alpha}}}^{\frac{k}{n}}\left(\int_{x-\frac{T}{n^{1-\alpha}}}^{s_1}\cdots\left(\int_{x-\frac{T}{n^{1-\alpha}}}^{s_{N-1}}f^{(N)}(s_N,s)ds_N\right)ds_{N-1}\cdots\right.\right.$$

$$ds_1,(FR)\int_{x-\frac{T}{n^{1-\alpha}}}^{\frac{k}{n}}\left(\int_{x-\frac{T}{n^{1-\alpha}}}^{s_1}\cdots\left(\int_{x-\frac{T}{n^{1-\alpha}}}^{s_{N-1}}f^{(N)}\left(x-\frac{T}{n^{1-\alpha}},s\right)ds_N\right)\right.$$

$$\left.\left.ds_{N-1}\cdots\right)ds_1\right)P(ds)$$

$$(\text{by Lemmas 1.13, 1.15, 18.1})\le\int_X\left(\int_{x-\frac{T}{n^{1-\alpha}}}^{\frac{k}{n}}\left(\int_{x-\frac{T}{n^{1-\alpha}}}^{s_1}\right.\right.$$

$$\left.\left.\cdots\left(\int_{x-\frac{T}{n^{1-\alpha}}}^{s_{N-1}}D\left(f^{(N)}(s_N,s),f^{(N)}\left(x-\frac{T}{n^{1-\alpha}},s\right)\right)ds_N\right)ds_{N-1}\cdots\right)ds\right.$$

$$P(ds)\qquad(\text{by Lemma 1.13, bounded}$$

convergence theorem, Proposition 18.14(i) and Fubini theorem)

$$=\int_{x-\frac{T}{n^{1-\alpha}}}^{\frac{k}{n}}\left(\int_{x-\frac{T}{n^{1-\alpha}}}^{s_1}\cdots\left(\int_{x-\frac{T}{n^{1-\alpha}}}^{s_{N-1}}\left(\int_X D\left(f^{(N)}(s_N,s),f^{(N)}\left(x-\frac{T}{n^{1-\alpha}},s\right)\right)\right.\right.\right.$$

$$\left.\left.\left.P(ds)\right)ds_N\right)ds_{N-1}\cdots\right)ds_1$$

$$\le\int_{x-\frac{T}{n^{1-\alpha}}}^{\frac{k}{n}}\left(\int_{x-\frac{T}{n^{1-\alpha}}}^{s_1}\cdots\left(\int_{x-\frac{T}{n^{1-\alpha}}}^{s_{N-1}}\Omega_1^{(\mathcal{F})}\left(f^{(N)},\left|s_N-x+\frac{T}{n^{1-\alpha}}\right|\right)_{L^1}ds_N\right)\right.$$

$$\left.\cdots ds_{N-1}\cdots\right)ds_1$$

$$\le\frac{\left(\frac{k}{n}-x+\frac{T}{n^{1-\alpha}}\right)^N}{N!}\cdot\Omega_1^{(\mathcal{F})}\left(f^{(N)},\frac{k}{n}-x+\frac{T}{n^{1-\alpha}}\right)_{L^1}$$

$$\le\frac{2^N T^N}{N!n^{N(1-\alpha)}}\cdot\Omega_1^{(\mathcal{F})}\left(f^{(N)},\frac{2T}{n^{1-\alpha}}\right)_{L^1}.$$

Thus, we got that

$$\int_X D\left(R_N\left(x-\frac{T}{n^{1-\alpha}},\frac{k}{n},s\right),f^{(N)}\left(x-\frac{T}{n^{1-\alpha}},s\right)\odot\frac{\left(\frac{k}{n}-x+\frac{T}{n^{1-\alpha}}\right)^N}{N!}\right)\cdot$$

$$P(ds)\le\frac{2^N T^N}{N!\cdot n^{N(1-\alpha)}}\Omega_1^{(\mathcal{F})}\left(f^{(N)},\frac{2T}{n^{1-\alpha}}\right)_{L^1}$$

$$\leq \frac{2^{N+1}T^N}{N!n^{N(1-\alpha)}}\Omega_1^{(\mathcal{F})}\left(f^{(N)},\frac{T}{n^{1-\alpha}}\right)_{L^1}.$$

Putting things together we derive

$$\int_X D\left(R_N\left(x-\frac{T}{n^{1-\alpha}},\frac{k}{n},s\right),\tilde{o}\right)P(ds)$$
$$\leq \frac{2^{N+1}T^N}{N!n^{N(1-\alpha)}}\Omega_1^{(\mathcal{F})}\left(f^{(N)},\frac{T}{n^{1-\alpha}}\right)_{L^1}$$
$$+\frac{2^N T^N}{N!n^{N(1-\alpha)}}\left(\int_X D(f^{(N)}(x,s),\tilde{o})P(ds)+\Omega_1^{(\mathcal{F})}\left(f^{(N)},\frac{T}{n^{1-\alpha}}\right)_{L^1}\right).$$

That is, we find

$$\int_X D\left(R_N\left(x-\frac{T}{n^{1-\alpha}},\frac{k}{n},s\right),\tilde{o}\right)P(ds) \tag{18.22}$$
$$\leq \frac{2^N T^N}{N!n^{N(1-\alpha)}}\left(\int_X D(f^{(N)}(x,s),\tilde{o})P(ds)+3\Omega_1^{(\mathcal{F})}\left(f^{(N)},\frac{T}{n^{1-\alpha}}\right)_{L^1}\right).$$

Also it holds

$$\sum_{k=\lceil nx-Tn^\alpha\rceil}^{\lceil nx+Tn^\alpha\rceil}\frac{b\left(n^{1-\alpha}\left(x-\frac{k}{n}\right)\right)}{In^\alpha}\cdot\int_X D\left(R_N\left(x-\frac{T}{n^{1-\alpha}},\frac{k}{n},s\right)\tilde{o}\right)P(ds)$$
$$\text{(by (18.15))} \leq \frac{2^N T^N b^*}{N!In^{N(1-\alpha)}}\left(2T+\frac{1}{n^\alpha}\right)\left[\int_X D(f^{(N)}(x,s),\tilde{o})P(ds)\right.$$
$$\left.+3\Omega_1^{(\mathcal{F})}\left(f^{(N)},\frac{T}{n^{1-\alpha}}\right)_{L^1}\right].$$

Putting things in $(*)$, see (18.21), etc.   □

Finally we present

Theorem 18.29. *All here are as in Assumptions 18.26. Then*

$$\int_X D((G_n f)(x, s), f(x, s)) P(ds) \le$$

$$\left| \sum_{k=\lceil nx-Tn^\alpha \rceil}^{\lceil nx+Tn^\alpha \rceil} \frac{S\left(n^{1-\alpha}\left(x - \frac{k}{n}\right)\right)}{I^* n^\alpha} - 1 \right|$$

$$\cdot \left[\int_X D(f(x, s), \tilde{o}) P(ds) + \Omega_1^{(\mathcal{F})}\left(f, \frac{T}{n^{1-\alpha}}\right)_{L^1} \right]$$

$$+ \Omega_1^{(\mathcal{F})}\left(f, \frac{T}{n^{1-\alpha}}\right)_{L^1}$$

$$+ \frac{S(T)}{I^*}\left(2T + \frac{1}{n^\alpha}\right) \cdot \left\{ \sum_{j=1}^{N-1} \frac{2^j T^j}{j! n^{(1-\alpha)j}} \right.$$

$$\left. \cdot \left[\int_X D(f^{(j)}(x, s), \tilde{o}) P(ds) + \Omega_1^{(\mathcal{F})}\left(f^{(j)}, \frac{T}{n^{1-\alpha}}\right)_{L^1} \right] \right\}$$

$$+ \frac{2^N T^N S(T)}{N! I^* n^{N(1-\alpha)}}\left(2T + \frac{1}{n^\alpha}\right) \tag{18.23}$$

$$\cdot \left[\int_X D(f^{(N)}(x, s), \tilde{o}) P(ds) + 3\Omega_1^{(\mathcal{F})}\left(f^{(N)}, \frac{T}{n^{1-\alpha}}\right)_{L^1} \right].$$

Remark 18.30. From Lemma 2.2, p. 79, [9], we have that

$$\lim_{n \to +\infty} \left| \sum_{k=\lceil nx-Tn^\alpha \rceil}^{\lceil nx+Tn^\alpha \rceil} \frac{S\left(n^{1-\alpha}\left(x - \frac{k}{n}\right)\right)}{I^* n^\alpha} - 1 \right| = 0, \quad \text{any } x \in \mathbb{R}.$$

Since $f^{(j)} \in C_{\mathcal{FR}}^{U1}(\mathbb{R})$, $j = 0, 1, \ldots, N$, we have that

$$\lim_{n \to +\infty} \Omega_1^{(\mathcal{F})}\left(f^{(j)}, \frac{T}{n^{1-\alpha}}\right)_{L^1} = 0.$$

Hence from (18.23) we get that

$$\lim_{n \to +\infty} \int_X D((G_n f)(x, s), f(x, s)) P(ds) = 0.$$

That is,

$$(G_n f)(x, s) \xrightarrow[n \to +\infty]{\text{“1-mean"}} f(x, s),$$

pointwise in $x \in \mathbb{R}$ with rates.

Proof (of Theorem 18.29). Basic facts here are justified as in the proof of Theorem 18.27. Again we see, that the terms in the sum (18.16) are non-zero iff $n^{1-\alpha}\left|x - \frac{k}{n}\right| \leq T$, iff

$$-\frac{T}{n^{1-\alpha}} \leq x - \frac{k}{n} \leq \frac{T}{n^{1-\alpha}}, \quad x \in \mathbb{R}.$$

Thus here $x - \frac{T}{n^{1-\alpha}} \leq \frac{k}{n}$. By the fuzzy Taylor formula (Theorem 18.3) we have

$$f\left(\frac{k}{n}, s\right) = \sum_{j=0}^{N-1}{}^{*} f^{(j)}\left(x - \frac{T}{n^{1-\alpha}}, s\right) \cdot \frac{\left(\frac{k}{n} - x + \frac{T}{n^{1-\alpha}}\right)^{j}}{j!} \oplus R_N\left(x - \frac{T}{n^{1-\alpha}}, \frac{k}{n}, s\right),$$

where

$$R_N\left(x - \frac{T}{n^{1-\alpha}}, \frac{k}{n}, s\right)$$

$$:= (FR) \int_{x - \frac{T}{n^{1-\alpha}}}^{\frac{k}{n}} \left(\int_{x - \frac{T}{n^{1-\alpha}}}^{s_1} \cdots \left(\int_{x - \frac{T}{n^{1-\alpha}}}^{s_{N-1}} f^{(N)}(s_N, s) ds_N \right) \right.$$
$$\left. ds_{N-1}\right) \ldots) \, ds_1.$$

Hence

$$\sum_{k=\lceil nx - Tn^{\alpha}\rceil}^{\lfloor nx + Tn^{\alpha}\rfloor}{}^{*} f\left(\frac{k}{n}, s\right) \odot \frac{S\left(n^{1-\alpha}\left(x - \frac{k}{n}\right)\right)}{I^* n^{\alpha}}$$

$$= \sum_{k=\lceil nx - Tn^{\alpha}\rceil}^{\lfloor nx + Tn^{\alpha}\rfloor}{}^{*} \sum_{j=0}^{N-1}{}^{*} f^{(j)}\left(x - \frac{T}{n^{1-\alpha}}, s\right)$$

$$\odot \frac{\left(\frac{k}{n} - x + \frac{T}{n^{1-\alpha}}\right)^{j}}{j!}$$

$$\odot \frac{S\left(n^{1-\alpha}\left(x - \frac{k}{n}\right)\right)}{I^* n^{\alpha}}$$

$$\oplus \sum_{k=\lceil nx - Tn^{\alpha}\rceil}^{\lfloor nx + Tn^{\alpha}\rfloor}{}^{*} R_N\left(x - \frac{T}{n^{1-\alpha}}, \frac{k}{n}, s\right)$$

$$\odot \frac{S\left(n^{1-\alpha}\left(x - \frac{k}{n}\right)\right)}{I^* n^{\alpha}}.$$

That is

$$(G_n f)(x, s) = \sum_{j=0}^{N-1}{}^* \sum_{k=\lceil nx-Tn^\alpha \rceil}^{\lfloor nx+Tn^\alpha \rfloor}{}^* f^{(j)}\left(x - \frac{T}{n^{1-\alpha}}, s\right)$$

$$\odot \frac{\left(\frac{k}{n} - x + \frac{T}{n^{1-\alpha}}\right)^j}{j!} \odot \frac{S\left(n^{1-\alpha}\left(x - \frac{k}{n}\right)\right)}{I^* n^\alpha}$$

$$\oplus \sum_{k=\lceil nx-Tn^\alpha \rceil}^{\lfloor nx+Tn^\alpha \rfloor}{}^* R_N\left(x - \frac{T}{n^{1-\alpha}}, \frac{k}{n}, s\right)$$

$$\odot \frac{S\left(n^{1-\alpha}\left(x - \frac{k}{n}\right)\right)}{I^* n^\alpha}.$$

Next we estimate

$$D((G_n f)(x, s), f(x, s))$$

$$\leq D\left((G_n f)(x, s), \sum_{k=\lceil nx-Tn^\alpha \rceil}^{\lfloor nx+Tn^\alpha \rfloor}{}^* f\left(x - \frac{T}{n^{1-\alpha}}, s\right) \odot \frac{S\left(n^{1-\alpha}\left(x - \frac{k}{n}\right)\right)}{I^* n^\alpha}\right)$$

$$+ D\left(f\left(x - \frac{T}{n^{1-\alpha}}, s\right) \odot \sum_{k=\lceil nx-Tn^\alpha \rceil}^{\lfloor nx+Tn^\alpha \rfloor} \frac{S\left(n^{1-\alpha}\left(x - \frac{k}{n}\right)\right)}{I^* n^\alpha},\right.$$

$$\left. f\left(x - \frac{T}{n^{1-\alpha}}, s\right) \odot 1\right) + D\left(f\left(x - \frac{T}{n^{1-\alpha}}, s\right), f(x, s)\right)$$

$$\text{(by Lemma 1.2)} \leq D\left(\sum_{j=1}^{N-1}{}^* \sum_{k=\lceil nx-Tn^\alpha \rceil}^{\lfloor nx+Tn^\alpha \rfloor}{}^* f^{(j)}\left(x - \frac{T}{n^{1-\alpha}}, s\right)\right.$$

$$\odot \frac{\left(\frac{k}{n} - x + \frac{T}{n^{1-\alpha}}\right)^j}{j!} \odot \frac{S\left(n^{1-\alpha}\left(x - \frac{k}{n}\right)\right)}{I^* n^\alpha}$$

$$\oplus \sum_{k=\lceil nx-Tn^\alpha \rceil}^{\lfloor nx+Tn^\alpha \rfloor}{}^* R_N\left(x - \frac{T}{n^{1-\alpha}}, \frac{k}{n}, s\right) \odot \frac{S\left(n^{1-\alpha}\left(x - \frac{k}{n}\right)\right)}{I^* n^\alpha}, \tilde{o}\right)$$

$$+ \left|\sum_{k=\lceil nx-Tn^\alpha \rceil}^{\lfloor nx+Tn^\alpha \rfloor} \frac{S\left(n^{1-\alpha}\left(x - \frac{k}{n}\right)\right)}{I^* n^\alpha} - 1\right| \cdot D\left(f\left(x - \frac{T}{n^{1-\alpha}}, s\right), \tilde{o}\right)$$

$$+ D\left(f\left(x - \frac{T}{n^{1-\alpha}}, s\right), f(x, s)\right)$$

$$\leq \sum_{j=1}^{N-1} \sum_{k=\lceil nx-Tn^\alpha \rceil}^{[nx+Tn^\alpha]} \frac{\left(\frac{k}{n}-x+\frac{T}{n^{1-\alpha}}\right)^j}{j!} \frac{S\left(n^{1-\alpha}\left(x-\frac{k}{n}\right)\right)}{I^* n^\alpha} \cdot$$

$$D\left(f^{(j)}\left(x-\frac{T}{n^{1-\alpha}},s\right),\tilde{o}\right)$$

$$+\sum_{k=\lceil nx-Tn^\alpha \rceil}^{[nx+Tn^\alpha]} \frac{S\left(n^{1-\alpha}\left(x-\frac{k}{n}\right)\right)}{I^* n^\alpha} D\left(R_N\left(x-\frac{T}{n^{1-\alpha}},\frac{k}{n},s\right),\tilde{o}\right)$$

$$+\left|\sum_{k=\lceil nx-Tn^\alpha \rceil}^{[nx+Tn^\alpha]} \frac{S\left(n^{1-\alpha}\left(x-\frac{k}{n}\right)\right)}{I^* n^\alpha}-1\right| \cdot D\left(f\left(x-\frac{T}{n^{1-\alpha}},s\right),\tilde{o}\right)$$

$$+D\left(f\left(x-\frac{T}{n^{1-\alpha}},s\right),f(x,s)\right).$$

Therefore we derive

$$\int_X D((G_nf)(x,s),f(x,s))P(ds)$$

$$\leq \sum_{j=1}^{N-1} \sum_{k=\lceil nx-Tn^\alpha \rceil}^{[nx+Tn^\alpha]} \frac{\left(\frac{k}{n}-x+\frac{T}{n^{1-\alpha}}\right)^j}{j!} \frac{S\left(n^{1-\alpha}\left(x-\frac{k}{n}\right)\right)}{I^* n^\alpha}$$

$$\cdot \int_X D\left(f^{(j)}\left(x-\frac{T}{n^{1-\alpha}},s\right),\tilde{o}\right)P(ds)$$

$$+\sum_{k=\lceil nx-Tn^\alpha \rceil}^{[nx+Tn^\alpha]} \frac{S\left(n^{1-\alpha}\left(x-\frac{k}{n}\right)\right)}{I^* n^\alpha} \int_X D\left(R_N\left(x-\frac{T}{n^{1-\alpha}},\frac{k}{n},s\right),\tilde{o}\right)P(ds)$$

$$+\left|\sum_{k=\lceil nx-Tn^\alpha \rceil}^{[nx+Tn^\alpha]} \frac{S\left(n^{1-\alpha}\left(x-\frac{k}{n}\right)\right)}{I^* n^\alpha}-1\right| \cdot \int_X D\left(f\left(x-\frac{T}{n^{1-\alpha}},s\right),\tilde{o}\right)P(ds)$$

$$+\int_X D\left(f\left(x-\frac{T}{n^{1-\alpha}},s\right),f(x,s)\right)P(ds).$$

That is

$$\int_X D((G_nf)(x,s),f(x,s))P(ds)$$

$$\leq \frac{S(T)}{I^*}\left(2T+\frac{1}{n^\alpha}\right)\cdot\left\{\sum_{j=1}^{N-1}\frac{2^j T^j}{j!n^{(1-\alpha)j}}\cdot\left[\int_X D(f^{(j)}(x,s),\tilde{o})P(ds)\right.\right.$$

$$\left.\left.+\Omega_1^{(\mathcal{F})}\left(f^{(j)},\frac{T}{n^{1-\alpha}}\right)_{L^1}\right]\right\}$$

$$+\sum_{k=\lceil nx-Tn^\alpha \rceil}^{[nx+Tn^\alpha]} \frac{S\left(n^{1-\alpha}\left(x-\frac{k}{n}\right)\right)}{I^* n^\alpha} \int_X D\left(R_N\left(x-\frac{T}{n^{1-\alpha}},\frac{k}{n},s\right),\tilde{o}\right)P(ds)$$

$$+\left|\sum_{k=\lceil nx-Tn^\alpha\rceil}^{[nx+Tn^\alpha]}\frac{S\left(n^{1-\alpha}\left(x-\frac{k}{n}\right)\right)}{I^*n^\alpha}-1\right| \tag{18.2.}$$

$$\cdot\left[\int_X D(f(x,s),\tilde{o})P(ds)+\Omega_1^{(\mathcal{F})}\left(f,\frac{T}{n^{1-\alpha}}\right)_{L^1}\right]+\Omega_1^{(\mathcal{F})}\left(f,\frac{T}{n^{1-\alpha}}\right)_{L^1}=:(*$$

Using (18.22) we find

$$\sum_{k=\lceil nx-Tn^\alpha\rceil}^{[nx+Tn^\alpha]}\frac{S\left(n^{1-\alpha}\left(x-\frac{k}{n}\right)\right)}{I^*n^\alpha}\cdot\int_X D\left(R_N\left(x-\frac{T}{n^{1-\alpha}},\frac{k}{n},s\right),\tilde{o}\right)P(ds)$$

$$(\text{by }(18.19))\leq\frac{2^N T^N S(T)}{N!I^*n^{N(1-\alpha)}}\left(2T+\frac{1}{n^\alpha}\right)\left[\int_X D(f^{(N)}(x,s),\tilde{o})P(ds)\right.$$

$$\left.+3\Omega_1^{(\mathcal{F})}\left(f^{(N)},\frac{T}{n^{1-\alpha}}\right)_{L^1}\right].$$

Putting things in (*), see (18.24), etc. □

19

$\mathcal{A}$-SUMMABILITY AND FUZZY KOROVKIN APPROXIMATION

The aim of this chapter is to present a fuzzy Korovkin-type approximation theorem by using a matrix summability method. We also study the rates of convergence of fuzzy positive linear operators. This chapter is based on [27].

19.1 Introduction

The study of the classical Korovkin type approximation theory is a well-established area of research, which deals with the problem of approximating a function f by means of a sequence $\{L_n(f)\}$ of positive linear operators. Most of the classical approximation operators tend to converge to the value of the function being approximated. However, at points of discontinuity, they often converge to the average of the left and right limits of the function. In this case, the matrix summability methods of the Cesáro type are applicable to correct the lack of convergence [45]. The main purpose of using summability theory has always been to make a nonconvergent sequence "converge". Some results regarding matrix summability for positive linear operators may be found in the papers [38], [39], [94]. In this chapter using a $\mathcal{A}$-summation process we study the approximation properties of sequence of positive linear operators defined on the space of all fuzzy continuous functions on the interval $[a, b]$.

G.A. Anastassiou: Fuzzy Mathematics: Approximation The., STUDFUZZ 251, pp. 373–384.
springerlink.com

Anastassiou and Duman [29] introduced the fuzzy analog of $A-$statistical convergence by using the metric D.

Let $(\mu_n)_{n \in \mathbb{N}}$ be a fuzzy number valued sequence and let $A = (a_{jn})$ is a non-negative regular summability matrix. Recall that the regularity conditions on a matrix A are known as Silverman-Toeplitz conditions in the functional analysis (see for details, [70]). Then, $(\mu_n)_{n \in \mathbb{N}}$ is $A-$statistically convergent to $\mu \in \mathbb{R}_\mathcal{F}$, which is denoted by $st_A - \lim\limits_{n \to \infty} D(\mu_n, \mu) = 0$ if for every $\varepsilon > 0$

$$\lim_{j \to \infty} \sum_{n:D(\mu_n,\mu) \geq \varepsilon} a_{jn} = 0$$

holds. Some basic results regarding $A-$statistical convergence for number sequences may be found in the papers [65], [77], [86]. Of course, the case of $A = C_1$, the Cesáro matrix of order one, immediately reduces to the statistical convergence of fuzzy valued sequences which defined by Nuray and Savaş [91]. Also, if A is replaced by the identity matrix, then we have the fuzzy convergence introduced by Matloka [85].

Let $\mathcal{A} := (A^n)_{n \geq 1}$, $A^n = \left(a_{kj}^n\right)_{k,j \in \mathbb{N}}$ be a sequence of infinite non-negative real matrices. For a sequence of real numbers, $x = (x_j)_{j \in \mathbb{N}}$, the double sequence

$$\mathcal{A}x := \{(Ax)_k^n : k, n \in \mathbb{N}\}$$

defined by $(Ax)_k^n := \sum\limits_{j=1}^{\infty} a_{kj}^n x_j$ is called the $\mathcal{A}-$transform of x whenever the series converges for all k and n. A sequence x is said to be $\mathcal{A}-$summable to L if

$$\lim_{k \to \infty} \sum_{j=1}^{\infty} a_{kj}^n x_j = L$$

uniformly in n ([42], [99]).

If $A^n = A$ for some matrix A, then $\mathcal{A}-$summability is the ordinary matrix summability by A. If, $a_{kj}^n = \frac{1}{k}$, for $n \leq j \leq k + n$, $(n = 1, 2, ...)$, and $a_{kj}^n = 0$ otherwise, then $\mathcal{A}-$summability reduces to almost convergence [82].

19.2   A Fuzzy Korovkin Type Theorem

In this section we prove a fuzzy Korovkin-type approximation theorem via the concept of $\mathcal{A}-$summation process.

Let $f : [a, b] \to \mathbb{R}_\mathcal{F}$ be fuzzy number valued functions. Then f is said to be fuzzy continuous at $x_0 \in [a, b]$ provided that whenever $x_n \to x_0$, then $D(f(x_n), f(x_0)) \to 0$ as $n \to \infty$. Also, we say that f is fuzzy continuous on $[a, b]$ if it is fuzzy continuous at every point $x \in [a, b]$. The set all fuzzy continuous functions on the interval $[a, b]$ is denoted by $C_\mathcal{F}[a, b]$ (see, for instance [18]). Notice that $C_{\mathcal{F}}[a,b]$ is only a cone not a vector space. Now let $L : C_{\mathcal{F}}[a,b] \to C_{\mathcal{F}}[a,b]$ be an operator. Then L is said to be fuzzy linear if, for every $\lambda_1, \lambda_2 \in \mathbb{R}$, $f_1, f_2 \in C_{\mathcal{F}}[a,b]$, and $x \in [a,b]$,

$$L(\lambda_1 \odot f_1 \oplus \lambda_2 \odot f_2; x) = \lambda_1 \odot L(f_1; x) \oplus \lambda_2 \odot L(f_2; x)$$

holds. Also L is called fuzzy positive linear operator if it is fuzzy linear and the condition $L(f; x) \preceq L(g; x)$ is satisfied for any $f, g \in C_{\mathcal{F}}[a,b]$ and all $x \in [a,b]$ with $f(x) \preceq g(x)$.

Throughout this chapter the symbol $\|f\|$ denotes the usual supremum norm of f.

A sequence $\{L_j\}$ of positive linear operators of $C[a,b]$ into itself is called an $\mathcal{A}$-summation process on $C[a,b]$ if $\{L_j(f)\}$ is $\mathcal{A}$-summable to f for every $f \in C[a,b]$, i.e.,

$$\lim_{k \to \infty} \left\| \sum_{j=1}^{\infty} a_{kj}^n L_j(f) - f \right\| = 0, \qquad \text{uniformly in } n, \qquad (19.1)$$

where it is assumed that the series in (19.1) converges for each k, n and f.

Let $\{L_j\}$ be sequence of positive linear operators of $C[a,b]$ into itself such that for each $k, n \in \mathbb{N}$

$$\sup_{n,k} \sum_{j=1}^{\infty} a_{kj}^n \|L_j(1)\| < \infty. \qquad (19.2)$$

Furthermore, for $k, n \in \mathbb{N}$ and $f \in C[a,b]$, let

$$B_k^n(f; x) := \sum_{j=1}^{\infty} a_{kj}^n L_j(f)$$

which is well defined by (19.2) and belongs to $C[a,b]$.

In this chapter, we establish a theorem of Korovkin type with respect to the convergence behavior (19.1) for a sequence of positive linear operators of $C_{\mathcal{F}}[a,b]$ into itself. So our results are extensions of type (19..1)

$$\lim_{k \to \infty} \left\| \sum_{j=1}^{\infty} a_{kj}^n L_j(f) - f \right\| = 0, \qquad \text{uniformly in } n.$$

If a sequence $\{L_j\}$ of positive linear operators of $C_{\mathcal{F}}[a,b]$ into itself is called an $\mathcal{A}$-summation process on $C_{\mathcal{F}}[a,b]$ if $\{L_j(f)\}$ is $\mathcal{A}$-summable to f for every $f \in C_{\mathcal{F}}[a,b]$; i.e.,

$$\lim_{k \to \infty} D^* \left(\sum_{j=1}^{\infty} a_{kj}^n L_j(f), f \right) = 0, \qquad \text{uniformly in } n.$$

Throughout the chapter we use the test functions $e_i(x) = x^i$ $(i = 0, 1, 2)$
Then Anastassiou [18] obtained the following fuzzy Korovkin theorem.

Theorem 19.1. ([18]) Let $\{L_n\}_{n \in \mathbb{N}}$ be a sequence of fuzzy positive linear operators from $C_{\mathcal{F}}[a, b]$ into itself. Assume that there exists a corresponding sequence $\left\{\tilde{L}_n\right\}_{n \in \mathbb{N}}$ of positive linear operators from $C[a, b]$ into itself with the property

$$\{L_n(f; x)\}_{\pm}^{(r)} = \tilde{L}_n\left(f_{\pm}^{(r)}; x\right) \tag{19.3}$$

for all $x \in [a, b]$, $r \in [a, b]$, $n \in \mathbb{N}$ and $f \in C_{\mathcal{F}}[a, b]$. Assume further that

$$\lim_{n \to \infty} \left\| \tilde{L}_n(e_i) - e_i \right\| = 0 \quad \text{for each } i = 0, 1, 2.$$

Then, for all $f \in C_{\mathcal{F}}[a, b]$, we have

$$\lim_{n \to \infty} D^*(L_n(f), f) = 0.$$

Recently the A-statistical analog of Theorem 19.1 has been studied by Anastassiou and Duman [29]. It will be read as follows.

Theorem 19.2. ([29]) Let $A = (a_{jn})$ be a non-negative regular summability matrix and let $\{L_n\}_{n \in \mathbb{N}}$ be a sequence of fuzzy positive linear operators from $C_{\mathcal{F}}[a, b]$ into itself. Assume that there exists a corresponding sequence $\left\{\tilde{L}_n\right\}_{n \in \mathbb{N}}$ of positive linear operators from $C[a, b]$ into itself with the property (19.3). Assume further that

$$st_A - \lim_{n \to \infty} \left\| \tilde{L}_n(e_i) - e_i \right\| = 0 \quad \text{for each } i = 0, 1, 2.$$

Then, for all $f \in C_{\mathcal{F}}[a, b]$, we have

$$st_A - \lim_{n \to \infty} D^*(L_n(f), f) = 0.$$

We now give the following generalization by using a matrix summability method.

Theorem 19.3. Let $\mathcal{A} = (A^n)_{n \geq 1}$ be a sequence of infinite non-negative real matrices such that

$$\sup_{n,k} \sum_{j=1}^{\infty} a_{kj}^n < \infty \tag{19.4}$$

and $\{L_j\}_{j \in \mathbb{N}}$ be a sequence of fuzzy positive linear operators from $C_{\mathcal{F}}[a, b]$ into itself. Assume that there exists a corresponding sequence $\left\{\tilde{L}_j\right\}_{j \in \mathbb{N}}$ of positive linear operators from $C[a, b]$ into itself with the property (19.3) and satisfying (19.2). Assume further that

$$\lim_{k \to \infty} \left\| \sum_{j=1}^{\infty} a_{kj}^n \tilde{L}_j(e_i) - e_i \right\| = 0 \text{ for each } i = 0, 1, 2 \qquad (19.5)$$

uniformly in n. Then, for all $f \in C_{\mathcal{F}}[a, b]$, we have

$$\lim_{k \to \infty} D^* \left(\sum_{j=1}^{\infty} a_{kj}^n L_j(f), f \right) = 0$$

uniformly in n.

Proof. Let $f \in C_{\mathcal{F}}[a, b]$, $x \in [a, b]$ and $r \in [0, 1]$. By the hypothesis, since $f_{\pm}^{(r)} \in C[a, b]$, we may write, for every $\varepsilon > 0$, that there exists a number $\delta > 0$ such that $\left| f_{\pm}^{(r)}(y) - f_{\pm}^{(r)}(x) \right| < \varepsilon$ holds for every $y \in [a, b]$ satisfying $|y - x| < \delta$. Then we get for all $y \in [a, b]$, that

$$\left| f_{\pm}^{(r)}(y) - f_{\pm}^{(r)}(x) \right| \le \varepsilon + 2 M_{\pm}^{(r)} \frac{(y - x)^2}{\delta^2}, \qquad (19.6)$$

where $M_{\pm}^{(r)} := \left\| f_{\pm}^{(r)} \right\|$. Now using the linearity and the positivity of the operators $\tilde{L}_j$ and considering inequality (19.6), we can write

$$\left| \sum_{j=1}^{\infty} a_{kj}^n \tilde{L}_j \left(f_{\pm}^{(r)}; x \right) - f_{\pm}^{(r)}(x) \right|$$

$$\le \sum_{j=1}^{\infty} a_{kj}^n \tilde{L}_j \left(\left| f_{\pm}^{(r)}(y) - f_{\pm}^{(r)}(x) \right|; x \right) + \left| f_{\pm}^{(r)}(x) \right| \left| \sum_{j=1}^{\infty} a_{kj}^n \tilde{L}_j(1; x) - 1 \right|$$

$$\le \sum_{j=1}^{\infty} a_{kj}^n \tilde{L}_j \left(\varepsilon + 2 M_{\pm}^{(r)} \frac{(y - x)^2}{\delta^2}; x \right) + \left\| f_{\pm}^{(r)} \right\| \left| \sum_{j=1}^{\infty} a_{kj}^n \tilde{L}_j(1; x) - 1 \right|$$

$$\le \varepsilon + \varepsilon \left| \sum_{j=1}^{\infty} a_{kj}^n \tilde{L}_j(1; x) - 1 \right| + \frac{2 M_{\pm}^{(r)}}{\delta^2} \sum_{j=1}^{\infty} a_{kj}^n \tilde{L}_j \left((y - x)^2; x \right)$$

$$+ M_{\pm}^{(r)} \left| \sum_{j=1}^{\infty} a_{kj}^n \tilde{L}_j(1; x) - 1 \right|$$

$$\le \varepsilon + \left(\varepsilon + M_{\pm}^{(r)} \right) \left| \sum_{j=1}^{\infty} a_{kj}^n \tilde{L}_j(1; x) - 1 \right| + \frac{2 M_{\pm}^{(r)}}{\delta^2} \sum_{j=1}^{\infty} a_{kj}^n \tilde{L}_j \left((y - x)^2; x \right)$$

$$= \quad \varepsilon + \left(\varepsilon + M_{\pm}^{(r)}\right)\left|\sum_{j=1}^{\infty} a_{kj}^{n} \widetilde{L}_{j}\left(1;x\right) - 1\right| + \frac{2M_{\pm}^{(r)}}{\delta^{2}}\left[\sum_{j=1}^{\infty} a_{kj}^{n} \widetilde{L}_{j}\left(y^{2};x\right)\right.$$

$$\left. -2x\sum_{j=1}^{\infty} a_{kj}^{n} \widetilde{L}_{j}\left(y;x\right) + x^{2}\sum_{j=1}^{\infty} a_{kj}^{n} \widetilde{L}_{j}\left(1;x\right)\right]$$

$$\leq \quad \varepsilon + \left(\varepsilon + M_{\pm}^{(r)}\right)\left|\sum_{j=1}^{\infty} a_{kj}^{n} \widetilde{L}_{j}\left(1;x\right) - 1\right|$$

$$+\frac{2M_{\pm}^{(r)}}{\delta^{2}}\left[\left|\sum_{j=1}^{\infty} a_{kj}^{n} \widetilde{L}_{j}\left(y^{2};x\right) - x^{2}\right|\right.$$

$$\left. +2\left|x\right|\left|\sum_{j=1}^{\infty} a_{kj}^{n} \widetilde{L}_{j}\left(y;x\right) - x\right| + x^{2}\left|\sum_{j=1}^{\infty} a_{kj}^{n} \widetilde{L}_{j}\left(1;x\right) - 1\right|\right]$$

$$\leq \quad \varepsilon + \left(\varepsilon + M_{\pm}^{(r)}\right)\left|\sum_{j=1}^{\infty} a_{kj}^{n} \widetilde{L}_{j}\left(1;x\right) - 1\right|$$

$$+\frac{2M_{\pm}^{(r)}}{\delta^{2}}\left|\sum_{j=1}^{\infty} a_{kj}^{n} \widetilde{L}_{j}\left(y^{2};x\right) - x^{2}\right| + \frac{4M_{\pm}^{(r)}c}{\delta^{2}}\left|\sum_{j=1}^{\infty} a_{kj}^{n} \widetilde{L}_{j}\left(y;x\right) - x\right|$$

$$+\frac{2M_{\pm}^{(r)}c^{2}}{\delta^{2}}\left|\sum_{j=1}^{\infty} a_{kj}^{n} \widetilde{L}_{j}\left(1;x\right) - 1\right|$$

$$\leq \quad \varepsilon + \left(\varepsilon + M_{\pm}^{(r)} + \frac{2M_{\pm}^{(r)}c^{2}}{\delta^{2}}\right)\left|\sum_{j=1}^{\infty} a_{kj}^{n} \widetilde{L}_{j}\left(e_{0};x\right) - e_{0}\right|$$

$$+\frac{4M_{\pm}^{(r)}c}{\delta^{2}}\left|\sum_{j=1}^{\infty} a_{kj}^{n} \widetilde{L}_{j}\left(e_{1};x\right) - e_{1}\right| + \frac{2M_{\pm}^{(r)}}{\delta^{2}}\left|\sum_{j=1}^{\infty} a_{kj}^{n} \widetilde{L}_{j}\left(e_{2};x\right) - e_{2}\right|$$

where $c := \max\left\{\left|a\right|,\left|b\right|\right\}.$ If we take

$$K_{\pm}^{(r)}\left(\varepsilon\right) := \max\left\{\varepsilon + M_{\pm}^{(r)} + \frac{2c^{2}M_{\pm}^{(r)}}{\delta^{2}}, \frac{4cM_{\pm}^{(r)}}{\delta^{2}}, \frac{2M_{\pm}^{(r)}}{\delta^{2}}\right\},$$

then we get

$$\left| \sum_{j=1}^{\infty} a_{kj}^n \tilde{L}_j \left(f_{\pm}^{(r)}; x \right) - f_{\pm}^{(r)}(x) \right| \leq \varepsilon + K_{\pm}^{(r)}(\varepsilon) \left\{ \left| \sum_{j=1}^{\infty} a_{kj}^n \tilde{L}_j (e_0; x) - e_0 \right| \right.$$

$$\left. + \left| \sum_{j=1}^{\infty} a_{kj}^n \tilde{L}_j (e_1; x) - e_1 \right| + \left| \sum_{j=1}^{\infty} a_{kj}^n \tilde{L}_j (e_2; x) - e_2 \right| \right\}.$$

Then we observe that

$$\sup_{r \in [0,1]} \max \left| \sum_{j=1}^{\infty} a_{kj}^n \tilde{L}_j \left(f_{\pm}^{(r)}; x \right) - f_{\pm}^{(r)}(x) \right|$$

$$\leq \varepsilon + \sup_{r \in [0,1]} \max K_{\pm}^{(r)}(\varepsilon) \left\{ \left| \sum_{j=1}^{\infty} a_{kj}^n \tilde{L}_j (e_0; x) - e_0 \right| + \left| \sum_{j=1}^{\infty} a_{kj}^n \tilde{L}_j (e_1; x) - e_1 \right| \right.$$

$$\left. + \left| \sum_{j=1}^{\infty} a_{kj}^n \tilde{L}_j (e_2; x) - e_2 \right| \right\}.$$

Taking $K := K(\varepsilon) := \sup_{r \in [0,1]} \max \left\{ K_{-}^{(r)}(\varepsilon), K_{+}^{(r)}(\varepsilon) \right\}$, we get

$$\sup_{x \in [a,b]} \sup_{r \in [0,1]} \max \left| \sum_{j=1}^{\infty} a_{kj}^n \tilde{L}_j \left(f_{\pm}^{(r)}; x \right) - f_{\pm}^{(r)}(x) \right|$$

$$\leq \varepsilon + K(\varepsilon) \left\{ \sup_{x \in [a,b]} \left| \sum_{j=1}^{\infty} a_{kj}^n \tilde{L}_j (e_0; x) - e_0 \right| + \sup_{x \in [a,b]} \left| \sum_{j=1}^{\infty} a_{kj}^n \tilde{L}_j (e_1; x) - e_1 \right| \right.$$

$$\left. + \sup_{x \in [a,b]} \left| \sum_{j=1}^{\infty} a_{kj}^n \tilde{L}_j (e_2; x) - e_2 \right| \right\}.$$

Consequently, we get

$$D^* \left(\sum_{j=1}^{\infty} a_{kj}^n L_j (f; x), f(x) \right) \leq \varepsilon + K \left\| \sum_{j=1}^{\infty} a_{kj}^n \tilde{L}_j (e_0) - e_0 \right\|$$

$$+ K \left\| \sum_{j=1}^{\infty} a_{kj}^n \tilde{L}_j (e_1) - e_1 \right\|$$

$$+ K \left\| \sum_{j=1}^{\infty} a_{kj}^n \tilde{L}_j (e_2) - e_2 \right\|.$$

By taking limit as $k \to \infty$ and by using (19.5), we obtain the desired result.
$\square$

If we take $A^n = I$, the identity matrix and $A^n = A$, for some matrix A in Theorem 19.3, we immediately get Theorem 19.1 and Theorem 19.2, respectively.

Remarks. We now present three examples of sequences of positive linear operators. The first one shows that approximation Theorem 19.3 works but the statistical version of Theorem 19.2 does not work. The second one shows that approximation Theorem 19.3 works but Theorem 19.1 does not work and the last one shows that statistical version of Theorem 19.2 works but Theorem 19.3 does not work.

Assume that $\mathcal{A} := (A^n)_{n \geq 1} = \left(a_{kj}^n \right)_{k, j \in \mathbb{N}}$ is a sequence of infinite matrices defined by $a_{kj}^n = \frac{1}{k}$ if $n \leq j \leq k + n$, $(n = 1, 2, ...)$, and $a_{kj}^n = 0$ otherwise, then A-summability reduces to almost convergence.

(i) Take $(u_j) = \left\{ (-1)^j \right\}$. Observe that u is almost convergent to zero, but it is not statistically convergent. Then consider the Fuzzy Bernstein-type polynomials as follows:

$$B_j^{(\mathcal{F})}(f; x) = (1 + u_j) \odot \sum_{k=0}^{j}{}^{*} \binom{j}{k} x^k (1 - x)^{j-k} \odot f\left(\frac{k}{j} \right),$$

where $f \in C_{\mathcal{F}}[0, 1]$, $x \in [0, 1]$ and $j \in \mathbb{N}$. So we write

$$\left\{ B_j^{(\mathcal{F})}(f; x) \right\}_{\pm}^{(r)} = \tilde{B}_j \left(f_{\pm}^{(r)}; x \right) = (1 + u_j) \sum_{k=0}^{j} \binom{j}{k} x^k (1 - x)^{j-k} f_{\pm}^{(r)}\left(\frac{k}{j} \right),$$

where $f_{\pm}^{(r)} \in C[0, 1]$. Then, we get

$$
\begin{aligned}
\tilde{B}_j(e_0; x) &= (1 + u_j) \\
\tilde{B}_j(e_1; x) &= (1 + u_j)\, x \\
\tilde{B}_j(e_2; x) &= (1 + u_j) \left(x^2 + \frac{(x - x^2)}{j} \right).
\end{aligned}
$$

Since u is almost convergent to zero we get

$$\lim_{k \to \infty} \left\| \sum_{j=1}^{\infty} a_{kj}^n \tilde{B}_j(e_i) - e_i \right\| = 0 \quad \text{for each } i = 0, 1, 2,$$

uniformly in n. So, by Theorem 19.3, we obtain, for all $f \in C_{\mathcal{F}}[0, 1]$, that

$$\lim_{k \to \infty} D^* \left(\sum_{j=1}^{\infty} a_{kj}^n B_j(f), f \right) = 0,$$

uniformly in n. However, since u is not statistically convergent to zero, we observe that $\{B_j\}$ satisfies the Theorem 19.3 but it does not satisfy the statistical version of Theorem 19.2.

(ii) Let $(v_j) = (1, 0, 1, 0, ...)$. We see that (v_j) is almost convergent to $\frac{1}{2}$, but not convergent. Then consider

$$L_j^{(\mathcal{F})}(f; x) = \left(\frac{1}{2} + v_j\right) \odot \sum_{k=0}^{j}{}^{*} \binom{j}{k} x^k (1 - x)^{j-k} \odot f\left(\frac{k}{j}\right),$$

for $f \in C_{\mathcal{F}}[0, 1]$, $x \in [0, 1]$ and $j \in \mathbb{N}$. Hence $\{L_j\}$ satisfies the Theorem 19.3 but it does not satisfy the Theorem 19.1.

(iii) Now consider the sequence of 0's and 1's defined as follows:

$$\underset{\to 100 \leftarrow}{0, ..., 0} \ , \ \underset{\to \ \ 10 \ \ \leftarrow}{1, ..., 1} \ , 0, ..., 0, 1, ...1, 1, ...$$

where the blocks of 0's are increasing by factors of 100 and the blocks of 1's are increasing by factors of 10. If we denote this sequence by (y_j), then (y_j) is not almost convergent but it is statistically convergent to zero [87]. Then we define

$$T_j^{(\mathcal{F})}(f; x) = (1 + y_j) \odot \sum_{k=0}^{j}{}^{*} \binom{j}{k} x^k (1 - x)^{j-k} \odot f\left(\frac{k}{j}\right)$$

for $f \in C_{\mathcal{F}}[0, 1]$, $x \in [0, 1]$ and $j \in \mathbb{N}$. We observe that $\{T_j\}_{j \in \mathbb{N}}$ satisfies the statistical version of Theorem 19.2 but it does not satisfy the Theorem 19.3.

19.3   Rate of Convergence

In this section we study the rates of convergence of the sequence of fuzzy positive linear operators examined in Theorem 19.3.

Let $f : [a, b] \to \mathbb{R}_{\mathcal{F}}$. Then the (first) fuzzy modulus of continuity of f, which is introduced by Gal [66] (see also [18]), is defined by

$$\omega_1^{(\mathcal{F})}(f, \delta) := \sup_{x, y \in [a, b]; |x - y| \leq \delta} D(f(x), f(y))$$

for any $0 < \delta \leq b - a$.

As in [15], we have

$$\omega_1^{(\mathcal{F})}(f, \delta) = \sup_{r \in [0, 1]} \max\left\{\omega_1\left(f_-^{(r)}, \delta\right), \omega_1\left(f_+^{(r)}, \delta\right)\right\}.$$

Then we give the following.

Theorem 19.4. Let $\mathcal{A} = (A^n)_{n\geq 1}$ be a sequence of infinite non-negative real matrices such that

$$\sup_{n,k} \sum_{j=1}^{\infty} a_{kj}^n < \infty$$

and $\{L_j\}_{j\in\mathbb{N}}$ be a sequence of fuzzy positive linear operators from $C_{\mathcal{F}}[a,b]$ into itself. Assume that there exists a corresponding sequence $\left\{\tilde{L}_j\right\}_{j\in\mathbb{N}}$ of positive linear operators from $C[a,b]$ into itself with the property (19.3) and satisfying (19.2). Suppose that $\tilde{L}_j$ satisfy the following conditions:

i. $\lim\limits_{k\to\infty} \left\| \sum\limits_{j=1}^{\infty} a_{kj}^n \tilde{L}_j (e_0) - e_0 \right\| = 0$ *uniformly in* n

ii. $\lim\limits_{k\to\infty} \omega_1^{(\mathcal{F})}(f, \mu_k^n) \left\| \sum\limits_{j=1}^{\infty} a_{kj}^n \tilde{L}_j (e_0) - e_0 \right\| = 0$ *uniformly in* n

iii. $\lim\limits_{k\to\infty} \omega_1^{(\mathcal{F})}(f, \mu_k^n) = 0$ *uniformly in* n, *where* $\mu_k^n = \sqrt{\left\| \sum\limits_{j=1}^{\infty} a_{kj}^n \tilde{L}_j (\varphi) \right\|}$

and $\varphi(y) = (y - x)^2$ *for each* $x \in [a,b]$.

Then for all $f \in C_{\mathcal{F}}[a,b]$ *we have*

$$\lim_{k\to\infty} D^* \left(\sum_{j=1}^{\infty} a_{kj}^n L_j(f), f \right) = 0$$

uniformly in n.

Proof. Using the linearity and the positivity of the operators $\tilde{L}_j$ we may write, for each $j \in \mathbb{N}$, that

$$\left| \sum_{j=1}^{\infty} a_{kj}^n \tilde{L}_j \left(f_{\pm}^{(r)}; x \right) - f_{\pm}^{(r)}(x) \right|$$

$$\leq \left| f_{\pm}^{(r)}(x) \right| \left| \sum_{j=1}^{\infty} a_{kj}^n \tilde{L}_j(1;x) - 1 \right| + \sum_{j=1}^{\infty} a_{kj}^n \tilde{L}_j \left(\left| f_{\pm}^{(r)}(y) - f_{\pm}^{(r)}(x) \right|; x \right)$$

$$\leq M_{\pm}^{(r)} \left| \sum_{j=1}^{\infty} a_{kj}^n \tilde{L}_j(e_0;x) - e_0 \right| + \sum_{j=1}^{\infty} a_{kj}^n \tilde{L}_j \left(\omega_1 \left(f_{\pm}^{(r)}, \frac{\delta(y-x)}{\delta} \right); x \right)$$

$$\leq M_{\pm}^{(r)} \left| \sum_{j=1}^{\infty} a_{kj}^n \tilde{L}_j(e_0;x) - e_0 \right|$$

$$+ \sum_{j=1}^{\infty} a_{kj}^n \tilde{L}_j \left(\left(1 + \left[\frac{|y-x|}{\delta}\right]\right) \omega_1 \left(f_{\pm}^{(r)}, \delta \right); x \right)$$

$$\leq M_{\pm}^{(r)} \left| \sum_{j=1}^{\infty} a_{kj}^n \tilde{L}_j(e_0;x) - e_0 \right|$$

$$+ \sum_{j=1}^{\infty} a_{kj}^n \tilde{L}_j \left(\left(1 + \frac{|y-x|^2}{\delta^2}\right) \omega_1 \left(f_{\pm}^{(r)}, \delta \right); x \right)$$

$$\leq M_{\pm}^{(r)} \left| \sum_{j=1}^{\infty} a_{kj}^n \tilde{L}_j(e_0;x) - e_0 \right| + \omega_1 \left(f_{\pm}^{(r)}, \delta \right) \sum_{j=1}^{\infty} a_{kj}^n \tilde{L}_j(e_0;x)$$

$$+ \frac{\omega_1 \left(f_{\pm}^{(r)}, \delta \right)}{\delta^2} \sum_{j=1}^{\infty} a_{kj}^n \tilde{L}_j \left((y-x)^2; x \right)$$

$$\leq M_{\pm}^{(r)} \left| \sum_{j=1}^{\infty} a_{kj}^n \tilde{L}_j(e_0;x) - e_0 \right| + \omega_1 \left(f_{\pm}^{(r)}, \delta \right) \left| \sum_{j=1}^{\infty} a_{kj}^n \tilde{L}_j(e_0;x) - e_0 \right|$$

$$+ \omega_1 \left(f_{\pm}^{(r)}, \delta \right) + \frac{\omega_1 \left(f_{\pm}^{(r)}, \delta \right)}{\delta^2} \sum_{j=1}^{\infty} a_{kj}^n \tilde{L}_j \left((y-x)^2; x \right)$$

where $M_{\pm}^{(r)} := \left\| f_{\pm}^{(r)} \right\|$. Then we observe that

$$\left\| \sum_{j=1}^{\infty} a_{kj}^n \tilde{L}_j \left(f_{\pm}^{(r)}; x \right) - f_{\pm}^{(r)}(x) \right\| \leq M_{\pm}^{(r)} \left\| \sum_{j=1}^{\infty} a_{kj}^n \tilde{L}_j(e_0;x) - e_0 \right\| +$$

$$+ \omega_1 \left(f_{\pm}^{(r)}, \delta \right) \left\| \sum_{j=1}^{\infty} a_{kj}^n \tilde{L}_j \left(e_0; x \right) - e_0 \right\|$$

$$+ \omega_1 \left(f_{\pm}^{(r)}, \delta \right) + \frac{\omega_1 \left(f_{\pm}^{(r)}, \delta \right)}{\delta^2} \left\| \sum_{j=1}^{\infty} a_{kj}^n \tilde{L}_j \left(\varphi; x \right) \right\|.$$

If we take $\delta = \mu_k^n$, we conclude that

$$\left\| \sum_{j=1}^{\infty} a_{kj}^n \tilde{L}_j \left(f_{\pm}^{(r)}; x \right) - f_{\pm}^{(r)} \left(x \right) \right\| \leq M_{\pm}^{(r)} \left\| \sum_{j=1}^{\infty} a_{kj}^n \tilde{L}_j \left(e_0; x \right) - e_0 \right\|$$

$$+ \omega_1 \left(f_{\pm}^{(r)}, \mu_k^n \right) \left\| \sum_{j=1}^{\infty} a_{kj}^n \tilde{L}_j \left(e_0; x \right) - e_0 \right\| + 2\omega_1 \left(f_{\pm}^{(r)}, \mu_k^n \right).$$

Then we get

$$D^* \left(\sum_{j=1}^{\infty} a_{kj}^n L_j \left(f \right), f \right) \leq M \left\| \sum_{j=1}^{\infty} a_{kj}^n \tilde{L}_j \left(e_0 \right) - e_0 \right\| \tag{19.7}$$

$$+ \left\| \sum_{j=1}^{\infty} a_{kj}^n \tilde{L}_j \left(e_0 \right) - e_0 \right\| \omega_1^{(\mathcal{F})} \left(f, \mu_k^n \right)$$

$$+ 2\omega_1^{(\mathcal{F})} \left(f, \mu_k^n \right)$$

where $M := \sup_{r \in [0,1]} \left\{ M_-^r, M_+^r \right\}$. Letting $k \to \infty$ and using the hypothesis (i), (ii) and (iii) we get

$$\lim_{k \to \infty} D^* \left(\sum_{j=1}^{\infty} a_{kj}^n L_j \left(f \right), f \right) = 0$$

uniformly in n, which completes the proof. $\square$

We note that the rate of convergence of the involved operators may be obtained from (19.7).

20

$\mathcal{A}$-SUMMABILITY AND FUZZY TRIGONOMETRIC KOROVKIN APPROXIMATION

The aim of this chapter is to present a fuzzy trigonometric Korovkin-type approximation theorem by using a matrix summability method. We also study the rates of convergence of fuzzy positive linear operators in trigonometric environment. This chapter is based on [28].

20.1 Introduction

Approximation theory which has a close relationship with other branches of mathematics has been used in the theory of polynomial approximation and various domains of functional analysis [3], in numerical solutions of differential and integral operators [78], in the studies of the interpolation operator of Hermit-Fejér [44], [45], [55] and the partial sums of Fourier series [79].

Most of the classical approximation operators tend to converge to the value of the function being approximated. However, at points of discontinuity, they often converge to the average of the left and right limits of the function. The main purpose of using summability theory has always been to make a nonconvergent sequence "converge". Some results regarding matrix summability for positive linear operators may be found in the paper [38], [39], [94]. In this chapter using a $\mathcal{A}$-summation process we study the

G.A. Anastassiou: Fuzzy Mathematics: Approximation The., STUDFUZZ 251, pp. 385–397.
springerlink.com   © Springer-Verlag Berlin Heidelberg 2010

approximation properties of sequence of positive linear operators on the space of all 2π-periodic and continuous functions on the whole real axis.

20.2   Background

Anastassiou and Duman [29] (see also [63]) introduced the fuzzy analog of $A-$statistical convergence by using the metric D.

Let $(\mu_n)_{n\in\mathbb{N}}$ be a fuzzy number valued sequence and let $A = (a_{jn})$ is a non-negative regular summability matrix. Recall that the regularity conditions on a matrix A are known as Silverman-Toeplitz conditions in the functional analysis (see for details, [70]). Then, $(\mu_n)_{n\in\mathbb{N}}$ is $A-$statistically convergent to $\mu \in \mathbb{R}_{\mathcal{F}}$, which is denoted by $st_A - \lim_{n\to\infty} D(\mu_n, \mu) = 0$ if for every $\varepsilon > 0$

$$\lim_{j\to\infty} \sum_{n:D(\mu_n,\mu)\geq\varepsilon} a_{jn} = 0$$

holds. Some basic results regarding $A-$statistical convergence for number sequences may be found in the papers [65], [77], [86]. Of course, the case of $A = C_1$, the Cesáro matrix of order one, immediately reduces to the statistical convergence of fuzzy valued sequences which defined by Nuray and Savaş [91]. Also, if A is replaced by the identity matrix, then we have the fuzzy convergence introduced by Matloka [85].

Let $\mathcal{A} := (A^n)_{n\geq 1}$, $A^n = \left(a_{kj}^n\right)_{k,j\in\mathbb{N}}$ be a sequence of infinite non-negative real matrices. For a sequence of real numbers, $x = (x_j)_{j\in\mathbb{N}}$, the double sequence

$$\mathcal{A}x := \{(Ax)_k^n : k, n \in \mathbb{N}\}$$

defined by $(Ax)_k^n := \sum_{j=1}^{\infty} a_{kj}^n x_j$ is called the $\mathcal{A}-$transform of x whenever the series converges for all k and n. A sequence x is said to be $\mathcal{A}-$summable to L if

$$\lim_{k\to\infty} \sum_{j=1}^{\infty} a_{kj}^n x_j = L$$

uniformly in n ([42], [99]).

If $A^n = A$ for some matrix A, then $\mathcal{A}-$summability is the ordinary matrix summability by A. If, $a_{kj}^n = \frac{1}{k}$, for $n \leq j \leq k + n$, $(n = 1, 2, ...)$, and $a_{kj}^n = 0$ otherwise, then $\mathcal{A}-$summability reduces to almost convergence [82].

20.3   A Fuzzy Korovkin Type Theorem

In this section we prove a fuzzy trigonometric Korovkin-type approximation theorem via the concept of $\mathcal{A}-$summation process.

Let $f : \mathbb{R} \to \mathbb{R}_{\mathcal{F}}$ be fuzzy number valued functions. Then f is said to be fuzzy continuous at $x_0 \in \mathbb{R}$ provided that whenever $x_n \to x_0$, then $D(f(x_n), f(x_0)) \to 0$ as $n \to \infty$. Also, we say that f is fuzzy continuous on $\mathbb{R}$ if it is fuzzy continuous at every point $x \in \mathbb{R}$. The set all fuzzy continuous functions on $\mathbb{R}$ is denoted by $C_{\mathcal{F}}(\mathbb{R})$ (see, for instance, [18], [32]). Notice that $C_{\mathcal{F}}(\mathbb{R})$ is only a cone not a vector space. By $C_{2\pi}^{(\mathcal{F})}(\mathbb{R})$ we mean the space of all fuzzy continuous and 2π-periodic functions on $\mathbb{R}$. Also the space of all real valued continuous and 2π-periodic functions is denoted by $C_{2\pi}(\mathbb{R})$.

Suppose that $f : [a, b] \to \mathbb{R}_{\mathcal{F}}$ be fuzzy number valued functions. Then, f is said to be fuzzy-Riemann integrable (or, FR-integrable) to $I \in \mathbb{R}_{\mathcal{F}}$ if, for given $\varepsilon > 0$, there exists a $\delta > 0$ such that, for any partition $P = \{[u, v]; \xi\}$ of $[a, b]$ with the norms $\Delta(P) < \delta$, we have

$$D\left(\sum_{P}^{*}(v - u) \odot f(\xi), I\right) < \varepsilon.$$

In this case, we write

$$I := (FR) \int_{a}^{b} f(x) \, dx.$$

By Corollary 13.2 of ([66], p. 644), we conclude that if $f \in C_{\mathcal{F}}[a, b]$ (fuzzy continuous on $[a, b]$), then f is FR-integrable on $[a, b]$.

Now let $L : C_{\mathcal{F}}(\mathbb{R}) \to C_{\mathcal{F}}(\mathbb{R})$ be an operator. Then L is said to be fuzzy linear if, for every $\lambda_1, \lambda_2 \in \mathbb{R}$, $f_1, f_2 \in C_{\mathcal{F}}(\mathbb{R})$, and $x \in \mathbb{R}$,

$$L(\lambda_1 \odot f_1 \oplus \lambda_2 \odot f_2; x) = \lambda_1 \odot L(f_1; x) \oplus \lambda_2 \odot L(f_2; x)$$

holds. Also L is called fuzzy positive linear operator if it is fuzzy linear and the condition $L(f; x) \preceq L(g; x)$ is satisfied for any $f, g \in C_{\mathcal{F}}(\mathbb{R})$ and all $x \in \mathbb{R}$ with $f(x) \preceq g(x)$.

A sequence $\{L_j\}$ of positive linear operators of $C(\mathbb{R})$ into itself is called an $\mathcal{A}$-summation process on $C(\mathbb{R})$ if $\{L_j(f)\}$ is $\mathcal{A}$-summable to f for every $f \in C(\mathbb{R})$, i.e.,

$$\lim_{k \to \infty}\left\|\sum_{j=1}^{\infty} a_{kj}^{n} L_j(f) - f\right\| = 0, \qquad \text{uniformly in } n, \qquad (20.1)$$

where it is assumed that the series in (20.1) converges for each k, n and f and $\|h\|$ denotes the usual supremum norm of $h \in C(\mathbb{R})$.

If a sequence $\{L_j\}$ of positive linear opeartors of $C_{\mathcal{F}}(\mathbb{R})$ into itself is called an $\mathcal{A}$-summation process on $C_{\mathcal{F}}(\mathbb{R})$ if $\{L_j(f)\}$ is $\mathcal{A}$-summable to f for every $f \in C_{\mathcal{F}}(\mathbb{R})$, i.e.,

$$\lim_{k \to \infty} D^* \left(\sum_{j=1}^{\infty} a_{kj}^n L_j(f), f \right) = 0, \qquad \text{uniformly in } n.$$

Some unification on Korovkin-type results through the use of a matrix summability method may be found in [3], [88], [89], [90], [100].

Let $\{L_j\}$ be a sequence of positive linear opeartors of $C(\mathbb{R})$ into itself such that for each $k, n \in \mathbb{N}$

$$\sup_{n,k} \sum_{j=1}^{\infty} a_{kj}^n \|L_j(1)\| < \infty. \tag{20.2}$$

Furthermore, for $k, n \in \mathbb{N}$ and $f \in C(\mathbb{R})$, let

$$B_k^n(f; x) := \sum_{j=1}^{\infty} a_{kj}^n L_j(f)$$

which is well defined by (20.2) and belong to $C(\mathbb{R})$.

In this chapter, we establish a theorem of Korovkin type with respect to the convergence behavior (20.1) for a sequence of positive linear operators of $C_{2\pi}^{(\mathcal{F})}(\mathbb{R})$ into itself. So the results of type (20.1) are extensions of

$$\lim_{k \to \infty} \left\| \sum_{j=1}^{\infty} a_{kj}^n L_j(f) - f \right\| = 0, \qquad \text{uniformly in } n.$$

Throughout the chapter the symbol $\|f\|$ denotes the usual supremum norm of $f \in C_{\mathcal{F}}(\mathbb{R})$ and we use the test functions f_i $(i = 0, 1, 2)$ defined by

$$f_0(x) = 1, \quad f_1(x) = \cos x, \quad f_2(x) = \sin x.$$

Then Anastassiou and Gal [32] obtained the following fuzzy Korovkin theorem.

Theorem 20.1. Let $\{L_n\}_{n \in \mathbb{N}}$ be a sequence of fuzzy positive linear operators defined on $C_{2\pi}^{(\mathcal{F})}(\mathbb{R})$. Suppose that there exists a corresponding sequence $\left\{ \tilde{L}_n \right\}_{n \in \mathbb{N}}$ of positive linear operators defined on $C_{2\pi}(\mathbb{R})$ with the property

$$\{L_n(f; x)\}_{\pm}^{(r)} = \tilde{L}_n \left(f_{\pm}^{(r)}; x \right) \tag{20.3}$$

for all $x \in [a, b]$, $r \in [0, 1]$, $n \in \mathbb{N}$ and $f \in C_{2\pi}^{\mathcal{F}}(\mathbb{R})$. Assume further that

$$\lim_{n \to \infty} \left\| \tilde{L}_n(f_i) - f_i \right\| = 0 \text{ for each } i = 0, 1, 2.$$

Then, for all $f \in C_{2\pi}^{\mathcal{F}}(\mathbb{R})$, we have

$$\lim_{n \to \infty} D^* (L_n(f), f) = 0.$$

Recently the A−statistical analog of Theorem 20.1 has been studied by Anastassiou and Duman [29]. It will be read as follows.

Theorem 20.2. ([29]) Let $A = (a_{jn})$ be a non-negative regular summability matrix and let $\{L_n\}_{n \in \mathbb{N}}$ be a sequence of fuzzy positive linear operators defined on $C_{2\pi}^{(\mathcal{F})}(\mathbb{R})$. Suppose that there exists a corresponding sequence $\left\{ \widetilde{L}_n \right\}_{n \in \mathbb{N}}$ of positive linear operators defined on $C_{2\pi}(\mathbb{R})$ with the property (20.3). Assume further that

$$st_A - \lim_{n \to \infty} \left\| \widetilde{L}_n(f_i) - f_i \right\| = 0 \quad \text{for each} \quad i = 0, 1, 2.$$

Then, for all $f \in C_{2\pi}^{(\mathcal{F})}(\mathbb{R})$, we have

$$st_A - \lim_{n \to \infty} D^* (L_n(f), f) = 0.$$

It is clear that if we replace the matrix A in Theorem 20.2 by the Cesáro matrix C_1, we immediately get the statistical fuzzy Korovkin result in the trigonometric case (see Corollary 2.2. of [29]).

We now give the following generalization by using a $\mathcal{A}$-summation process

.

Theorem 20.3. Let $\mathcal{A} = (A^n)_{n \geq 1}$ be a sequence of infinite non-negative real matrices such that

$$\sup_{n,k} \sum_{j=1}^{\infty} a_{kj}^n < \infty \tag{20.4}$$

and let $\{L_n\}_{n \in \mathbb{N}}$ be a sequence of fuzzy positive linear operators defined on $C_{2\pi}^{(\mathcal{F})}(\mathbb{R})$. Suppose that there exists a corresponding sequence $\left\{ \widetilde{L}_n \right\}_{n \in \mathbb{N}}$ of positive linear operators defined on $C_{2\pi}(\mathbb{R})$ with the property (20.3) and satisfying (20.2). Suppose further that

$$\lim_{k \to \infty} \left\| \sum_{j=1}^{\infty} a_{kj}^n \widetilde{L}_j(f_i) - f_i \right\| = 0 \quad \text{for each} \quad i = 0, 1, 2. \tag{20.5}$$

uniformly in n. Then, for all $f \in C_{2\pi}^{(\mathcal{F})}(\mathbb{R})$, we have

$$\lim_{k \to \infty} D^* \left(\sum_{j=1}^{\infty} a_{kj}^n L_j(f), f \right) = 0$$

uniformly in n.

Proof. Assume that I is a closed bounded interval with length 2π of $\mathbb{R}$. Let $f \in C_{2\pi}^{(\mathcal{F})} (\mathbb{R})$, $x \in I$ and $r \in [0, 1]$. Taking $[f(x)]^{(r)} = \left[f_-^{(r)}(x), f_+^{(r)}(x)\right]$ we get $f_\pm^{(r)}(x) \in C_{2\pi}(\mathbb{R})$. Then, we immediately see from ([78], p.7) that, for every $\varepsilon > 0$, there exists a $\delta > 0$ such that

$$\left| f_\pm^{(r)}(y) - f_\pm^{(r)}(x) \right| \le \varepsilon + 2M_\pm^{(r)} \frac{\sin^2\left(\frac{y-x}{2}\right)}{\sin^2 \frac{\delta}{2}} \tag{20.6}$$

holds for all $y \in \mathbb{R}$ and $x \in I$ where $M_\pm^{(r)} := \left\| f_\pm^{(r)} \right\|$. Now using the linearity and the positivity of the operators $\tilde{L}_j$ and considering inequality (20.6), we can write

$$\left| \sum_{j=1}^{\infty} a_{kj}^n \tilde{L}_j \left(f_\pm^{(r)}; x \right) - f_\pm^{(r)}(x) \right|$$

$$\le \sum_{j=1}^{\infty} a_{kj}^n \tilde{L}_j \left(\left| f_\pm^{(r)}(y) - f_\pm^{(r)}(x) \right|; x \right) + \left| f_\pm^{(r)}(x) \right| \left| \sum_{j=1}^{\infty} a_{kj}^n \tilde{L}_j (1; x) - 1 \right|$$

$$\le \sum_{j=1}^{\infty} a_{kj}^n \tilde{L}_j \left(\varepsilon + 2M_\pm^{(r)} \frac{\sin^2\left(\frac{y-x}{2}\right)}{\sin^2 \frac{\delta}{2}}; x \right) + \left\| f_\pm^{(r)} \right\| \left| \sum_{j=1}^{\infty} a_{kj}^n \tilde{L}_j (1; x) - 1 \right|$$

$$\le \varepsilon \sum_{j=1}^{\infty} a_{kj}^n \tilde{L}_j (1; x) + \frac{2M_\pm^{(r)}}{\sin^2 \frac{\delta}{2}} \sum_{j=1}^{\infty} a_{kj}^n \tilde{L}_j \left(\sin^2\left(\frac{y-x}{2}\right); x \right)$$

$$+ M_\pm^{(r)} \left| \sum_{j=1}^{\infty} a_{kj}^n \tilde{L}_j (1; x) - 1 \right|$$

$$\le \varepsilon + \left(\varepsilon + M_\pm^{(r)} \right) \left| \sum_{j=1}^{\infty} a_{kj}^n \tilde{L}_j (1; x) - 1 \right|$$

$$+ \frac{2M_\pm^{(r)}}{\sin^2 \frac{\delta}{2}} \sum_{j=1}^{\infty} a_{kj}^n \tilde{L}_j \left(\sin^2\left(\frac{y-x}{2}\right); x \right)$$

$$\le \varepsilon + \left(\varepsilon + M_\pm^{(r)} \right) \left| \sum_{j=1}^{\infty} a_{kj}^n \tilde{L}_j (1; x) - 1 \right| + \frac{M_\pm^{(r)}}{\sin^2 \frac{\delta}{2}} \left\{ \left(\sum_{j=1}^{\infty} a_{kj}^n \tilde{L}_j (1; x) - 1 \right) \right.$$

$$\left. - \cos x \left(\sum_{j=1}^{\infty} a_{kj}^n \tilde{L}_j (\cos y; x) - \cos x \right) \right.$$

$$- \sin x \left(\sum_{j=1}^{\infty} a_{kj}^n \tilde{L}_j \left(\sin y; x \right) - \sin x \right) \Bigg\}$$

$$\leq \quad \varepsilon + \left(\varepsilon + M_{\pm}^{(r)} + \frac{M_{\pm}^{(r)}}{\sin^2 \frac{\delta}{2}} \right) \left| \sum_{j=1}^{\infty} a_{kj}^n \tilde{L}_j \left(1; x \right) - 1 \right|$$

$$+ \left| \cos x \right| \frac{M_{\pm}^{(r)}}{\sin^2 \frac{\delta}{2}} \left| \sum_{j=1}^{\infty} a_{kj}^n \tilde{L}_j \left(\cos y; x \right) - \cos x \right|$$

$$+ \left| \sin x \right| \frac{M_{\pm}^{(r)}}{\sin^2 \frac{\delta}{2}} \left| \sum_{j=1}^{\infty} a_{kj}^n \tilde{L}_j \left(\sin y; x \right) - \sin x \right|.$$

If we take $K_{\pm}^{(r)} \left(\varepsilon \right) := \varepsilon + M_{\pm}^{(r)} + \frac{M_{\pm}^{(r)}}{\sin^2 \frac{\delta}{2}}$, then we get

$$\left| \sum_{j=1}^{\infty} a_{kj}^n \tilde{L}_j \left(f_{\pm}^{(r)}; x \right) - f_{\pm}^{(r)} \left(x \right) \right| \quad \leq \quad \varepsilon + K_{\pm}^{(r)} \left(\varepsilon \right) \Bigg\{ \left| \sum_{j=1}^{\infty} a_{kj}^n \tilde{L}_j \left(f_0; x \right) - f_0 \right|$$

$$+ \left| \sum_{j=1}^{\infty} a_{kj}^n \tilde{L}_j \left(f_1; x \right) - f_1 \right|$$

$$+ \left| \sum_{j=1}^{\infty} a_{kj}^n \tilde{L}_j \left(f_2; x \right) - f_2 \right| \Bigg\}.$$

Then we observe that

$$\sup_{r \in [0,1]} \max \left| \sum_{j=1}^{\infty} a_{kj}^n \tilde{L}_j \left(f_{\pm}^{(r)}; x \right) - f_{\pm}^{(r)} \left(x \right) \right| \leq \varepsilon$$

$$+ \sup_{r \in [0,1]} \max K_{\pm}^{(r)} \left(\varepsilon \right) \Bigg\{ \left| \sum_{j=1}^{\infty} a_{kj}^n \tilde{L}_j \left(f_0; x \right) - f_0 \right|$$

$$+ \left| \sum_{j=1}^{\infty} a_{kj}^n \tilde{L}_j \left(f_1; x \right) - f_1 \right|$$

$$+ \left| \sum_{j=1}^{\infty} a_{kj}^n \tilde{L}_j \left(f_2; x \right) - f_2 \right| \Bigg\}.$$

Taking $K := K(\varepsilon) := \sup\limits_{r\in[0,1]} \max\left\{K_-^{(r)}(\varepsilon), K_+^{(r)}(\varepsilon)\right\}$, we get

$$\sup\limits_{x\in\mathbb{R}}\sup\limits_{r\in[0,1]} \max\left|\sum\limits_{j=1}^{\infty}a_{kj}^n\tilde{L}_j\left(f_{\pm}^{(r)};x\right) - f_{\pm}^{(r)}(x)\right| \le$$

$$\varepsilon + K(\varepsilon)\left\{\sup\limits_{x\in\mathbb{R}}\left|\sum\limits_{j=1}^{\infty}a_{kj}^n\tilde{L}_j(f_0;x) - f_0\right|\right.$$

$$+ \sup\limits_{x\in\mathbb{R}}\left|\sum\limits_{j=1}^{\infty}a_{kj}^n\tilde{L}_j(f_1;x) - f_1\right|$$

$$\left.+ \sup\limits_{x\in\mathbb{R}}\left|\sum\limits_{j=1}^{\infty}a_{kj}^n\tilde{L}_j(f_2;x) - f_2\right|\right\}.$$

Consequently, we obtain

$$D^*\left(\sum\limits_{j=1}^{\infty}a_{kj}^nL_j(f),f\right) \le \varepsilon + K\left\|\sum\limits_{j=1}^{\infty}a_{kj}^n\tilde{L}_j(f_0;x) - f_0\right\|$$

$$+K\left\|\sum\limits_{j=1}^{\infty}a_{kj}^n\tilde{L}_j(f_1;x) - f_1\right\|$$

$$+K\left\|\sum\limits_{j=1}^{\infty}a_{kj}^n\tilde{L}_j(f_2;x) - f_2\right\|.$$

By taking limit as $k \to \infty$ and by using (20.5) we obtain the desired result.
□

Remarks. We now present three examples of sequences of positive linear operators. The first one shows that approximation Theorem 20.3 works but the statistical version of Theorem 20.2 does not work. The second one shows that approximation Theorem 20.3 works but Theorem 20.1 does not work and the last one shows that statistical version of Theorem 20.2 works but Theorem 20.3 does not work.

Assume that $\mathcal{A} := (A^n)_{n\ge1} = \left(a_{kj}^n\right)_{k,j\in\mathbb{N}}$ is a sequence of infinite matrices defined by $a_{kj}^n = \frac{1}{k}$ if $n \le j \le k+n$, $(n = 1,2,...)$, and $a_{kj}^n = 0$ otherwise, then A−summability reduces to almost convergence. Now define the fuzzy Fejer operators F_j as follows:

$$F_j(f;x) = \frac{1}{j\pi} \odot \left\{(FR)\int\limits_{-\pi}^{\pi} f(y) \odot \frac{\sin^2\left(\frac{j}{2}(y-x)\right)}{2\sin^2\left[\left(\frac{y-x}{2}\right)\right]}dy\right\},$$

where $j \in \mathbb{N}$, $f \in C_{2\pi}^{(\mathcal{F})}(\mathbb{R})$ and $x \in \mathbb{R}$. Then observe that the operators F_j are fuzzy positive linear. So we write

$$\{F_j\,(f;x)\}_{\pm}^{(r)} = \widetilde{F}_j\left(f_{\pm}^{(r)};x\right) := \frac{1}{j\pi}\int\limits_{-\pi}^{\pi} f_{\pm}\,(y)\,\frac{\sin^2\left(\frac{j}{2}\,(y-x)\right)}{2\sin^2\left[\left(\frac{y-x}{2}\right)\right]}\,dy$$

where $f_{\pm}^{(r)} \in C_{2\pi}(\mathbb{R})$ and $r \in [0,1]$. Then, we get (see [78])

$$\begin{aligned}
\widetilde{F}_j\,(f_0;x) &= 1 \\
\widetilde{F}_j\,(f_1;x) &= \frac{n-1}{n}\,\cos x \\
\widetilde{F}_j\,(f_2;x) &= \frac{n-1}{n}\,\sin x.
\end{aligned}$$

(i) Take $(u_j) = \left\{(-1)^j\right\}$. Observe that u is almost convergent to zero, but it is not statistically convergent. Now using the sequence (u_j) and the fuzzy Fejer operators F_j, we introduce the following fuzzy positive linear operators defined on the space $C_{2\pi}^{(\mathcal{F})}(\mathbb{R})$:

$$B_j\,(f;x) = (1+u_j) \odot F_j\,(f;x),$$

where $j \in \mathbb{N}$, $f \in C_{2\pi}^{(\mathcal{F})}(\mathbb{R})$ and $x \in \mathbb{R}$. So, the corresponding real positive linear operators are given by

$$\widetilde{B}_j\left(f_{\pm}^{(r)};x\right) := \frac{1+u_j}{j\pi}\int\limits_{-\pi}^{\pi} f_{\pm}\,(y)\,\frac{\sin^2\left(\frac{j}{2}\,(y-x)\right)}{2\sin^2\left[\left(\frac{y-x}{2}\right)\right]}\,dy,$$

where $f_{\pm}^{(r)} \in C_{2\pi}(\mathbb{R})$. Then we get, for all $j \in \mathbb{N}$ and $x \in \mathbb{R}$, that

$$\begin{aligned}
\left\|\widetilde{B}_j\,(f_0) - f_0\right\| &= u_j \\
\left\|\widetilde{B}_j\,(f_1) - f_1\right\| &\leq u_j + \frac{1+u_j}{j} \\
\left\|\widetilde{B}_j\,(f_2) - f_2\right\| &\leq u_j + \frac{1+u_j}{j}.
\end{aligned}$$

Since u is almost convergent to zero we get

$$\lim_{k\to\infty}\left\|\frac{1}{k}\sum_{j=n}^{n+k}\widetilde{B}_j\,(f_i) - f_i\right\| = 0 \quad \text{for each } i = 0,1,2$$

uniformly in n. So, by Theorem 20.3, we obtain, for all $f \in C_{2\pi}^{(\mathcal{F})}(\mathbb{R})$, that

$$\lim_{k\to\infty} D^*\left(\sum_{j=1}^{\infty} a_{kj}^n B_j\,(f),f\right) = 0$$

uniformly in n. However, since u is not statistically convergent to zero, we observe that $\{B_j\}$ satisfies the Theorem 20.3 but it does not satisfy the statistical version of Theorem 20.2.

(ii) Let $(v_j) = (1, 0, 1, 0, ...)$. We see that (v_j) is almost convergent to $\frac{1}{2}$, but not convergent (in the ordinary sense). Then consider

$$L_j(f;x) = \left(\frac{1}{2} + v_j\right) \odot F_j(f;x),$$

for $f \in C_{2\pi}^{(\mathcal{F})}(\mathbb{R})$, $x \in \mathbb{R}$ and $j \in \mathbb{N}$. Hence $\{L_j\}$ satisfies the Theorem 20.3 but it does not satisfy the Theorem 20.1.

(iii) Now consider the sequence of 0's and 1's defined as follows:

$$\underset{\substack{\to 100 \ \leftarrow}}{0, ..., 0} \ , \ \underset{\substack{\to \ 10 \ \leftarrow}}{1, ..., 1} \ , 0, ..., 0, 1, ..., 1, 1, ...$$

where the blocks of 0's are increasing by factors of 100 and the blocks of 1's are increasing by factors of 10. If we denote this sequence by (y_j), then (y_j) is not almost convergent but it is statistically convergent to zero ([87]). Then we define

$$T_j(f;x) = (1 + y_j) \odot F_j(f;x)$$

for $f \in C_{2\pi}^{(\mathcal{F})}(\mathbb{R})$, $x \in \mathbb{R}$ and $j \in \mathbb{N}$. We observe that $\{T_j\}_{j \in \mathbb{N}}$ satisfies the statistical version of Theorem 20.2 but it does not satisfy Theorem 20.3.

20.4 Rate of Convergence

In this section we study the rates of convergence of the sequence of fuzzy positive linear operators examined in Theorem 20.3.

Let $f \in C_{2\pi}^{(\mathcal{F})}(\mathbb{R})$. Then the (first) fuzzy modulus of continuity of f, which is introduced by Gal [66] (see also [18], [32]), is defined by

$$\omega_1^{(\mathcal{F})}(f, \delta) := \sup_{x, y \in \mathbb{R}; |x-y| \leq \delta} D(f(x), f(y))$$

for any $\delta > 0$.

It is easy to see that, for any $c > 0$ and all $f_{\pm}^{(r)} \in C_{2\pi}(\mathbb{R})$

$$\omega_1^{(\mathcal{F})}\left(f_{\pm}^{(r)}, c\delta\right) \leq (1 + [c]) \omega_1^{(\mathcal{F})}\left(f_{\pm}^{(r)}, \delta\right)$$

where $[c]$ is defined to be greatest integer less than or equal to c.

Lemma 20.4. [18] Let $f \in C_{2\pi}^{(\mathcal{F})}(\mathbb{R})$. then it holds.

$$\omega_1^{(\mathcal{F})}(f, \delta) = \sup_{r \in [0,1]} \max\left\{\omega_1\left(f_-^{(r)}, \delta\right), \omega_1\left(f_+^{(r)}, \delta\right)\right\},$$

for any $\delta > 0$.

Then we have the following.

Theorem 20.5. Let $\mathcal{A} = (A^n)_{n \geq 1}$ be a sequence of infinite non-negative real matrices such that

$$\sup_{n,k} \sum_{j=1}^{\infty} a_{kj}^n < \infty$$

and $\{L_j\}_{j \in \mathbb{N}}$ be a sequence of fuzzy positive linear operators defined on $C_{2\pi}^{(\mathcal{F})}(\mathbb{R})$. Assume that there exists a corresponding sequence $\left\{ \tilde{L}_j \right\}_{j \in \mathbb{N}}$ of positive linear operators defined on $C_{2\pi}(\mathbb{R})$ with the property (20.3). Suppose that $\tilde{L}_j$ satisfy the following conditions:

i. $\lim\limits_{k \to \infty} \left\| \sum\limits_{j=1}^{\infty} a_{kj}^n \tilde{L}_j (f_0) - f_0 \right\| = 0$ *uniformly in* n,

ii. $\lim\limits_{k \to \infty} \omega_1^{(\mathcal{F})} (f, \mu_k^n) = 0$ *uniformly in* n,

iii. $\lim\limits_{k \to \infty} \omega_1^{(\mathcal{F})} (f, \mu_k^n) \left\| \sum\limits_{j=1}^{\infty} a_{kj}^n \tilde{L}_j (f_0) - f_0 \right\| = 0$ *uniformly in* n,

where $\mu_k^n = \pi \sqrt{\left\| \sum\limits_{j=1}^{\infty} a_{kj}^n \tilde{L}_j (\varphi) \right\|}$ *and* $\varphi(y) = \sin^2 \left(\frac{y-x}{2} \right)$ *for each* $x \in \mathbb{R}$.

Then for all $f \in C_{2\pi}^{(\mathcal{F})}(\mathbb{R})$, *we have*

$$\lim_{k \to \infty} D^* \left(\sum_{j=1}^{\infty} a_{kj}^n L_j (f), f \right) = 0$$

uniformly in n.

Proof. Using the linearity and the positivity of the operators $\widetilde{L}_j$ and using property (20.4) by [97], we may write, for each $j \in \mathbb{N}$, $\delta > 0$, that

$$\left| \sum_{j=1}^{\infty} a_{kj}^n \widetilde{L}_j \left(f_{\pm}^{(r)}; x \right) - f_{\pm}^{(r)}(x) \right|$$

$$\leq \left| f_{\pm}^{(r)}(x) \right| \left| \sum_{j=1}^{\infty} a_{kj}^n \widetilde{L}_j (1; x) - 1 \right| + \sum_{j=1}^{\infty} a_{kj}^n \widetilde{L}_j \left(\left| f_{\pm}^{(r)}(y) - f_{\pm}^{(r)}(x) \right|; x \right)$$

$$\leq M_{\pm}^{(r)} \left| \sum_{j=1}^{\infty} a_{kj}^n \widetilde{L}_j (f_0; x) - f_0 \right|$$

$$+ \sum_{j=1}^{\infty} a_{kj}^n \widetilde{L}_j \left(\left(1 + \left(\frac{\pi}{\delta} \right)^2 \sin^2 \left(\frac{y-x}{2} \right) \right) \omega_1 \left(f_{\pm}^{(r)}, \delta \right); x \right)$$

$$\leq M_{\pm}^{(r)} \left| \sum_{j=1}^{\infty} a_{kj}^n \widetilde{L}_j (f_0; x) - f_0 \right| + \omega_1 \left(f_{\pm}^{(r)}, \delta \right) \sum_{j=1}^{\infty} a_{kj}^n \widetilde{L}_j (f_0; x)$$

$$+ \left(\frac{\pi}{\delta} \right)^2 \omega_1 \left(f_{\pm}^{(r)}, \delta \right) \sum_{j=1}^{\infty} a_{kj}^n \widetilde{L}_j \left(\sin^2 \left(\frac{y-x}{2} \right); x \right)$$

$$\leq M_{\pm}^{(r)} \left| \sum_{j=1}^{\infty} a_{kj}^n \widetilde{L}_j (f_0; x) - f_0 \right| + \omega_1 \left(f_{\pm}^{(r)}, \delta \right) \left| \sum_{j=1}^{\infty} a_{kj}^n \widetilde{L}_j (f_0; x) - f_0 \right|$$

$$+ \left(\frac{\pi}{\delta} \right)^2 \omega_1 \left(f_{\pm}^{(r)}, \delta \right) \sum_{j=1}^{\infty} a_{kj}^n \widetilde{L}_j \left(\sin^2 \left(\frac{y-x}{2} \right); x \right) + \omega_1 \left(f_{\pm}^{(r)}, \delta \right)$$

where $M_{\pm}^{(r)} := \left\| f_{\pm}^{(r)} \right\|$. Taking supremum over $x \in \mathbb{R}$, we easily see that

$$\left\| \sum_{j=1}^{\infty} a_{kj}^n \widetilde{L}_j \left(f_{\pm}^{(r)}; x \right) - f_{\pm}^{(r)}(x) \right\|$$

$$\leq M_{\pm}^{(r)} \left\| \sum_{j=1}^{\infty} a_{kj}^n \widetilde{L}_j (f_0; x) - f_0 \right\| + \omega_1 \left(f_{\pm}^{(r)}, \delta \right) \left\| \sum_{j=1}^{\infty} a_{kj}^n \widetilde{L}_j (f_0; x) - f_0 \right\|$$

$$+ \left(\frac{\pi}{\delta} \right)^2 \omega_1 \left(f_{\pm}^{(r)}, \delta \right) \left\| \sum_{j=1}^{\infty} a_{kj}^n \widetilde{L}_j \left(\sin^2 \left(\frac{y-x}{2} \right); x \right) \right\| + \omega_1 \left(f_{\pm}^{(r)}, \delta \right).$$

Now putting $\delta := \mu_k^n$, we conclude that

$$\left\| \sum_{j=1}^{\infty} a_{kj}^n \widetilde{L}_j \left(f_{\pm}^{(r)}; x \right) - f_{\pm}^{(r)} \left(x \right) \right\|$$

$$\leq\ M_{\pm}^{(r)} \left\| \sum_{j=1}^{\infty} a_{kj}^n \widetilde{L}_j \left(f_0; x \right) - f_0 \right\| + \omega_1 \left(f_{\pm}^{(r)}, \mu_k^n \right) \left\| \sum_{j=1}^{\infty} a_{kj}^n \widetilde{L}_j \left(f_0; x \right) - f_0 \right\|$$

$$+ 2\omega_1 \left(f_{\pm}^{(r)}, \mu_k^n \right).$$

Then we get from Lemma 20.4

$$D^* \left(\sum_{j=1}^{\infty} a_{kj}^n L_j \left(f \right), f \right) \tag{20.7}$$

$$\leq\ M \left\| \sum_{j=1}^{\infty} a_{kj}^n \widetilde{L}_j \left(f_0 \right) - f_0 \right\|$$

$$+ \omega_1^{(\mathcal{F})} \left(f, \mu_k^n \right) \left\| \sum_{j=1}^{\infty} a_{kj}^n \widetilde{L}_j (f_0) - f_0 \right\| + 2\omega_1^{(\mathcal{F})} \left(f, \mu_k^n \right), \tag{20.1}$$

where $M := \sup_{r \in [0,1]} \max \left\{ M_+^{(r)}, M_-^{(r)} \right\}$. If we take limit as $k \to \infty$ on the both sides of inequality (20.7) and use the hypotesis (i), (ii) and (iii), we immediately see that

$$\lim_{k \to \infty} D^* \left(\sum_{j=1}^{\infty} a_{kj}^n L_j \left(f \right), f \right) = 0.$$

Inequality (20.7) gives the above convergence with rates.   □

21

UNIFORM REAL AND FUZZY ESTIMATES FOR DISTANCES BETWEEN WAVELET TYPE OPERATORS AT REAL AND FUZZY ENVIRONMENT

The basic fuzzy wavelet type operators A_k, B_k, C_k, D_k, $k \in \mathbb{Z}$ were studied in [14], [16], see also Chapters 12, 16, for their pointwise and uniform convergence with rates to the fuzzy unit operator. Also they were studied in [23], see also Chapter 13, in terms of estimating their fuzzy differences and giving their pointwise convergence with rates to zero.

For prior related and similar study of convergence to the unit of real analogs of these wavelet type operators see [9], Section II.

Here in Section 21.1 we present the complete study of finding uniform estimates for the distances between the real Wavelet type operators A_k, B_k, C_k, D_k, $k \in \mathbb{Z}$.

Their differences converge to zero with rates. This is done via elegant tight Jackson type inequalities involving the modulus of continuity of the higher order derivative of the engaged real function. Based on these real analysis results in Section 21.2 we establish the corresponding fuzzy results regarding uniform estimates for the fuzzy differences between the fuzzy wavelet type operators. These fuzzy diferences converge to zero with rates give via fuzzy Jackson type tight inequalities. The last inequalities involve the fuzzy modulus of continuity of the higher order fuzzy derivative of the engaged fuzzy function.

The defining all these operators real scaling function is not assumed to be orthogonal and is of compact support.

G.A. Anastassiou: Fuzzy Mathematics: Approximation The., STUDFUZZ 251, pp. 399–429.
springerlink.com   © Springer-Verlag Berlin Heidelberg 2010

21.1   Estimates for Distances of Real Wavelet type Operators

The real wavelet type operators A_k, B_k, C_k, D_k, $k \in \mathbb{Z}$ we study here converge to the unit operator I, and in that respect were studied extensively in [9], Section II.

We need

Definition 21.1. Let $f : \mathbb{R} \to \mathbb{R}$ be continuous function. We define the first modulus of continuity of f by

$$\omega_1 (f, \delta) = \sup_{\substack{x,y \\ |x-y| \le \delta}} |f(x) - f(y)|, \delta > 0.$$

We give

Theorem 21.2. Let $f \in C^N (\mathbb{R})$, $N \ge 1$, $x \in \mathbb{R}$, and $k \in \mathbb{Z}$. Let φ be a bounded function of compact support $\subseteq [-a, a]$, $a > 0$ such that $\sum_{j=-\infty}^{\infty} \varphi(x - j) = 1$, all $x \in \mathbb{R}$. Suppose $\varphi \ge 0$. Put

$$(B_k f)(x) = \sum_{j=-\infty}^{\infty} f\left(\frac{j}{2^k}\right) \varphi\left(2^k x - j\right),$$

$$(D_k f)(x) = \sum_{j=-\infty}^{\infty} \delta_{kj}(f) \varphi\left(2^k x - j\right),$$

where

$$\delta_{kj}(f) = \sum_{r=0}^{n} w_r f\left(\frac{j}{2^k} + \frac{r}{2^k n}\right),$$

$n \in \mathbb{N}$, $w_r \ge 0$, $\sum_{r=0}^{n} w_r = 1$. Then

$$E_{1k}(x) = \left| (D_k f)(x) - (B_k f)(x) - \sum_{i=1}^{N} \frac{\left(B_k f^{(i)}\right)(x)}{2^{ki} n^i i!} \left(\sum_{r=1}^{n} w_r r^i\right) \right|$$

$$\le \frac{\sum_{r=1}^{n} w_r r^N \omega_1\left(f^{(N)}, \frac{r}{2^k n}\right)}{2^{kN} n^N N!}. \tag{21.1}$$

Remark. (i) Given that $f^{(N)}$ is continuous and bounded or uniformly continuous we have that $\omega_1\left(f^{(N)}, \frac{r}{2^k n}\right) < \infty$, and $E_{1k}(x) \to 0$, $x \in \mathbb{R}$, as $k \to \infty$.

(ii) One also has

$$\|E_{1k}\|_\infty = \left\| D_k f - B_k f - \sum_{i=1}^N \frac{B_k f^{(i)}}{2^{ki} n^i i!} \left(\sum_{r=1}^n w_r r^i \right) \right\|_\infty$$

$$\leq \frac{\sum_{r=1}^n w_r r^N \omega_1\left(f^{(N)}, \frac{r}{2^k n}\right)}{2^{kN} n^N N!}.$$

Under the assumptions of (i) we get also $\|E_{1k}\|_\infty \to 0$, as $k \to \infty$.
(iii) By (21.1) we get

$$|(D_k f)(x) - (B_k f)(x)| \leq \sum_{i=1}^N \frac{\left|B_k f^{(i)}(x)\right|}{2^{ki} n^i i!} \left(\sum_{r=1}^n w_r r^i \right)$$

$$+ \frac{\sum_{r=1}^n w_r r^N \omega_1\left(f^{(N)}, \frac{r}{2^k n}\right)}{2^{kN} n^N N!}. \qquad (21.2)$$

Given that $\left\|f^{(i)}\right\|_\infty < \infty$, for $i = 1, \ldots, N$, we obtain $\left|B_k f^{(i)}(x)\right| \leq \left\|f^{(i)}\right\|_\infty$, and

$$|(B_k f)(x) - (D_k f)(x)| \leq \sum_{i=1}^N \frac{\left\|f^{(i)}\right\|_\infty}{2^{ki} n^i i!} \left(\sum_{r=1}^n w_r r^i \right)$$

$$+ \frac{\sum_{r=1}^n w_r r^N \omega_1\left(f^{(N)}, \frac{r}{2^k n}\right)}{2^{kN} n^N N!}. \qquad (21.3)$$

Clearly we get

$$|(B_k f)(x) - (D_k f)(x)| \leq \sum_{i=1}^N \frac{\left\|f^{(i)}\right\|_\infty}{2^{ki} i!} + \frac{\omega_1\left(f^{(N)}, \frac{1}{2^k}\right)}{2^{kN} N!} =: T_1$$

and

$$\|B_k f - D_k f\|_\infty \leq T_1. \qquad (21.4)$$

So as $k \to \infty$, we have $\|B_k f - D_k f\|_\infty \to 0$.
Proof of Theorem 21.2. Because $f \in C^N(\mathbb{R})$, $N \geq 1$ we have

$$\sum_{r=0}^n w_r f\left(\frac{j}{2^k} + \frac{r}{2^k n}\right) = f\left(\frac{j}{2^k}\right) + \sum_{i=1}^N \frac{f^{(i)}\left(\frac{j}{2^k}\right)}{i!} \sum_{r=0}^n w_r \left(\frac{r^i}{2^{ki} n^i}\right)$$

$$+ \sum_{r=0}^n w_r \int_{j/2^k}^{(j/2^k)+\frac{r}{2^k n}} \left(f^{(N)}(t) - f^{(N)}(j/2^k)\right) \frac{\left(\frac{j}{2^k} + \frac{r}{2^k n} - t\right)^{N-1}}{(N-1)!} dt.$$

Hence we get

$$\sum_{j=-\infty}^{\infty} \delta_{kj}\left(f\right) \varphi\left(2^k x - j\right) = \sum_{j=-\infty}^{\infty} f\left(\frac{j}{2^k}\right) \varphi\left(2^k x - j\right)$$

$$+\sum_{i=1}^{N} \frac{\sum_{j=-\infty}^{\infty} f^{(i)}\left(\frac{j}{2^k}\right) \varphi\left(2^k x - j\right)}{2^{ki} i! n^i} \left(\sum_{r=0}^{n} w_r r^i\right)$$

$$+\sum_{r=0}^{n} w_r \sum_{j=-\infty}^{\infty} \varphi\left(2^k x - j\right) \int_{j/2^k}^{(j/2^k)+\frac{r}{2^k n}} \left(f^{(N)}\left(t\right) - f^{(N)}\left(j/2^k\right)\right) \cdot$$

$$\frac{\left(\frac{j}{2^k} + \frac{r}{2^k n} - t\right)^{N-1}}{(N-1)!} dt.$$

So, we observe that

$$\left(D_k f\right)(x) - \left(B_k f\right)(x) - \sum_{i=1}^{N} \frac{\left(B_k f^{(i)}\right)(x)}{2^{ki} n^i i!} \left(\sum_{r=0}^{n} w_r r^i\right) = \mathcal{R}_1,$$

where

$$\mathcal{R}_1 = \sum_{r=1}^{n} w_r \sum_{j=-\infty}^{\infty} \varphi\left(2^k x - j\right) \int_{j/2^k}^{(j/2^k)+\frac{r}{2^k n}} \left(f^{(N)}\left(t\right) - f^{(N)}\left(j/2^k\right)\right) \cdot$$

$$\frac{\left(\frac{j}{2^k} + \frac{r}{2^k n} - t\right)^{N-1}}{(N-1)!} dt.$$

Set

$$\Gamma_{jr} = \left| \int_{j/2^k}^{(j/2^k)+\frac{r}{2^k n}} \left(f^{(N)}\left(t\right) - f^{(N)}\left(j/2^k\right)\right) \frac{\left(\frac{j}{2^k} + \frac{r}{2^k n} - t\right)^{N-1}}{(N-1)!} dt \right|.$$

So that

$$|\mathcal{R}_1| \leq \sum_{r=1}^{n} w_r \sum_{j=-\infty}^{\infty} \varphi\left(2^k x - j\right) \Gamma_{jr}.$$

Next we see that

$$\Gamma_{jr} \leq \int_{j/2^k}^{(j/2^k)+\frac{r}{2^k n}} \left| f^{(N)}\left(t\right) - f^{(N)}\left(j/2^k\right) \right| \frac{\left(\frac{j}{2^k} + \frac{r}{2^k n} - t\right)^{N-1}}{(N-1)!} dt$$

$$\leq \int_{j/2^k}^{(j/2^k)+\frac{r}{2^k n}} \omega_1\left(f^{(N)}, |t - (j/2^k)|\right) \frac{\left(\frac{j}{2^k} + \frac{r}{2^k n} - t\right)^{N-1}}{(N-1)!} dt$$

$$\leq \omega_1\left(f^{(N)}, \frac{r}{2^k n}\right) \int_{j/2^k}^{(j/2^k)+\frac{r}{2^k n}} \frac{\left(\frac{j}{2^k} + \frac{r}{2^k n} - t\right)^{N-1}}{(N-1)!} dt$$

$$= \omega_1\left(f^{(N)}, \frac{r}{2^k n}\right) \frac{\left(\frac{r}{2^k n}\right)^{N}}{N!}.$$

That is

$$\Gamma_{jr} \le \omega_1 \left(f^{(N)}, \frac{r}{2^k n} \right) \frac{\left(\frac{r}{2^k n} \right)^N}{N!}.$$

So we have found that

$$|R_1| \le \frac{\sum\limits_{r=1}^{n} w_r r^N \omega_1 \left(f^{(N)}, \frac{r}{2^k n} \right)}{2^{kN} n^N N!}.$$

The proof of the theorem is complete.   □

We continue with

Theorem 21.4. Let $f \in C^N (\mathbb{R})$, $N \ge 1$, $x \in \mathbb{R}$, and $k \in \mathbb{Z}$. Let φ be a bounded function of compact support $\subseteq [-a, a]$, $a > 0$ such that $\sum\limits_{j=-\infty}^{\infty} \varphi (x - j) = 1$, all $x \in \mathbb{R}$. Suppose $\varphi \ge 0$. Put

$$(B_k f)(x) = \sum\limits_{j=-\infty}^{\infty} f\left(\frac{j}{2^k} \right) \varphi (2^k x - j),$$

$$(C_k f)(x) = \sum\limits_{j=-\infty}^{\infty} \gamma_{kj} (f) \varphi (2^k x - j),$$

where

$$\gamma_{kj} (f) = 2^k \int_{2^{-k} j}^{2^{-k} (j+1)} f (t) \, dt = 2^k \int_{0}^{2^{-k}} f\left(t + \frac{j}{2^k} \right) dt.$$

Then

$$E_{2k} (x) = \left| (C_k f)(x) - (B_k f)(x) - \sum\limits_{i=1}^{N} \frac{\left(B_k f^{(i)} \right) (x)}{2^{ki} (i+1)!} \right|$$

$$\le \frac{\omega_1 \left(f^{(N)}, \frac{1}{2^k} \right)}{2^{kN} (N+1)!}. \tag{21.5}$$

Remark 21.5. (i) Given that $f^{(N)}$ is continuous and bounded or uniformly continuous we have that $\omega_1 \left(f^{(N)}, \frac{1}{2^k} \right) < \infty$, and $E_{2k} (x) \to 0$, $x \in \mathbb{R}$, as $k \to \infty$.

(ii) One also has

$$\|E_{2k}\|_{\infty} = \left\| C_k f - B_k f - \sum\limits_{i=1}^{N} \frac{B_k f^{(i)}}{2^{ki} (i+1)!} \right\|_{\infty}$$

$$\le \frac{\omega_1 \left(f^{(N)}, \frac{1}{2^k} \right)}{2^{kN} (N+1)!}.$$

Under the assumptions of (i) we get also $\|E_{2k}\|_\infty \to 0$, as $k \to \infty$.
(iii) By (21.5) we get

$$|(B_k f)(x) - (C_k f)(x)| \le \sum_{i=1}^{N} \frac{\left|\left(B_k f^{(i)}\right)(x)\right|}{2^{ki}(i+1)!} + \frac{\omega_1\left(f^{(N)}, \frac{1}{2^k}\right)}{2^{kN}(N+1)!}. \qquad (21.6)$$

Given that $\left\|f^{(i)}\right\|_\infty < \infty$, for $i = 1, \dots, N$, we obtain

$$|(B_k f)(x) - (C_k f)(x)| \le \sum_{i=1}^{N} \frac{\left\|f^{(i)}\right\|_\infty}{2^{ki}(i+1)!} + \frac{\omega_1\left(f^{(N)}, \frac{1}{2^k}\right)}{2^{kN}(N+1)!} =: T_2,$$

and

$$\|B_k f - C_k f\|_\infty \le T_2. \qquad (21.7)$$

So as $k \to \infty$, we get $\|B_k f - C_k f\|_\infty \to 0$.
Proof of Theorem 21.4. Because $f \in C^N(\mathbb{R})$, $N \ge 1$ we have

$$
\begin{aligned}
f\left(t + \frac{j}{2^k}\right) = {}& f\left(\frac{j}{2^k}\right) + \sum_{i=1}^{N} \frac{f^{(i)}\left(\frac{j}{2^k}\right)}{i!} t^i \\
& + \int_{j/2^k}^{t+(j/2^k)} \left(f^{(N)}(s) - f^{(N)}\left(j/2^k\right)\right) \frac{\left(t + \frac{j}{2^k} - s\right)^{N-1}}{(N-1)!} ds.
\end{aligned}
$$

Hence we get

$$
\begin{aligned}
\gamma_{kj}(f) = {}& 2^k \int_0^{2^{-k}} f\left(t + \frac{j}{2^k}\right) dt = f\left(\frac{j}{2^k}\right) + \sum_{i=1}^{N} \frac{f^{(i)}\left(\frac{j}{2^k}\right)}{2^{ki}(i+1)!} \\
& + 2^k \int_0^{2^{-k}} \left(\int_{j/2^k}^{t+(j/2^k)} \left(f^{(N)}(s) - f^{(N)}\left(j/2^k\right)\right) \frac{\left(t + \frac{j}{2^k} - s\right)^{N-1}}{(N-1)!} ds\right) dt.
\end{aligned}
$$

Hence we get

$$
\begin{aligned}
\sum_{j=-\infty}^{\infty} \gamma_{kj}(f)\, \varphi\left(2^k x - j\right) = {}& \sum_{j=-\infty}^{\infty} f\left(\frac{j}{2^k}\right) \varphi\left(2^k x - j\right) + \\
& \sum_{i=1}^{N} \sum_{j=-\infty}^{\infty} \frac{f^{(i)}\left(\frac{j}{2^k}\right) \varphi\left(2^k x - j\right)}{2^{ki}(i+1)!} + \\
& \sum_{j=-\infty}^{\infty} \varphi\left(2^k x - j\right) 2^k \int_0^{2^{-k}} \left(\int_{j/2^k}^{t+(j/2^k)} \left(f^{(N)}(s) - f^{(N)}\left(j/2^k\right)\right) \right. \\
& \left. \frac{\left(t + \frac{j}{2^k} - s\right)^{N-1}}{(N-1)!} ds\right) dt.
\end{aligned}
$$

So, we see that

$$(C_k(f))(x) - (B_k(f))(x) - \sum_{i=1}^{N} \frac{\left(B_k\left(f^{(i)}\right)\right)(x)}{2^{ki}(i+1)!} = \mathcal{R}_2,$$

where

$$\mathcal{R}_2 = \sum_{j=-\infty}^{\infty} \varphi\left(2^k x - j\right) 2^k \cdot$$

$$\int_0^{2^{-k}} \left(\int_{j/2^k}^{t+(j/2^k)} \left(f^{(N)}(s) - f^{(N)}\left(j/2^k\right)\right) \frac{\left(t + \frac{j}{2^k} - s\right)^{N-1}}{(N-1)!} ds \right) dt.$$

Set

$$\Gamma_j(t) = \left| \int_{j/2^k}^{t+(j/2^k)} \left(f^{(N)}(s) - f^{(N)}\left(j/2^k\right)\right) \frac{\left(t + \frac{j}{2^k} - s\right)^{N-1}}{(N-1)!} ds \right|.$$

So that

$$|\mathcal{R}_2| \le \sum_{j=-\infty}^{\infty} \varphi\left(2^k x - j\right) 2^k \int_0^{2^{-k}} \Gamma_j(t) \, dt.$$

Next we observe that

$$
\begin{aligned}
\Gamma_j(t) &\le \int_{j/2^k}^{t+(j/2^k)} \left| f^{(N)}(s) - f^{(N)}\left(j/2^k\right) \right| \frac{\left(t + \frac{j}{2^k} - s\right)^{N-1}}{(N-1)!} ds \\
&\le \int_{j/2^k}^{t+(j/2^k)} \omega_1\left(f^{(N)}, \left|s - (j/2^k)\right|\right) \frac{\left(t + \frac{j}{2^k} - s\right)^{N-1}}{(N-1)!} ds \\
&\le \omega_1\left(f^{(N)}, t\right) \int_{j/2^k}^{t+(j/2^k)} \frac{\left(t + \frac{j}{2^k} - s\right)^{N-1}}{(N-1)!} ds \\
&= \omega_1\left(f^{(N)}, t\right) \frac{t^N}{N!}.
\end{aligned}
$$

I.e. we get

$$\Gamma_j(t) \le \omega_1\left(f^{(N)}, t\right) \frac{t^N}{N!},$$

and

$$
\begin{aligned}
2^k \int_0^{2^{-k}} \Gamma_j(t) \, dt &\le 2^k \int_0^{2^{-k}} \omega_1\left(f^{(N)}, t\right) \frac{t^N}{N!} dt \\
&\le \frac{2^k}{N!} \omega_1\left(f^{(N)}, \frac{1}{2^k}\right) \int_0^{2^{-k}} t^N dt \\
&= \frac{2^k}{(N+1)!} \omega_1\left(f^{(N)}, \frac{1}{2^k}\right) 2^{-k(N+1)} \\
&= \frac{1}{2^{kN}(N+1)!} \omega_1\left(f^{(N)}, \frac{1}{2^k}\right).
\end{aligned}
$$

That is we obtain

$$2^k \int_0^{2^{-k}} \Gamma_j(t)\, dt \le \frac{1}{2^{kN}(N+1)!} \omega_1\left(f^{(N)}, \frac{1}{2^k}\right).$$

It is clear that

$$|\mathcal{R}_2| \le \frac{\omega_1\left(f^{(N)}, \frac{1}{2^k}\right)}{2^{kN}(N+1)!}.$$

The proof of the theorem is now finished.   □

We continue with

Theorem 21.6. Let $f \in C^N(\mathbb{R})$, $N \ge 1$, $x \in \mathbb{R}$, and $k \in \mathbb{Z}$. Let φ be a bounded function of compact support $\subseteq [-a, a]$, $a > 0$ such that $\sum_{j=-\infty}^{\infty} \varphi(x-j) = 1$, all $x \in \mathbb{R}$. Suppose $\varphi \ge 0$. Put

$$(C_k f)(x) = \sum_{j=-\infty}^{\infty} \gamma_{kj}(f)\, \varphi\left(2^k x - j\right),$$

where

$$\gamma_{kj}(f) = 2^k \int_{2^{-k}j}^{2^{-k}(j+1)} f(t)\, dt = 2^k \int_0^{2^{-k}} f\left(t + \frac{j}{2^k}\right) dt,$$

and

$$(D_k f)(x) = \sum_{j=-\infty}^{\infty} \delta_{kj}(f)\, \varphi\left(2^k x - j\right),$$

where

$$\delta_{kj}(f) = \sum_{r=0}^n w_r f\left(\frac{j}{2^k} + \frac{r}{2^k n}\right),$$

$n \in \mathbb{N}$, $w_r \ge 0$, $\sum_{r=0}^n w_r = 1$. Then

$$
\begin{aligned}
E_{3k}(x) \;=\; & \left| (C_k f)(x) - (D_k f)(x) - \sum_{i=1}^N \frac{\left(D_k f^{(i)}\right)(x)}{2^{ki}(i+1)!} \right.\\
& \left. \left[\left(1 - \frac{r}{n}\right)^{i+1} - (-1)^{i+1}\left(\frac{r}{n}\right)^{i+1}\right] \right|
\end{aligned}
\tag{21.8}
$$

$$\le \frac{\left(\sum_{r=0}^n w_r \left[\left(\frac{r}{n}\right)^{N+1} + \left(1-\frac{r}{n}\right)^{N+1}\right]\right)}{2^{kN}(N+1)!} \omega_1\left(f^{(N)}, \frac{1}{2^k}\right).$$

Remark 21.7 (i) Given that $f^{(N)}$ is continuous and bounded or uniformly continuous we have that $\omega_1\left(f^{(N)}, \frac{1}{2^k}\right) < \infty$, and $E_{3k}(x) \to 0, x \in \mathbb{R}$, as $k \to \infty$.

(ii) One also has

$$
\begin{aligned}
\|E_{3k}\|_\infty &= \left\|C_k f - D_k f - \sum_{i=1}^N \frac{D_k f^{(i)}}{2^{ki}(i+1)!} \cdot \right. \\
&\qquad \left.\left[\left(1-\frac{r}{n}\right)^{i+1} - (-1)^{i+1}\left(\frac{r}{n}\right)^{i+1}\right]\right\|_\infty \\
&\leq \frac{\left(\sum_{r=0}^n w_r \left[\left(\frac{r}{n}\right)^{N+1} + \left(1-\frac{r}{n}\right)^{N+1}\right]\right)}{2^{kN}(N+1)!}\omega_1\left(f^{(N)}, \frac{1}{2^k}\right).
\end{aligned}
$$

Under the assumption of (i) we get also $\|E_{3k}\|_\infty \to 0$, as $k \to \infty$.

(iii) By (21.8) we get

$$
\begin{aligned}
|(C_k f)(x) - (D_k f)(x)| &\leq \sum_{i=1}^N \frac{\left|\left(D_k f^{(i)}\right)(x)\right|}{2^{ki-1}(i+1)!} \\
&+ \frac{\left(\sum_{r=0}^n w_r \left[\left(\frac{r}{n}\right)^{N+1} + \left(1-\frac{r}{n}\right)^{N+1}\right]\right)}{2^{kN}(N+1)!}\omega_1\left(f^{(N)}, \frac{1}{2^k}\right).
\end{aligned}
\tag{21.9}
$$

Clearly then

$$
|(C_k f)(x) - (D_k f)(x)| \leq \sum_{i=1}^N \frac{\left|\left(D_k f^{(i)}\right)(x)\right|}{2^{ki-1}(i+1)!} + \frac{\omega_1\left(f^{(N)}, \frac{1}{2^k}\right)}{2^{kN-1}(N+1)!}.
\tag{21.10}
$$

Given that $\left\|f^{(i)}\right\|_\infty < \infty$, for $i = 1, \ldots, N$, we get

$$
\left|\left(D_k f^{(i)}\right)(x)\right| \leq \left\|f^{(i)}\right\|_\infty,
$$

and

$$
|(C_k f)(x) - (D_k f)(x)| \leq \sum_{i=1}^N \frac{\left\|f^{(i)}\right\|_\infty}{2^{ki-1}(i+1)!} + \frac{\omega_1\left(f^{(N)}, \frac{1}{2^k}\right)}{2^{kN-1}(N+1)!} =: T_3
$$

$$
\tag{21.11}
$$

and

$$
\|(C_k f)(x) - (D_k f)(x)\|_\infty \leq T_3.
\tag{21.12}
$$

So as $k \to \infty$, we get $\|C_k f - D_k f\|_\infty \to 0$.

Proof of Theorem 21.6. Because $f \in C^N(\mathbb{R})$, $N \geq 1$ we have

$$f\left(t + \frac{j}{2^k}\right) = f\left(\frac{j}{2^k} + \frac{r}{2^k n}\right) + \sum_{i=1}^{N} \frac{f^{(i)}\left(\frac{j}{2^k} + \frac{r}{2^k n}\right)}{i!}\left(t - \frac{r}{2^k n}\right)^i$$

$$+ \int_{(j/2^k)+\frac{r}{2^k n}}^{t+(j/2^k)} \left(f^{(N)}(s) - f^{(N)}\left(\frac{j}{2^k} + \frac{r}{2^k n}\right)\right) \frac{\left(t + \frac{j}{2^k} - s\right)^{N-1}}{(N-1)!} ds.$$

Hence we get

$$\gamma_{kj}(f) = 2^k \int_0^{2^{-k}} f\left(t + \frac{j}{2^k}\right) dt = \sum_{r=0}^{n} w_r f\left(\frac{j}{2^k} + \frac{r}{2^k n}\right)$$

$$+ \sum_{i=1}^{N} \sum_{r=0}^{n} \frac{w_r f^{(i)}\left(\frac{j}{2^k} + \frac{r}{2^k n}\right)}{i!} 2^k \int_0^{2^{-k}} \left(t - \frac{r}{2^k n}\right)^i dt + \sum_{r=0}^{n} w_r 2^k \cdot$$

$$\int_0^{2^{-k}} \left(\int_{(j/2^k)+\frac{r}{2^k n}}^{t+(j/2^k)} \left(f^{(N)}(s) - f^{(N)}\left(\frac{j}{2^k} + \frac{r}{2^k n}\right)\right) \frac{\left(t + \frac{j}{2^k} - s\right)^{N-1}}{(N-1)!} ds\right) dt.$$

That is we have

$$\sum_{j=-\infty}^{\infty} \gamma_{kj}(f) \varphi\left(2^k x - j\right) = \sum_{j=-\infty}^{\infty} \delta_{kj}(f) \varphi\left(2^k x - j\right)$$

$$+ \sum_{i=1}^{N} \frac{\sum_{j=-\infty}^{\infty} \delta_{kj}\left(f^{(i)}\right) \varphi\left(2^k x - j\right)}{2^{ki}(i+1)!} \cdot$$

$$\left[\left(1 - \frac{r}{n}\right)^{i+1} - (-1)^{i+1}\left(\frac{r}{n}\right)^{i+1}\right]$$

$$+ \sum_{j=-\infty}^{\infty} \varphi\left(2^k x - j\right) \sum_{r=0}^{n} w_r 2^k$$

$$\cdot \int_0^{2^{-k}} \left(\int_{(j/2^k)+\frac{r}{2^k n}}^{t+(j/2^k)} \left(f^{(N)}(s) - f^{(N)}\left(\frac{j}{2^k} + \frac{r}{2^k n}\right)\right)\right.$$

$$\left. \frac{\left(t + \frac{j}{2^k} - s\right)^{N-1}}{(N-1)!} ds\right) dt.$$

Consequently we get

$$(C_k f)(x) - (D_k f)(x) - \sum_{i=1}^{N} \frac{\left(D_k f^{(i)}\right)(x)}{2^{ki}(i+1)!}\left[\left(1 - \frac{r}{n}\right)^{i+1} - (-1)^{i+1}\left(\frac{r}{n}\right)^{i+1}\right] = \mathcal{R}_3,$$

where

$$R_3 = \sum_{j=-\infty}^{\infty} \varphi\left(2^k x - j\right) \sum_{r=0}^{n} w_r 2^k$$

$$\cdot \int_0^{2^{-k}} \left(\int_{(j/2^k)+\frac{r}{2^k n}}^{t+(j/2^k)} \left(f^{(N)}(s) - f^{(N)}\left(\frac{j}{2^k} + \frac{r}{2^k n}\right) \right) \frac{\left(t + \frac{j}{2^k} - s\right)^{N-1}}{(N-1)!} ds \right) dt.$$

Set

$$\Gamma_{jr}(t) = \left| \int_{(j/2^k)+\frac{r}{2^k n}}^{t+(j/2^k)} \left(f^{(N)}(s) - f^{(N)}\left(\frac{j}{2^k} + \frac{r}{2^k n}\right) \right) \frac{\left(t + \frac{j}{2^k} - s\right)^{N-1}}{(N-1)!} ds \right|.$$

So that

$$|R_3| \le \sum_{j=-\infty}^{\infty} \varphi\left(2^k x - j\right) \sum_{r=0}^{n} w_r 2^k \int_0^{2^{-k}} \Gamma_{jr}(t)\, dt.$$

Next we observe that
(i) Case of $\frac{r}{2^k n} \le t$. So we have

$$\Gamma_{jr}(t) \le \int_{(j/2^k)+\frac{r}{2^k n}}^{t+(j/2^k)} \left| f^{(N)}(s) - f^{(N)}\left(\frac{j}{2^k} + \frac{r}{2^k n}\right) \right| \frac{\left(t + \frac{j}{2^k} - s\right)^{N-1}}{(N-1)!} ds$$

$$\le \int_{(j/2^k)+\frac{r}{2^k n}}^{t+(j/2^k)} \omega_1\left(f^{(N)}, \left| s - \frac{j}{2^k} - \frac{r}{2^k n} \right|\right) \frac{\left(t + \frac{j}{2^k} - s\right)^{N-1}}{(N-1)!} ds$$

$$\le \omega_1\left(f^{(N)}, t - \frac{r}{2^k n}\right) \int_{(j/2^k)+\frac{r}{2^k n}}^{t+(j/2^k)} \frac{\left(t + \frac{j}{2^k} - s\right)^{N-1}}{(N-1)!} ds$$

$$\le \omega_1\left(f^{(N)}, t\right) \frac{\left(t - \frac{r}{2^k n}\right)^{N}}{N!}$$

$$\le \omega_1\left(f^{(N)}, \frac{1}{2^k}\right) \frac{\left(t - \frac{r}{2^k n}\right)^{N}}{N!}.$$

So we obtain

$$\Gamma_{jr}(t) \le \omega_1\left(f^{(N)}, \frac{1}{2^k}\right) \frac{\left(t - \frac{r}{2^k n}\right)^{N}}{N!}.$$

(ii) Case of $\frac{r}{2^k n} \geq t$. We have

$$
\begin{aligned}
\Gamma_{jr}(t) &= \left| \int_{t+(j/2^k)}^{(j/2^k)+\frac{r}{2^k n}} \left(f^{(N)}(s) - f^{(N)}\left(\frac{j}{2^k} + \frac{r}{2^k n} \right) \right) \right. \\
&\qquad \left. \frac{\left(s - \left(t + \frac{j}{2^k} \right) \right)^{N-1}}{(N-1)!} ds \right| \\
&\leq \int_{t+(j/2^k)}^{(j/2^k)+\frac{r}{2^k n}} \left| f^{(N)}(s) - f^{(N)}\left(\frac{j}{2^k} + \frac{r}{2^k n} \right) \right| \frac{\left(s - \left(t + \frac{j}{2^k} \right) \right)^{N-1}}{(N-1)!} ds \\
&\leq \int_{t+(j/2^k)}^{(j/2^k)+\frac{r}{2^k n}} \omega_1 \left(f^{(N)}, \left(\frac{j}{2^k} + \frac{r}{2^k n} - s \right) \right) \frac{\left(s - \left(t + \frac{j}{2^k} \right) \right)^{N-1}}{(N-1)!} ds \\
&\leq \omega_1 \left(f^{(N)}, \frac{r}{2^k n} - t \right) \int_{t+(j/2^k)}^{(j/2^k)+\frac{r}{2^k n}} \frac{\left(s - \left(t + \frac{j}{2^k} \right) \right)^{N-1}}{(N-1)!} ds \\
&= \omega_1 \left(f^{(N)}, \frac{r}{2^k n} - t \right) \frac{\left(\frac{r}{2^k n} - t \right)^{N}}{N!} \\
&\leq \omega_1 \left(f^{(N)}, \frac{1}{2^k} \right) \frac{\left(\frac{r}{2^k n} - t \right)^{N}}{N!}.
\end{aligned}
$$

So we derive

$$
\Gamma_{jr}(t) \leq \omega_1 \left(f^{(N)}, \frac{1}{2^k} \right) \frac{\left(\frac{r}{2^k n} - t \right)^{N}}{N!}.
$$

Therefore we have found

$$
\Gamma_{jr}(t) \leq \frac{\omega_1 \left(f^{(N)}, \frac{1}{2^k} \right)}{N!} \left| t - \frac{r}{2^k n} \right|^{N}.
$$

Furthermore we see that

$$
\begin{aligned}
\int_0^{2^{-k}} \Gamma_{jr}(t)\, dt &\leq \frac{\omega_1 \left(f^{(N)}, \frac{1}{2^k} \right)}{N!} \int_0^{2^{-k}} \left| t - \frac{r}{2^k n} \right|^{N} dt \\
&= \frac{\omega_1 \left(f^{(N)}, \frac{1}{2^k} \right)}{N!} \cdot \\
&\qquad \left[\int_0^{\frac{r}{2^k n}} \left(\frac{r}{2^k n} - t \right)^{N} dt + \int_{\frac{r}{2^k n}}^{2^{-k}} \left(t - \frac{r}{2^k n} \right)^{N} dt \right] \\
&= \frac{\omega_1 \left(f^{(N)}, \frac{1}{2^k} \right)}{(N+1)!} \left[\left(\frac{r}{2^k n} \right)^{N+1} + \left(\frac{1}{2^k} - \frac{r}{2^k n} \right)^{N+1} \right] \\
&= \frac{\omega_1 \left(f^{(N)}, \frac{1}{2^k} \right)}{(N+1)!} \frac{1}{2^{k(N+1)}} \left[\left(\frac{r}{n} \right)^{N+1} + \left(1 - \frac{r}{n} \right)^{N+1} \right].
\end{aligned}
$$

Thus

$$2^k \int_0^{2^{-k}} \Gamma_{jr}(t)\, dt \le \frac{\omega_1\left(f^{(N)}, \frac{1}{2^k}\right)}{2^{kN}(N+1)!} \left[\left(\frac{r}{n}\right)^{N+1} + \left(1 - \frac{r}{n}\right)^{N+1}\right].$$

Finally we derive that

$$|\mathcal{R}_3| \le \left(\frac{\omega_1\left(f^{(N)}, \frac{1}{2^k}\right)}{2^{kN}(N+1)!}\right) \sum_{r=0}^n w_r \left[\left(\frac{r}{n}\right)^{N+1} + \left(1 - \frac{r}{n}\right)^{N+1}\right],$$

proving the theorem. □
We continue with

Theorem 21.8. Let $f \in C^N(\mathbb{R})$, $N \ge 1$, $x \in \mathbb{R}$ and $k \in \mathbb{Z}$, also $\left\|f^{(i)}\right\|_\infty < \infty$, $i = 1, \dots, N$. Let φ be a bounded function of compact support $\subseteq [-a, a]$, $a > 0$ such that $\sum_{j=-\infty}^{\infty} \varphi(x - j) = 1$ all $x \in \mathbb{R}$. Suppose $\varphi \ge 0$ and φ is Lebesgue measurable (then $\int_{-\infty}^{\infty} \varphi(x)\, dx = 1$). Define

$$\varphi_{kj}(x) := 2^{k/2} \varphi\left(2^k x - j\right) \text{ all } k, j \in \mathbb{Z},$$

$$\langle f, \varphi_{kj} \rangle = \int_{-\infty}^{\infty} f(t)\, \varphi_{kj}(t)\, dt,$$

and

$$(A_k f)(x) = \sum_{j=-\infty}^{\infty} \langle f, \varphi_{kj} \rangle\, \varphi_{kj}(x)$$

$$= \sum_{j=-\infty}^{\infty} \left(\int_{-\infty}^{\infty} f\left(\frac{u}{2^k}\right) \varphi(u - j)\, du\right) \varphi\left(2^k x - j\right),$$

also define

$$(B_k f)(x) = \sum_{j=-\infty}^{\infty} f\left(\frac{j}{2^k}\right) \varphi\left(2^k x - j\right).$$

Then

$$|(A_k f)(x) - (B_k f)(x)| \le \|A_k f - B_k f\|_\infty \qquad (21.13)$$

$$\le \sum_{i=1}^N \frac{\left\|f^{(i)}\right\|_\infty}{2^{ki} i!} a^i + \frac{a^N}{2^{kN} N!} \omega_1\left(f^{(N)}, \frac{a}{2^k}\right),$$

$x \in \mathbb{R}$.
So as $k \to \infty$, we get $\|A_k f - B_k f\|_\infty \to 0$.

Proof. By $\sum_{j=-\infty}^{\infty} \varphi \left(2^k x - j\right) = 1$ we have that $\int_{-\infty}^{\infty} \varphi \left(u - j\right) du = 1$.

Notice that

$$
\begin{aligned}
(A_k f)(x) - (B_k f)(x) &= \sum_{j=-\infty}^{\infty} \langle f, \varphi_{kj} \rangle \varphi_{kj}(x) - \sum_{j=-\infty}^{\infty} f\left(\frac{j}{2^k}\right) \varphi\left(2^k x - j\right) \\
&= \sum_{j=-\infty}^{\infty} \left(\int_{-\infty}^{\infty} f\left(\frac{u}{2^k}\right) \varphi(u-j) du \right) \varphi\left(2^k x - j\right) \\
&\quad - \sum_{j=-\infty}^{\infty} f\left(\frac{j}{2^k}\right) \varphi\left(2^k x - j\right) \\
&= \sum_{j=-\infty}^{\infty} \left[\int_{-\infty}^{\infty} f\left(\frac{u}{2^k}\right) \varphi(u-j) du - f\left(\frac{j}{2^k}\right) \right] \cdot \\
&\quad \varphi\left(2^k x - j\right) \\
&= \sum_{j=-\infty}^{\infty} \left[\int_{-\infty}^{\infty} \left(f\left(\frac{u}{2^k}\right) - f\left(\frac{j}{2^k}\right) \right) \varphi(u-j) du \right] \cdot \\
&\quad \varphi\left(2^k x - j\right).
\end{aligned}
$$

That is

$$
\begin{aligned}
(A_k f)(x) - (B_k f)(x) &= \sum_{j=-\infty}^{\infty} \left[\int_{-\infty}^{\infty} \left(f\left(\frac{u}{2^k}\right) - f\left(\frac{j}{2^k}\right) \right) \varphi(u-j) du \right] \cdot \\
&\quad \varphi\left(2^k x - j\right).
\end{aligned}
$$

By *supp* $\varphi \subseteq [-a, a]$ we have that $\varphi(u-j)$ is nonzero when $-a \le u-j \le a$, that is when $j - a \le u \le j + a$. Hence

$$
(A_k f)(x) - (B_k f)(x) =
$$
$$
\sum_{j=-\infty}^{\infty} \left[\int_{j-a}^{j+a} \left(f\left(\frac{u}{2^k}\right) - f\left(\frac{j}{2^k}\right) \right) \varphi(u-j) du \right] \varphi\left(2^k x - j\right).
$$

Next we see that

$$
f\left(\frac{u}{2^k}\right) - f\left(\frac{j}{2^k}\right) = \sum_{i=1}^{N} \frac{f^{(i)}\left(\frac{j}{2^k}\right) (u-j)^i}{i! \, 2^{ki}} + \mathcal{R}_4,
$$

where

$$
\mathcal{R}_4 = \int_{j/2^k}^{u/2^k} \left(f^{(N)}(t) - f^{(N)}\left(\frac{j}{2^k}\right) \right) \frac{\left(\frac{u}{2^k} - t\right)^{N-1}}{(N-1)!} dt.
$$

(i) Case of $j \leq u$. We have

$$
\begin{aligned}
|\mathcal{R}_4| &\leq \int_{j/2^k}^{u/2^k} \left| f^{(N)}(t) - f^{(N)}\left(\frac{j}{2^k}\right) \right| \frac{\left(\frac{u}{2^k} - t\right)^{N-1}}{(N-1)!} dt \\
&\leq \int_{j/2^k}^{u/2^k} \omega_1\left(f^{(N)}, \left(t - \frac{j}{2^k}\right)\right) \frac{\left(\frac{u}{2^k} - t\right)^{N-1}}{(N-1)!} dt \\
&\leq \omega_1\left(f^{(N)}, \frac{(u-j)}{2^k}\right) \int_{j/2^k}^{u/2^k} \frac{\left(\frac{u}{2^k} - t\right)^{N-1}}{(N-1)!} dt \\
&= \omega_1\left(f^{(N)}, \frac{(u-j)}{2^k}\right) \frac{(u-j)^N}{2^{kN} N!}.
\end{aligned}
$$

(ii) Case of $j \geq u$. We have

$$
\begin{aligned}
|\mathcal{R}_4| &= \left| \int_{u/2^k}^{j/2^k} \left(f^{(N)}(t) - f^{(N)}\left(\frac{j}{2^k}\right) \right) \frac{\left(t - \frac{u}{2^k}\right)^{N-1}}{(N-1)!} dt \right| \\
&\leq \int_{u/2^k}^{j/2^k} \left| f^{(N)}(t) - f^{(N)}\left(\frac{j}{2^k}\right) \right| \frac{\left(t - \frac{u}{2^k}\right)^{N-1}}{(N-1)!} dt \\
&\leq \int_{u/2^k}^{j/2^k} \omega_1\left(f^{(N)}, \left(\frac{j}{2^k} - t\right)\right) \frac{\left(t - \frac{u}{2^k}\right)^{N-1}}{(N-1)!} dt \\
&\leq \omega_1\left(f^{(N)}, \frac{j-u}{2^k}\right) \int_{u/2^k}^{j/2^k} \frac{\left(t - \frac{u}{2^k}\right)^{N-1}}{(N-1)!} dt \\
&= \omega_1\left(f^{(N)}, \frac{j-u}{2^k}\right) \frac{(j-u)^N}{2^{kN} N!}.
\end{aligned}
$$

So we have proved that

$$
|\mathcal{R}_4| \leq \omega_1\left(f^{(N)}, \frac{|u-j|}{2^k}\right) \frac{|u-j|^N}{2^{kN} N!} \leq \omega_1\left(f^{(N)}, \frac{a}{2^k}\right) \frac{a^N}{2^{kN} N!},
$$

i.e.

$$
|\mathcal{R}_4| \leq \frac{a^N}{2^{kN} N!} \omega_1\left(f^{(N)}, \frac{a}{2^k}\right).
$$

Furthermore we observe that

$$
\begin{aligned}
&\int_{j-a}^{j+a} \left(f\left(\frac{u}{2^k}\right) - f\left(\frac{j}{2^k}\right) \right) \varphi(u-j) \, du \\
&= \sum_{i=1}^{N} \frac{f^{(i)}(j/2^k)}{2^{ki} i!} \int_{j-a}^{j+a} (u-j)^i \varphi(u-j) \, du + \int_{j-a}^{j+a} \mathcal{R}_4 \varphi(u-j) \, du.
\end{aligned}
$$

Therefore

$$\left| \int_{j-a}^{j+a} \left(f\left(\frac{u}{2^k}\right) - f\left(\frac{j}{2^k}\right) \right) \varphi\left(u-j\right) du \right| \leq \sum_{i=1}^{N} \frac{\left\|f^{(i)}\right\|_\infty}{2^{ki} i!} a^i +$$

$$\frac{a^N}{2^{kN} N!} \omega_1\left(f^{(N)}, \frac{a}{2^k}\right).$$

The last proves the theorem.   □

We continue with

Theorem 21.9. Let $f \in C^N(\mathbb{R})$, $N \geq 1$, $x \in \mathbb{R}$ and $k \in \mathbb{Z}$, also $\left\|f^{(i)}\right\|_\infty < \infty$, $i = 1, \ldots, N$. Let φ be a bounded function of compact support $\subseteq [-a, a]$, $a > 0$ such that $\sum_{j=-\infty}^{\infty} \varphi(x-j) = 1$ all $x \in \mathbb{R}$. Suppose $\varphi \geq 0$ and φ is Lebesgue measurable (then $\int_{-\infty}^{\infty} \varphi(x)\,dx = 1$). Define

$$\varphi_{kj}(x) := 2^{k/2} \varphi\left(2^k x - j\right) \text{ all } k, j \in \mathbb{Z},$$

$$\langle f, \varphi_{kj} \rangle = \int_{-\infty}^{\infty} f(t)\,\varphi_{kj}(t)\,dt,$$

and

$$\begin{aligned} (A_k f)(x) &= \sum_{j=-\infty}^{\infty} \langle f, \varphi_{kj} \rangle \varphi_{kj}(x) \\ &= \sum_{j=-\infty}^{\infty} \left(\int_{-\infty}^{\infty} f\left(\frac{u}{2^k}\right) \varphi(u-j)\,du \right) \varphi\left(2^k x - j\right). \end{aligned}$$

Also define

$$(D_k f)(x) = \sum_{j=-\infty}^{\infty} \delta_{kj}(f)\,\varphi\left(2^k x - j\right),$$

where

$$\delta_{kj}(f) = \sum_{r=0}^{n} w_r f\left(\frac{j}{2^k} + \frac{r}{2^k n}\right),$$

$n \in \mathbb{N}$, $w_r \geq 0$, $\sum_{r=0}^{n} w_r = 1$. Then

$$|(A_k f)(x) - (D_k f)(x)| \leq \|A_k f - D_k f\|_\infty \tag{21.14}$$

$$\leq \sum_{i=1}^{N} \frac{\left\|f^{(i)}\right\|_\infty}{2^{ki} i!} (a+1)^i +$$

$$\frac{(a+1)^N}{N! 2^{kN}} \omega_1\left(f^{(N)}, \frac{(a+1)}{2^k}\right).$$

So as $k \to \infty$, we get $\|A_k f - D_k f\|_\infty \to 0$.

Proof. By $\sum_{j=-\infty}^{\infty} \varphi\left(2^k x - j\right) = 1$ we have that $\int_{-\infty}^{\infty} \varphi\left(u - j\right) du = 1$.
Notice that

$$
\begin{aligned}
\left(A_k f\right)(x) - \left(D_k f\right)(x) &= \sum_{j=-\infty}^{\infty} \langle f, \varphi_{kj}\rangle \varphi_{kj}(x) - \sum_{j=-\infty}^{\infty} \delta_{kj}(f) \varphi\left(2^k x - j\right) \\
&= \sum_{j=-\infty}^{\infty} \left(2^{k/2} \langle f, \varphi_{kj}\rangle - \delta_{kj}(f)\right) \varphi\left(2^k x - j\right) \\
&= \sum_{j=-\infty}^{\infty} \left(\int_{-\infty}^{\infty} f\left(\frac{u}{2^k}\right) \varphi\left(u - j\right) du - \right. \\
&\qquad \left. \sum_{r=0}^{n} w_r f\left(\frac{j}{2^k} + \frac{r}{2^k n}\right)\right) \varphi\left(2^k x - j\right) \\
&= \sum_{j=-\infty}^{\infty} \left[\sum_{r=0}^{n} w_r \left(\int_{-\infty}^{\infty} f\left(\frac{u}{2^k}\right) \varphi\left(u - j\right) du - \right.\right. \\
&\qquad \left.\left. f\left(\frac{j}{2^k} + \frac{r}{2^k n}\right)\right)\right] \varphi\left(2^k x - j\right) \\
&= \sum_{j=-\infty}^{\infty} \left[\sum_{r=0}^{n} w_r \left(\int_{-\infty}^{\infty} \left(f\left(\frac{u}{2^k}\right) - f\left(\frac{j}{2^k} + \frac{r}{2^k n}\right)\right)\right.\right. \\
&\qquad \left.\left. \varphi\left(u - j\right) du\right)\right] \varphi\left(2^k x - j\right).
\end{aligned}
$$

That is, by the compact support of φ we have

$$
\left(A_k f\right)(x) - \left(D_k f\right)(x) =
$$

$$
\sum_{j=-\infty}^{\infty} \left[\sum_{r=0}^{n} w_r \left(\int_{j-a}^{j+a} \left(f\left(\frac{u}{2^k}\right) - f\left(\frac{j}{2^k} + \frac{r}{2^k n}\right)\right) \varphi\left(u - j\right) du\right)\right]
$$
$$
\cdot \varphi\left(2^k x - j\right).
$$

Next we see that

$$
f\left(\frac{u}{2^k}\right) - f\left(\frac{j}{2^k} + \frac{r}{2^k n}\right) = \sum_{i=1}^{N} \frac{f^{(i)}\left(\frac{j}{2^k} + \frac{r}{2^k n}\right) \left(u - j - \frac{r}{n}\right)^i}{i!} \cdot \frac{1}{2^{ki}} + \mathcal{R}_5,
$$

where

$$
\mathcal{R}_5 = \int_{\frac{j}{2^k} + \frac{r}{2^k n}}^{\frac{u}{2^k}} \left(f^{(N)}(t) - f^{(N)}\left(\frac{j}{2^k} + \frac{r}{2^k n}\right)\right) \frac{\left(\frac{u}{2^k} - t\right)^{N-1}}{(N-1)!} dt.
$$

(i) Case $j + \frac{r}{n} \leq u$. We have easily that

$$|\mathcal{R}_5| \leq \frac{\omega_1\left(f^{(N)}, \frac{1}{2^k}\left(u - j - \frac{r}{n}\right)\right)}{N! 2^{kN}} \left(u - j - \frac{r}{n}\right)^N.$$

(ii) Case $j + \frac{r}{n} \geq u$. We have that

$$
\begin{aligned}
|\mathcal{R}_5| &= \left| \int_{\frac{u}{2^k}}^{\frac{j}{2^k} + \frac{r}{2^k n}} \left(f^{(N)}\left(\frac{j}{2^k} + \frac{r}{2^k n}\right) - f^{(N)}(t) \right) \frac{\left(t - \frac{u}{2^k}\right)^{N-1}}{(N-1)!} dt \right| \\
&\leq \int_{\frac{u}{2^k}}^{\frac{j}{2^k} + \frac{r}{2^k n}} \left| f^{(N)}\left(\frac{j}{2^k} + \frac{r}{2^k n}\right) - f^{(N)}(t) \right| \frac{\left(t - \frac{u}{2^k}\right)^{N-1}}{(N-1)!} dt \\
&\leq \frac{\omega_1\left(f^{(N)}, \frac{1}{2^k}\left(j + \frac{r}{n} - u\right)\right)}{N! 2^{kN}} \left(j + \frac{r}{n} - u\right)^N.
\end{aligned}
$$

So we have proved that

$$|\mathcal{R}_5| \leq \frac{\omega_1\left(f^{(N)}, \frac{1}{2^k}\left|j + \frac{r}{n} - u\right|\right)}{N! 2^{kN}} \left|j + \frac{r}{n} - u\right|^N \leq \frac{\omega_1\left(f^{(N)}, \frac{a+1}{2^k}\right)}{N! 2^{kN}} (a+1)^N,$$

i.e.

$$|\mathcal{R}_5| \leq \frac{(a+1)^N}{N! 2^{kN}} \omega_1\left(f^{(N)}, \frac{a+1}{2^k}\right).$$

Furthermore we observe that

$$\int_{j-a}^{j+a} \left(f\left(\frac{u}{2^k}\right) - f\left(\frac{j}{2^k} + \frac{r}{2^k n}\right) \right) \varphi(u - j) \, du =$$

$$\sum_{i=1}^{N} \frac{f^{(i)}\left(\frac{j}{2^k} + \frac{r}{2^k n}\right)}{i! 2^{ki}} \int_{j-a}^{j+a} \left(u - j - \frac{r}{n}\right)^i \varphi(u - j) \, du + \int_{j-a}^{j+a} \mathcal{R}_5 \varphi(u - j) \, du.$$

Therefore

$$\left| \int_{j-a}^{j+a} \left(f\left(\frac{u}{2^k}\right) - f\left(\frac{j}{2^k} + \frac{r}{2^k n}\right) \right) \varphi(u - j) \, du \right| \leq$$

$$\sum_{i=1}^{N} \frac{\|f^{(i)}\|_\infty}{2^{ki} i!} (a+1)^i + \frac{(a+1)^N}{N! 2^{kN}} \omega_1\left(f^{(N)}, \frac{(a+1)}{2^k}\right).$$

The last proves the theorem. $\square$

We continue with

Theorem 21.10. Let $f \in C^N(\mathbb{R})$, $N \geq 1$, $x \in \mathbb{R}$ and $k \in \mathbb{Z}$, also $\|f^{(i)}\|_\infty < \infty$, $i = 1, \ldots, N$. Let φ be a bounded function of compact support $\subseteq [-a, a]$, $a > 0$ such that $\sum_{j=-\infty}^{\infty} \varphi (x - j) = 1$ all $x \in \mathbb{R}$. Suppose $\varphi \geq 0$ and φ is Lebesgue measurable (then $\int_{-\infty}^{\infty} \varphi(x) \, dx = 1$). Define

$$\varphi_{kj} (x) := 2^{k/2} \varphi \left(2^k x - j \right) \text{ all } k, j \in \mathbb{Z},$$

$$\langle f, \varphi_{kj} \rangle = \int_{-\infty}^{\infty} f(t) \, \varphi_{kj}(t) \, dt,$$

and

$$
\begin{aligned}
(A_k f)(x) &= \sum_{j=-\infty}^{\infty} \langle f, \varphi_{kj} \rangle \, \varphi_{kj}(x) \\
&= \sum_{j=-\infty}^{\infty} \left(\int_{-\infty}^{\infty} f \left(\frac{u}{2^k} \right) \varphi(u - j) \, du \right) \varphi \left(2^k x - j \right).
\end{aligned}
$$

Also define

$$(C_k f)(x) = \sum_{j=-\infty}^{\infty} \gamma_{kj} (f) \, \varphi \left(2^k x - j \right),$$

where

$$\gamma_{kj} (f) = 2^k \int_{2^{-k} j}^{2^{-k} (j+1)} f(t) \, dt = 2^k \int_0^{2^{-k}} f \left(t + \frac{j}{2^k} \right) dt.$$

Then

$$|(A_k f)(x) - (C_k f)(x)| \leq \|A_k f - C_k f\|_\infty \tag{21.15}$$

$$\leq \sum_{i=1}^{N} \frac{\|f^{(i)}\|_\infty}{i! 2^{ki}} (a + 1)^i + \frac{(a+1)^N}{2^{kN} N!} \omega_1 \left(f^{(N)}, \frac{(a+1)}{2^k} \right).$$

So as $k \to \infty$, we get $\|A_k f - C_k f\|_\infty \to 0$.

Proof. By $\sum_{j=-\infty}^{\infty} \varphi \left(2^k x - j \right) = 1$ we have that $\int_{-\infty}^{\infty} \varphi(u - j) \, du = 1$.

Notice that

$$(A_k f)(x) - (C_k f)(x) = \sum_{j=-\infty}^{\infty} \langle f, \varphi_{kj} \rangle \, \varphi_{kj}(x) - \sum_{j=-\infty}^{\infty} \gamma_{kj} (f) \, \varphi \left(2^k x - j \right)$$

$$= \sum_{j=-\infty}^{\infty} \left(2^{k/2} \langle f, \varphi_{kj} \rangle - \gamma_{kj}(f) \right) \varphi \left(2^k x - j \right)$$

$$= \sum_{j=-\infty}^{\infty} \left(\int_{-\infty}^{\infty} f\left(\frac{u}{2^k} \right) \varphi(u-j)\, du - 2^k \int_0^{2^{-k}} f\left(t + \frac{j}{2^k} \right) dt \right) \varphi \left(2^k x - j \right)$$

$$= \sum_{j=-\infty}^{\infty} \left[2^k \int_0^{2^{-k}} \left(\int_{-\infty}^{\infty} f\left(\frac{u}{2^k} \right) \varphi(u-j)\, du \right) dt - 2^k \int_0^{2^{-k}} f\left(t + \frac{j}{2^k} \right) dt \right] \cdot$$
$$\varphi \left(2^k x - j \right)$$

$$= \sum_{j=-\infty}^{\infty} \left[2^k \int_0^{2^{-k}} \left[\left(\int_{-\infty}^{\infty} f\left(\frac{u}{2^k} \right) \varphi(u-j)\, du \right) - f\left(t + \frac{j}{2^k} \right) \right] dt \right] \cdot$$
$$\varphi \left(2^k x - j \right)$$

$$= \sum_{j=-\infty}^{\infty} \left[2^k \int_0^{2^{-k}} \left[\int_{-\infty}^{\infty} \left(f\left(\frac{u}{2^k} \right) - f\left(t + \frac{j}{2^k} \right) \right) \varphi(u-j)\, du \right] dt \right] \cdot$$
$$\varphi \left(2^k x - j \right).$$

That is, by the compact support of φ we have

$$(A_k f)(x) - (C_k f)(x) =$$
$$\sum_{j=-\infty}^{\infty} \left[2^k \int_0^{2^{-k}} \left[\int_{j-a}^{j+a} \left(f\left(\frac{u}{2^k} \right) - f\left(t + \frac{j}{2^k} \right) \right) \varphi(u-j)\, du \right] dt \right] \cdot$$
$$\varphi \left(2^k x - j \right).$$

Next we see that

$$f\left(\frac{u}{2^k} \right) - f\left(t + \frac{j}{2^k} \right) = \sum_{i=1}^{N} \frac{f^{(i)}\left(t + \frac{j}{2^k} \right)}{i!} \left(\frac{u}{2^k} - t - \frac{j}{2^k} \right)^i + \mathcal{R}_6,$$

where

$$\mathcal{R}_6 = \int_{t+\frac{j}{2^k}}^{\frac{u}{2^k}} \left(f^{(N)}(s) - f^{(N)}\left(t + \frac{j}{2^k} \right) \right) \frac{\left(\frac{u}{2^k} - s \right)^{N-1}}{(N-1)!}\, ds.$$

(i) Case $t + \frac{j}{2^k} \leq \frac{u}{2^k}$. We have easily that

$$|\mathcal{R}_6| \leq \omega_1 \left(f^{(N)}, \left(\frac{u}{2^k} - t - \frac{j}{2^k} \right) \right) \frac{\left(\frac{u}{2^k} - t - \frac{j}{2^k} \right)^N}{N!}.$$

(ii) Case $t + \frac{j}{2^k} \geq \frac{u}{2^k}$. We have that

$$
\begin{aligned}
|\mathcal{R}_6| &= \left| \int_{\frac{u}{2^k}}^{t + \frac{j}{2^k}} \left(f^{(N)} \left(t + \frac{j}{2^k} \right) - f^{(N)}(s) \right) \frac{\left(s - \frac{u}{2^k} \right)^{N-1}}{(N-1)!} ds \right| \\
&\leq \int_{\frac{u}{2^k}}^{t + \frac{j}{2^k}} \left| f^{(N)} \left(t + \frac{j}{2^k} \right) - f^{(N)}(s) \right| \frac{\left(s - \frac{u}{2^k} \right)^{N-1}}{(N-1)!} ds \\
&\leq \omega_1 \left(f^{(N)}, t + \frac{j}{2^k} - \frac{u}{2^k} \right) \frac{\left(t + \frac{j}{2^k} - \frac{u}{2^k} \right)^N}{N!}.
\end{aligned}
$$

So we have proved that

$$
\begin{aligned}
|\mathcal{R}_6| &\leq \omega_1 \left(f^{(N)}, \left| t + \frac{(j-u)}{2^k} \right| \right) \frac{\left| t + \frac{(j-u)}{2^k} \right|^N}{N!} \\
&\leq \omega_1 \left(f^{(N)}, \frac{(a+1)}{2^k} \right) \frac{(a+1)^N}{2^{kN} N!}.
\end{aligned}
$$

I.e. we found that

$$
|\mathcal{R}_6| \leq \frac{(a+1)^N}{2^{kN} N!} \omega_1 \left(f^{(N)}, \frac{(a+1)}{2^k} \right).
$$

Furthermore we observe that

$$
\int_{j-a}^{j+a} \left(f \left(\frac{u}{2^k} \right) - f \left(t + \frac{j}{2^k} \right) \right) \varphi(u-j) \, du =
$$

$$
\sum_{i=1}^{N} \frac{f^{(i)} \left(t + \frac{j}{2^k} \right)}{i!} \int_{j-a}^{j+a} \left(\frac{(u-j)}{2^k} - t \right)^i \varphi(u-j) \, du + \int_{j-a}^{j+a} \mathcal{R}_6 \varphi(u-j) \, du.
$$

Therefore

$$
\left| \int_{j-a}^{j+a} \left(f \left(\frac{u}{2^k} \right) - f \left(t + \frac{j}{2^k} \right) \right) \varphi(u-j) \, du \right| \leq
$$

$$
\sum_{i=1}^{N} \frac{\|f^{(i)}\|_\infty}{i! 2^{ki}} (a+1)^i + \frac{(a+1)^N}{2^{kN} N!} \omega_1 \left(f^{(N)}, \frac{(a+1)}{2^k} \right).
$$

The last proves the theorem. $\qquad\square$

We give

Theorem 21.11. Let $f \in C^N(\mathbb{R})$, $N \geq 1$, $x \in \mathbb{R}$ and $k \in \mathbb{Z}$. Let φ be a bounded function of compact support $\subseteq [-a, a]$, $a > 0$ such that

$\sum_{j=-\infty}^{\infty} \varphi(x-j) = 1$ all $x \in \mathbb{R}$. Suppose $\varphi \geq 0$. Assume further $\left\|f^{(i)}\right\|_{\infty} < \infty$, $i = 1, \ldots, N$. Put

$$(B_k f)(x) = \sum_{j=-\infty}^{\infty} f\left(\frac{j}{2^k}\right) \varphi\left(2^k x - j\right),$$

$$(C_k f)(x) = \sum_{j=-\infty}^{\infty} \gamma_{kj}(f) \varphi\left(2^k x - j\right),$$

where

$$\gamma_{kj}(f) = 2^k \int_0^{2^{-k}} f\left(t + \frac{j}{2^k}\right) dt,$$

and

$$(D_k f)(x) = \sum_{j=-\infty}^{\infty} \delta_{kj}(f) \varphi\left(2^k x - j\right),$$

where

$$\delta_{kj}(f) = \sum_{r=0}^{n} w_r f\left(\frac{j}{2^k} + \frac{r}{2^k n}\right),$$

$n \in \mathbb{N}$, $w_r \geq 0$, $\sum_{r=0}^{n} w_r = 1$. Then

(i)

$$|(B_k f)(x) - (D_k f)(x)| \leq \|B_k f - D_k f\|_{\infty} \qquad (21.16)$$
$$\leq \sum_{i=1}^{N} \frac{\left\|f^{(i)}\right\|_{\infty}}{2^{ki} i!} + \frac{\omega_1\left(f^{(N)}, \frac{1}{2^k}\right)}{2^{kN} N!},$$

(ii)

$$|(B_k f)(x) - (C_k f)(x)| \leq \|B_k f - C_k f\|_{\infty} \qquad (21.17)$$
$$\leq \sum_{i=1}^{N} \frac{\left\|f^{(i)}\right\|_{\infty}}{2^{ki}(i+1)!} + \frac{\omega_1\left(f^{(N)}, \frac{1}{2^k}\right)}{2^{kN}(N+1)!},$$

(iii)

$$|(C_k f)(x) - (D_k f)(x)| \leq \|C_k f - D_k f\|_{\infty} \qquad (21.18)$$
$$\leq \sum_{i=1}^{N} \frac{\left\|f^{(i)}\right\|_{\infty}}{2^{ki-1}(i+1)!} + \frac{\omega_1\left(f^{(N)}, \frac{1}{2^k}\right)}{2^{kN-1}(N+1)!}.$$

So as $k \to \infty$, we get

$$\|B_k f - D_k f\|_{\infty} \to 0, \|B_k f - C_k f\|_{\infty} \to 0, \|C_k f - D_k f\|_{\infty} \to 0.$$

Proof. By Theorems 21.2, 21.4, 21.6 and especially use of (21.4), (21.7) and (21.12). $\qquad \square$

21.2 Estimates for Distances of Fuzzy Wavelet type Operators

We need some background which follows.

Remark 21.12 ([15]). Here $r \in [0, 1]$, $x_i^{(r)}, y_i^{(r)} \in \mathbb{R}$, $i = 1, \ldots, m \in \mathbb{N}$. Suppose that

$$\sup_{r \in [0,1]} \max \left(x_i^{(r)}, y_i^{(r)} \right) \in \mathbb{R}, \text{ for } i = 1, \ldots, m.$$

Then one sees easily that

$$\sup_{r \in [0,1]} \max \left(\sum_{i=1}^m x_i^{(r)}, \sum_{i=1}^m y_i^{(r)} \right) \leq \sum_{i=1}^m \sup_{r \in [0,1]} \max \left(x_i^{(r)}, y_i^{(r)} \right).$$

Definition 21.13. Let $f : \mathbb{R} \to \mathbb{R}_{\mathcal{F}}$, we define the (first) fuzzy modulus of continuity of f by

$$\omega_1^{(\mathcal{F})}(f, \delta) = \sup_{\substack{x,y \in \mathbb{R} \\ |x-y| \leq \delta}} D(f(x), f(y)), \delta > 0.$$

We define $C_{\mathcal{F}}^U(\mathbb{R})$ the space of uniformly continuous functions from $\mathbb{R} \to \mathbb{R}_{\mathcal{F}}$, also $C(\mathbb{R}, \mathbb{R}_{\mathcal{F}})$ the space of fuzzy continuous functions on $\mathbb{R}$.

Proposition 21.14 ([15]). Let $f \in C_{\mathcal{F}}^U(\mathbb{R})$. Then $\omega_1^{(\mathcal{F})}(f, \delta) < \infty$, any $\delta > 0$.

Proposition 21.15 ([15]). It holds $\lim_{\delta \to 0} \omega_1^{(\mathcal{F})}(f, \delta) = \omega_1^{(\mathcal{F})}(f, 0) = 0$, iff $f \in C_{\mathcal{F}}^U(\mathbb{R})$.

Proposition 21.16 ([15]). Let $f : \mathbb{R} \to \mathbb{R}_{\mathcal{F}}$ be a fuzzy real number valued function. Assume that $\omega_1^{(\mathcal{F})}(f, \delta)$, $\omega_1 \left(f_-^{(r)}, \delta \right)$, $\omega_1 \left(f_+^{(r)}, \delta \right)$ are finite, for $\delta > 0$, all $r \in [0, 1]$. Then

$$\omega_1^{(\mathcal{F})}(f, \delta) = \sup_{r \in [0,1]} \max \left\{ \omega_1 \left(f_-^{(r)}, \delta \right), \omega_1 \left(f_+^{(r)}, \delta \right) \right\}.$$

Note. It is clear from Propositions 21.15, 21.16 that if $f \in C_{\mathcal{F}}^U(\mathbb{R})$, then $f_\pm^{(r)} \in C_U(\mathbb{R})$ (uniformly continuous on $\mathbb{R}$).

We denote by $C^N(\mathbb{R}, \mathbb{R}_{\mathcal{F}})$, $N \geq 1$, the space of all N−times continuously fuzzy differentiable functions from $\mathbb{R}$ into $\mathbb{R}_{\mathcal{F}}$.

We mention

Theorem 21.17 ([71]). Let $f : [a, b] \subseteq \mathbb{R} \to \mathbb{R}_{\mathcal{F}}$ be H−fuzzy differentiable. Let $t \in [a, b]$, $0 \leq r \leq 1$. Clearly

$$[f(t)]^r = \left[(f(t))^{(r)}_-, (f(t))^{(r)}_+\right] \subseteq \mathbb{R}.$$

Then $(f(t))^{(r)}_\pm$ are differentiable and

$$[f'(t)]^r = \left[\left((f(t))^{(r)}_-\right)', \left((f(t))^{(r)}_+\right)'\right].$$

I.e. $(f')^{(r)}_\pm = \left(f^{(r)}_\pm\right)'$, $\forall r \in [0, 1]$.

Remark 21.18 ([15]). Let $f \in C^N(\mathbb{R}, \mathbb{R}_{\mathcal{F}})$, $N \geq 1$. Then by Theorem 21.17 we obtain $\left[f^{(i)}(t)\right]^r = \left[\left((f(t))^{(r)}_-\right)^{(i)}, \left((f(t))^{(r)}_+\right)^{(i)}\right]$, for $i = 0, 1, 2, \ldots, N$, and in particular we have that

$$\left(f^{(i)}\right)^{(r)}_\pm = \left(f^{(r)}_\pm\right)^{(i)},$$

for any $r \in [0, 1]$.

Note 21.19. (i) Let $f : \mathbb{R} \to \mathbb{R}_{\mathcal{F}}$ fuzzy continuous, then $f^{(r)}_\pm : \mathbb{R} \to \mathbb{R}$ are continuous, $\forall r \in [0, 1]$.
(ii) Let $f \in C^N(\mathbb{R}, \mathbb{R}_{\mathcal{F}})$, $N \geq 1$. Then by Theorem 21.17, we have $f^{(r)}_\pm \in C^N(\mathbb{R})$, for any $r \in [0, 1]$.

We need also
Definition 21.20. Denote by $C^{NB}_{\mathcal{F}}(\mathbb{R}) := \{f : \mathbb{R} \to \mathbb{R}_{\mathcal{F}}|$ such that all fuzzy derivatives $f^{(i)} : \mathbb{R} \to \mathbb{R}_{\mathcal{F}}$, $i = 0, 1, \ldots, N$ exist and are fuzzy continuous and furthermore $D^*\left(f^{(i)}, \tilde{0}\right) < \infty$, for $i = 1, \ldots, N\}$, $N \geq 1$.

Notice here that

$$
\begin{aligned}
D^*\left(f^{(i)}, \tilde{0}\right) &= \sup_{r \in [0,1]} \max \left(\left\|\left(f^{(i)}\right)^{(r)}_-\right\|_\infty, \left\|\left(f^{(i)}\right)^{(r)}_+\right\|_\infty\right) \\
&= \sup_{r \in [0,1]} \max \left(\left\|\left(f^{(r)}\right)^{(i)}_-\right\|_\infty, \left\|\left(f^{(r)}\right)^{(i)}_+\right\|_\infty\right), \quad i = 1, \ldots, N.
\end{aligned}
$$

Notice also that

$$D^*\left(f^{(i)}, \tilde{0}\right) < \infty, \text{ implies } \left\|\left(f^{(i)}\right)^{(r)}_\pm\right\|_\infty < \infty, i = 1, \ldots, N, \forall r \in [0, 1].$$

We mention

Theorem 21.21 ([67]). Let $f : [a,b] \to \mathbb{R}_{\mathcal{F}}$ be fuzzy continuous. Then $(FR) \int_a^b f(x) \, dx$ exists and belongs to $\mathbb{R}_{\mathcal{F}}$, furthermore it holds

$$\left[(FR) \int_a^b f(x) \, dx \right]^r = \left[\int_a^b (f)_-^{(r)}(x) \, dx, (f)_+^{(r)}(x) \, dx \right],$$

$\forall r \in [0,1]$.

Clearly $f_{\pm}^{(r)} : [a,b] \to \mathbb{R}$ are continuous functions.

In this section we study the fuzzy corresponding analogs of real wavelet type operators A_k, B_k, C_k, D_k, $k \in \mathbb{Z}$, of first section. For simplicity we keep the same notation at the fuzzy level. So, depending on the context we understand accordingly whether our operator is real of fuzzy, that is whether is operating on real valued functions or on fuzzy valued functions.

We present the next main fuzzy wavelet type result.

Theorem 21.22. Let $f \in C_{\mathcal{F}}^{NB}(\mathbb{R})$, $N \geq 1$, $x \in \mathbb{R}$, and $k \in \mathbb{Z}$. Let φ be a bounded real valued function of compact support $\subseteq [-a,a]$, $a > 0$ such that $\sum_{j=-\infty}^{\infty} \varphi(x-j) = 1$, all $x \in \mathbb{R}$. Suppose $\varphi \geq 0$. Put

$$(B_k f)(x) = \sum_{j=-\infty}^{\infty}{}^* f\left(\frac{j}{2^k}\right) \odot \varphi\left(2^k x - j\right),$$

$$(C_k f)(x) = \sum_{j=-\infty}^{\infty}{}^* \left(2^k \odot (FR) \int_0^{2^{-k}} f\left(t + \frac{j}{2^k}\right) dt\right) \odot \varphi\left(2^k x - j\right),$$

and

$$(D_k f)(x) = \sum_{j=-\infty}^{\infty}{}^* \delta_{kj}(f) \odot \varphi\left(2^k x - j\right),$$

where

$$\delta_{kj}(f) = \sum_{\tilde{r}=0}^{n}{}^* w_{\tilde{r}} \odot f\left(\frac{j}{2^k} + \frac{\tilde{r}}{2^k n}\right),$$

$n \in \mathbb{N}$, $w_{\tilde{r}} \geq 0$, $\sum_{\tilde{r}=0}^{n}{}^* w_{\tilde{r}} = 1$.

Then

(i)

$$D\left((B_k f)(x), (D_k f)(x)\right) \leq D^*(B_k f, D_k f) \tag{21.19}$$

$$\leq \sum_{i=1}^{N} \frac{D^*\left(f^{(i)}, \tilde{0}\right)}{2^{ki} i!} + \frac{\omega_1^{(\mathcal{F})}\left(f^{(N)}, \frac{1}{2^k}\right)}{2^{kN} N!},$$

(ii)

$$D\left(\left(B_k f\right)(x), \left(C_k f\right)(x)\right) \;\leq\; D^*\left(B_k f, C_k f\right) \tag{21.20}$$

$$\leq\; \sum_{i=1}^{N} \frac{D^*\left(f^{(i)}, \widetilde{0}\right)}{2^{ki}\,(i+1)!} + \frac{\omega_1^{(\mathcal{F})}\left(f^{(N)}, \frac{1}{2^k}\right)}{2^{kN}\,(N+1)!},$$

and
(iii)

$$D\left(\left(C_k f\right)(x), \left(D_k f\right)(x)\right) \;\leq\; D^*\left(C_k f, D_k f\right) \tag{21.21}$$

$$\leq\; \sum_{i=1}^{N} \frac{D^*\left(f^{(i)}, \widetilde{0}\right)}{2^{ki-1}\,(i+1)!} + \frac{\omega_1^{(\mathcal{F})}\left(f^{(N)}, \frac{1}{2^k}\right)}{2^{kN-1}\,(N+1)!}.$$

Note. We see that

$$D\left(f^{(N)}(x), f^{(N)}(y)\right) \;\leq\; D\left(f^{(N)}(x), \widetilde{0}\right) + D\left(f^{(N)}(y), \widetilde{0}\right)$$

$$\leq\; 2D^*\left(f^{(N)}, \widetilde{0}\right) < \infty.$$

Thus $\omega_1^{(\mathcal{F})}\left(f, \frac{1}{2^k}\right) < \infty, \forall k \in \mathbb{Z}$.
Consequently as $k \to \infty$ we obtain

$$D^*\left(B_k f, D_k f\right), D^*\left(B_k f, C_k f\right), D^*\left(C_k f, D_k f\right) \to 0$$

with rates.
Proof. (i) We observe the following

$$
\begin{aligned}
D\left(\left(B_k f\right)(x), \left(D_k f\right)(x)\right) \;&=\; \sup_{r\in[0,1]} \max\left\{ \left|\left(\left(B_k f\right)(x)\right)_-^{(r)} - \left(\left(D_k f\right)(x)\right)_-^{(r)}\right|, \right.\\
&\qquad\qquad\qquad \left.\left|\left(\left(B_k f\right)(x)\right)_+^{(r)} - \left(\left(D_k f\right)(x)\right)_+^{(r)}\right|\right\}\\
&=\; \sup_{r\in[0,1]} \max\left\{ \left|\left(B_k\left(f_-^{(r)}\right)\right)(x) - \left(D_k\left(f_-^{(r)}\right)\right)(x)\right|, \right.\\
&\qquad\qquad\qquad \left.\left|\left(B_k\left(f_+^{(r)}\right)\right)(x) - \left(D_k\left(f_+^{(r)}\right)\right)(x)\right|\right\}\\
&\leq\; \sup_{r\in[0,1]} \max\left\{ \left\|B_k\left(f_-^{(r)}\right) - D_k\left(f_-^{(r)}\right)\right\|_\infty, \right.\\
&\qquad\qquad\qquad \left.\left\|B_k\left(f_+^{(r)}\right) - D_k\left(f_+^{(r)}\right)\right\|_\infty\right\}
\end{aligned}
$$

$$
\begin{aligned}
(21.16) \\
\leq \sup_{r\in[0,1]} \max \Bigg\{ &\sum_{i=1}^{N} \frac{\left\| \left(f_-^{(r)} \right)^{(i)} \right\|_\infty}{2^{ki}i!} \\
&+ \frac{\omega_1 \left(\left(f_-^{(r)} \right)^{(N)}, \frac{1}{2^k} \right)}{2^{kN}N!}, \\
&\sum_{i=1}^{N} \frac{\left\| \left(f_+^{(r)} \right)^{(i)} \right\|_\infty}{2^{ki}i!} + \frac{\omega_1 \left(\left(f_+^{(r)} \right)^{(N)}, \frac{1}{2^k} \right)}{2^{kN}N!} \Bigg\}
\end{aligned}
$$

$$
\begin{aligned}
= \sup_{r\in[0,1]} \max \Bigg\{ &\sum_{i=1}^{N} \frac{\left\| \left(f^{(i)} \right)_-^{(r)} \right\|_\infty}{2^{ki}i!} + \frac{\omega_1 \left(\left(f^{(N)} \right)_-^{(r)}, \frac{1}{2^k} \right)}{2^{kN}N!}, \\
&\sum_{i=1}^{N} \frac{\left\| \left(f^{(i)} \right)_+^{(r)} \right\|_\infty}{2^{ki}i!} + \frac{\omega_1 \left(\left(f^{(N)} \right)_+^{(r)}, \frac{1}{2^k} \right)}{2^{kN}N!} \Bigg\}
\end{aligned}
$$

$$
\begin{aligned}
\leq \sum_{i=1}^{N} \frac{1}{2^{ki}i!} \sup_{r\in[0,1]} \max \left\{ \left\| \left(f^{(i)} \right)_-^{(r)} \right\|_\infty, \left\| \left(f^{(i)} \right)_+^{(r)} \right\|_\infty \right\} \\
+ \frac{1}{2^{kN}N!} \sup_{r\in[0,1]} \max \left\{ \omega_1 \left(\left(f^{(N)} \right)_-^{(r)}, \frac{1}{2^k} \right), \omega_1 \left(\left(f^{(N)} \right)_+^{(r)}, \frac{1}{2^k} \right) \right\}
\end{aligned}
$$

$$
= \sum_{i=1}^{N} \frac{1}{2^{ki}i!} D^* \left(f^{(i)}, \tilde{0} \right) + \frac{1}{2^{kN}N!} \omega_1^{(\mathcal{F})} \left(f^{(N)}, \frac{1}{2^k} \right),
$$

proving the theorem's (21.19).

(ii) We observe the following

$$
\begin{aligned}
D\left((B_k f)(x), (C_k f)(x) \right) &= \sup_{r\in[0,1]} \max \left\{ \left| ((B_k f)(x))_-^{(r)} - ((C_k f)(x))_-^{(r)} \right|, \right. \\
&\qquad\qquad\qquad \left. \left| ((B_k f)(x))_+^{(r)} - ((C_k f)(x))_+^{(r)} \right| \right\} \\
&= \sup_{r\in[0,1]} \max \left\{ \left| \left(B_k \left(f_-^{(r)} \right) \right)(x) - \left(C_k \left(f_-^{(r)} \right) \right)(x) \right|, \right. \\
&\qquad\qquad\qquad \left. \left| \left(B_k \left(f_+^{(r)} \right) \right)(x) - \left(C_k \left(f_+^{(r)} \right) \right)(x) \right| \right\} \\
&\leq \sup_{r\in[0,1]} \max \left\{ \left\| B_k \left(f_-^{(r)} \right) - C_k \left(f_-^{(r)} \right) \right\|_\infty, \right.
\end{aligned}
$$

$$\left\| B_k\left(f_+^{(r)}\right) - C_k\left(f_+^{(r)}\right)\right\|_\infty\Big\}$$

$$\overset{(21.17)}{\leq} \ \sup_{r\in[0,1]}\max\left\{\sum_{i=1}^N \frac{\left\|\left(f_-^{(r)}\right)^{(i)}\right\|_\infty}{2^{ki}(i+1)!} +\right.$$

$$\frac{\omega_1\left(\left(f_-^{(r)}\right)^{(N)},\frac{1}{2^k}\right)}{2^{kN}(N+1)!},$$

$$\left.\sum_{i=1}^N \frac{\left\|\left(f_+^{(r)}\right)^{(i)}\right\|_\infty}{2^{ki}(i+1)!} + \frac{\omega_1\left(\left(f_+^{(r)}\right)^{(N)},\frac{1}{2^k}\right)}{2^{kN}(N+1)!}\right\}$$

$$= \ \sup_{r\in[0,1]}\max\left\{\sum_{i=1}^N \frac{\left\|\left(f^{(i)}\right)_-^{(r)}\right\|_\infty}{2^{ki}(i+1)!} + \frac{\omega_1\left(\left(f^{(N)}\right)_-^{(r)},\frac{1}{2^k}\right)}{2^{kN}(N+1)!},\right.$$

$$\left.\sum_{i=1}^N \frac{\left\|\left(f^{(i)}\right)_+^{(r)}\right\|_\infty}{2^{ki}(i+1)!} + \frac{\omega_1\left(\left(f^{(N)}\right)_+^{(r)},\frac{1}{2^k}\right)}{2^{kN}(N+1)!}\right\}$$

$$\leq \ \sum_{i=1}^N \frac{1}{2^{ki}(i+1)!}\sup_{r\in[0,1]}\max\left\{\left\|\left(f^{(i)}\right)_-^{(r)}\right\|_\infty,\left\|\left(f^{(i)}\right)_+^{(r)}\right\|_\infty\right\}$$

$$+\frac{1}{2^{kN}(N+1)!}\sup_{r\in[0,1]}\max\left\{\omega_1\left(\left(f^{(N)}\right)_-^{(r)},\frac{1}{2^k}\right),\omega_1\left(\left(f^{(N)}\right)_+^{(r)},\frac{1}{2^k}\right)\right\}$$

$$= \ \sum_{i=1}^N \frac{1}{2^{ki}(i+1)!}D^*\left(f^{(i)},\tilde{0}\right) + \frac{1}{2^{kN}(N+1)!}\omega_1^{(\mathcal{F})}\left(f^{(N)},\frac{1}{2^k}\right),$$

proving the theorem's (21.20).

(iii) We observe the next

$$D\left((C_kf)(x),(D_kf)(x)\right) = \sup_{r\in[0,1]}\max\left\{\left|((C_kf)(x))_-^{(r)} - ((D_kf)(x))_-^{(r)}\right|,\right.$$

$$\left.\left|((C_kf)(x))_+^{(r)} - ((D_kf)(x))_+^{(r)}\right|\right\}$$

$$= \sup_{r\in[0,1]}\max\left\{\left|\left(C_k\left(f_-^{(r)}\right)\right)(x) - \left(D_k\left(f_-^{(r)}\right)\right)(x)\right|,\right.$$

$$\left.\left|\left(C_k\left(f_+^{(r)}\right)\right)(x) - \left(D_k\left(f_+^{(r)}\right)\right)(x)\right|\right\}$$

$$\leq \sup_{r\in[0,1]} \max\left\{\left\|C_k\left(f_-^{(r)}\right) - D_k\left(f_-^{(r)}\right)\right\|_\infty,\right.$$

$$\left.\left\|C_k\left(f_+^{(r)}\right) - D_k\left(f_+^{(r)}\right)\right\|_\infty\right\}$$

$$\overset{(21.18)}{\leq} \sup_{r\in[0,1]} \max\left\{\sum_{i=1}^{N} \frac{\left\|\left(f_-^{(r)}\right)^{(i)}\right\|_\infty}{2^{ki-1}(i+1)!} + \right.$$

$$\frac{\omega_1\left(\left(f_-^{(r)}\right)^{(N)}, \frac{1}{2^k}\right)}{2^{kN-1}(N+1)!},$$

$$\left.\sum_{i=1}^{N} \frac{\left\|\left(f_+^{(r)}\right)^{(i)}\right\|_\infty}{2^{ki-1}(i+1)!} + \frac{\omega_1\left(\left(f_+^{(r)}\right)^{(N)}, \frac{1}{2^k}\right)}{2^{kN-1}(N+1)!}\right\}$$

$$= \sup_{r\in[0,1]} \max\left\{\sum_{i=1}^{N} \frac{\left\|\left(f^{(i)}\right)_-^{(r)}\right\|_\infty}{2^{ki-1}(i+1)!} + \frac{\omega_1\left(\left(f^{(N)}\right)_-^{(r)}, \frac{1}{2^k}\right)}{2^{kN-1}(N+1)!},\right.$$

$$\left.\sum_{i=1}^{N} \frac{\left\|\left(f^{(i)}\right)_+^{(r)}\right\|_\infty}{2^{ki-1}(i+1)!} + \frac{\omega_1\left(\left(f^{(N)}\right)_+^{(r)}, \frac{1}{2^k}\right)}{2^{kN-1}(N+1)!}\right\}$$

$$\leq \sum_{i=1}^{N} \frac{1}{2^{ki-1}(i+1)!} \sup_{r\in[0,1]} \max\left\{\left\|\left(f^{(i)}\right)_-^{(r)}\right\|_\infty, \left\|\left(f^{(i)}\right)_+^{(r)}\right\|_\infty\right\}$$

$$+ \frac{1}{2^{kN-1}(N+1)!} \sup_{r\in[0,1]} \max\left\{\omega_1\left(\left(f^{(N)}\right)_-^{(r)}, \frac{1}{2^k}\right),\right.$$

$$\left.\omega_1\left(\left(f^{(N)}\right)_+^{(r)}, \frac{1}{2^k}\right)\right\}$$

$$= \sum_{i=1}^{N} \frac{1}{2^{ki-1}(i+1)!} D^*\left(f^{(i)}, \tilde{0}\right) + \frac{1}{2^{kN-1}(N+1)!} \omega_1^{(\mathcal{F})}\left(f^{(N)}, \frac{1}{2^k}\right),$$

proving the theorem's (21.21).
   Above we need that ([16])

$$(B_k f)_\pm^{(r)} = B_k\left(f_\pm^{(r)}\right),$$

$$(C_k f)_\pm^{(r)} = C_k\left(f_\pm^{(r)}\right), \text{ and}$$

$$(D_k f)_\pm^{(r)} = D_k\left(f_\pm^{(r)}\right),$$

$\forall r \in [0, 1]$. $\qquad\qquad\qquad\qquad\qquad\qquad\qquad\qquad\qquad\qquad\qquad\qquad$ $\square$

Denote by $C_b(\mathbb{R}, \mathbb{R}_{\mathcal{F}})$ the space of bounded fuzzy continuous functions on $\mathbb{R}$ with respect to metric D.

We finish with the following fuzzy wavelet type main result

Theorem 21.23. Let $f \in C_{\mathcal{F}}^{NB}(\mathbb{R}) \cap C_b(\mathbb{R}, \mathbb{R}_{\mathcal{F}})$, $N \geq 1$, $x \in \mathbb{R}$, and $k \in \mathbb{Z}$. Let the scaling function $\varphi(x)$ a real valued function with *supp* $\varphi(x) \subseteq [-a, a]$, $0 < a < +\infty$, φ is continuous on $[-a, a]$, $\varphi(x) \geq 0$, such that $\sum_{j=-\infty}^{\infty} \varphi(x - j) = 1$ on $\mathbb{R}$ (then $\int_{-\infty}^{\infty} \varphi(x) \, dx = 1$).
Define

$$\varphi_{kj}(t) := 2^{k/2} \varphi(2^k t - j) \quad \text{all } k, j \in \mathbb{Z}, t \in \mathbb{R}$$

$$\langle f, \varphi_{kj} \rangle := (FR) \int_{\frac{j-a}{2^k}}^{\frac{j+a}{2^k}} f(t) \odot \varphi_{kj}(t) \, dt,$$

and the fuzzy wavelet type operator

$$(A_k f)(x) = \sum_{j=-\infty}^{\infty}{}^* \langle f, \varphi_{kj} \rangle \odot \varphi_{kj}(x), \quad x \in \mathbb{R}.$$

The fuzzy wavelet type operators B_k, C_k, D_k are as in Theorem 21.22.
Then
(i)

$$D((A_k f)(x), (B_k f)(x)) \leq D^*(A_k f, B_k f)$$

$$\leq \sum_{i=1}^{N} \frac{D^*\left(f^{(i)}, \tilde{0}\right)}{2^{ki} i!} a^i + \frac{a^N}{2^{kN} N!} \omega_1^{(\mathcal{F})}\left(f^{(N)}, \frac{a}{2^k}\right), \qquad (21.22)$$

(ii)

$$D((A_k f)(x), (C_k f)(x)) \leq D^*(A_k f, C_k f)$$

$$\leq \sum_{i=1}^{N} \frac{D^*\left(f^{(i)}, \tilde{0}\right)}{i! 2^{ki}} (a+1)^i + \frac{(a+1)^N}{2^{kN} N!} \omega_1^{(\mathcal{F})}\left(f^{(N)}, \frac{(a+1)}{2^k}\right), \qquad (21.23)$$

and
(iii)

$$D((A_k f)(x), (D_k f)(x)) \leq D^*(A_k f, D_k f) \qquad\qquad\qquad (21.24)$$

$$\leq \sum_{i=1}^{N} \frac{D^*\left(f^{(i)}, \tilde{0}\right)}{2^{ki} i!} (a+1)^i + \frac{(a+1)^N}{N! 2^{kN}} \omega_1^{(\mathcal{F})}\left(f^{(N)}, \frac{(a+1)}{2^k}\right).$$

Notice that $D^* (A_k f, B_k f), D^* (A_k f, C_k f), D^* (A_k f, D_k f) \to 0$ as $k \to \infty$ with rates.

Proof. Similar to the proof of Theorem 21.22. Also notice here (see also [16]) that $(A_k f)_{\pm}^{(r)} = A_k \left(f_{\pm}^{(r)} \right), \forall r \in [0, 1]$.
It is based on (21.13), (21.14) and (21.15).   □

Note 21.24. In [14] we proved, see also Chapter 12, for $f \in C_{\mathcal{F}}^U (\mathbb{R})$ as $k \to \infty$, we get uniformly that A_k, B_k, C_k, $D_k \to I$ unit operator with rates in the D metric. In the case of A_k we need also f be fuzzy bounded.

References

[1]   C. D. Aliprantis, O. Burkinshaw, Positive Operators, Academic Press, New York and London, 1985.

[2]   C. D. Aliprantis, O. Burkinshaw, Principles of Real Analysis, Third edition, Academic Press, Inc., San Diego, CA, 1998.

[3]   F. Altomare , M. Campiti, Korovkin type approximation theory and its applications, Walter de Gruyter Publ. Berlin, 1994.

[4]   G.A. Anastassiou, *Moments in Probability and Approximation Theory*, Pitman/Longman, #287, UK, 1993.

[5]   G.A. Anastassiou, Ostrowski type inequalities, Proc. Amer. Math. Soc. 123 (1995), No. 12, 3775-3781.

[6]   G.A. Anastassiou, G.A., Rate of convergence of some neural network operators to the unit-univariate case, *J. Math. Anal. Appl.* **212**, (1997), 237–262.

[7]   G.A. Anastassiou, Shape and probability preserving univariate wavelet type operators, *Commun. Appl. and Anal.*, **1**, No. 3 (1997), 303–314.

[8]   G.A. Anastassiou, Higher order univariate wavelet type approximation, *Approximation Theory*, in Memory of A.K. Varma, Marcel Dekker, New York, 1998, pp. 43–60.

[9]   G.A. Anastassiou, *Quantitative Approximations*, Chapman & Hall/CRC, 2001, Boca Raton, New York.

[10]   G.A. Anastassiou, On *H*-fuzzy differentiation, *Mathematica Balkanica*, New Series, Vol. 16, Fasc. 1–4 (2002), 153–193.

[11]   G.A. Anastassiou, Rate of convergence of fuzzy neural network operators, univariate case, *Journal of Fuzzy Mathematics*, **10**, No. 3 (2002), 755–780.

[12]   G.A. Anastassiou, High order approximation by univariate shift-invariant integral operators, in *Nonlinear Analysis and Applications*, edited by Ravi Agarwal and Donal O'Regan, Vol. I, 141–164, Kluwer, 2003.

[13]   G.A. Anastassiou, Fuzzy Ostrowski type inequalities, Comput. Appl. Math. 22 (2003), No. 2, 279-292.

[14]   G.A. Anastassiou, Fuzzy wavelet type operators, *Nonlinear Functional Analysis and Applications*, **9**, No. 2 (2004), 251–269.

[15]   G.A. Anastassiou, Fuzzy approximation by fuzzy Convolution type operators, Computers & Mathematics with Applications, 48 (2004), 1369-1386.

[16]   G.A. Anastassiou, *High Order Fuzzy Approximation by Fuzzy Wavelet type and Neural Network Operators,* Computers and Math. with Appl., special issue edited by G. Anastassiou, Vol. 48, 2004, 1387-1401.

[17]   G.A. Anastassiou, Univariate fuzzy-random neural network approximation operators, *Computers and Mathematics with Applications*, Special issue/Proceedings edited by G.Anastassiou of special session "Comutational Methods in Analysis", AMS meeting in Orlando,Florida, November 2002, Vol.48 (2004), 1263–1283.

[18]   G.A. Anastassiou, On basic Fuzzy Korovkin Theory, *Studia Univ. Babes - Bolyai, Mathematica*, Vol.L, No.4, 2005, 3-10.

[19]   G.A. Anastassiou, Fuzzy Taylor formulae, *Cubo*, Vol.7, No.3, 2005, 1-13.

[20]   G.A. Anastassiou, Higher order Fuzzy Korovkin Theory via Inequalities, Communications in Applied Analysis, 10 (2006), no. 2, 359-392.

[21]   G.A. Anastassiou, On Fuzzy Global Smoothness Preservation, J. Fuzzy Math., vol 15, no. 1, 2007, 219-231.

[22]   G.A. Anastassiou, Fuzzy Random Korovkin Theory and Inequalities, Mathematical Inequalities & Applications, Vol. 10, No. 1, 2007, 63-94.

[23] G.A. Anastassiou, *Quantitative estimates for Distance between Fuzzy Wavelet type operators,* Journal of Concrete and Applicable Mathematics, Vol 5, No. 1 (2007), 25-52.

[24] G.A. Anastassiou, Fuzzy Korovkin Theorems and Inequalities, J. Fuzzy Mathematics, Vol. 15, No.1 (2007), 169-205.

[25] G.A. Anastassiou, C. Cottin, and H.H. Gonska, Global smoothness preservation by multivariate approximation operators, in *Israel Mathematical Conference Proceedings* (S. Baron and D. Leviatan, eds.), Weizmann Science Press, Vol. IV, 1991, pp. 31–44.

[26] G.A. Anastassiou, C. Cottin, and H.H. Gonska, Global smoothness of approximating functions, *Analysis,* **11** (1991), 43–57.

[27] G.A. Anastassiou, K. Demirci, S. Karakus, $\mathcal{A}$-summability and fuzzy Korovkin-type approximation, submitted 2009.

[28] G.A. Anastassiou, K. Demirci, S. Karakus, $\mathcal{A}$-summability and fuzzy trigonometric Korovkin-type approximation, submitted 2009.

[29] G.A. Anastassiou and O. Duman, Statistical fuzzy approximation by fuzzy positive linear operators, Comput. Math. Appl. 55 (2008), 573-580.

[30] G.A. Anastassiou and S. G. Gal, Approximation Theory. Moduli of Continuity and Global Smoothness Preservation, Birkhäuser, Boston, Basel, Berlin, 2000.

[31] G.A. Anastassiou and S. Gal, On a fuzzy trigonometric approximation theorem of Weierstrass-type, *Journal of Fuzzy Mathematics,* **9,** No. 3 (2001), Los Angeles, 701–708.

[32] G.A. Anastassiou, S.G. Gal, On fuzzy trigonometric Korovkin theory, Nonlinear Funct. Anal. Appl. 11 (2006), 385-395.

[33] G.A. Anastassiou and H.H. Gonska, On stochastic global smoothness, *Rev. Acad. Cienc. Zaragoza,* **49** (1994), 119–136.

[34] G.A. Anastassiou and H. H. Gonska, On some shift-invariant integral operators, multivariate case, in *Proc. Int. Conf. Approx. Prob. and Related Fields,* U.C.S.B., Santa Barbara, CA, May 20–22, 1993 (Anastassiou, G.A. and Rachev, S.T., eds.), Plenum Press, New York, 1994, 41–64.

[35] G.A. Anastassiou and H. H. Gonska, On some shift invariant integral operators, univariate case, *Ann. Polon. Math.,* **LXI,** 3 (1995), 225–243.

[36]   G.A. Anastassiou and X.M. Yu, Monotone and probabilistic wavelet approximation, *Stochastic Anal. Appl.*, **10**(3) (1992), 251–264.

[37]   Tom M. Apostol, Mathematical Analysis, Addison-Wesley Publ. Company, Reading, Massachusetts, 1957.

[38]   Ö.G. Atlıhan, C. Orhan, Matrix summability and positive linear operators, Positivity 11 (2007), 387-389.

[39]   Ö.G. Atlıhan, C. Orhan, Summation process of positive linear operators, Computers & Mathematics with Applications, 56 (2008) 1188-1195.

[40]   B. Bede and S. G. Gal, Quadrature rules for integrals of fuzzy-number-valued functions, Fuzzy Sets and Systems, 145 (2004), no.3, 359-380.

[41]   B. Bede and S. G. Gal, Best Approximation and Jackson-Type estimates by generalized fuzzy polynomials, J. of Concrete and Applicable Mathematics, Vol.2 (2004), No. 3, 213-232.

[42]   H.T. Bell, Order summability and almost convergence, Proc. Am. Math. Soc. 38 (1973) 548-553.

[43]   E.K. Blum, Numerical Analysis and Computation Theory and Practice, Addison-Wesley Series in Mathematics, Addison-Wesley Publishing Company, XII, 1972.

[44]   R. Bojanic and F. Cheng. Estimates for the rate of approximation of functions of bounded variation by Hermite-Fejér polynomials, Proceedings of the conference of Canadian Math. Soc. 3 (1983), 5-17.

[45]   R. Bojanic and M. K. Khan. Summability of Hermite-Fejér interpolation for functions of bounded variation, J. Nat. Sci. Math. 32 No. 1 (1992), 5-10.

[46]   D. Butnariu, measurability concepts of fuzzy mappings, *Fuzzy Sets and Systems*, **31** (1989), 77–82.

[47]   P. Cardaliaguet and G. Euvrard, Approximation of a function and its derivative with a neural network, *Neural Networks*, **5** (1992), 207–220.

[48]   A. Colubi, J.S. Dominguez-Menchero, M. Lopez-Diaz, and D. Ralescu, A $D_E[0,1]$ representation of random upper semicontinuous functions, *Proc. A.M.S.*, **130**, No. 11 (2002), 3237–3242.

[49]   Wu Congxin and Liu Danhong, A fuzzy Weierstrass approximation theorem, *J. Fuzzy Mathematics* **7**, No. 1 (1999), 101–104.

[50] Congxin Wu and Ming M a. On embedding problem of fuzzy number space: Part 1, *Fuzzy Sets and Systems*, **44** (1991), 33—38.

[51] Congxin Wu and Ming Ma. On embedding problem of fuzzy number space: Part 2, *Fuzzy Sets and Systems*, **45**, 1992, 189–202.

[52] Wu Congxin and Gong Zengtai, On Henstock integral of fuzzy-number-valued functions, I, *Fuzzy Sets and Systems*, **115**, No. 3 (2000), 377–391.

[53] Congxin W u and Zengtai G o n g. On Henstock integral of fuzzy number valued functions (I), *Fuzzy Sets and Systems*, **120**, No. 3, 2001, 523–532.

[54] R. Cristescu, Ordered structures in ordered vector spaces, *Ed. St. Enciclop.*, Bucharest, 1983 (in Romanian).

[55] R. A. DeVore, the Approximation of Continuous Functions by Positive Linear Operators, Lecture Notes in Mathematics, Springer-Verlag, 293 (1972), Berlin.

[56] Ronald A. DeVore and George G. Lorentz, *Constructive Approximation*, Vol. 303, Springer-Verlag, Berlin, New York, 1993.

[57] P. Diamond, P. Kloeden, Metric Spaces of Fuzzy Sets, World Scientific, New Jersey, 1994.

[58] P. Diamond and P. Kloeden, Metric topology of fuzzy numbers and fuzzy analysis, Chapter 11 in *Fundamentals of Fuzzy Sets, Handbook of Fuzzy Sets Series*, 7, Kluwer, Boston, 2000.

[59] P. Diamond, P. Kloeden, A. Vladimirov, Spikes, broken planes and the approximation of convex sets, Fuzzy Sets and Systems 99(1999), 225-232.

[60] P. Diamond and A. Ramer, Approximation of knowledge and fuzzy sets, *Cybernetics and Systems: An International Journal*, **24** (1993), 407–417.

[61] M. Lopez-Diaz and M. Angeles Gil, Approximating integrably bounded fuzzy random variables in terms of the "Generalized" Hausdorff metric, *Information Sciences*, **104** (1998), 279–291.

[62] R. Dudley, *Real Analysis and Probability*, Wadsworth & Brooks/Cole Mathematics Series, Pacific Grove, CA, 1989.

[63] O. Duman and G.A. Anastassiou, On statistical fuzzy trigonometric Korovkin theory, J. Comput. Anal. Appl. 10 (2008), 333-344.

[64] A.M. Fink, Bounds on the deviation of a function from its averages, *Czechoslavak Math. J.*, **42** (117) (1992), 289–310.

[65] A.R. Freedman, J.J. Sember, Densities and summability, Pacific J. Math. 95 (1981) 293–305.

[66] S. Gal, Approximation Theory in Fuzzy Setting, Chapter 13 in *Handbook of Analytic-Computational Methods in Applied Mathematics*, pp. 617-666, edited by G. Anastassiou, Chapman & Hall/CRC, 2000, Boca Raton, New York.

[67] R. Goetschel, Jr. and W. Voxman, *Elementary fuzzy calculus*, Fuzzy Sets and Systems, 18 (1986), 31-43.

[68] W.J. Gordon and J.A. Wixom, Shepard's method of metric interpolation to bivariate and multivariate interpolation, *Math. Comp.*, **32** (1978), 253–264.

[69] C.W. Groetsch, *Elements of Applicable Functional Analysis*, Marcel Dekker, Inc., New York and Basel, 1980.

[70] G. H. Hardy, Divergent series, Oxford Uni. Press, London, 1949.

[71] O. Kaleva, *Fuzzy differential equations,* Fuzzy Sets and Systems, 24 (1987), 301-317.

[72] O. Kaleva, Interpolation of fuzzy data, Fuzzy Sets and Systems 61(1994), 63-70.

[73] L.V. Kantorovich, On the moment problem for a finite interval, *Dokl. Akad. Nauk SSSR*, **14** (1937), 531–537 (in Russian).

[74] J.H.B. Kemperman, The general moment problem, a geometric approach, *The Annals of Mathematical Statistics*, **39**, No. 1 (1968), 93–122.

[75] Yun Kyong Kim and Byung Moon Ghil, Integrals of fuzzy-number-valued functions, *Fuzzy Sets and Systems*, **86** (1997), 213–222.

[76] J.P. King, J.J. Swetits, Positive linear operators and summability, J. Aust. Math. Soc. 11 (1970) 281–290.

[77] E. Kolk, Matrix summability of statistically convergent sequences, Analysis 13 (1993), 77–83.

[78] P.P. Korovkin, *Linear Operators and Approximation Theory*, Hindustan Publ. Corp., Delhi, India, 1960.

[79] B. Kuttner. On the Gibbs phenomenon for Riesz means, J. London Math. Soc. 19 (1944), 153-161.

[80] Puyin Liu, Analysis of approximation of continuous fuzzy functions by multivariate fuzzy polynomials, Fuzzy Sets and Systems 127(2002), 299-313.

[81] W. Lodwick, J. Santos, Constructing consistent fuzzy surfaces from fuzzy data, Fuzzy Sets and Systems, 135(2003), 259-277.

[82] G.G. Lorentz, A contribution to the theory of divergent sequences, Acta Math. 80 (1948) 167–190.

[83] G.G. Lorentz, Appproximation of Functions, Chelsea Publishing Company, New York, 1986 (second edition).

[84] R. Lowen, A fuzzy Lagrange interpolation theorem, Fuzzy Sets and Systems 34(1990) 33-38.

[85] M. Matloka, Sequences of fuzzy numbers, BUSEFAL 28 (1986) 28–37.

[86] H.I. Miller, A measure theoretical subsequence characterization of statistical convergence, Trans. Amer. Math. Soc. 347 (1995) 1811–1819.

[87] H.I. Miller, C. Orhan, On almost convergent and statistically convergent subsequences. Acta Math. Hungar. 93 (2001), 135–151.

[88] R.N. Mohapatra, Quantitative results on almost convergence of a sequence of positive linear operators, J. Approx. Theory 20 (1977) 239–250.

[89] T. Nishishiraho, Quantitative theorems on linear approximation processes of convolution operators in Banach spaces, Tôhoku Math. J. 33(1981) 109–126.

[90] T. Nishishiraho, Convergence of positive linear approximation processes, Tôhoku Math. J. 35 (1983) 441–458.

[91] F. Nuray, E. Savaş, Statistical convergence of sequences of fuzzy numbers, Math. Slovaca 45 (1995), 269–273.

[92] A. Ostrowski, Über die Absolutabweichung einer differentiebaren Funktion von ihrem Integralmittelwert, *Comment. Math. Helv.*, **10** (1938), 226–227.

[93] M. L. Puri and D. A. Ralescu, Differentials of fuzzy functions, *J. of Math. Analysis and Appl.*, **91**, 1983, 552–558.

[94] C. Radu, $\mathcal{A}$–summability and approximation of continuous periodic functions, Studia Univ. Babeş-Bolyai Math. 52 (2007) 155-161.

[95] D. Shepard, A two-dimensional interpolation function for irregularly-spaced data, in *Proc. 1968 ACM National Conference*, 1968, 517–524.

[96] O. Shisha and B. Mond, The degree of convergence of sequences of linear positive operators, *Nat. Acad. of Sci. U.S.*, **60** (1968), 1196–1200.

[97] O. Shisha and B. Mond, The degree of approximation to periodic functions by linear positive operators, *J. Approx. Th.*, **1** (1968), 335–339.

[98] Murray R. Spiegel, *Advanced Calculus*, Schaum's Outline Series, McGraw-Hill Book Company, New York. 1963.

[99] M. Steiglitz, Eine verallgemeinerung des begriffs der fastkonvergenz, Math. Japon. 18 (1973) 53-70.

[100] J.J. Swetits, On summability and positive linear operators, J. Approx. Theory 25 (1979) 186–188.

[101] J. Szabados, On a problem of R. DeVore, Acta Math. Hungar., 27 (1-2)(1976) 219-223.

[102] P. Teran and M. Lopez-Diaz, Remarks on Korovkin-type approximation of fuzzy random variables, in *Statistical Modeling, Analysis and Management of Fuzzy Data*, edited by C. Bertoluzza, M. Gil, D. Ralescu, Physica-Verlag, 2002, pp. 90–103.

[103] L. A. Zadeh, Fuzzy sets, *Information and Control*, **8**, 1965, 338–353.

List of Symbols

Index